SOCIAL PROBLEMS

SELECTIONS FROM CQ RESEARCHER

$SAGE | PINE FORGE

Los Angeles | London | New Delhi
Singapore | Washington DC

For information:

Pine Forge Press
An Imprint of SAGE Publications, Inc.
2455 Teller Road
Thousand Oaks, California 91320
E-mail: order@sagepub.com

SAGE Publications Ltd.
1 Oliver's Yard
55 City Road
London EC1Y 1SP
United Kingdom

SAGE Publications India Pvt. Ltd.
B 1/I 1 Mohan Cooperative Industrial Area
Mathura Road, New Delhi 110 044
India

SAGE Publications Asia-Pacific Pte. Ltd.
33 Pekin Street #02-01
Far East Square
Singapore 048763

Printed in the United States of America

Library of Congress Cataloging-in-Publication Data

Social problems : selections from CQ Researcher.
    p. cm.
Includes bibliographical references and index.
ISBN 978-1-4129-7862-0 (pbk.)
    1. United States—Social policy—1993- 2. United States—Economic conditions—2001- 3. United States—Politics and government—2001-2009. I. Congressional Quarterly, inc. II. CQ researcher.

HN65.S57523 2010
361.0973—dc22   2009026406

This book is printed on acid-free paper.

10   11   12   13   10   9   8   7   6   5   4   3   2

Acquisitions Editor:	David Repetto
Editorial Assistant:	Nancy Scrofano
Production Editor:	Laureen Gleason
Typesetter:	C&M Digitals (P) Ltd.
Cover Designer:	Candice Harman
Marketing Manager:	Jennifer Reed Banando

Contents

Work and the Economy

Health and Medicine

The Media

OUR SOCIAL AND PHYSICAL WORLDS

Alcohol and Drug Abuse

INDIVIDUAL ACTION AND SOCIAL CHANGE

Social Problems and Social Action

Annotated Contents

THE BASES OF INEQUALITY
Social Class and Poverty
Global Food Crisis: What's Causing the Rising Prices?

Food prices have spiked around the world over the past year, bringing hunger and unrest to many developing countries, along with pain at the checkout counter for lower-income American families. In North Korea, for example, where 35 percent of the population is undernourished, the price of the major food staple, rice, soared 186 percent, and overall food prices rose 70 percent. With 2.1 billion people around the world living on less than $2 a day, such price increases may plunge hundreds of millions into malnutrition and starvation. Drought, high oil prices that make food transport pricey and diversion of corn for use as a biofuel all contribute to the price spike. The effect of globalization — which has led poor countries to abandon domestic food crops in favor of commodity crops for export — also has been blamed. The crisis also has sparked international tension over the impact of wealthy nations' farm subsidies and meat-heavy diets, which take many more resources to produce than grain- or legume-based diets.

Race and Ethnicity
Affirmative Action: Is It Time to End Racial Preferences?

Since the 1970s, affirmative action has played a key role in helping minorities get ahead. But many Americans say school and job candidates should be chosen on merit, not race. This November, ballot initiatives in Colorado and Nebraska would eliminate race as a selection criterion for job or school candidates but would allow preferences for those trying to struggle out of poverty, regardless of their race. It's an approach endorsed by foes of racial affirmative action. Big states, meanwhile, including California and Texas, are still struggling to reconcile restrictions on the use of race in college admissions designed to promote diversity. Progress toward that goal has been slowed by a major obstacle: Affirmative action hasn't lessened the stunning racial disparities in academic performance plaguing elementary and high school education. Still, the once open hostility to affirmative action of decades ago has faded. Even some race-preference critics don't want to eliminate it entirely but seek ways to keep diversity without eroding admission and hiring standards.

Gender

Gender Pay Gap: Are Women Paid Fairly in the Workplace?

More than four decades after Congress passed landmark anti-discrimination legislation —including the Equal Pay Act of 1963 — a debate continues to rage over whether women are paid fairly in the workplace. Contending that gender bias contributes to a significant "pay gap," reformists support proposed federal legislation aimed at bringing women's wages more closely in line with those of men. Others say new laws are not needed because the wage gap largely can be explained by such factors as women's choices of occupation and the amount of time they spend in the labor force. Meanwhile, a class-action suit charging Wal-Mart Stores with gender bias in pay and promotions —the biggest sex-discrimination lawsuit in U.S. history — may be heading for the Supreme Court. Some women's advocates argue that a controversial high-court ruling last year makes it more difficult to sue over wage discrimination.

Sexual Orientation

Gay Marriage Showdowns: Will Voters Bar Marriage for Same-Sex Couples?

The California Supreme Court gave gay rights advocates a major victory in May, ruling the state's constitution guarantees same-sex couples the same marriage rights as opposite-sex pairs. Thousands of same-sex couples from California and around the country have already taken advantage of the decision to obtain legal recognition from California for their unions. Opponents, however, have placed on the state's Nov. 4 ballot a constitutional amendment that would deny marriage rights to same-sex couples by defining marriage as the union of one man and one woman. Similar proposals are on the ballot in Arizona and Florida. The ballot-box showdowns come as nationwide polls indicate support for some legal protection for same-sex couples, but not necessarily marriage equality. In California, one early poll showed support for the ballot measure, but more recently it has been trailing. Opposing groups expect to spend about $20 million each before the campaign ends.

Age and Aging

Aging Baby Boomers: Will the 'Youth Generation' Redefine Old Age?

In January, the oldest baby boomers will turn 62 — and become eligible to collect Social Security benefits. For the next 18 years, a member of the baby boom generation —the 78 million Americans born between 1946 and 1964 — will reach that age every eight seconds. Boomers have long been famous for their desire to stay or at least act young. What will they be like as seniors? Many predict they will reshape the nation's view of old age, as healthier boomers continue to work and stay active longer than their parents. Others worry that the vast expansion of the nation's senior population will put unaffordable strains on government entitlement programs like Social Security and Medicare. Still others worry boomers could upset the economy as they begin spending down their assets all at once. Boomers have left their imprint on every stage of American life they've passed through, and there's no reason to think that the senior years will be any exception.

OUR SOCIAL INSTITUTIONS
Families

Future of Marriage: Is Traditional Matrimony Going Out of Style?

In the past 40 years, the nation's marriage rate has dropped from three-quarters of American households to slightly over half. Moreover, nearly 50 percent of all U.S.

marriages now end in divorce, and the number of households with unmarried couples has risen dramatically. Some scholars say that although traditional marriage will not disappear entirely, it will never again be the nation's pre-eminent social arrangement. In the future, they say, the United States will look more like Europe, where couples increasingly are opting to cohabit rather than marry. But other experts argue that the recent decrease in the divorce rate and other positive trends point to a brighter future for marriage. Meanwhile, actions by a number of state courts and local officials in favor of same-sex unions have helped ignite a debate over the issue and prompted conservatives to push for a constitutional amendment banning gay marriage.

Education

Student Aid: Will Many
Low-Income Students Be Left Out?

With a record number of students hoping to attend college next year — and fees higher than ever — finding a way to pay the bills will be tough for many. Congress and the Bush administration made common cause in 2007 to increase federal Pell Grants for students and reduce some student-loan interest rates. Nevertheless, critics say the increases won't go far enough. To help middle-class families, states increasingly offer merit-based grants for college aid. But with merit scholarships replacing need-based aid, low-income and minority students — who often don't have the grades for scholarships — are finding their college dreams harder to realize. Meanwhile, longtime concern that private lenders rake in excess profits from their high-interest student loans has reached new heights. Investigations of student lending are being conducted in several states, even as universities and lenders settle allegations of loan fraud with New York's attorney general.

Work and the Economy

Public-Works Projects: Do They Stimulate
the Economy More Than Tax Cuts?

To battle the Great Depression, President Franklin D. Roosevelt put millions of unemployed Americans to work on New Deal projects such as repairing roads and building cabins in national parks. To stimulate today's ailing economy, Congress has enacted a $787 billion package that includes tax cuts and spending on infrastructure, including expanding highway and rail systems and weatherizing buildings. But many conservatives argue that government spending does not create jobs and merely diverts money from the private sector, which they call the only true engine of job creation. Meanwhile, infrastructure experts worry that if federal public-works dollars are spent too quickly, the money will go to eco-unfriendly projects, such as additional highway lanes that encourage fossil-fuel use and suburban sprawl, rather than to more future-oriented "green" initiatives like expanding rail and public transit and upgrading the electrical grid to accommodate alternative power sources.

Health and Medicine

Universal Coverage: Will All Americans
Finally Get Health Insurance?

Some 45 million Americans lacked health insurance in 2005 — a number that has been climbing for two decades. Every month, about 2 million Americans become uninsured, at least temporarily, as lower-paying service jobs with minimal benefits replace union-dominated manufacturing jobs with health benefits — undercutting the nation's employer-based coverage system. Health costs — rising faster than wages or inflation — also push employers to drop coverage. Past legislative proposals for universal coverage relied heavily on government management, drawing fatal opposition from physicians and insurance companies. But now consensus may be forming around proposals requiring most Americans to buy private insurance with public assistance. Republican governors in California and Massachusetts back such plans, as does former Sen. John Edwards, the first presidential hopeful to announce what's expected to be a slew of universal-coverage proposals in the coming 2008 election.

The Media

Television's Future: Will TV Remain
the Dominant Mass Medium?

Television is changing rapidly, and so is the TV audience. Viewers are ignoring broadcast schedules and watching programs via Internet "streams" and iPod downloads. Or they are "time-shifting" and skipping the commercials, using digital video recorders, such as TiVo, or video-on-demand television. Millions are also spending time watching user-generated video on sites such as YouTube. Many TV executives are wondering whether television can sustain its traditional approach to making money — "renting out" millions of viewers at a time to advertisers. For all the ferment, however, Americans are watching more television than ever, using the new devices not to avoid traditional TV but to catch up on shows they otherwise would have missed. There's an atmosphere of experimentation and uncertainty in the industry reminiscent of the dot-com boom, but television and advertising executives insist that the future of TV is bright.

OUR SOCIAL AND PHYSICAL WORLDS
Alcohol and Drug Abuse
Drinking on Campus: Have Efforts to Reduce Alcohol Abuse Failed?
Tremendous media attention has been focused on heavy college drinking during the past decade, but drinking habits have changed little. Alcohol is still the drug of choice among young people, especially on college campuses. Each year, some 1,700 students die due to drunken driving or other alcohol-related incidents. This year, the Duke University lacrosse team made the news when an exotic dancer accused three of its members of raping her at an alcohol-fueled party. Studies have found that rates of binge drinking and its often-devastating outcomes have remained remarkably stable over time, despite various attempts to reduce alcohol consumption on campus. According to the National Institute of Alcohol Abuse and Alcoholism, too many college alcohol programs are not supported by research. As a result, there is considerable debate about what colleges and surrounding communities can do to reduce excessive drinking among students.

Crime and Criminal Justice
Juvenile Justice: Are Sentencing Policies Too Harsh?
As many as 200,000 youths charged with crimes today are tried in adult courts, where judges tend to be tougher and punishments harsher —including sentencing to adult prisons. But with juvenile crime now on the decline, youth advocates are seizing the moment to push for major changes in iron-fisted juvenile justice systems nationwide. Above all, they want to roll back harsh state punishments — triggered by the crack cocaine-fueled crime wave of the late 1980s and early '90s — that sent thousands of adolescents to adult courts and prisons. Many prosecutors say the get-tough approach offers society the best protection. But critics say young people often leave prison more bitter and dangerous than when they went in. Moreover, recent brain studies show weak impulse control in young people under age 18, prompting some states to reconsider their tough punishments. Prosecutors respond that even immature adolescents know right from wrong.

Cities and Suburbs
Rebuilding New Orleans: Should Flood-Prone Areas Be Redeveloped?
Five months after Hurricane Katrina flooded most of New Orleans, some 80 percent of the "Crescent City" remains unrepaired. Damage is estimated at $35 billion. Most schools and businesses are still closed, and two-thirds of the 460,000 residents have moved out. How many will return remains troublingly uncertain. Municipal leaders only this month began setting up a process to decide which of the city's 73 neighborhoods can be resettled and which would be left uninhabited to soak up future floodwaters. Questions about who will help the city's poorer residents — many of them African-American — hang over the city, along with concern about how much of New Orleans' storied popular culture will survive. Meanwhile, as a new hurricane season approaches, efforts to repair and strengthen the protective system of levees, canals and pumps lag behind schedule.

The Environment
Regulating Toxic Chemicals: Do We Know Enough About Chemical Risks?
Chemicals are integral to many everyday products, from electronics and toys to building materials and household goods. But environmental, health and consumer advocates say the agencies responsible for protecting Americans from exposure to harmful chemicals are

allowing too many dangerous substances into the market without testing them for toxicity. Some goods, such as medicines, are tested for safety before they can be sold, but many common products do not go through premarket safety screening. Many concerns focus on infants and young children, who are especially sensitive to toxic hazards. Chemical manufacturers say the existing regulatory system works effectively and can be tightened to address new concerns, but critics argue that a precautionary approach — which would require producers to show that materials are safe before they can be marketed — would protect consumers more fully.

War and Terrorism

Cost of the Iraq War: Are Economic Woes a Casualty of Unexpectedly High Costs?

The fifth anniversary of the Iraq War hit just as the subprime mortgage crisis and rising unemployment in the United States were turning the economic situation bleak. Against this backdrop, a Nobel laureate economist and a federal budget expert linked the economic downturn to the war and calculated its eventual total financial cost at $3 trillion and possibly even more, plus the tens of thousands of Americans and Iraqis killed or wounded. President George W. Bush dismisses the linkage argument, contending the war creates job opportunities at home and that military spending in Iraq and Afghanistan amounts to only a "modest fraction" of the U.S. economy. But even Republican lawmakers have been asking why taxpayers are funding much of the rebuilding of oil-rich Iraq while it reaps billions in profits thanks to record-high oil prices. For its part, the administration says Iraq is now starting to bear more of the reconstruction costs.

INDIVIDUAL ACTION AND SOCIAL CHANGE
Social Problems and Social Action

The Obama Presidency: Can Barack Obama Deliver the Change He Promises?

As the 44th president of the United States, Barack Hussein Obama confronts a set of challenges more daunting perhaps than any chief executive has faced since the Great Depression and World War II. At home, the nation is in the second year of a recession that Obama warns may get worse before the economy starts to improve. Abroad, he faces the task of withdrawing U.S. forces from Iraq, reversing the deteriorating conditions in Afghanistan and trying to ease the Israeli-Palestinian conflict. Still, Obama begins his four years in office with the biggest winning percentage of any president in 20 years and a strong Democratic majority in both houses of Congress. In addition, as the first African-American president, Obama starts with a reservoir of goodwill from Americans and people and governments around the world. But he began encountering criticism and opposition from Republicans in his first days in office as he filled in the details of his campaign theme: "Change We Can Believe In."

Preface

Will voters bar marriage for same-sex couples? Are women paid fairly in the workplace? Is it time to end racial preferences? Will all Americans finally get health insurance? These questions and many more are addressed in a unique selection of articles for debate focused on social problems offered exclusively through *CQ Researcher*, CQ Press and SAGE. This collection aims to promote in-depth discussion, facilitate further research and help students formulate their own positions on crucial issues.

This first edition includes sixteen up-to-date reports by *CQ Researcher,* an award-winning weekly policy brief that brings complicated issues down to earth. Each report chronicles and analyzes current social problems. This collection was carefully crafted to cover a range of issues including the global food crisis, gay marriage, affirmative action, aging baby boomers, the Obama Presidency and much more. All in all, this reader will help your students become better versed on current social problems and gain a deeper, more critical perspective of timely and important issues.

CQ RESEARCHER

CQ Researcher was founded in 1923 as *Editorial Research Reports* and was sold primarily to newspapers as a research tool. The magazine was renamed and redesigned in 1991 as *CQ Researcher.* Today, students are its primary audience. While still used by hundreds of journalists and newspapers, many of which reprint portions of the reports, the *Researcher's* main subscribers are now high school, college and public libraries. In 2002, *Researcher* won the American Bar Association's coveted Silver Gavel award for magazine excellence for a series of nine reports on civil liberties and other legal issues.

Researcher staff writers—all highly experienced journalists—sometimes compare the experience of writing a *Researcher* report to drafting a college term paper. Indeed, there are many similarities. Each report is as long as many term papers—about 11,000 words—and is written by one person without any significant outside help. One of the key differences is that writers interview leading experts, scholars and government officials for each issue.

Like students, staff writers begin the creative process by choosing a topic. Working with the *Researcher's* editors, the writer identifies a controversial subject that has important public policy implications. After a topic is selected, the writer embarks on one to two weeks of intense research. Newspaper and magazine articles are clipped or downloaded, books are ordered and information is gathered from a wide variety of sources, including interest groups, universities and the government. Once the writers are well informed, they develop a detailed outline, and begin the interview process. Each report requires a minimum of ten to fifteen interviews with academics, officials, lobbyists and people working in the field. Only after all interviews are completed does the writing begin.

CHAPTER FORMAT

Each issue of *CQ Researcher,* and therefore each selection in this book, is structured in the same way. Each begins with an overview, which briefly summarizes the areas that will be explored in greater detail in the rest of the chapter. The next section chronicles important and current debates on the topic under discussion and is structured around a number of key questions. These questions are usually the subject of much debate among practitioners and scholars in the field. Hence, the answers presented are never conclusive but detail the range of opinion on the topic.

Next, the "Background" section provides a history of the issue being examined. This retrospective covers important legislative measures, executive actions and court decisions that illustrate how current policy has evolved. Then the "Current Situation" section examines contemporary policy issues, legislation under consideration and legal action being taken. Each selection concludes with an "Outlook" section, which addresses possible regulation, court rulings and initiatives from Capitol Hill and the White House over the next five to ten years.

Each report contains features that augment the main text: two to three sidebars that examine issues related to the topic at hand, a pro versus con debate between two experts, a chronology of key dates and events and an annotated bibliography detailing major sources used by the writer.

ACKNOWLEDGMENTS

We wish to thank many people for helping to make this collection a reality. Tom Colin, managing editor of *CQ Researcher,* gave us his enthusiastic support and cooperation as we developed this edition. He and his talented staff of editors and writers have amassed a first-class library of *Researcher* reports, and we are fortunate to have access to that rich cache. We also wish to thank our colleagues at CQ Press, a division of SAGE and a leading publisher of books, directories, research publications and Web products on U.S. government, world affairs and communications. They have forged the way in making these readers a useful resource for instruction across a range of undergraduate and graduate courses.

Some readers may be learning about *CQ Researcher* for the first time. We expect that many readers will want regular access to this excellent weekly research tool. For subscription information or a no-obligation free trial of *CQ Researcher,* please contact CQ Press at www.cqpress .com or toll-free at 1-866-4CQ-PRESS (1-866-427-7737).

We hope that you will be pleased by this edition of *Social Problems: Selections From CQ Researcher.* We welcome your feedback and suggestions for future editions. Please direct comments to David Repetto, Sr. Acquisitions Editor, Pine Forge Press, an Imprint of SAGE Publications, 2455 Teller Road, Thousand Oaks, CA 91320, or david.repetto@sagepub.com.

—The Editors of SAGE

Contributors

Thomas J. Billitteri is a *CQ Researcher* staff writer based in Fairfield, Pennsylvania, who has more than 30 years' experience covering business, nonprofit institutions and public policy for newspapers and other publications. He has written previously for *CQ Researcher* on "Domestic Poverty," "Curbing CEO Pay" and "Mass Transit." He holds a BA in English and an MA in journalism from Indiana University.

Marcia Clemmitt is a veteran social-policy reporter who previously served as editor in chief of *Medicine & Health* and staff writer for *The Scientist*. She has also been a high-school math and physics teacher. She holds a liberal arts and sciences degree from St. John's College, Annapolis, and a master's degree in English from Georgetown University. Her recent reports include "Climate Change," "Health Care Costs," "Cyber Socializing" and "Student Aid."

Alan Greenblatt is a staff writer at *Governing* magazine. He previously covered elections, agriculture and military spending for *CQ Weekly*, where he won the National Press Club's Sandy Hume Award for political journalism. He graduated from San Francisco State University in 1986 and received a master's degree in English literature from the University of Virginia in 1988. His recent *CQ Researcher* reports include "Sex Offenders" and "Pension Crisis."

Kenneth Jost graduated from Harvard College and Georgetown University Law Center. He is the author of the *Supreme Court*

Yearbook and editor of *The Supreme Court from A to Z* (both CQ Press). He was a member of the *CQ Researcher* team that won the 2002 ABA Silver Gavel Award. His previous reports include "Gays on Campus" and "Gay Marriage."

Peter Katel is a *CQ Researcher* staff writer who previously reported on Haiti and Latin America for *Time* and *Newsweek* and covered the Southwest for newspapers in New Mexico. He has received several journalism awards, including the Bartolomé Mitre Award for coverage of drug trafficking, from the Inter-American Press Association. He holds an AB in university studies from the University of New Mexico. His recent reports include "Oil Jitters," "Race and Politics" and "Rise in Counterinsurgency."

Barbara Mantel is a freelance writer in New York City whose work has appeared in *The New York Times*, the *Journal of Child and Adolescent Psychopharmacology* and *Mamm Magazine*. She is a former correspondent and senior producer for National Public Radio and has won several journalism awards, including the National Press Club's Best Consumer Journalism Award and Lincoln University's Unity Award. She holds a BA in history and economics from the University of Virginia and an MA in economics from Northwestern University.

David Masci specializes in science, religion and foreign policy issues. Before joining *The CQ Researcher* in 1996, he was a reporter at Congressional Quarterly's *Daily Monitor* and *CQ Weekly*. He holds a law degree from The George Washington University and a BA in medieval history from Syracuse University. His recent reports include "Rebuilding Iraq" and "Human Trafficking and Slavery."

Jennifer Weeks is a *CQ Researcher* contributing writer in Watertown, Massachusetts, who specializes in energy and environmental issues. She has written for *The Washington Post*, *The Boston Globe Magazine* and other publications, and has 15 years' experience as a public-policy analyst, lobbyist and congressional staffer. She has an AB degree from Williams College and master's degrees from the University of North Carolina and Harvard.

1

Global Food Crisis

What's Causing the Rising Prices?

Marcia Clemmitt

AFP/Getty Images/David Greedy

Flooded corn crops throughout the Midwest are contributing to rising food prices in the United States and abroad, where higher prices already have plunged millions of poor people into malnutrition and starvation. Drought, high transportation costs stemming from higher oil prices and a growing diversion of corn for use as a biofuel also have contributed to the price hikes.

From *CQ Researcher*,
June 27, 2008.

Spiking food prices have brought pain at supermarket checkout counters for millions of American families this past year, but in many developing countries, the situation is far more severe:

- In Somalia, people who can no longer afford food in markets try to stave off starvation with a watery soup made from the mashed branches of thorn trees.[1]
- In North Korea, where more than a third of the population is undernourished, the price of rice, the major food staple, soared 186 percent between 2007 and 2008, and overall food prices rose 70 percent.
- In Yemen, where 36 percent of the population is undernourished, wheat prices doubled.[2]
- In tiny Burundi, where about half the population is desperately poor, the price tripled for the landlocked nation's food staple, farine noir, a mixture of black flour and ground cassava root.[3]

With 2.1 billion people worldwide living on less than $2 a day and another 880 million living on less than $1 a day, price increases of such magnitude have plunged hundreds of millions into malnutrition and starvation.[4]

The price spikes have several causes, including drought and bad harvests in major food-exporting countries, high oil prices that make food more expensive to chemically fertilize and transport and a growing diversion of corn for use as a biofuel.

Some critics also blame the impact of globalization and the continued use of farm subsidies by industrialized nations, which they say undercut prices in poor countries.

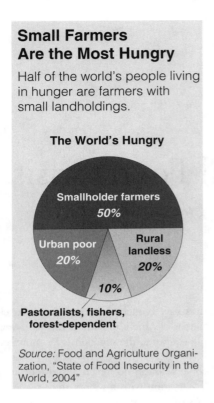

Small Farmers Are the Most Hungry

Half of the world's people living in hunger are farmers with small landholdings.

The World's Hungry

Smallholder farmers
50%

Urban poor
20%

Rural landless
20%

10%

Pastoralists, fishers, forest-dependent

Source: Food and Agriculture Organization, "State of Food Insecurity in the World, 2004"

With harvests expected to improve and more land being brought into cultivation, prices are expected to drop somewhat next year, according to the United Nations Food and Agriculture Organization (FAO) and the Organization for Economic Cooperation and Development (OECD), representing industrialized nations.[5] Nevertheless, experts warn, serious pressure on the world's food supply poses a long-term threat.

"The era of cheap food may be over," as rising oil prices drive the cost of food production and transport upward, said Haruhiko Kuroda, president of the Asian Development Bank.[6] Over the past several decades, the world's food system has been transformed from local production to a global market, where many countries produce large quantities of just a few crops each, mainly for export, while depending on imports for much of their own food supply.

"A core problem is that 35 countries don't produce enough food to give their residents a 2,000-calorie-per-day diet, even if all their production was being distributed equally" among citizens, says Cornell University Professor of Applied Economics and Management Christopher B. Barrett.

Furthermore, most of the world's population growth now occurs in the very developing nations that are currently unable to produce enough food to feed themselves, Barrett says.

Readjusting the global food system to avoid future crises will require fundamental rethinking of how and where food is produced and how it's allocated, analysts say.

"We're running up against this brick wall called finite resources," mainly the fertile soil and ample water needed to sustain good harvests, says Randall Doyle, an assistant professor of history at Central Michigan University.

"I always tell [food] producers that this whole thing is not rocket science — it's far more complicated," says Jerry L. Hatfield, supervisory plant physiologist at the U.S. Department of Agriculture's (USDA) National Soil Tilth Research Laboratory in Ames, Iowa. To manage water resources successfully, for example, "you have to look at the whole landscape."

"Feeding 6 billion people is really hard," says Curt Ellis, a filmmaker in Portland, Ore., whose documentary on American farming, "King Corn," aired recently on PBS. "I don't think we've figured out the right way to do it."

"We've got to increase the supply," says Mark Alley, a professor of agriculture at Virginia Tech and president-elect of the American Society of Agronomists.

That is especially difficult "in Europe, the United States and Australia, where our ability to exponentially increase food production is quite limited," says Doyle. This means that the most attention must be spent on increasing agriculture yields in developing countries, especially in Africa, where agriculture is least advanced, he says.

However, development experts say, there's no consensus on how future farming should look — what balance should be struck between large-scale industrial farming for export and smaller farms that produce food for local consumption.

"There's no consensus in the global development community about agriculture," says Peter Gubbels, vice president for international programs at World Neighbors, a nonprofit development organization in Oklahoma City that helps poor farmers in developing countries become self-supporting. Nevertheless, "there are growing movements in every country" to return to more local production, he says. "Some call that food sovereignty, and now we're even beginning to see the U.N. and the World Bank" talking about it.

The food crisis has sparked international tension over the rich diets enjoyed by industrialized nations and the fear that, as developing countries add more animal products to their menus, food crises will increase.

"There's still plenty of food for everyone, but only if everyone eats a grain and legume-based diet," said Peter Timmer, a fellow at the Washington-based Center for Global Development. "If the diet includes large . . . amounts of animal protein (not to mention biofuels for our SUVs), food demand is running ahead of global production," he said.[7]

In India, the "middle class is larger than our entire population," said President George W. Bush in May, and "when you start getting wealth, you start demanding better nutrition and better food," including meat, which increases global food demand and "causes prices to go up."[8]

But Indians reacted with outrage to Bush's implication that their diets have fueled food-price spikes. "Bush is shifting the blame to hide the truth," said Devinder Sharma, chair of the New Dehli-based Forum for Biotechnology and Food Security. "We all know that the food crisis is an outcome of the American policy of diverting huge land area from food to fuel production," under a congressional mandate to increase use of biofuels, mainly corn-based ethanol.[9]

While greater consumption of meat in developing countries is a long-term trend, it's not a factor in current price spikes, says Brian Wright, professor of agricultural and resource economics at the University of California at Berkeley. For example, he says, Indians consume only 37 eggs a year per person, and "meat consumption is almost not on the charts."

Other analysts argue that developing countries' farm sectors have been crippled because the United States and other wealthy nations shut out poor nations' farm exports while subsidizing their own farmers to sell abroad below cost.

"The U.S. and the European Union in particular have preached free markets but have been in blatant disregard" of trade rules, which they repeatedly tweak to their own advantage, says Thomas Dobbs, a professor emeritus of economics at South Dakota State University in Brookings. "We produce too much of the wrong kind of thing," then "dump it on Third World markets and remove [those countries'] incentive for local production," he says.[10]

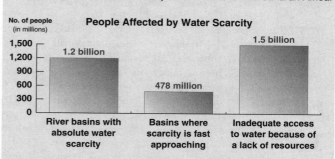

Billions Suffer From Water Scarcity

Three billion people — almost half the world's population — face serious actual or potential water shortages. Most are in the Middle East and North Africa, followed by South Asia and sub-Saharan Africa.

People Affected by Water Scarcity

No. of people (in millions)

- River basins with absolute water scarcity: 1.2 billion
- Basins where scarcity is fast approaching: 478 million
- Inadequate access to water because of a lack of resources: 1.5 billion

Sources: International Water Management Institute, in "World Development Report 2008: Agriculture for Development," The World Bank, 2007

But many U.S. policy makers hotly defend the subsidies. By and large, the United States has not constructed overwhelming trade barriers against agricultural products, said former U.S. Secretary of Agriculture Ann M. Veneman. U.S. farm subsidies "haven't changed market access into this country. At least 91 percent of African produce comes into this country duty free."[11]

Addressing these contentious issues will be difficult because "the poor are voiceless," says Cornell's Barrett. "The loudest and often the shrillest voices are those who aren't paying attention to the billion or so people who are living on a dollar or so a day."

As farmers, scientists and policy makers seek ways to feed a changing planet and expanding population, here are some of the questions that are being asked:

Can enough food be sustainably produced to feed the global population?

Environmentalists warn that water and soil resources soon may be outstripped by growing dietary demands.

"As the economy grows, its demands are outgrowing the Earth," said Lester R. Brown, founder of the Earth Policy Institute, which supports sustainable economic policies. "While the world economy multiplied sevenfold in just 50 years, the Earth's natural life-support systems remained essentially the same. Water use tripled, but the capacity . . . to produce fresh water through evaporation changed little. The demand for seafood increased fivefold, but the sustainable yield of oceanic fisheries was unchanged," he said.[12]

Hungry Countries Face Double Whammy

High food prices and climate change trigger unrest.

The recent worldwide price spike for grains and other food staples has left vulnerable populations in many developing countries at risk for malnutrition and starvation. And the growing hunger crisis leaves unstable nations open to "an emerging security problem as well," says Cornell University Professor of Applied Economics and Management Christopher B. Barrett.

But the very instability of many developing nations also leaves them vulnerable to yet another threat — climate change.

Local circumstances largely determine which populations are the most vulnerable, says Johan Selle, director of operations at iJET, an international risk-management consultancy in Annapolis, Md. In rural Kenya, for example, former nomadic peoples "are really struggling," says Selle. "Ten years ago they would have produced their own food," but today many are urban dwellers without enough income to buy imported food at skyrocketing global prices, he says.

The food-price crisis has triggered riots and strikes in more than 30 countries, mostly places where existing political unease has made populations ripe for protest.

Countries like Jordan, which have enough financial stability to subsidize food for their poorest citizens, don't see the unrest, Selle says. But "wherever there's instability, food shortages are the final straw," says Frederic Ngoga Gateretse, iJET regional manager for Africa.

"The capacity of pressure groups to organize" in a country and a population's "history of taking to the streets" largely determine whether the food crisis has triggered unrest, Gateretse says.

Guinea, in West Africa, for example, "is pretty much a failed state, and trade unions there have the capacity to mobilize and get on the streets" after a 10-year struggle of opposition parties trying to remove the current president from office, he says. Guinea has experienced four union-led food strikes, says Gateretse. Trade unions also have been involved in protests in the West African nation of Cameroon, where unrest over food prices lasted for five days in February and left many people dead or injured, he says.

Lack of a stable government or an economic infrastructure also paved the way for food strikes in Haiti, where the unrest was a response to the government's long-term inability to take care of the people, says Selle.

Urban populations have been more likely to riot than rural populations "because the foods they buy to eat are more likely to be affected by price hikes, since they require fuel [to grow, produce and transport], like bread," says Gateretse.

In some countries, riots are unlikely — even with extreme food stress — because citizens fear reprisals. Zimbabwe is suffering from severe shortages, for example, but people know that the military would crush any active rebellion "so they have yet to see protests on the street," says Selle.

In any event, the impact of climate change on harvests worldwide likely means unrest over food supplies will long outlast the current price spike in many developing countries.

Many of the threatened countries lie near the equator and are at high risk for desertification and water shortages, says Gateretse. "Many governments in poor countries do not have the ability to anticipate or handle

"The bottom line is that it is now more difficult for farmers to keep up with the growing demand for grain," said Brown. "Food insecurity may soon eclipse terrorism as the overriding concern of national governments."[13]

The current American diet, in particular, may not be sustainable, many commentators say.

"Perhaps three Earths would be required to support the current human population if everyone lived the over-consumptive North American lifestyle," noted the environmentalist Web site OilEmpire.us.[14]

A factor in recent grain price hikes "is the amount that's being used to increase the meat and milk supply," as more people consume more such foods, says Virginia Tech's Alley. "Can we produce all the food we need? Yes. All we want? Not necessarily."

Furthermore, "there is no way to produce more food without occupying more land and taking down more trees" in the rainforest, said Blairo Maggi, owner of the soybean-producing company Andre Maggi Group and governor of Brazil's Mato Grosso state.[15]

climate-change-related crises, and no one is training the governments to do so," he says.

Wealthy nations, including the United States, have not offered the help that developing countries need to develop more stable food systems, said Jacques Diouf, director-general of the U.N. Food and Agriculture Organization (FAO). "The developing countries did, in fact, forge policies, strategies and programs that — if they had received appropriate funding — would have given us world food security," but the industrialized nations spent the money on subsidies for their own farmers instead, Diouf said.[1]

But many analysts say that even if aid from wealthy nations had been available, unstable regimes in many nations can't use the aid effectively.

"A lot of developmental solutions are undermined by corruption" in developing countries' governments, says John Walton, a professor of sociology at the University of California at Davis.

"Many of the countries in most trouble are failed states," says Brian Wright, a professor of agricultural and resource economics at the University of California at Berkeley. Without stable and functional governments, countries have "no local research capabilities," without which agricultural science is ineffective, he says. "You can't just take a plant from another country and stick it in the ground." Local researchers must take research findings from elsewhere and figure out how to adapt them to local conditions.

Even the most well-intentioned assistance can easily fail if local government is unstable or corrupt, says Josh N. Ruxin, an assistant professor of public health at New York's Columbia University and director of its Millennium Village Project in Rwanda. "Stability in government is extremely important. It means that interventions can be rolled out large scale, and faster," he says.

"That's one of the reasons I live and work in Rwanda, because now there's a lot of transparency and stability in government," Ruxin says. "If you look at a place that's extremely corrupt, like Zimbabwe, with bad agriculture policies, you're just hitting your head against the wall."

Farming requires infrastructure, such as roads for farmers to get their crops to market, says Ray Cesca, president of the World Agricultural Forum, which seeks to improve world agriculture. "It's up to the government to build the road, but sometimes the money disappears," he says.

Unfortunately, stable governments are important for moving a population out of poverty and hunger, but poverty and hunger themselves act against development of stable government, says Thomas Dobbs, professor emeritus of economics at South Dakota State University in Brookings. "It's a vicious cycle, and difficult to get out of."

For example, a sustainable system to give a country food security would include "more domestic production, more garden plots" tended over the years right where people live, says Dobbs. "But when there's a civil war and people are moved into camps, that system all falls apart."

There's plenty that industrialized countries like the United States, foundations and other donors can do to help, says Ruxin. For example, "very few governments in sub-Saharan Africa have sufficient agricultural extension [education] to reach all their farmers, and donor funds can have tremendous impact" by helping support such efforts, he says.

And the "failed-state" excuse for not offering assistance doesn't hold water, says Cornell's Barrett. "Of course, nothing productive is going to happen in Zimbabwe, but there are other places with elected governments that don't get support" from wealthy nations. "Donors fiddle" while the future ebbs away, he says.

---

[1] Quoted in Elisabeth Rosenthal and Andrew Martin, "U.N. Issues Warning on Food Crisis," *The New York Times*, June 4, 2008, p. A6.

A growing number of analysts argue that today's industrial-style agriculture, which depends heavily on fossil fuels for fertilizer manufacture and long-distance transport, cannot be sustained.

A 2008 report for the intergovernmental group International Assessment of Agricultural Science and Technology for Development, based on input from private- and public-sector participants from developed and developing nations, concluded that "the dominant practice of industrial, large-scale agriculture is unsustainable, mainly because of the dependence of such farming on cheap oil, its negative effects on ecosystems and growing water scarcity."[16]

As fossil-fuel supplies run out, farm products like corn are increasingly called into service as biofuel, and the heightened demand translates into higher food prices.

"I worry that with biofuels, the food market will end up looking like the worldwide pharmaceutical market. The rich will get theirs, and the poor will die," says

Food Prices Jumped 50 Percent

Prices for several food staples have risen dramatically in recent years. Most have increased by 50 percent, while skim milk powder has more than doubled. Prices are expected to drop in the future, but not back to earlier levels.

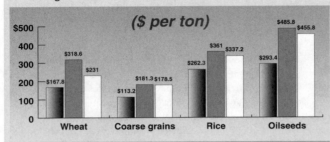

Average World Prices for Selected Commodities, 2002-2013

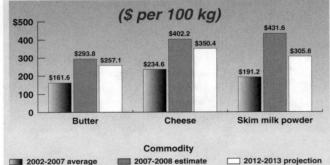

Source: "OECD-FAO Agricultural Outlook 2008-2017," Organisation for Economic Cooperation and Development, 2008

Success depends on investing in more agricultural research and then making sure it's implemented, says Wright. "The economic returns on crop research are the highest" returns on any research, he says.

"Earth will have 9 billion people by 2050, and we need a 33 percent jump in food productivity," says Doyle at Central Michigan University. 'We need another Green Revolution," he says, referring to the period from the late 1940s through the 1970s, when Western researchers spurred huge crop-yield improvements in developing countries, especially in Asia, using new high-yield grain varieties.

"Do we have the technology to increase the supply? Absolutely," Alley says. "We have better varieties of seeds, methods to control pests," and more. "The issue is implementation of these technologies."

China provides evidence that yields can improve, Alley says. In the early 1990s, some predicted that China's swelling population would soon require it to import virtually all the grain produced on the planet. But within a decade, China had implemented farm improvements that made the country a net grain exporter until as recently as last year, he says.

Developing countries have plenty of unfarmed arable — farmable — land, although barriers exist to its being brought into use, says Cesca. The greatest amounts of potentially arable land that can be used — not counting rainforests — are in Africa, followed by Asia and Latin America, often in the most food-deprived nations, he says. Currently, however, poverty, unstable governments and a lack of business-friendly policies — such as local barriers to setting up banks or getting loans — block development of the land for farming, Cesca says.

Harvests can be improved even in very difficult environments if farmers get help to improve their practices, says Gubbels at World Neighbors.

In many places, such as the Sahel, a dry, semi-tropical belt of shrub- and grasslands that runs across Africa, farmers cannot get into [large-scale] industrial agriculture because "fertilizer costs three to four times as much" as in the United States because of transport

Berkeley's Wright. "I would abandon the grain-based biofuels."

Nevertheless, Wright and others say that human effort has vastly increased food production in the past and can continue to do so.

The improvements in agriculture made during the 20th century greatly increased food yields, says Wright. Today there is "some sign that the rate of increase of yields is declining, but there's no evidence that we're going to reach a limit. We still have fairly good yield increases."

"Potentially, can we feed everybody? Of course we can," says Ray Cesca, president of the World Agricultural Forum, which aims to bring together public- and private-sector expertise and resources to improve world agriculture.

costs, Gubbels says. However, small farms can greatly expand their harvests if farmers get help to learn and adopt the best techniques, he says. "A lot of it is building on farmers' own knowledge," but beefing up agricultural-extension programs to teach best practices is vital, he says.

In mountainous areas, for example, few farmers have dug trenches at proper intervals to stem water and soil loss, partly because "it's very labor intensive. But we got people to do this, and now all their neighbors are doing it," with better crops as a result, says Josh N. Ruxin, assistant professor of public health at New York's Columbia University and director of the University's Millennium Village Project in Rwanda.

"A lot of subsistence farmers have just scattered seed and prayed it would produce," but encouraging underutilized practices like sowing seeds in rows to prevent young plants from crowding each other out can greatly increase harvests, even in the most unpromising regions, Ruxin says.

Do U.S. farm and trade policies harm poor people in developing countries?

Critics say U.S. policies that give hefty subsidies to American growers encourage developing countries to favor U.S.-produced food at below-market prices. As a result, say critics, farmers in developing countries lose their local markets and grow poorer and developing nations themselves lose their food-production capability, risking famine if imports are unavailable or spike in price.

Other analysts argue, however, that American farmers need financial protection against fluctuating harvests and prices and that, ultimately, global trade is best for everyone.

Trade globalization helps developing countries create modern economies, rather than keeping them mired in a subsistence lifestyle, according to conservative British columnist Janet Daley. "What developing countries need is to develop, not to have their present conditions of life and work preserved like a museum exhibit," she wrote. Replacing globalization with more local food production merely means "sustaining agricultural activity that would not otherwise be sustainable in the global marketplace."[17]

Opening developing-world markets to imports can help those nations in the long run, says Berkeley's Wright. There can be some value in keeping a modest

local ability to produce farm staples like grains, but "globalization is what stops famines," he says. Bad weather or bad farm-sector decision-making can mean severe food shortages, "but if you have another place to get food, you have a safety net."

The argument for importing food rather than striving for agricultural self-sufficiency goes back to Scottish economist Adam Smith, whose 1776 classic *The Wealth of Nations* first described the logic of free-market capitalism. Global free trade in food is the international counterpart of Smith's key "proposition that people within a national economy will all be better off if they all specialize at what they do best instead of trying to be self-sufficient," said Jagdish Bhagwati, a professor of economics and political science at New York's Columbia University.[18]

In fact, the U.S. food market "tends to be fairly open" to developing-country products, said Bob Young, chief economist for the American Farm Bureau, which represents farmers. "The average tariff faced by countries trying to land agricultural products here is around 12 percent. The average tariff faced by our farmers is around 62 percent," making U.S. trade barriers minor compared to those put up by countries in the rest of the world, Young said.[19]

U.S. farmers need subsidies to offset the much stricter food-safety and environmental regulations they face, said Young. U.S. subsidies amount to "compensation to help level the playing field" for the more heavily regulated U.S. farmer, said Young. "We provide protection to other sectors of the economy when they face unfair competition. Why should agriculture be any different?"[20]

The key to determining conditions for the poor is not so much globalization itself but how governments respond to it, say some economists.

"Opening the economy to trade . . . need not make the poor worse off if appropriate domestic policies and institutions are in place," wrote Pranab Bardhan, a professor of economics at the University of California, Berkeley. Cases in point are the developing economies of Mauritius, South Korea and Botswana, whose citizens have prospered in recent years while the lot of the poor stagnated in other countries with similar resources, such as Jamaica, the Philippines and Angola, said Bardhan.[21]

But many commentators criticize U.S. farm and trade policy as harmful and hypocritical.

"The idea that you should import everything that you can buy cheaper from abroad means that you lose your ability to provide your own staples," a dangerous state that threatens nations' food security, says South Dakota State University's Dobbs. "Ironically, this is something the United States would never do" but persists in urging on others, he says.

The United States, Europe and Japan spend about $350 billion a year subsidizing their own farmers, "and at the same time they negotiate trade agreements with other countries requiring them to drop their own farm subsidies," says Gubbels at World Neighbors. "As a result, governments [in developing countries] tell their people, 'Move to the cities,' " thus shrinking their own farm sectors and driving up poverty. He describes a farmer in Ghana who raises chickens but can't sell them profitably in local markets because Europe dumps its excess chicken production in Africa at below-market prices.

A few decades ago, Haiti grew all the rice it needed and was one of the world's largest exporters of sugar and tropical produce, wrote William P. Quigley, a professor of law at Loyola University in New Orleans. But in April, food riots over high prices that threatened to push some families into starvation claimed the lives of six Haitians.[22]

Quigley blames the United States — and the U.S.-backed World Bank and International Monetary Fund — for pushing Haitian leaders to open the way to imports that decimated the country's farms. "In the 1980s, imported rice poured into Haiti, below the cost of what our farmers could produce," said a Haitian priest quoted by Quigley. "Farmers lost their businesses. . . . After a few years of cheap imported rice, local production went way down."[23]

Today Haiti, with an annual per capita income of $400, is the third-largest importer of U.S. rice, writes Quigley. Meanwhile, U.S. rice farmers have been supported by three different government subsidies that averaged more than $1 billion a year since 1998, Quigley said.[24]

In Mexico, U.S.-subsidized corn, imported under the North American Free Trade Agreement (NAFTA), is "swamping small farmers," wrote Conn Hallinan, an analyst for Foreign Policy in Focus, a Washington think tank that seeks diplomatic solutions to international problems. "Some 2 million farmers have left the land, and 18 million subsist on less than $2 a day,

accelerating rural poverty and helping to fuel" emigration, he said.[25]

"The extremely high level of U.S. government payments to farmers, while simultaneously encouraging other countries to reduce domestic agricultural supports," is an "egregious example of hypocrisy and double-speak," wrote analysts from the University of Tennessee's Agricultural Policy Analysis Center.[26]

Under World Trade Organization (WTO) agreements, the United States committed to reduce payments to American farmers but found ways around those commitments, which "have risen dramatically since 1996 and stand as a testament to U.S. admonitions to 'do as I say, not as I do,' when it comes to trade liberalization," said the center. "Our farm policy directly affects the livelihoods and sustainability of small farmers around the world."[27]

Do U.S. food aid policies harm people in developing countries?

Historically, the United States has been the largest donor of food to developing nations, tiding numerous countries over crises.[28] But critics of U.S. aid programs argue that much U.S. aid may benefit American producers, including multinational corporations, more than hungry people abroad. Indeed, they say most development aid completely bypasses the poor in rural areas who need help the most.

America's oldest and biggest food aid program, Food for Peace, was signed into law by President Dwight D. Eisenhower in 1954. Since then, it and other programs "have brought together governments, businesses, multilateral institutes such as the U.N. World Food Programme (WFP), and American voluntary organizations in a valuable public-private partnership intended to reduce hunger," wrote Cornell's Barrett and Daniel G. Maxwell, an associate professor of development economics at Tufts University and a former regional director for the charity CARE International.

In its first 50 years, Food for Peace alone "contributed more than 340 million metric tons of food aid to save and improve the lives of many hundreds of millions of poor and hungry people," according to Barrett and Maxwell.[29]

Food aid from the U.S. and other wealthy nations "clearly had a significant role in reducing loss of life during food emergencies in such countries as Ethiopia, Sudan, Somalia, Afghanistan, Rwanda and Haiti," according to analysts at the USDA's Economic Research Service.[30]

During Somalia's 1992-1993 civil war, food aid contributed about 70 percent of Somalians' food consumption and about half of Eritrea's between 2000 and 2004, they wrote.[31]

"The United States has traditionally been the major provider of emergency food aid during international humanitarian disasters," according to the Congressional Research Service (CRS), Congress' nonpartisan research agency. Historically, the United States "has been the world's largest provider of food aid, both emergency and non-emergency," accounting for just over 55 percent of total food aid in the 1990s.[32]

The resumption this May of U.S. food aid to North Korea is a "very timely" and "hugely significant contribution" that could head off the famine that threatens that country as food prices soar, said Jennifer Parmalee, a spokesperson for the World Food Programme. The U.S. suspended food aid to ccommunist North Korea in 2006 but over the long term has been the country's "biggest historical donor."[33]

A May announcement of a $40 million, three-year food-aid package to Bangladesh will provide "a means and incentive for children to stay in school so Bangladesh can prepare the next generation of leaders," since a large proportion of the aid will feed schoolchildren and pregnant and breastfeeding mothers, said James F. Moriarty, U.S. ambassador to Bangladesh. In 2007, floods and a cyclone destroyed crops in Bangladesh.[34]

But the current structure of aid also causes problems, many analysts say.

"There's an immediate need to reform the way we provide food aid," says Gubbels of World Neighbors. The United States is now "the only developed country that obliges people to buy U.S. grain," thus "tying its food aid to support" for large agribusiness companies like Decatur, Ill.-based Archer Daniels Midland and Minneapolis-based Cargill. Canada and Europe have stopped that practice. The U.S. method is "expensive

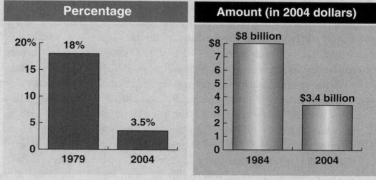

Agricultural Assistance on the Decline

Aid to agriculture amounted to only 3.5 percent of total development assistance in 2004, less than one-fifth the percentage 25 years earlier. Aid to agriculture in 2004 totalled $3.4 billion, less than half of the amount in 1984.

Official Development Assistance for Agriculture

Percentage	Amount (in 2004 dollars)
1979: 18% — 2004: 3.5%	1984: $8 billion — 2004: $3.4 billion

Source: "World Development Report 2008: Agriculture for Development," The World Bank, 2007

because you have to ship it all the way over there," says Gubbels.

In addition, "these often are not the foods people are used to," and bringing in tons of food from abroad ends up "degrading local agriculture," Gubbels says. "If you really want to help, give them cash or try to buy the food locally," he says.

The cost to taxpayers of food aid that must be bought from producers in the donor country and then shipped "has been shown to be significantly higher — in many cases 30-50 percent higher" — than aid from alternative sources, said the charity CARE USA.[35]

Moreover, the timing and extent of aid often has more to do with market conditions in wealthy donor countries than with developing-countries' needs, said the international charity Oxfam International. For example, in 1973, many developing countries faced food shortages, but that same year U.S. Food for Peace shipments dropped to less than a tenth of 1960s levels, according to the charity. The reason? Cereal prices were high around the world so grain producers' revenues from "commercial sales made surplus disposal" of food as aid to developing nations "unnecessary," said the group.[36]

Critics also blast "monetization" — the U.S. practice of shipping American grain to charities in a developing country, which then sell the grain locally and use the proceeds to finance their work. In 2007, CARE announced it would stop accepting monetized food aid by 2009, saying that the process is inefficient and delivers food not to the hungry but only to people who can afford to buy it.[37]

U.S. food aid has focused too much on addressing emergency situations "at the expense of addressing the chronic hunger and poverty that makes these crises so serious," said the charity Catholic Relief Services (CRS). To help countries avoid future food crises, food aid should be accompanied by other assistance, "such as investments in agricultural development aimed at small-scale producers," said CRS.[38]

Only 4 percent of development aid goes to small farmers, even though about 75 percent of those who survive on under a dollar a day live in rural areas, says Gubbels. Instead, most agricultural-development assistance goes to areas near coasts, where climate, land and location make it easier to produce large yields and export harvests efficiently into world trade, he says. A better policy for actually relieving hunger would be to "help the small farmers feed themselves," Gubbels says.

Aid programs should also begin paying for new vitamin- and mineral-fortified ready-to-eat food supplements that developing countries can produce locally, Milton Tectonidis, chief nutritionist for the charity Doctors Without Borders, told CBS News' "60 Minutes." For example, small factories in three African countries now produce Plumpynut — a nutritionist-invented mix of peanut butter, powdered milk, sugar, vitamins and minerals — that can keep otherwise malnourished toddlers healthy for about $1 a day and doesn't require scarce refrigeration or clean water to use.

"In three weeks, we can cure a kid that looked like they're half-dead," Tectonidis said. "There's many countries in Africa now saying, 'We want a factory. We want a factory.' Well let's give it to them," by redirecting some U.S. and European food aid to such projects, he said.[39]

BACKGROUND

Food Riots and Famine

"A hungry man is an angry man," runs a proverb common to nations from Zimbabwe to Scotland. Like hunger itself, protests over food shortages have occurred throughout history.

"America has a history of food riots, though the last were in the Depression," and "I wouldn't be surprised to see them again here," says Rose Hayden-Smith, a fellow in the foundation-funded Food and Society Policy program in Columbia, Mo., and a youth- and community-garden adviser for the University of California's Master Gardener Program.[40] The Civil War saw food riots in the United States, as did the so-called Gilded Age of the 1890s, an era marked by corporate corruption and a rising income gap between rich and poor, she says.

Some food protests are spontaneous revolts by hungry citizens, while others are organized by activists like unions and linked to broader political events.

In Northern France in 1911 rising prices and a meat shortage due to hoof-and-mouth disease spurred women to march "to the markets in protest," demanding "lower prices and dump[ing] carts of eggs and butter if and when their demands were not met," wrote Lynne Taylor, a professor of history at the University of Waterloo in Ontario, Canada.[41]

In Barcelona, Spain, in 1918, women's groups led rioting to protest inflationary food prices during the country's post-World War I economic collapse. Rioting women "attacked bread shops and coal wagons and took over a ship laden with fish," wrote Taylor. When police tried to break up the crowds, "the women turned on them, stripping some officers of their pants" and "thrashing them."[42]

In the 1970s and '80s a "global wave" of food riots in developing countries was sparked by "structural-adjustment" policies imposed by the International Monetary Fund and World Bank as a condition for developing nations to receive loans. These free-market adjustments included "incentives to go from small-farm agriculture to industrial, export-driven agriculture" and requirements to cut government subsidies for items like food staples and bus fare, says John Walton, a professor of sociology at the University of California at Davis.

Often, the result was "prices doubling overnight," triggering protests seen in numerous countries over the decades, he says. "Some countries had 10 or 12 instances of strikes and demonstrations." Governments had a variety of reactions, with some working out softer ways of making the prescribed changes, such as making them temporary, says Walton. At the extremes, consequences were harsh for governments and citizens, he says. "Sudan's government fell. Some governments just plowed ahead and repressed the protesters," killing 50 in Morocco and 40 in Cairo, Egypt.

Food crises generally are more political and economic events than natural ones, scholars say.

"Famines themselves, when looked at historically, have turned out not to be about food supply so much as about food distribution," says Walton.

Economic and political inequality lie behind famine, according to Indian economists Jean Dreze, of the Dehli School of Economics, and Amartya Sen, a professor at Harvard University who won the 1998 Nobel Prize in economics. "The developing of modern economic relations and of extensive interdependences even between distant parts of the economy" has created "many new ways in which different sections of the population can see . . . their command over food shift violently and suddenly," they wrote.[43]

For example, the shift to a world economy where virtually every country makes the bulk of its income through trade and where most cheap staple foods are imported has increased the risk of hunger for many, wrote Sen and Dreze. "Pastoralist nomads can be reduced to starvation if the relative price of animal products falls in relation to that of staple food, since their subsistence depends on their ability to sell . . . animal products . . . to buy enough calories from . . . grain. Fishermen may go hungry if the price of fish fails to keep up with that of . . . rice."[44]

The world's growing numbers of wage laborers who own no land that they can farm to tide them over tough times are "particularly vulnerable" to famine in the modern era, Dreze and Sen noted.[45]

Modern-day food crises are most acute in developing nations, many of which were operated for centuries as colonies of European nations, and "the colonial powers invested little in the food production systems" of their colonies, according to the Washington, D.C.-based International Food Policy Research Institute (IFPRI).[46]

As a result of such policies, "by the mid-1960s, hunger and malnutrition were widespread" in the developing world, "which increasingly depended on food aid from rich countries," said IFPRI.[47]

Increasing the Harvest

While the political world has yet to figure out how to ensure that food is distributed fairly, agricultural

scientists — boosted by large government investments in research — have been busy, developing farm techniques that increased harvests exponentially.

Beginning in the 1890s, Germany, the United Kingdom and the United States launched major agricultural-research enterprises, says the University of California's Wright. At the time, rising demand from growing populations was forcing farms to expand into less fertile land, but research on higher-yield plants and better growing methods kept yields increasing, says Wright.

"The story of English wheat is typical," according to the IFPRI. "It took nearly 1,000 years for wheat yields to increase from 0.5 to two metric tons per hectare, but only 40 years to climb from two to six metric tons per hectare."[48]

Agricultural disasters themselves often spurred improvements.

In the 1930s, for example, the Great Plains of the United States and Canada — especially Kansas, Oklahoma and the Texas panhandle — turned into a veritable "Dust Bowl." Drought combined with decades of intense, single-crop farming allowed wind to blow the once-rich topsoil away in huge dust clouds, darkening skies all the way to the East Coast and rendering millions of acres of farmland useless.

Out of the disaster came better soil conservation, says the USDA's Hatfield. In fact, "one of the places where we've done the best at conserving soil is the Great Plains," he says, by means such as reduced plowing, which helps keep soil from drying out.

But while the industrialized world increased its farm yields through science in the first half of the 20th century, developing countries made little progress, even as their populations soared. By the 1960s, hunger in the developing world, especially Asia, could no longer be ignored. A 1967 report of the President's Science Advisory Committee noted that "the scale, severity and duration of the world food problem are so great that a massive, long-range, innovative effort unprecedented in human history will be required to master it."[49]

At the time, American entomologist and population scientist Paul Ehrlich predicted that hundreds of millions of people would starve to death during the next few decades because Earth was incapable of sustaining the population explosion, says Wright.

"These were very confident predictions in those days," Wright says. "Why didn't it happen? Because of the increase

CHRONOLOGY

1910s-1960s *Famines hit China, Russia and Africa. U.S. farm-support programs are created during the Great Depression.*

1917 War Department's School Garden program urges students to plant vegetable gardens.

1928 Drought-driven famine kills 3 million people in northern China.

1943 Famine in British colony of Bengal — now Bangladesh and western India — kills up to 3 million.

1945 The number of Los Angeles children tending schoolyard gardens for the war effort hits 13,000.

1948 In the Green Revolution's first victory, Mexico produces enough grain to feed itself, using high-yield wheat varieties developed by Rockefeller Foundation-backed research.

1968 Four-year drought hits Africa's Sahel.

1970s *Drought and government mismanagement ruin harvests in several countries. World Bank and International Monetary Fund require developing countries to drop farm trade barriers in return for loans.*

1970 American plant-breeding pioneer Norman Borlaug wins the Nobel Peace Prize for improving crop yields in developing countries.

1973 Ethiopian government of Haile Selassie falls after its inaction allows drought to trigger a famine.

1977 India becomes self-sufficient in rice — another victory for the Green Revolution.

1980s-1990s *Large agribusinesses proliferate. Globalization of food-trading systems squeezes out small farms and local food processors. Commodity prices drop worldwide.*

1995 Congress approves Community Food Projects grants to help charities feed hungry people with locally produced food.

1996 President Bill Clinton signs agriculture reform bill to wean farmers off government subsidies over seven years. . . . Floods, drought and the cutoff of food aid from China and the former Soviet Union bring famine to North Korea.

1998 Congress backs off subsidy reform, approving $5.9 billion in emergency farm supports. . . . War and drought cause famine in Ethiopia.

2000s *Congress continues increasing farm subsidies, over objections from the World Trade Organization (WTO). Food shortages hit Zimbabwe after strife over who should control farmlands cripples agriculture in the country, once a major food exporter. Number of U.S. farm households falls to a few hundred thousand from 5 million in the 1930s, while farm output is 10 times larger.*

2001 President George W. Bush calls for cutbacks in farm subsidies to prevent overproduction.

2002 Congress passes farm bill that increases subsidies.

2006 U.S. farm-research budget totals $2.8 billion, down from $6 billion in 1980. . . . U.S. support for developing-country agriculture totals $624 million, down from $2.3 billion in 1980. . . . WTO's Doha Round of international trade talks collapses over disagreement on U.S. and European farm subsidies.

2007 World grain stocks drop to historically low levels. . . . Congress increases requirements for biofuel in the gasoline supply, increasing U.S. demand for corn.

2008 U.N. Food and Agriculture Organization says 850 million people are undernourished. . . . U.N. warns that solving global food problems could cost $30 billion a year. . . . Svalbard Global Seed Vault is opened in Arctic Norway to preserve food-plant biodiversity. . . . International Assessment of Agricultural Knowledge concludes that chemical fertilizer-based industrial farming has depleted soil and water and must be combined with organic and small-farm techniques to keep agriculture sustainable. . . . Congress passes farm bill rejecting Bush administration pleas to end shipment of U.S.-produced food as aid and buy food in developing countries' local markets.

More Than Seeds Needed to Improve Harvests

Poor farmers need access to credit, decent roads.

Worldwide anxiety over food prices could be a good thing if it finally focuses attention on building a sustainable food system for developing countries, agriculture experts say.

"Years ago, if we had the kind of concern we're seeing now, we wouldn't have these problems today," says Ray Cesca, president of the World Agricultural Forum, which seeks to improve world agriculture. For example, the United Nations' Millennium Development Goals include cutting poverty in half by 2015, "but we're going in the opposite direction," Cesca says.

Improving harvests takes investment in more than just seeds and fertilizers. Farmers in developing countries desperately need access to credit to help them through the inevitable ups and downs of farm production, says Peter Gubbels, vice president for international programs at World Neighbors, a nonprofit development group in Oklahoma City.

"When small-scale farmers run out of food, they go to a rich landlord and borrow a sack of grain," says Gubbels. "Then three months later they have to pay it back with interest, which may mean they have to leave their own fields and work as a laborer to pay off the debt." World Neighbors contributes seed money for local farmers' co-ops to create small banks "so that when they fall into need they can borrow from themselves and avoid the cycle of debt and slavery," he says.

Even to sell food in local markets, farmers need basic infrastructure such as roads.

To get a country on the right track, "feeding your own people has got to be the top priority," but many countries "haven't built the infrastructure for it," says Cesca. For example, in U.S. cities there are "central distribution points where everyone can go to buy and sell food, but that takes investment, and somebody needs the foresight to say, 'We'll have to create one of these,'" he says.

Similarly, in many developing countries, "most of the productivity rots in the field because there's no technology in cold storage" to preserve it, says Cesca.

Sub-Saharan Africa exports around $10 billion in food-related products annually, "but the local markets have the potential for around $135 billion" in sales, "so why wouldn't you pay attention to the local markets?" Cesca asks.

In the past, wealthy nations have pushed developing countries to commit most of their agriculture to commodities for export, like coffee or sugar. But in the long run that's a losing strategy, says Josh N. Ruxin, an assistant professor of public health at New York's Columbia University and director of the university's Millennium Village Project in Rwanda. The current price spike aside, since the 19th century prices of basic commodities — like coffee, corn and sugar — "have all gone down," says Ruxin. "Over time, producing only commodities makes people poorer, even if they become more efficient and productive," he says.

A better plan is agricultural diversity, says Ruxin. "We ask, 'What different crops can we produce for both local and national markets?'" says Ruxin. This needs to be thought out carefully, he says. For example, Africa can't export mangoes to Europe because storage and transport are too expensive. "But they can grow mangoes and export dried organic mango strips processed locally."

And mango strips are only the beginning, says Ruxin. On 32 hectares of formerly abandoned land, a farm coop recently planted the first pomegranate trees in Rwandan history in coordination with the federal U.S. African Development Foundation and a Los Angeles-based company, POM Wonderful. "Israel and Turkey can only produce so many pomegranates," and there's room for developing nations to get in on the game, says Ruxin. "The answer lies in asking, 'What crops will do well?'"

In the current price crisis, the hungry need immediate aid, Ruxin says. But solving long-term food problems requires patience.

"One reason African farmers don't plant fruit trees is because they take five to eight years to mature, and that doesn't work when you're hungry today," he explains. That's where focused research can help. A new variety of mango tree will produce a 50-fruit crop in two years, increasing to 500 fruits in five years, Ruxin says. "Five hundred mangoes that sell for a dollar apiece" can mean an unheard-of level of wealth for a subsistence farmer, he says.

in crop yields, and that didn't happen by accident" but through investments in research, he says. "Indians have a life expectancy in the 60s today, up from the 40s to 50s several decades ago," and much of that comes from better nutrition. "There is about a third more food available per person in the world today than when Ehrlich was writing."

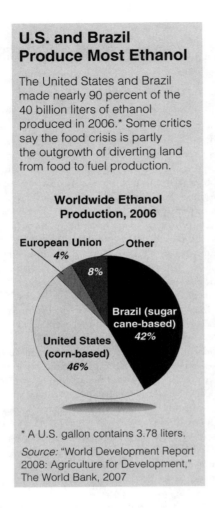

U.S. and Brazil Produce Most Ethanol

The United States and Brazil made nearly 90 percent of the 40 billion liters of ethanol produced in 2006.* Some critics say the food crisis is partly the outgrowth of diverting land from food to fuel production.

Worldwide Ethanol Production, 2006

European Union 4%
Other 8%
Brazil (sugar cane-based) 42%
United States (corn-based) 46%

* A U.S. gallon contains 3.78 liters.

Source: "World Development Report 2008: Agriculture for Development," The World Bank, 2007

Beginning in the mid 1940s and accelerating in the '60s, the Rockefeller and Ford foundations began promulgating new farming products and techniques in developing countries, including improved irrigation, better crop-management techniques and, first and foremost, new varieties of wheat, rice and other grains that were more responsive to modern chemical fertilizers.[50]

This so-called Green Revolution provided "the highest return on a public investment in recorded history," says Cornell's Barrett. For an amount variously estimated at between several hundred million and 1 billion dollars, "we kept a billion people from falling into poverty and moved a billion out of poverty. That's a remarkable accomplishment."

By the 1990s, nearly 75 percent of Asian rice came from Green Revolution seeds, along with about half the wheat planted in Africa, Latin America and Asia and about 70 percent of the world's corn, according to Food First, an Oakland, Calif.-based anti-hunger think tank.[51]

More recently, however, the limits of the Green Revolution's achievements have become clear.

For one thing, "we basically became complacent" about agriculture's achievements, and "investments in research have slowed radically," says Barrett. Furthermore, "we took for granted that agricultural yields were going to continue to improve," partly because of continued heavy use of fertilizers. "But now — with worries about the environment growing — fertilizer use has dropped off," meaning that continual yield increases are no longer assured, he says.

Finally, the focus on agricultural yield does nothing to change economic and political problems that are root causes of most famines, said Food First. "If the poor don't have the money to buy food, increased production is not going to help them."[52]

Subsidies and Trade

Congress enacted its first farm-subsidy legislation in 1933, during the Great Depression, to ensure farmers had a stable enough income to keep farming, even in tough times.[53]

"Since then, in various ways and to varying degrees, Congress has sought to raise farm income through a combination of commodity-specific price supports and supply controls, mainly import restrictions," according to trade analysts at the Cato Institute, a libertarian, Washington-based think tank.[54]

Subsidies "originally were intended to raise farm prices and incomes from Depression levels and to provide greater income stability," wrote South Dakota State University's Dobbs. "The goal was to maintain a nation of moderate-sized family farms."[55]

In the second half of the 20th century, however, the food-supply web changed dramatically. International trade in food commodities swelled, huge food-processing companies gained control of international food markets and giant agribusinesses squeezed out small and medium-sized family farms. These developments led critics to argue that wealthy nations' farm subsidies should be scrapped.

"Only 8 percent of producers receive 78 percent of subsidies," with 80 percent of farmers getting an average subsidy of under $1,000, wrote Keith Ashdown, vice president for policy at the Washington advocacy group

Taxpayers for Common Sense. "Farm payments are based on production levels, so the bigger the farm, the bigger the government check," and the family farmers whom subsidies originally were designed for see virtually no assistance today.[56]

Intense constituent pressure on farm-state politicians has kept subsidies in place, however.

For example, U.S. guarantees of a minimum price to sugar producers and trade barriers to keep foreign sugar out have long raised concerns, and the U.S. House in 1996 came within five votes of abolishing the sugar supports, wrote Mark A. Groombridge, a research fellow at the Cato Institute and now a special assistant in the State Department's Office of Arms Control. Since 1996, however, Congress has actually increased sugar supports, despite widespread criticism.[57]

Beginning in 1995, the WTO has worked toward international consensus on rules to lower trade barriers worldwide. Since 2001, the trade organization has focused on reducing agricultural trade barriers in the so-called Doha Development Round of negotiations.

The Doha Round has stalled, however, mainly because wealthy nations refuse to alter their subsidy programs while insisting that developing nations abandon their own trade barriers and farm supports. For example, both U.S. cotton subsidies and the European Union's sugar programs have been found in violation of WTO rules, but the programs have not been revised, said the Cato Institute analysts.[58]

In 2003, at a street protest against WTO talks being held in Cancun, Mexico, Lee Kyung Hai, a South Korean farmer and former director of the Korean Advanced Farmers Union, committed suicide by stabbing himself in the heart; he was carrying a banner that read "WTO Kills Farmers."[59]

Earlier, Lee had spoken of another farmer who committed suicide by swallowing poison, after falling into serious debt. "I was powerless to do anything but hear the howling of my friend's wife," he said.[60]

CURRENT SITUATION

Price Spike

Skyrocketing prices for food staples are fueling protests around the world this year.

"In 2006, no one was predicting this current price boom," says David Orden, a senior research fellow at the International Food Policy Research Institute and a professor of economics at Virginia Tech. A constellation of supply-and-demand factors came together to drive prices skyward, and though "good crops are forecasted this year, they are not harvested yet, so people are still nervous" enough about future food availability to keep demand high, he says.

High demand and a historically low supply of commodities like grain are the main drivers of the price spike. For example, "we're at historic lows of grain inventories," says Cornell's Barrett. "In the U.S. we have less than one month's supply when there are usually three or four."

A long-term drought that has slashed Australia's grain exports has also contributed to a scare that caused some other rice-producing countries to ban their exports, out of fear they would not have enough to feed their own people, says Berkeley's Wright.

Most recently, a disastrous May 2 cyclone devastated crops in Myanmar, which had been among the few countries expected to export rice in 2008.[61]

Demand pressures also are driving up prices.

Notably, in its 2005 energy bill, Congress mandated adding 4 billion gallons of renewable fuel — mostly corn-based ethanol — to the U.S. gasoline supply in 2006, with increased amounts added in succeeding years.[62]

The mandate drives up demand and prices for corn, says Barrett. And in our processed-food world, corn demand is already much higher than many realize, he says. "Corn feeds chickens and sweetens soft drinks" to name just a few uses, Barrett says. Rising corn prices also drive up prices for meat and other sweeteners like sugar cane, he says.

Investors are also driving the price spike. Many have been dabbling in the grain markets as other investment sectors, such as housing, have become unstable in the past year.[63] The new investors also have encouraged "some speculators to withhold grain from the market in hopes of selling it for a higher price later," says Central Michigan University's Doyle.

But while speculation has been a factor in recent price spikes, it "will not in the long term keep prices high, because there's only so much wheat or corn . . . that someone can hold off the market . . . because it is a perishable product," said Dean Baker, co-director of the Center for Economic and Policy Research in Washington.

To some extent, the demand-supply equation will come into balance naturally and food prices will drop again, says

Is America's Food System Ripe for Change?

Return to locally grown food is advocated.

Americans spend an average of only 10 percent of their disposable income on food, compared with 50 percent in Indonesia and 30 percent in China. U.S. spending today is also low by historical standards: In 1933, Americans spent about a quarter of their income on food.[1]

Nevertheless, with food prices spiking, more Americans are feeling pain at the supermarket checkout. "Our [monthly] food bills are $600, $700," up from around $400 a year or two ago, said Jomarie Ortiz, the mother of four teenage sons in Bloomfield, N.J. "The cereal [price] was astronomical."[2]

Food experts say the rising prices are but one indication of serious, long-term problems in the American food system, which relies too much on fuel-guzzling long-distance shipping and single-crop, commodity-based agriculture that damages the environment and stints human nutrition needs.

"In the last half-century we made a profound choice to have the bulk of our food system in commodity crops, like corn, soybeans and rice," says Curt Ellis, the Portland, Ore.-based filmmaker of "King Corn," a documentary on American farming that aired recently on PBS television. "These dominate our diets in unrecognizable forms," from the fast-food hamburger made from corn-fed beef to soda sweetened with corn syrup rather than with sugar or honey, he says.

"All across our food shelves are products that are seemingly diverse but actually are not" when you read labels carefully, Ellis says. The result is a meat- and simple-carbohydrate-heavy diet that sees "one-in-three first-graders on the road to developing Type 2 diabetes" and "young people in America today — my generation — potentially having shorter lifespans than their parents," says the 28-year-old Ellis.

In addition, environmentally sustainable farming methods aren't yet the order of the day in American agriculture, despite the real possibility that depleted soils and fluctuating rainfall — partly brought on by climate change — may threaten future harvests.

"Even with our productive crop yields, we've got spots in the heartland with really poor yields because" of soil erosion, depletion of soil nutrients and water problems, says Jerry L. Hatfield, supervisory plant physiologist at the U.S. Department of Agriculture's National Soil Tilth Research Laboratory in Ames, Iowa.

"We need to focus on how reliant we are on a good natural-resource base" of rich soil and sufficient water, Hatfield says. "We can work on this genetic stuff" — breeding or bioengineering higher-yield or hardier plant varieties — "but at a certain point environment" is the determining factor for harvests, he says. "I don't want to be an alarmist, but I still believe that we're on a path where our soils are

Barrett. To meet high demand, "farmers will put more land into cultivation and use more fertilizers," he says. But the poorest countries will have the most trouble making adequate adjustments, he says. "Poor farmers can't get credit" to put in irrigation systems, for example, he says.

Columbia University's Ruxin says the price surge could actually lead to a new era of smarter agricultural development in developing countries.

"The overall direction over the decades" has been to steer African nations toward large-scale industrial farming of products for export to Europe and the United States, but "overnight that policy has been stood on its head," Ruxin says. Today, "if you're a poor, subsistence-level farmer and eat what you produce and maybe trade a little, and

prices go up 60 percent, your food is suddenly extremely valuable, and your ability to get into the market is greater," which could help build local agriculture, he says.

"We're asking how can we help smaller farms cooperate" to produce for local and regional markets, where buyers have been hit exceptionally hard by the price of imported grain. Thanks in part to the price spike, "the jury is clearly coming down on the side of producing more locally, and that was not the picture five years ago," says Ruxin.

Farm Bill

A flurry of attention followed April statements by U.N. food officials branding recent food-price spikes as signs of a long-term disaster in the making.

continuing to degrade, and when I look at our practices, we could do a lot more" to conserve soil, Hatfield says.

"When you choose to plant 90 million acres with just one crop — corn — it's a thirsty crop, a nitrogen-thirsty crop," so it depletes soil resources and "doesn't make efficient use of the limited water and nitrogen that we have," says Ellis.

Environmentally sustainable farming generally requires more crop diversity, says Thomas Dobbs, professor emeritus of economics at South Dakota State University in Brookings. "You can't just do corn/soybean cropping," as farms from Ohio to North Dakota mostly do today, he says. "You also have to have small grains like wheat or oats and some plants like clover or alfalfa," which take nitrogen from the air and return it to the soil where it acts as a fertilizer, he says.

Some farmers argue that they can't diversify their fields because under current law they can only get government subsidies by keeping many acres in corn and soybeans, says Hatfield. "But if we focused on food security, we'd think about changing to a more diverse system of crops."

The American food system has been nationalized, with little food grown close to where it's eaten, and changing that is job No. 1 in the eyes of many food analysts, as fossil-fuel prices rise.

"In a petroleum-limited world, the best solutions are the ones you can find close to home," says Ellis. "It's unfathomable to me that Iowa can feed the whole country with corn but grows so few fruits and vegetables," even though "fruits and vegetables used to grow just fine there."

"In Virginia 10 or 12 years ago, we developed a good system with a nice niche for growing fall broccoli, but we couldn't compete with California's competitive advantage because of their climate and [higher] yield," says Mark Alley, a professor of agriculture at Virginia Tech. But with fuel prices rising "that's changing" back in favor of more local growing, he says.

But switching to locally grown produce will require rebuilding a lost food infrastructure.

"We've created a national food system, stripping rural areas of food-processing capabilities," says Rose Hayden-Smith, a youth- and community-garden adviser for the University of California's Master Gardener Program. Processing and distribution infrastructure will need to be rebuilt from the ground up, she says.

Furthermore, urban farming and home and city-lot community gardening should be brought back in a big way, says Hayden-Smith. "We need the government to introduce 'victory gardens' again," she says. "They were phenomenally successful, especially in World War II, when America was able to export much more food to our starving European allies" because the "citizen gardener" — including schoolchildren — produced so much of the food eaten in U.S. homes, says Hayden-Smith.

"I'd like to see the White House put in a garden," just like first lady Eleanor Roosevelt did, she says.

[1] "Americans Spend Less than 10 Percent of Disposable Income on Food," *Salem-News* [Salem, Ore.], July 19, 2006, www.salem-news.com.

[2] Andrew Martin and Michael M. Grynbaum, "Costs Surge for Stocking the Pantry," *The New York Times*, March 15, 2008, p. A1.

High food prices are "creating the biggest challenge that the [World Food Programme] has faced in its 45-year history, a silent tsunami threatening to plunge more than 100 million people on every continent into hunger," said WFP Executive Director Josette Sheeran.[64]

"We must take immediate action in a concerted way," said U.N. Secretary-General Ban Ki-moon.[65]

But food and farm politics are intensely national and too easily override international concerns. A case in point, say critics, is the new U.S. farm bill, which Congress overwhelmingly approved in May, 318-106 in the House and 85-15 in the Senate, overcoming an earlier presidential veto.[66]

The bill trims some farm subsidies and increases the percentage of funding for U.S. nutrition programs, such

as food stamps and the Fresh Fruit and Vegetable Snack Program for schools, said House Agriculture Committee Chairman Collin Peterson, D-Minn., a supporter of the bill.[67]

"Today's overwhelming vote was an indication that we produced a strong, bipartisan farm bill that is good for our farmers and ranchers and essential to our state," said Sen. Amy Klobuchar, D-Minn., a member of the Senate Agriculture Committee.[68]

But the legislation comes in for harsh criticism from others.

The farm bill "is a travesty, it's disgusting" says Berkeley's Wright. "When farmers are richer than ever and the poor can't eat, they gave $10 billion more to farmers."

Grain and rice vendors wait for customers at a market in Beijing, China, in July 2007. Farm improvements made China a net exporter of grain until last year, when high prices began putting food staples out of the reach of many Chinese.

Singled out for special censure are the legislation's protections for U.S.-produced sugar, particularly championed by Peterson, whose home state of Minnesota is a major sugar-beet producer.

The bill raises subsidies for U.S. sugar producers and requires imported sugar to be used for ethanol production, not food, "virtually locking in an 85 percent share of the U.S. market for domestic . . . growers, even though a number of foreign countries can grow sugar more cheaply," wrote trade analyst Daniel Griswold of the Cato Institute. The sugar provisions interfere with free trade and drive up prices for American consumers, he insisted.[69]

"This farm bill just heads in the wrong direction in terms of our international obligations," which is why President Bush vetoed it, said Deputy Secretary of Agriculture Chuck Conner. "We would expect [nations that trade with the United States] to protest in every way they can."[70]

During the more than year-long battle over the new bill, the legislation's architects also remained at odds with President Bush and many anti-poverty groups over changes the White House wanted in food-aid programs. U.S. food aid currently comes in the form of U.S.-produced food that U.S. transport companies ferry to developing countries, where it is either handed out or sold in local markets by charities that use the proceeds to provide various kinds of assistance.

The White House pushed to turn a quarter of the American aid — about $400 million annually — into cash to buy food in developing countries' own regional markets, a move that Bush and several charities, including CARE and Oxfam, say would provide more food more quickly and cost less by cutting transport times and costs.[71]

Congress' nonpartisan auditing agency, the Government Accountability Office (GAO), agreed. The amount of food actually delivered as U.S. aid actually dropped by 52 percent between 2002 and 2007 because transport and other associated costs rose, said the GAO.[72]

But the Bush plan runs afoul of U.S. food producers and shippers. Accordingly, Congress approved only a small pilot program — totaling $60 million over four years — to test overseas purchasing.[73]

Gloria Tosi, a lobbyist for companies that transport American food aid, called even the tiny pilot program "a bad idea."[74]

OUTLOOK
Environmental Concerns

The science of farming traditionally has focused on increasing harvests, but today concerns have shifted to making farming environmentally sustainable. The change was prompted by soaring prices for fossil fuels that power farm machinery and are used in fertilizer manufacturing, plus dawning awareness that soil and water resources are rapidly being depleted.

In the United States, "One thing we need to focus on is how reliant we are on a good natural-resource base" — mainly soil and water — says the USDA's Hatfield. "Even with our productive yields we've still got spots in the heartland with really poor yields" due to erosion and depletion of soil nutrients, he says.

In India and Pakistan, farmers "are mining the groundwater deeper and deeper," depleting future stores, says South Dakota State University's Dobbs.

Climate change is adding to the troubles. In China, Australia and Africa, deserts are encroaching on once-fertile lands, says Doyle at Central Michigan University. "Where the hell is all the food going to be grown?"

Transport of food in a globalized marketplace is "a huge issue with the price of diesel fuel as it is today" and oil supplies dwindling, says Virginia Tech's Alley, president-elect of the American Society of Agronomists.

Farmers must soon reevaluate many practices in light of environmental needs.

AT ISSUE

Is the U.S. ethanol fuel program worsening the world food crisis?

YES

Gawain Kripke
Senior Policy Adviser, Oxfam America

From testimony before the House Energy and Commerce
Subcommittee on Energy and Air Quality, May 6, 2008

Diversion of corn to ethanol is playing a significant role in reducing corn supplies for food and feed. In 2008 the U.S. Department of Agriculture estimates that 3.1 million bushels of U.S. corn will be used to produce biofuels. That's an increase of nearly 50 percent over the 2.1 million bushels last year and close to twice the 1.6 million bushels of 2006.

In 2008 the U.S. will convert approximately one-quarter (23.7 percent) of our corn production into biofuels. We're rapidly diverting larger portions of our corn supply to fuel, leaving less for food.

For about 1.2 billion people around the world, corn is the preferred staple cereal. Consider that the U.S. produces more than 40 percent of the world's corn supply. Dedicating 3.1 million bushels of corn for ethanol this year will take more than one-tenth of the global corn supply off the market for food and feed. Furthermore, the U.S. exports nearly twice as much corn as all the other exporters combined. So, reduced supply and/or higher prices in the U.S. corn market have significant implications for the rest of the world.

Although ethanol mandates and subsidies directly impact corn prices, they also have cascading impacts on other agricultural commodities. Higher corn prices are encouraging farmers to commit more acreage and agricultural inputs to corn. This leaves less for other crops, especially soybeans, which are often planted in alternate years with corn. As a result, production for other commodities is lower and prices are higher.

Higher corn prices also lead consumers to choose other, cheaper cereals to substitute for food or feed. Over time, this increased demand increases the prices for other commodities.

"Biofuel demand has propelled the prices not only for corn but also for other grains, meat, poultry and dairy through cost push and crop and demand substitution effects," according to the [International Monetary Fund's] World Economic Outlook.

The International Food Policy Research Institute (IFPRI), one of the premier organizations tracking food and hunger issues, estimates that biofuels will drive up corn prices by between 27 percent and 72 percent by 2020, depending on the scenario. Other commodities such as oil seeds used for biodiesel fuel would rise by 18 percent to 44 percent. "In general, subsidies for biofuels that use agricultural-production resources are extremely anti-poor, because they implicitly act as a tax on basic food, which represents a large share of poor people's expenditures," the IFPRI said.

NO

Rick Tolman
CEO, National Corn Growers Association

From testimony before the House Energy and Commerce
Subcommittee on Energy and Air Quality, May 6, 2008

Recently, the media and ethanol critics have demonized corn ethanol and attempted to solely blame rising food costs on higher commodity costs and government policies promoting renewable fuel.

In attempting to justify their opposition to ethanol expansion and the Renewable Fuels Standard (RFS) enacted by Congress in 2007, opponents continue to claim that higher corn prices are causing higher retail food prices. A look at the facts surrounding food prices simply doesn't support that logic.

More so, the effects of $120-per-barrel oil have far-reaching effects on the consumer price for food. A recent study by the Oregon Department of Agriculture details the factors affecting food price: a growing middle class in Latin America and Asia; drought in Australia; low worldwide wheat stocks; increases in labor costs; a declining U.S. dollar; regional pests, diseases, droughts and frost; and marginal impacts from ethanol demand for corn and sugarcane.

One recent study found that a $1-per-gallon increase in the price of gas has three times the impact on food prices as does a $1-per-bushel increase in the price of corn.

In fact, just 19 cents of every consumer dollar can be attributed to the actual cost of farm products like grains, oilseeds and meat. Retail food products like cereals, snack foods and beverage corn sweeteners contain very little corn. Consider that even when corn is priced at $5 per bushel, a standard box of corn flakes contains less than eight cents' worth of corn.

Corn is a more significant ingredient for meat, dairy and egg production. Still, corn represents a relatively small share of these products from a retail price perspective. As an example, according to the National Cattlemen's Beef Association, it takes about three pounds of corn to produce one pound of beef. This equates to 27 cents' worth of corn in a pound of beef when corn is $5 per bushel.

Because corn and other grains constitute such a small portion of retail food products, higher grain prices are unlikely to have any significant impact on overall food inflation, according to a number of experts. According to [U.S. Department of Agriculture] economist Ephraim Liebtag, a 50 percent increase in corn prices translates to an overall increase of retail food prices of less than 1 percent. Similarly, a recent analysis by Informa Economics found that higher corn prices "explain" only 4 percent of the increase in retail food prices.

"About two years ago, I was in northwestern Iowa to talk about reducing tillage" — plowing fields less to reduce soil erosion and depletion, recalls Hatfield. "I asked, 'What are you going to do when diesel fuel goes above $3 a gallon?' And they looked at me as if I were from Mars," he says. But today, "Farmers are asking different questions, like 'How do I reduce my tilling? How do I use manure' " to replace fossil-fuel-based fertilizer?

The debate is heating up right now about the best means to maximize harvests in sub-Saharan Africa and other challenging environments, and industrial, fossil-fuel-based farming is not necessarily winning, says Dobbs. A large study by Great Britain's University of Essex found that "sustainable practices . . . led to a 96 percent increase in per-hectare food production," showing that "there's potential for a lot of increased productivity without abandoning small-farm agriculture," he says.

But others say that tough times may lead to increased acceptance of high-tech methods, including so-called genetically modified foods — crop varieties created by gene manipulation in the biotechnology lab — of which many consumers are wary.[75]

In Australia and New Zealand, for example, there was a great deal of resistance to transgenic crops — which may someday be cheaper and drought-resistant — but as food prices go up, the opposition is starting to be muffled, says Doyle.

Food shortages and the resulting unrest "are not going to go away tomorrow by any stretch of the imagination," says Johan Selle, an international risk-management expert, in Annapolis, Md.

"When a businessman is getting $5 for a loaf of bread, he will not bring it down to $2. We are going to see this for another five, six, seven years," he says. "Unless someone — the U.S., the U.N., the World Trade Organization — takes the lead" to create a more stable system, "we're going to struggle for a long time."

NOTES

1. Jeffrey Gettleman, 'Famine Looms as Wars Rend Horn of Africa," *The New York Times*, May 17, 2008, p. A1.

2. "The List: The World's Most Dangerous Food Crises," *Foreign Policy*, April 2008, www.foreignpolicy.com.

3. Maria Bartiromo, "Food Emergency: On the Front Line With the U.N.'s Josette Sheeran," Facetime, *Business Week*, May 1, 2008, www.businessweek.com.

4. For background, see "Agriculture for Development, World Development Report 2008," World Bank, Oct. 19, 2007, http://econ.worldbank.org/WBSITE/EXTERNAL/EXTDEC/EXT-RESEARCH/EXTWDRS/EXTWDR2008/0,,contentMDK:21410054~menuPK:3149676~pagePK:64167689~piPK:64167673~theSitePK:2795143, 00.html.

5. "Agricultural Outlook 2008-2017," United Nations Food and Agriculture Organization (FAO) and the Organization for Economic Cooperation and Development (OECD).

6. Quoted in Laurie Garrett, "Food Failures and Futures," Council on Foreign Relations working paper, May 15, 2008, www.cfr.org/content/publications/attachments/CGS_WorkingPaper_2.pdf.

7. "Asian Rice Crisis puts 10 Million or More at Risk: Q&A With Peter Timmer," Center for Global Development, April 21, 2008, www.cgdev.org.

8. Quoted in Rama Lakshimi, "Bush Comment on Food Crisis Brings Anger, Ridicule in India," *The Washington Post*, May 8, 2008, p. A18.

9. Quoted in *ibid*.

10. For background on globalization and world trade, see Samuel Loewenberg, "Anti-Americanism," *CQ Global Researcher*, March 2007, pp. 51-74; Brian Hansen, "Globalization Backlash," *CQ Researcher*, Sept. 28, 2001, pp. 761-784, and Mary H. Cooper, "World Trade," *CQ Researcher*, June 9, 2000, pp. 497-520.

11. Quoted in Elizabeth Becker, "U.S. Defends Its Farm Subsidies Against Rising Foreign Criticism," *The New York Times*, June 27, 2002.

12. Lester R. Brown, *Outgrowing the Earth: The Food Security Challenge in an Age of Falling Water Tables and Rising Temperatures* (2005), p. 3.

13. *Ibid.*, p. 8.

14. "Peak Grain: Feeding Nine Billion After Peak Oil and Climate Change," OilEmpire.us Web site, www.oilempire.us/peak-grain.html.

15. Quoted in Tom Philpott, "Food Crisis Resolved!" "Gristmill blog," *Grist*, http://grist.org, April 28, 2008.

16. Stephen Leahy, "Africa: Reinventing Agriculture," Inter Press Service, April 15, 2008, http://allafrica .com. For background see Peter Behr, "Looming Water Crisis," *CQ Global Researcher*, February 2008, pp. 27-56.

17. Janet Daley, "Forget Fairtrade — Only Free Trade Can Help Poor," *Daily Telegraph* (United Kingdom), Feb. 25, 2008, www.telegraph.co.uk.

18. Jagdish Bhagwati, "Protectionism," *The Concise Encyclopedia of Economics*, Library of Economics and Liberty Web site, www.econlib.org.

19. Daniel T. Griswold and Bob Young, "Online Debate: Should the United States Cut Its Farm Subsidies?" Council on Foreign Relations, April 27, 2007, www.cfr.org.

20. *Ibid.*

21. Pranab Bardhan, "Does Globalization Help or Hurt the World's Poor?" *Scientific American*, March 26, 2006, www.sciam.com.

22. Bill Quigley, "30 Years Ago Haiti Grew All the Rice It Needed. What Happened?" *Counterpunch*, April 21, 2008, www.counterpunch.org/quigley04212008.html.

23. *Ibid.*

24. *Ibid.* For background see Peter Katel, "Haiti's Dilemma," *CQ Researcher*, Feb. 18, 2005, pp. 149-172.

25. Conn Hallinan, "The Devil's Brew of Poverty Relief," *Foreign Policy in Focus*, July 19, 2006, www .fpif.org.

26. Daryll E. Ray, Daniel G. De La Torre Ugarte and Kelly J. Tiller, "Rethinking U.S. Agricultural Policy: Changing Course to Secure Farmer Livelihoods Worldwide," Agricultural Policy Analysis Center, University of Tennessee, September 2003, p. 2.

27. *Ibid.*

28. For background, see Mary H. Cooper, "Foreign Aid After Sept. 11," *CQ Researcher*, April 26, 2002, pp. 361-392.

29. Christopher B. Barrett and Daniel G. Maxwell, "Recasting Food Aid's Role," *Policy Brief*, Cornell University Department of Applied Economics and Management, August 2004, www.aem.cornell.edu/ faculty_sites/cbb2/papers/BM_policybrief.pdf.

30. Shahla Shapouri and Stacey Rosen, "Fifty Years of U.S. Food Aid and Its Role in Reducing World Hunger," *Amber Waves*, U.S. Department of Agriculture Economic Research Service, September 2004, www.ers.usda.gov.

31. *Ibid.*

32. Charles E. Hanrahan, "Indian Ocean Earthquake and Tsunami: Food Aid Needs and the U.S Response," Congressional Research Service, April 8, 2005.

33. Quoted in "North Korea Welcomes Resumption of U.S. Food Aid," Agence France-Presse, May 17, 2008, http://news.yahoo.com. For background, see Kenneth Jost, "Future of Korea," *CQ Researcher*, May 19, 2000, pp. 425-448.

34. Quoted in "U.S. Giving Bangladesh $40 million in Food Aid," The Associated Press, May 4, 2008, www.msnbc.msn.com.

35. Quoted in Matthew Bolton and Michael Manske, "Non-Emergency Food Aid: A Resource in the Fight Against Hunger and a Tool for Development," policy statement, Counterpart International, www.hunger-center.org/international/documents/Michael%20 Manske%20Non%20Emergency%20Food%20 Aid.pdf.

36. "Food Aid or Hidden Dumping: Separating Wheat from Chaff," *Oxfam Briefing Paper 71*, March 2005, www.oxfam.org/en/files/bp71_food_aid_240305 .pdf.

37. For background, see Eben Harrell, "CARE Turns Down U.S. Food Aid," *Time*, Aug. 15, 2007, www .time.com/time/nation/article/0,8599,1653360,00. html.

38. "Food Aid for Food Security," Catholic Relief Services, http://crs.org/public-policy/food_aid .cfm.

39. Quoted in "A Life Saver Called 'Plumpynut,' " "60 Minutes," June 22, 2008, www.cbsnews.com/ stories/2007/10/19/60minutes/main3386661 .shtml.

40. For background on food riots, see B.P. Garnett, "Mob Disturbances in the United States," *Editorial Research Reports*, Oct. 20, 1931. Available at *CQ Researcher Plus Archive*.

41. Lynne Taylor, "Food Riots Revisited," *Journal of Social History*, winter 1996, p. 483.

42. *Ibid.*

43. Jean Dreze and Amartya Sen, *Hunger and Public Action* (1989), p. 4.

44. *Ibid.*, p. 5.

45. *Ibid.*, p. 6.

46. "Green Revolution: Curse or Blessing?" International Food Policy Research Institute, 2002, www.ifpri .org.

47. *Ibid.*

48. *Ibid.*

49. Quoted in *ibid.*

50. For background, see "The Green Revolution & Dr. Norman Borlaug," The Norman Borlaug Institute for Crop Improvement, www.nbipsr.org.

51. "Lessons from the Green Revolution," Food First Web site, April 8, 2000, www.foodfirst.org.

52. *Ibid.*

53. For recent background, see David Hosansky, "Farm Subsidies," *CQ Researcher*, May 17, 2002, pp. 433-456, and Brian Hansen, "Crisis on the Plains," *CQ Researcher*, May 9, 2003. For historical background, several articles appear in *Editorial Research Reports*, including B. W. Patch, "Government Subsidies to Private Industry," April 26, 1933, and Charles E. Noyes, "Government Payments to Farmers," Sept. 3, 1941.

54. Daniel Griswold, Stephen Slivinski and Christopher Preble, "Ripe for Reform: Six Good Reasons to Reduce U.S. Farm Subsidies and Trade Barriers," Center for Trade Policy Studies, Cato Institute, Sept. 14, 2005.

55. Thomas Dobbs, "Is It Too Late for Progressive Farm and Food Policy Reforms? Critical Decisions Facing Congress," *Dailykos.com* Web site, Jan. 26, 2008, www.dailykos.com.

56. Keith Ashdown, "Congress Continues with Corruption and Failure: Handouts to Large Agribusiness Corporations," *The Progress Report* Web site, www.progress.org/2005/tcs176.htm.

57. Mark A. Groombridge, "America's Bittersweet Sugar Policy," Center for Trade Policy Studies, Cato Institute, Dec. 4, 2001.

58. Griswold, *et al.*, *op. cit.*

59. John Ross, "Bridging the Distance," *San Francisco Bay Guardian*, Sept. 17, 2003.

60. Quoted in *ibid.*

61. For background, see "Myanmar Cyclone Aftermath: Rice Shortage," *The Daily Green* Web site, May 7, 2008, www.thedailygreen.com.

62. For background, see Ben Lieberman, "The Ethanol Mandate Should Not Be Expanded," *Heritage Foundation Backgrounder #2020*, March 28, 2007, www.heritage.org/Research/energyandenvironment/ bg2020.cfm.

63. For background, see Marcia Clemmitt, "Mortgage Crisis," *CQ Researcher*, Nov. 2, 2007, pp. 913-936, and Kenneth Jost. "Financial Crisis," *CQ Researcher*, May 9, 2008, pp. 409-432.

64. Quoted in Thalif Deen, "A Silent Tsunami Threatening the Global Population," *Daily News* [Sri Lanka], April 30, 2008, www.dailynews.lk.

65. Quoted in *ibid.*

66. For background see "The 2008 Farm Bill Enacted Into Law," American Farmland Trust Web site, www .farmland.org.

67. Quoted in Peter Shinn, "Farm Bill Heads for Congressional Passage Next Week," Brownfield Network Web site, May 8, 2008, www.brown-fieldnetwork.com.

68. Quoted in Pamela Brogan, "Coleman, Klobuchar Vote to Override Bush Veto," *Saint Cloud* (Minnesota) *Times*, May 23, 2008, www.sctimes.com/apps/pbcs .dll/article?AID=/20080523/NEWS01/105220105/0/ archives.

69. Daniel Griswold, "Ag Committee Chair Demands Higher Food Prices," Cato-at-liberty blog, Cato Institute, May 5, 2008, www.cato-at-liberty.org.

70. Quoted in Greg Hitt, "Farm Bill May Hinder Trade Talks," *The Wall Street Journal*, May 14, 2008, p. A2.

71. For background, see Missy Ryan, "Congress Spurns Bush's Call for Food Aid Switch," Reuters, from Yahoo! News Canada Web site, May 8, 2008, http:// ca.news.yahoo.com; Michael Janofsky and Christopher Swann, "Congress Resists Speeding U.S. Food Aid, Benefiting Archer, APL," Bloomberg News, May 19, 2008, www.bloomberg.com.

72. "Foreign Assistance: Various Challenges Impede the Efficiency and Effectiveness of U.S. Food Aid," Government Accountability Office, April 2007, www.gao.gov/new.items/d07560.pdf.

73. *Ibid.*

74. Quoted in Janofsky and Swann, *op. cit.*

75. For background, see David Hosansky, "Food Safety," *CQ Researcher*, Nov. 1, 2002, pp. 897-920, and David Hosansky, "Biotech Foods," *CQ Researcher*, March 30, 2001, pp. 249-272.

BIBLIOGRAPHY

Books

Brown, Lester B., *Outgrowing the Earth: The Food Security Challenge in an Age of Falling Water Tables and Rising Temperatures*, W. W. Norton, 2005.
An environmentalist argues that climate change and resource depletion demand major changes in the global food system.

Federico, Giovanni, *An Economic History of World Agriculture, 1800-2000*, Princeton University Press, 2005.
A professor of economic history at the European University Institute in Florence, Italy, describes how farming remade itself over two centuries to feed a growing world.

Pollan, Michael, *The Omnivore's Dilemma: A Natural History of Four Meals*, Penguin, 2007.
A journalist chronicles the path of common foods through America's food-production systems.

Sen, Amartya, *Poverty and Famines: An Essay on Entitlement and Deprivation*, Oxford University Press, 1983.
A Nobel Prize-winning economist argues that there is enough food to feed the world, but many poor people can't get enough money to buy the food.

Vernon, James, *Hunger: A Modern History*, Harvard University Press, 2007.
A history professor at the University of California, Berkeley, argues that views about hunger shifted in the 20th century.

Articles

"The New Face of Hunger," *The Economist*, April 19, 2008, pp. 32-34.
Soaring food prices may be a sign of a world food system that needs large-scale overhaul to be sustainable.

Akl, Aida F., "Market Speculation Drives Food Prices," *Voice of America News*, May 16, 2008, www.voanews.com.
A surge of commercial and private investment in grain markets is a small but significant contributing cause of rising food prices.

Bello, Walden, "Manufacturing a Food Crisis," *The Nation*, May 15, 2008, www.thenation.com.
A professor of sociology at the University of the Philippines and a critic of globalization argues that contemporary trade policies have crippled developing-country food markets.

Hedges, Stephen J., "Grain Prices Grow, But So Do Risks," *Chicago Tribune*, May 26, 2008, www.chicagotribune.com/news/nationworld/chi-grain_hedgesmay27,0,1989503.story.
Many grain farmers aren't benefiting from skyrocketing commodity prices.

Janofsky, Michael, and Christopher Swann, "Congress Resists Speeding U.S. Food Aid, Benefiting Archer, APL," *Bloomberg.com*, May 29, 2008, www.bloomberg.com.
American food-aid policies benefit industrial food producers and shipping companies and limit the amount of aid taxpayers' dollars can buy.

Philpott, Tom, "Sticker Shock" Gristmill blog, *Grist*, April 25, 2008, http://gristmill.grist.org.
A writer for an environmental magazine explains the causes of the food-price spike.

Reports and Studies

"Foreign Assistance: Various Challenges Impede the Efficiency and Effectiveness of U.S. Food Aid," *Government Accountability Office*, April 2007.
Congress' nonpartisan auditing arm finds that U.S. food-aid practices reduce the amount, timeliness and quality of the food assistance the U.S. provides to hungry populations.

"Signing Away the Future," *Oxfam International*, March 2007, www.oxfam.org/en/files/bp101_regional_trade_agreements_0703/download.
An international charity argues that trade agreements are cutting developing countries out of the global economy.

Constantin, Anne Laure, "A Time of High Prices: An Opportunity for the Rural Poor," *Institute for Agriculture and Trade Policy*, April 2008, www.iatp.org.
A nonprofit group that promotes family farming argues that current food-price spikes may provide a window for countries to beef up their local farm production.

Griswold, Daniel, "Grain Drain: The Hidden Cost of U.S. Rice Subsidies," *Cato Institute*, November 2006, www.freetrade.org/node/539.

An analyst for a libertarian think tank argues that U.S. supports to rice producers distort markets and hurt farmers and consumers in the United States and abroad.

Riedl, Brian M., "Seven Reasons to Veto the Farm Bill," Backgrounder #2134, *Heritage Foundation*, May 12, 2008, www.heritage.org/Research/Agriculture/bg2134.cfm.
A conservative think tank argues that U.S. farm subsidies benefit wealthy large-scale farmers but not struggling family farms.

Schnepf, Randy, "High Agricultural Commodity Prices: What Are the Issues?" *Congressional Research Service*, May 6, 2008, www.iatp.org/tradeobservatory/library.cfm?refid=102843.
Congress' nonpartisan research office provides a primer on the food-price spike.

For More Information

Agribusiness Accountability Initiative, www.agribusinessaccountability.org. Think tank and advocacy group that supports alternatives to large agribusinesses; sponsored by the Des Moines, Iowa-based National Catholic Rural Life Conference.

American Farm Bureau, 600 Maryland Ave., S.W., Suite 1000W, Washington, DC 20024; (202) 406-3600; www.fb.org. Independent organization representing the interests of U.S. farmers.

Center for Trade Policy Studies, Cato Institute, 1000 Massachusetts Ave., N.W., Washington, DC 20001-5403; (202) 842-0200; www.freetrade.org. Libertarian think tank analyzes farm policies as an advocate of global free markets.

Food First — Institute for Food and Development Policy, 98 60th St., Oakland, CA 94618; (510) 654-4400; www.foodfirst.org. Think tank and advocacy group that works to eliminate hunger.

Grist, 710 Second Ave., Suite 860, Seattle, WA 98104; www.grist.org. Environmental journalism group provides information on sustainable farming and other green issues.

International Assessment of Agricultural Knowledge, Science and Technology for Development, www.agassessment.org. International intergovernmental group that has produced expert assessments of future global food and agriculture needs.

International Food Policy Research Institute, 2033 K St., N.W., Washington, DC 20006-1002; (202) 862-5600; www.ifpri.org. Part an international network of food-policy groups seeking sustainable solutions to hunger.

King Corn, www.pbs.org/independentlens/kingcorn. Web site for a public-television documentary on the dominant role of corn in U.S. farming and food supply.

La Via Campesina, Jl. Mampang Prapatan XIV No. 5, Jakarta Selatan, DKI Jakarta, Indonesia 12790; +62-21-7991890; www.viacampesina.org. International organization that represents the interests of peasant farmers, small landholders, farmworkers and rural, landless people.

Organization for Competitive Markets, P.O. Box 6486, Lincoln, NE 68506; www.competitivemarkets.com. Think tank and advocacy group that promotes government-regulated free markets in agriculture.

World Agricultural Forum, One Metropolitan Square Plaza, Suite 1300, Saint Louis, MO 63102; (314) 206-3218; www.worldagforum.org. International group that brings together government, business, academic and non-governmental groups to seek solutions for global food problems.

World Neighbors, 4127 N.W. 122nd St., Oklahoma City, OK 73120; (405) 752-9700; www.wn.org. Provides training and support for rural communities in developing countries to improve their farm yields and make other community improvements.

2

Affirmative Action

Is it Time to End Racial Preferences?

Peter Katel

Law student Jessica Peck Corry, executive director of the Colorado Civil Rights Initiative, supports Constitutional Amendment 46, which would prohibit all government entities in Colorado from discriminating for or against anyone because of race, ethnicity or gender. Attorney Melissa Hart counters that the amendment would end programs designed to reach minority groups.

From *CQ Researcher*, October 17, 2008.

N o white politician could have gotten the question George Stephanopoulos of ABC News asked Sen. Barack Obama. "You said . . . that affluent African-Americans, like your daughters, should probably be treated as pretty advantaged when they apply to college," he began. "How specifically would you recommend changing affirmative action policies so that affluent African-Americans are not given advantages and poor, less affluent whites are?"[1]

The Democratic presidential nominee, speaking during a primary election debate in April, said his daughters' advantages should weigh more than their skin color. "You know, Malia and Sasha, they've had a pretty good deal."[2]

But a white applicant who has overcome big odds to pursue an education should have those circumstances taken into account, Obama said. "I still believe in affirmative action as a means of overcoming both historic and potentially current discrimination," Obama said, "but I think that it can't be a quota system and it can't be something that is simply applied without looking at the whole person, whether that person is black, or white or Hispanic, male or female."[3]

Supporting affirmative action on the one hand, objecting to quotas on the other — Obama seemed to know he was threading his way through a minefield. Decades after it began, affirmative action is seen by many whites as nothing but a fancy term for racial quotas designed to give minorities an unfair break. Majority black opinion remains strongly pro-affirmative action, on the grounds that the legacy of racial discrimination lives on. Whites and blacks are 30 percentage points apart on the issue, according to a 2007 national survey by the nonpartisan Pew Research Center.[4]

Americans Support Boost for Disadvantaged

A majority of Americans believe that individuals born into poverty can overcome their disadvantages and that society should be giving them special help (top poll). Fewer, however, endorse race-based affirmative action as the way to help (bottom).

	Agree	Disagree
We should help people who are working hard to overcome disadvantages and succeed in life.	93%	6
People who start out with little and work their way up are the real success stories.	91	7
Some people are born poor, and there's nothing we can do about that.	26	72
We shouldn't give special help at all, even to those who started out with more disadvantages than most.	16	81

If there is only one seat available, which student would you admit to college, the high-income student or the low-income student?

	Percentage selecting:	
	Low-income student	High-income student
If both students get the same admissions test score?	63%	3%
If low-income student gets a slightly lower test score?	33	54
If the low-income student is also black, and the high-income student is white?	36	39
If the low-income student is also Hispanic, and the high-income student is not Hispanic?	33	45

Source: Anthony P. Carnevale and Stephen J. Rose, "Socioeconomic Status, Race/ Ethnicity, and Selective College Admissions," The Century Foundation, March 2003

Now, with the candidacy of Columbia University and Harvard Law School graduate Obama turning up the volume on the debate, voters in two states will be deciding in November whether preferences should remain in effect in state government hiring and state college admissions.

Originally, conflict over affirmative action focused on hiring. But during the past two decades, the debate has shifted to whether preference should be given in admissions to top-tier state schools, such as the University of California at Los Angeles (UCLA) based on race, gender or ethnic background. Graduating from such schools is

seen as an affordable ticket to the good life, but there aren't enough places at these schools for all applicants, so many qualified applicants are rejected.

Resentment over the notion that some applicants got an advantage because of their ancestry led California voters in 1996 to ban affirmative action in college admissions. Four years later, the Florida legislature, at the urging of then-Gov. Jeb Bush, effectively eliminated using race as an admission standard for colleges and universities. And initiatives similar to the California referendum were later passed in Washington state and then in Michigan, in 2006.

Race is central to the affirmative action debate because the doctrine grew out of the civil rights movement and the Civil Rights Act of 1964, which outlawed discrimination based on race, ethnicity or gender. The loosely defined term generally is used as a synonym for advantages — "preferences" — that employers and schools extend to members of a particular race, national origin or gender.

"The time has come to pull the plug on race-based decision-making," says Ward Connerly, a Sacramento, Calif.-based businessman who is the lead organizer of the Colorado and Nebraska ballot initiative campaigns, as well as earlier ones elsewhere. "The Civil Rights Act of 1964 talks about treating people equally without regard to race, color or national origin. When you talk about civil rights, they don't just belong to black people."

Connerly, who is black, supports extending preferences of some kind to low-income applicants for jobs — as long as the beneficiaries aren't classified by race or gender.

But affirmative action supporters say that approach ignores reality. "If there are any preferences in operation in our society, they're preferences given to people with white skin and who are men and who have financial and other advantages that come with that," says Nicole Kief, New York-based state strategist for the American Civil

Liberties Union's racial justice program, which is opposing the Connerly-organized ballot initiative campaigns.

Yet, of the 38 million Americans classified as poor, whites make up the biggest share: 17 million people. Blacks account for slightly more than 9 million and Hispanics slightly less. Some 576,000 Native Americans are considered poor. Looking beyond the simple numbers, however, reveals that far greater percentages of African-Americans and Hispanics are likely to be poor: 25 percent of African-Americans and 20 percent of Hispanics live below the poverty line, but only 10 percent of whites are poor.[5]

In 2000, according to statistics compiled by *Chronicle of Higher Education* Deputy Editor Peter Schmidt, the average white elementary school student attended a school that was 78 percent white, 9 percent black, 8 percent Hispanic, 3 percent Asian and 30 percent poor. Black or Hispanic children attended a school in which 57 percent of the student body shared their race or ethnicity and about two-thirds of the students were poor.[6]

These conditions directly affect college admissions, according to The Century Foundation. The liberal think tank reported in 2003 that white students account for 77 percent of the students at high schools in which the greatest majority of students go on to college. Black students account for only 11 percent of the population at these schools, and Hispanics 7 percent.[7]

A comprehensive 2004 study by the Urban Institute, a nonpartisan think tank, found that only about half of black and Hispanic high school students graduate, compared to 75 and 77 percent, respectively, of whites and Asians.[8]

Politically conservative affirmative action critics cite these statistics to argue that focusing on college admissions and hiring practices rather than school reform was a big mistake. The critics get some support from liberals who want to keep affirmative action — as long as it's

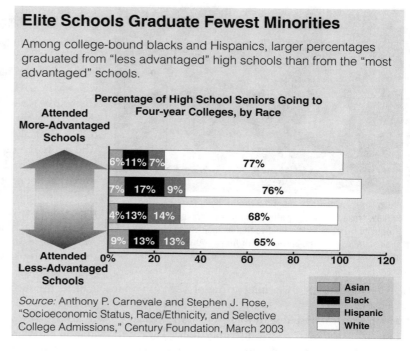

Elite Schools Graduate Fewest Minorities

Among college-bound blacks and Hispanics, larger percentages graduated from "less advantaged" high schools than from the "most advantaged" schools.

Percentage of High School Seniors Going to Four-year Colleges, by Race

Attended More-Advantaged Schools

6% 11% 7% 77%
7% 17% 9% 76%
4% 13% 14% 68%
9% 13% 13% 65%

Attended Less-Advantaged Schools

0%   20   40   60   80   100   120

Asian
Black
Hispanic
White

Source: Anthony P. Carnevale and Stephen J. Rose, "Socioeconomic Status, Race/Ethnicity, and Selective College Admissions," Century Foundation, March 2003

based on socioeconomic status instead of race. "Affirmative action based on race was always kind of a cheap and quick fix that bypassed the hard work of trying to develop the talents of low-income minority students generally," says Richard D. Kahlenberg, a senior fellow at The Century Foundation.

Basing affirmative action on class instead of race wouldn't exclude racial and ethnic minorities, Kahlenberg argues, because race and class are so closely intertwined.

President Lyndon B. Johnson noted that connection in a major speech that laid the philosophical foundations for affirmative action programs. These weren't set up for another five years, a reflection of how big a change they represented in traditional hiring and promotion practices, where affirmative action began. "You do not take a person who, for years, has been hobbled by chains and liberate him, bring him up to the starting line of a race and then say, 'You are free to compete with all the others,' and still justly believe that you have been completely fair," Johnson said in "To Fulfill These Rights," his 1965 commencement speech at Howard University in Washington, D.C., one of the country's top historically black institutions.[9]

Liasion/Lara Joe Regan

Asian-American enrollment at the University of California at Berkeley rose dramatically after California voters in 1996 approved Proposition 209, a ballot initiative that banned affirmative action at all state institutions. Enrollment of African-American, Hispanic and Native American students, however, plunged.

By the late 1970s, a long string of U.S. Supreme Court decisions began setting boundaries on affirmative action, partly in response to white job and school applicants who sued over "reverse discrimination." The court's bottom line: Schools and employers could take race into account, but not as a sole criterion. Setting quotas based on race, ethnicity or gender was prohibited. (The prohibition of gender discrimination effectively ended the chances for passage of the proposed Equal Rights Amendment [ERA], which feminist organizations had been promoting since 1923. The Civil Rights Act, along with other legislation and court decisions, made many supporters of women's rights "lukewarm" about the proposed amendment, Roberta W. Francis, then chair of the National Council of Women's Organizations' ERA task force, wrote in 2001.)[10]

The high court's support for affirmative action has been weakening through the years. Since 1991 the court has included Justice Clarence Thomas, the lone black member and a bitter foe of affirmative action. In his 2007 autobiography, Thomas wrote that his Yale Law School degree set him up for rejection by major law firm interviewers. "Many asked pointed questions unsubtly suggesting that they doubted I was as smart as my grades indicated," he wrote. "Now I knew what a law degree from Yale was worth when it bore the taint of racial preference."[11]

Some of Thomas' black classmates dispute his view of a Yale diploma's worth. "Had he not gone to a school like Yale, he would not be sitting on the Supreme Court," said William Coleman III, a Philadelphia attorney who was general counsel to the U.S. Army in the Clinton administration.[12]

But that argument does not seem to impress Thomas, who was in a 5-4 minority in the high court's most recent affirmative action ruling, in which the justices upheld the use of race in law-school admissions at the University of Michigan. But even Justice Sandra Day O'Connor, who wrote the majority opinion, signaled unease with her position. In 25 years, she wrote, affirmative action would "no longer be necessary."[13]

Paradoxically, an Obama victory on Nov. 4 might be the most effective anti-affirmative action event of all.

"The primary rationale for affirmative action is that America is institutionally racist and institutionally sexist," Connerly, an Obama foe, told The Associated Press. "That rationale is undercut in a major way when you look at the success of Sen. [Hillary Rodham] Clinton and Sen. Obama."

Asked to respond to Connerly's remarks, Obama appeared to draw some limits of his own on affirmative action. "Affirmative action is not going to be the long-term solution to the problems of race in America," he told a July convention of minority journalists, "because, frankly, if you've got 50 percent of African-American or Latino kids dropping out of high school, it doesn't really matter what you do in terms of affirmative action; those kids are not getting into college."[14]

As critics and supporters discuss the future of affirmative action, here are some of the questions being debated:

Has affirmative action outlived its usefulness?

In the United States of the late 1960s and '70s, even some outright opponents of race-based affirmative action conceded that it represented an attempt to deal with the consequences of longstanding, systematic racial discrimination, which had legally ended only shortly before.

But ever since opposition to affirmative action began growing in the 1980s, its opponents themselves have invoked the very principles that the civil rights movement had embraced in its fight to end discrimination. Taking a job or school applicant's race or ethnicity into account is immoral, opponents argue, even for supposedly benign purposes. And a policy of racial/ethnic preferences, by definition, cannot lead to equality.

In today's United States, critics say, minority applicants don't face any danger that their skin color or ethnic heritage will hold them back. Instead, affirmative-action beneficiaries face continuing skepticism from others — and even from themselves, that they somehow were given an advantage that their academic work didn't entitle them to receive.

Meanwhile, opponents and supporters readily acknowledge that a disproportionate share of black and Latino students receive substandard educations, starting in and lasting through high school. Affirmative action hasn't eliminated the link between race/ethnicity and poverty and academic deprivation, they agree.

Few Poor Students Attend Top Schools

Nearly three-quarters of students entering tier 1 colleges and universities come from the wealthiest families, but only 3 percent of students from the bottom quartile enter top schools. Far more students from poorer backgrounds enroll in less prestigious schools, and even more in community colleges.

Socioeconomic Status of Entering College Classes

School prestige level	First quartile (lowest)	Second quartile	Third quartile	Fourth quartile (highest)
Tier 1	3%	6%	17%	74%
Tier 2	7	18	29	46
Tier 3	10	19	36	35
Tier 4	16	21	28	35
Community Colleges	21	30	27	22

Source: Anthony P. Carnevale and Stephen J. Rose, "Socioeconomic Status, Race/ Ethnicity, and Selective College Admissions," The Century Foundation, March 2003

Critics of race preferences, however, say they haven't narrowed the divide that helped to trigger affirmative action in the first place. Affirmative action advocates favor significantly reforming K-12 education while simultaneously giving a leg up to minorities who managed to overcome their odds at inadequate public schools.

And some supporters say affirmative action is important for other reasons, which transcend America's racial history. Affirmative action helps to ensure continuation of a democratic political culture, says James E. Coleman Jr., a professor at Duke University Law School.

"It's not just about discrimination or past discrimination," says Coleman, who attended all-black schools when growing up and then graduated from Harvard College and Columbia Law School in the early 1970s, during the early days of affirmative action. "It's in our self-interest. We want leaders of all different backgrounds, all different races; we ought to educate them together."

But Connerly, the California businessman behind anti-affirmative action ballot initiatives, says that race and gender preferences are the wrong tool with which to promote diversity, because they effectively erode academic standards. "Excellence can be achieved by any group of people," says Connerly, a former member of the University of California Board of Regents. "So we will keep the standards where they ought to be, and we will expect people to meet those standards."

But legislators interested in a "quick fix" have found it simpler to mandate diversity than to devise ways to improve schools. "There are times when someone has to say, 'This isn't right. We're going to do something about it,' " Connerly says. "But in the legislative process, I can find no evidence of leadership anywhere."

Like others, Connerly also cites the extraordinary academic achievements of Asian-American students — who haven't benefited from affirmative action. Affirmative action supporters don't try to dispute that point. "At the University of California at Berkeley, 40 percent of the students are Asian," says Terry H. Anderson, a history professor at Texas A&M University in College Station. "What does that say about family structure? It makes a big statement. Family structure is so important, and it's something that affirmative action can't help at all."

But if encouraging minority-group enrollment at universities doesn't serve as a social and educational cure-all, says Anderson, who has written a history of affirmative

Few Poor Students Score High on SAT

Two-thirds of students who scored at least 1300 on the SAT came from families ranking in the highest quartile of socioeconomic status, compared with only 3 percent of students from the lowest-income group. Moreover, more than one-fifth of those scoring under 1000 — and 37 percent of non-test-takers — come from the poorest families.

SAT Scores by Family Socioeconomic Status*

Score	First Quartile (lowest)	Second Quartile	Third Quartile	Fourth Quartile (highest)
>1300	3%	10%	22%	66%
1200–1300	4	14	23	58
1100–1200	6	17	29	47
1000–1100	8	24	32	36
<1000	21	25	30	24
Non-taker	37	30	22	10

* The maximum score is 1600

Note: Percentages do not add to 100 due to rounding.

Source: Anthony P. Carnevale and Stephen J. Rose, "Socioeconomic Status, Race/ Ethnicity, and Selective College Admissions," The Century Foundation, March 2003

Does race-based affirmative action still face powerful public opposition?

At the state and federal level, affirmative action has generated enormous conflict over the decades, played out in a long chain of lawsuits and Supreme Court decisions, as well as the hard-fought ballot initiatives this year in Arizona, Missouri and Oklahoma — all three of which ended in defeat for race, ethnic and gender preferences.

But today's political agenda — dominated by the global financial crisis, the continuing downward slide of real estate prices, the continuing conflict in Iraq and escalated combat in Afghanistan — would seem to leave little space for a reignited affirmative action conflict.

Nevertheless, supporters and opponents of affirmative action fought hard in five states over proposed ballot initiatives, two of which will go before voters in November.

Nationally, the nonpartisan Pew Research Center reported last year that black and white Americans are divided by a considerable margin on whether minority group members should get preferential treatment. Among blacks, 57 answered yes, but only 27 percent of whites agreed. That gap was somewhat bigger in 1991, when 68 percent of blacks and only 17 percent of whites favored preferences.[15]

Obama's statement to ABC News' Stephanopoulos that his daughters shouldn't benefit from affirmative action reflected awareness of majority sentiment against race preference.[16]

Still, the exchange led to some predictions that it would resurface. "The issue of affirmative action is likely to dog Sen. Obama on the campaign trail as he seeks to win over white, blue-collar voters in battleground states like Michigan," The Wall Street Journal predicted in June.[17]

Just two and a half weeks before the election, that forecast hadn't come to pass. However, earlier in the year interest remained strong enough that campaigners for

action, the policy still serves a valuable purpose. "It's become part of our culture. On this campus, it's been 'out' to be racist for years and years. I'm looking at kids born in 1990; they just don't feel self-conscious about race or gender, they just expect to be treated equally."

Standing between the supporters and the enemies of affirmative action's racial/ethnic preferences are the affirmative action reformers. "I don't think it's time to completely abolish all forms of affirmative action," says the Century Foundation's Kahlenberg. "But it's clear there are strong legal, moral and political problems with relying solely on race."

And at the practical level, race isn't the only gauge of hardship that some students must overcome, even to be capable of competing for admission to a top-tier school. "There are students from low-income backgrounds," Kahlenberg says, "who aren't given the same opportunities as wealthier students are given, and they deserve a leg up in admissions. Someone's test scores and grades are a reflection not only of how hard they work and how talented they are, but what sorts of opportunities they've had."

state ballot initiatives were able to gather 136,589 signatures in Nebraska and about 130,000 in Colorado to require that the issue be put before voters in those states.

Meanwhile, the initiative efforts in Arizona, Missouri and Oklahoma were doomed after the validity of petition signatures was challenged in those states. Connerly, the chief organizer of the initiatives, blames opponents' tactics and, in Oklahoma, an unusually short, 90-day window during which signatures must be collected. But once initiatives get on ballots, he says, voters approve them. "There is something about the principle of fairness that most people understand."

Without congressional legislation prohibiting preferences, Connerly says, the initiatives are designed to force state governments "to abide by the moral principle that racial discrimination — whether against a white or black or Latino or Native American — is just wrong."

But reality can present immoral circumstances as well, affirmative action defenders argue. "Racial discrimination and gender discrimination continue to present obstacles to people of color and women," says the American Civil Liberties Union's (ACLU) Kief. "Affirmative action is a way to chip away at some of these obstacles."

Kief says the fact that Connerly has played a central role in all of the initiatives indicates that true grassroots opposition to affirmative action is weak in states where initiatives have passed or are about to be voted on.

However, The Century Foundation's Kahlenberg points out that pro-affirmative action forces work hard to block ballot initiatives, because when such initiatives have gone before voters they have been approved. And the most recent successful ballot initiative, in Michigan in 2006, passed by a slightly bigger margin — 57 percent to 43 percent — than its California counterpart in 1996, which was approved by 54-46.[18]

Further evidence that anti-affirmative action initiatives are hard to fight surfaced this year in Colorado, where the group Coloradans for Equal Opportunity failed to round up enough signatures to put a pro-affirmative action initiative on the ballot.

Kahlenberg acknowledges that affirmative action politics can be tricky. Despite abiding public opposition to preferences, support among blacks is so strong that Republican presidential campaigns tend to downplay affirmative action, for fear of triggering a huge turnout among black voters, who vote overwhelmingly Democratic. In 1999,

Democratic presidential candidate Sen. Barack Obama, speaking in Philadelphia on Oct. 11, 2008, represents the new face of affirmative action in the demographically changing United States: His father was Kenyan and a half-sister is half-Indonesian.

then-Florida Gov. Jeb Bush kept a Connerly-sponsored initiative out of that state largely in order to lessen the chances of a major black Democratic mobilization in the 2000 presidential election, in which his brother would be running.[19]

"When you have an initiative on the ballot," Kahlenberg says, "some Republicans think that it increases minority turnout, so they're not sure whether these initiatives play to their party or not." Republican opposition to affirmative action goes back to the Reagan administration. Reagan, however, passed up a chance to ban affirmative action programs throughout the federal government, displaying a degree of GOP ambivalence. However, Connerly is an outspoken Republican.[20]

Nevertheless, an all-out Republican push against affirmative action during the past decade failed to catch on at the national level. In 1996, former Republican Senate Majority Leader Bob Dole of Kansas was running for president, and the affirmative action initiative was on the same ballot in California. "The initiative passed, but there was no trickle-down help for Bob Dole," says Daniel A. Smith, a political scientist at the University of Florida who has written on affirmative action politics.

This year, to be sure, anxieties growing out of the financial crisis and economic slowdown could rekindle passions over preferences. But Smith argues the economic environment makes finger-pointing at minorities less likely. "Whites are not losing jobs to African-Americans," he says. "Whites and African-Americans are losing jobs to the Asian subcontinent — they're going to

Bangalore. The global economy makes it more difficult to have a convenient domestic scapegoat for lost jobs."

Has affirmative action diverted attention from the poor quality of K-12 education in low-income communities?

If there's one point on which everyone involved in the affirmative action debate agrees, it's that public schools attended by most low-income students are worsening.

"The educational achievement gap between racial groups began growing again in the 1990s," Gary Orfield, a professor of education and social policy at Harvard University, wrote. "Our public schools are becoming increasingly segregated by race and income, and the segregated schools are, on average, strikingly inferior in many important ways, including the quality and experience of teachers and the level of competition from other students. . . . It is clear that students of different races do not receive an equal chance for college."[21]

The decline in education quality has occurred at the same time various race-preference policies have governed admission to the nation's best colleges and universities. The policies were designed to provide an incentive for schools and students alike to do their best, by ensuring that a college education remains a possibility for all students who perform well academically.

But the results have not been encouraging. In California alone, only 36 percent of all high school students in 2001 had taken all the courses required for admission to the state university system, according to a study by the Civil Rights Project at Harvard University. Among black students, only 26 percent had taken the prerequisites, and only 24 percent of Hispanics. Meanwhile, 41 percent of white students and 54 percent of Asians had taken the necessary courses.[22]

In large part as a result of deficient K-12 education, decades of race-preference affirmative action at top-tier colleges and universities have yielded only small percentages of black and Hispanic students. In 1995, according to an exhaustive 2003 study by The Century Foundation, these students accounted for 6 percent of admissions to the 146 top-tier institutions.[23]

Socioeconomically, the picture is even less diverse. Seventy-four percent of students came from families in the wealthiest quarter of the socioeconomic scale; 3 percent came from families in the bottom quarter.[24]

For race-preference opponents, the picture demonstrates that efforts at ensuring racial and ethnic diversity in higher education would have been better aimed at improving K-12 schools across the country.

"If you've tried to use race for 40-some years, and you still have this profound gap," Connerly says, "yet cling to the notion that you have given some affirmative action to black and Latino and American Indian students — though Asians, without it, are outstripping everybody — maybe the way we've been doing it wasn't the right way to do it."

Meanwhile, he says, making a point that echoes through black, conservative circles, "Historically black colleges and universities (HBCUs) — if you look at doctors and pharmacists across our nation, you'll find them coming from schools that are 90 percent black. These schools are not very diverse, but they put a premium on quality."

But not all HBCUs are in that class, affirmative action supporters point out. "A lot of people who come out with a degree in computer science from minority-serving institutions know absolutely no mathematics," says Richard Tapia, a mathematics professor at Rice University and director of the university's Center for Equity and Excellence in Education. "I once went to a historically black university and had lunch with a top student who was going to do graduate work at Purdue, but when I talked to her I realized that her knowledge of math was on a par with that of a Rice freshman. The gap is huge."

Tapia, who advocates better mentoring for promising minority students at top-flight institutions, argues that the effect of relegating minority students to a certain defined group of colleges and universities, including historically black institutions, limits their chances of advancement in society at large. "From the elite schools you're going to get leadership."

Still, a question remains as to whether focusing on preferential admissions has helped perpetuate the very conditions that give rise to preferences in the first place.

"At the K-12 level you could argue that affirmative action has led to stagnation," says Richard Sander, a professor of law at UCLA Law School. "There's very little forward movement, very little closing of the black-white gap of the past 20 to 30 years."

Coleman of Duke University agrees that public education for most low-income students needs help. But that issue has nothing to do with admissions to top-drawer universities and professional schools, he says. "Look at minority students who get into places like that," he says. "For the most part, they haven't gone to the weakest high schools; they've often gone to the best."

Yet the affirmative action conflict focuses on black students, who are assumed to be academically under-qualified, Coleman says, while white students' place at the best schools isn't questioned. The classroom reality differs, he says. "We have a whole range of students with different abilities. All of the weak students are not minority students; all of the strong students are not white students."

BACKGROUND

Righting Wrongs

The civil rights revolution of the 1950s and '60s forced a new look at the policies that had locked one set of Americans out of most higher-education institutions and higher-paying jobs.

As early as 1962, the Congress of Racial Equality (CORE), one of the most active civil-rights organizations, advocated hiring practices that would make up for dis-crimination against black applicants. "We are approaching employers with the proposition that they have effectively excluded Negroes from their work force a long time, and they now have a responsibility and obligation to make up for their past sins," the organization said in a statement from its New York headquarters.[25]

Facing CORE-organized boycotts, a handful of compa-nies in New York, Denver, Detroit, Seattle and Baltimore changed their hiring procedures to favor black applicants.

In July 1964, President Lyndon B. Johnson pushed Congress to pass the landmark Civil Rights Act, which had been championed by President John F. Kennedy since his 1960 presidential election campaign.

The law's Title VII, which prohibits racial, religious or sexual discrimination in hiring, said judges enforcing the law could order "such affirmative action as may be appropriate" to correct violations.[26]

Title VII didn't specify what kind of affirmative action could be decreed. But racial preferences were openly discussed in the political arena as a tool to equalize opportunities. Official working definitions of affirmative action didn't emerge until the end of the 1960s, under President Richard M. Nixon.

In 1969, the administration approved the "Philadelphia Plan," which set numerical goals for black and other minority employment on federally financed construction jobs. One year later, the plan was expanded to cover all businesses with 50 or more employees and federal contracts of at least $50,000. The contracts were to set hiring goals and timetables designed to match up a firm's minority representation with the workforce demographics in its area. The specified minorities were: "Negro, Oriental, American Indian and Spanish Surnamed Americans."[27]

The sudden change in the workplace environment prompted a wave of lawsuits. In the lead, a legal challenge by 13 black electric utility workers in North Carolina led to one of the most influential U.S. Supreme Court deci-sions on affirmative action, the 1971 *Griggs v. Duke Power Co.* case.[28]

In a unanimous decision, the high court concluded that an aptitude test that was a condition of promotion for the workers violated the Civil Rights Act. Duke Power may not have intended the test to weed out black appli-cants, Chief Justice Warren E. Burger wrote in the decision. But, he added, "Congress directed the thrust of the Act to the consequences of employment practices, not simply the motivation."[29]

If the point of the Civil Rights Act was to ensure that the consequences of institutions' decisions yielded balanced workforces, then goals and timetables to lead to that outcome were consistent with the law as well. In other words, eliminating racial discrimination could mean paying attention to race in hiring and promotions.

That effort would produce a term that captured the frustration and anger among white males who were com-peting with minority-group members for jobs, promotions or school admissions: "reverse discrimination."

The issue went national with a challenge by Allan Bakke, a white, medical school applicant, to the University of California. He'd been rejected two years in a row while minority-group members — for whom 16 slots in the 100-member class had been set aside — were admitted with lower qualifying scores.

After the case reached the Supreme Court, the justices in a 5-4 decision in 1978 ordered Bakke admitted and prohibited the use of racial quotas. But they allowed race to be considered along with other criteria. Representing the University of California was former Solicitor General Archibald Cox, the Watergate special prosecutor who was

CHRONOLOGY

1960s *Enactment of civil rights law opens national debate on discrimination.*

1964 Civil Rights Act of 1964 bars discrimination in employment and at federally funded colleges.

1965 President Lyndon B. Johnson calls for a massive national effort to create social and economic equality.

1969 Nixon administration approves "Philadelphia Plan" setting numerical goals for minority employment on all federally financed building projects.

1970s–1980s *Affirmative action expands throughout the country, prompting legal challenges and growing voter discontent, leading to new federal policy.*

1971 The U.S. Supreme Court's landmark *Griggs v. Duke Power Co.* decision, growing out of a challenge by 13 black electric utility workers in North Carolina, is seen as authorizing companies and institutions to set out goals and timetables for minority hiring.

1978 Supreme Court's decision in *University of California Regents v. Bakke,* arising from a medical-school admission case, rules out racial quotas but allows race to be considered with other factors.

1980 Ronald W. Reagan is elected president with strong support from white males who see affirmative action as a threat.

1981-1983 Reagan administration reduces affirmative action enforcement.

1985 Attorney General Edwin Meese III drafts executive order outlawing affirmative action in federal government; Reagan never signs it.

1987 Supreme Court upholds job promotion of a woman whose advancement was challenged by a male colleague claiming higher qualifications.

1990s *Ballot initiatives banning race and gender preferences prompt President Bill Clinton to acknowledge faults in affirmative action.*

1994 White voter discontent energizes the "Republican revolution" that topples Democrats from control of Congress.

1995 Supreme Court rules in *Adarand Constructors v. Peña* that affirmative action programs must be "narrowly tailored" for cases of extreme discrimination. . . . Clinton concedes that affirmative action foes have some valid points but concludes, "Mend it, but don't end it." . . . Senate votes down anti-affirmative action bill.

1996 California voters pass nation's first ballot initiative outlawing racial, ethnic and gender preferences. . . . 5th U.S. Circuit Court of Appeals rules that universities can't take race into account in evaluating applicants.

1998 Washington state voters pass ballot initiative identical to California's.

2000s *Affirmative action in university admissions stays on national agenda, leading to major Supreme Court ruling; Sen. Barack Obama's presidential candidacy focuses more attention on the issue.*

2003 Supreme Court's *Gratz v. Bollinger* ruling rejects University of Michigan undergraduate admission system for awarding extra points to minority applicants, but simultaneous *Grutter v. Bollinger* decision upholds UM law school admissions policy, which includes race as one factor among many. . . . Justice Sandra Day O'Connor writes in 5-4 majority opinion in *Grutter* that affirmative action won't be necessary in 25 years. . . . Century Foundation study finds strong linkage between socioeconomic status, race and chances of going to college.

2006 Michigan passes nation's third ballot initiative outlawing racial, ethnic and gender preferences.

2008 Opponents of affirmative action in Arizona, Missouri and Oklahoma fail to place anti-affirmative action initiatives on ballot, but similar campaigns succeed in Colorado and Nebraska. . . . U.S. Civil Rights Commission opens study of minority students majoring in science and math. . . . Saying his daughters are affluent and shouldn't benefit from race preferences, Obama endorses affirmative action for struggling, white college applicants.

fired on orders of President Nixon in 1973. Cox's grand-daughter, Melissa Hart, helps lead the opposition to an anti-affirmative action ballot initiative in Colorado (*see p. 39*).[30]

In 1979 and 1980, the court upheld worker training and public contracting policies that included so-called set-asides for minority-group employees or minority-owned companies. But in the latter case, the deciding opinion specified that only companies that actually had suffered discrimination would be eligible for those contracts.[31]

Divisions within the Supreme Court reflected growing tensions in the country as a whole. A number of white people saw affirmative action as injuring the educational and career advancement of people who hadn't themselves caused the historical crimes that gave rise to affirmative action.

Reversing Course

President Ronald W. Reagan took office in 1981 with strong support from so-called "Reagan Democrats" — white, blue-collar workers who had turned against their former party on issues including affirmative action.[32]

Initially, Reagan seemed poised to fulfill the hopes of those who wanted him to ban all preferences based on race, ethnicity and gender. The latter category followed an upsurge of women fighting to abolish limits on their education and career possibilities.

Yet Reagan's appointees were divided on the issue, and the president himself never formalized his rejection of quotas and related measures. Because no law required the setting of goals and timetables, Reagan could have banned them by executive order. During Reagan's second term, Attorney General Edwin Meese III drafted such an order. But Reagan never signed it.

Nevertheless, the Reagan administration did systematically weaken enforcement of affirmative action. In Reagan's first term he cut the budgets of the Equal Employment Opportunity Commission and the Office of Federal Contract Compliance — the two front-line agencies on the issue — by 12 and 34 percent, respectively, between 1981 and 1983. As a result, the compliance office blocked only two contractors during Reagan's two terms, compared with 13 that were barred during President Jimmy Carter's term.

The Justice Department also began opposing some affirmative action plans. In 1983, Justice won a partial court reversal of an affirmative action plan for the New Orleans Police Department. In a police force nearly devoid of black supervisors, the plan was designed to expand the number — a move considered vital in a city whose population was nearly one-half black.

Affirmative action cases kept moving through the Supreme Court. In 1984-1986, the court overturned plans that would have required companies doing layoffs to disregard the customary "first hired, last fired" rule, because that custom endangered most black employees, given their typically short times on the job.

And in 1987, a 5-4 Supreme Court decision upheld an Alabama state police plan requiring that 50 percent of promotions go to black officers. The same year, the court upheld 6-3 the promotion of a woman employee of Santa Clara County, Calif., who got promoted over a male candidate who had scored slightly higher on an assessment. The decision marked the first court endorsement of affirmative action for women.

In the executive branch, divided views persisted in the administration of Reagan's Republican successor, George H. W. Bush. In 1990 Bush vetoed a pro-affirmative action bill designed to reverse recent Supreme Court rulings, one of which effectively eased the way for white men to sue for reverse discrimination.

The legislation would have required "quotas," Bush said, explaining his veto. But the following year, he signed a compromise, the Civil Rights Act of 1991.[33] Supported by the civil rights lobby, the bill wrote into law the *Griggs v. Duke Power* requirement that an employer prove that a job practice — a test, say — is required for the work in question. A practice that failed that test could be shown to result in discrimination, even if that hadn't been the intention.

Bush also reversed a directive by his White House counsel that would have outlawed all quotas, set-asides and related measures. The administration's ambivalence reflected divided views in American society. Local government and corporate officials had grown appreciative of affirmative action for calming racial tensions. In 1985, the white Republican mayor of Indianapolis refused a Justice Department request to end affirmative action in the police department. Mayor William Hudnut said that the "white majority has accepted the fact that we're making a special effort for minorities and women."[34]

Yet among white males, affirmative action remained a very hot-button issue. "When we hold focus groups," a Democratic pollster said in 1990, "if the issue of

'Percent Plans' Offer Alternative to Race-Based Preferences

But critics say approach fails to level playing field.

In recent years, voters and judges have blocked race and ethnicity preferences in university admissions in three big states with booming minority populations — California, Florida and Texas. Nonetheless, lawmakers devised a way to ensure that public universities remain open to black and Latino students.

The so-called "percent plans" promise guaranteed admission based on a student's high school class standing, not on skin tone. That, at least is the principle.

But the man who helped end racial affirmative action preferences in two of the states involved argues affirmative action is alive and well, simply under another name. Moreover, says Ward Connerly, a black businessman in Sacramento, Calif., who has been a leader in organizing anti-affirmative action referendums, the real issue — the decline in urban K-12 schools — is being ignored.

"Legislatures and college administrators lack the spine to say, 'Let's find the problem at its core,' " says Connerly, a former member of the University of California Board of Regents. "Instead, they go for a quick fix they believe will yield the same number of blacks and Latinos as before."

Even Connerly's opponents agree "percent plans" alone don't put high schools in inner cities and prosperous suburbs on an equal footing. "In some school districts in Texas, 50 percent of the graduates could make it here easily," says Terry H. Anderson, a history professor at Texas A&M University in College Station. "Some school districts are so awful that not one kid could graduate here, I don't care what race you're talking about."

All the plans — except at selective schools — ignore SAT or ACT scores (though students do have to present their scores). The policy troubles Richard D. Kahlenberg, a senior fellow at The Century Foundation, who champions "class-based" affirmation action. "The grade of A in one high school is very different from the grade of A in another," he says.

Texas lawmakers originated the percent plan concept after a 5th U.S. Circuit Court of Appeals decision in 1996 (*Hopwood v. Texas*) prohibited consideration of race in college admissions. Legislators proposed guaranteeing state university admissions to the top 10 percent of graduates of the state's public and private high schools. Then-Gov. George W. Bush signed the bill, which includes automatic admission to the flagship campuses, the University of Texas at Austin and Texas A&M.[1]

In California, the impetus was the 1996 voter approval of Proposition 209, which prohibited racial and ethnic preferences by all state entities. Borrowing the Texas idea, California lawmakers devised a system in which California high school students in the top 4 percent of their classes are eligible for the California system, but not necessarily to attend the two star institutions, UC Berkeley and UCLA. (Students in the top 4 percent-12.5 percent range are admitted to community colleges and can transfer to four-year institutions if they maintain 2.4 grade-point averages.)[2]

Connerly was active in the Proposition 209 campaign and was the key player — but involuntarily — in Florida's adoption of a percent plan. In 1999, Connerly was preparing to mount an anti-affirmative action initiative in Florida. Then-Gov. Jeb Bush worried it could hurt his party's standing with black voters — with possible repercussions on his brother

affirmative action comes up, you can forget the rest of the session. That's all . . . that's talked about."[35]

Mending It

From the early 1990s to 2003 race-based affirmative action suffered damage in the political arena and the courts.

In 1994, white male outrage at preferences for minority groups and women was a key factor in congressional elections that toppled Democrats from control of both houses. As soon as the Congress changed hands, its new leaders targeted affirmative action. "Sometimes the best-qualified person does not get the job because he or she

George's presidential campaign. Instead Gov. Bush launched "One Florida," a percent plan approved by the legislature.

In Florida, the top 20 percent of high school graduates are guaranteed admission to the state system. To attend the flagship University of Florida at Gainesville they must meet tougher standards. All three states also require students to have completed a set of required courses.

Percent plan states also have helped shape admissions policies by experimenting with ways to simultaneously keep academic standards high, while ensuring at least the possibility that promising students of all socioeconomic circumstances have a shot at college.

In Florida, the consequences of maintaining high admissions standards at UF were softened by another program, "Bright Futures," which offers tuition reductions of 75 percent — or completely free tuition — depending on completion of AP courses and on SAT or ACT scores.

The effect, says University of Florida political scientist Daniel A. Smith, is to ensure a plentiful supply of top students of all races and ethnicities. "We have really talented minorities — blacks, Latinos, Asian-Americans — because 'One Florida' in combination with 'Bright Futures' has kept a lot of our talented students in the state. We have students who turned down [partial] scholarships to Duke and Harvard because here they're going for free."

At UCLA, which also has maintained rigorous admission criteria, recruiters spread out to high schools in low-income areas in an effort to ensure that the school doesn't become an oasis of privilege. The realities of race and class

"The time has come to pull the plug on race-based decision-making," says Ward Connerly, a Sacramento, Calif., businessman who spearheaded anti-affirmative action ballot initiatives in Colorado, Nebraska and other states.

mean that some of that recruiting work takes place in mostly black or Latino high schools.

"It's the fallacy of [Proposition] 209 that you can immediately move to a system that doesn't take account of race and that treats everybody fairly," said Tom Lifka, a UCLA assistant vice chancellor in charge of admissions. He said the new system meets legal standards.[3]

Consciously or not, Lifka was echoing the conclusion of the most thorough analysis of the plans' operations in the three states. The 2003 study, sponsored by Harvard University's Civil Rights Project, concluded that the states had largely succeeded in maintaining racial and ethnic diversity on their campuses.

But the report added that aggressive recruitment, academic aid to high schools in low-income areas and similar measures played a major role.

"Without such support," wrote Catherine L. Horn, an education professor at the University of Houston, and Stella M. Flores, professor of public policy and higher education at Vanderbilt, "the plans are more like empty shells, appearing to promise eligibility, admission and enrollment for previously excluded groups but actually doing very little."[4]

---

[1] Catherine L. Horn and Stella M. Flores, "Percent Plans in College Admissions: A Comparative Analysis of Three States' Experiences," Civil Rights Project, Harvard University, 2003, pp. 20-23, www.civilrightsproject.ucla.edu/research/affirmativeaction/tristate.pdf.

[2] *Ibid.*

[3] Quoted in David Leonhardt, "The New Affirmative Action," *The New York Times Magazine*, Sept. 30, 2007, p. 76.

[4] Horn and Flores, *op. cit.*, pp. 59-60.

---

may be one color," Majority Leader Dole said in a television interview. "That may not be the way it should be in America."[36]

The following year, the U.S. Supreme Court imposed limits on the use of preferences, ruling on a white, male contractor's challenge to a federal program that

encouraged general contractors to favor minority subcontractors. Justice O'Connor wrote in the 5-4 majority opinion in *Adarand Constructors v. Peña* that any racial or ethnic preferences had to be "narrowly tailored" to apply only to "pervasive, systematic and obstinate discriminatory conduct."[37]

Getty Images/Bill Pugliano

Supporters of affirmative action in Lansing, Mich., rally against a proposed statewide anti-affirmative action ballot initiative in September 2006; voters approved the proposal that November. The initiative followed a 2003 U.S. Supreme Court ruling upholding the use of race in law-school admissions at the University of Michigan. Justice Sandra Day O'Connor, who wrote the majority 5-4 opinion, predicted, however, that in 25 years affirmative action would "no longer be necessary."

Some justices had wanted all preferences overturned. Though that position failed to win a majority, the clear unease that O'Connor expressed added to the pressure on politicians who supported affirmative action.

In that climate, President Bill Clinton gave a 1995 speech at the National Archives in Washington in which he acknowledged that critics had a point. He said he didn't favor "the unjustified preference of the unqualified over the qualified of any race or gender." But affirmative action was still needed because discrimination persisted, Clinton added. His bottom line: "Mend it, but don't end it."[38]

The slogan seemed to match national politicians' mood. One day after Clinton's speech, the Senate voted down a bill to abolish all preferences, with 19 Republicans siding with Democrats in a 61-36 vote.

But in California, one of the country's major affirmative action laboratories, the "end it" argument proved more popular. Racial/ethnic preferences had become a major issue in a state whose minority population was booming. California's higher-education system also included two of the nation's top public institutions: the University of California at Berkeley (UCB) and UCLA.

Among many white, Anglo Californians, affirmative action had come to be seen as a system under which black and Latino applicants were getting into those two schools at the expense of whites or Asians with higher grades and SAT scores.

By 1996, the statewide university system's majority-Republican Board of Regents voted to end all race, ethnic and gender preferences in admissions. The board did allow universities to take applicants' socioeconomic circumstances into account.

And in the same year, California voters approved Proposition 209, which outlawed all race, ethnicity and gender preferences by all state entities. Connerly helped organize that referendum and followed up with successful campaigns in Washington state in 1998 and in Michigan in 2003.

Meanwhile, the "reverse discrimination" issue that had been decided in the *Bakke* case flared up in Texas, where Cheryl Hopwood and two other white applicants to the University of Texas law school challenged their rejections, pointing to the admissions of minority students with lower grades and test scores. In 1996, the 5th U.S. Circuit Court of Appeals decided for the plaintiffs, ruling that universities couldn't take race into account when assessing applicants.

The appeals judges had overruled the *Bakke* decision, at least in their jurisdiction of Texas, Mississippi and Louisiana, yet the Supreme Court refused to consider the case.

But in 2003, the justices ruled on two separate cases, both centering on admissions to another top-ranked public higher education system: the University of Michigan. One case arose from admissions procedures for the undergraduate college, the other from the system for evaluating applicants to the university's law school.[39]

The Supreme Court decided against the undergraduate admissions policy because it automatically awarded 20 extra points on the university's 150-point evaluation scale to blacks, Latinos and American Indians. By contrast, the law school took race into account in what Justice O'Connor, in the majority opinion in the 5-4 decision, called a "highly individualized, holistic review" of each candidate aimed at producing a diverse student population.[40]

CURRENT SITUATION

'Formal Equality'

In the midst of war and the Wall Street meltdown, affirmative action may not generate as many headlines as it used to. But the issue still packs enough punch to have put anti-affirmative action legislation up for popular vote in Colorado and Nebraska this year.

"This is a progressive approach," said Jessica Peck Corry, executive director of the Colorado Civil Rights Initiative, which is campaigning for proposed Constitutional Amendment 46. The amendment would prohibit all state government entities from discriminating for or against anyone because of race, ethnicity or gender. "America is too diverse to put into stagnant race boxes," she says.

Melissa Hart, a co-chair of "No On 46," counters that the amendment would require "formal equality" that shouldn't be confused with the real thing. She likens the proposal to "a law that says both the beggar and the king may sleep under a bridge." In the real world, she says, only one of them will spend his nights in a bedroom.

Unlike California, Michigan and Washington — the states where voters have approved initiatives of this type over the past 12 years — the Colorado campaign doesn't follow a major controversy over competition for university admissions.

To be sure, Corry — a libertarian Republican law student, blogger and past failed candidate for state Senate — has publicly opposed affirmative action for several years.[41] But Corry, who is also a policy analyst at the Denver-based Independence Institute, a libertarian think tank, acknowledges that the referendum campaign in Colorado owes its start to Connerly. He began taking the ballot initiative route in the 1990s, after concluding that neither state legislatures nor Congress would ever touch the subject.

"They just seem to lack the stomach to do what I and the majority of Americans believe should be done," Connerly says. "Clearly, there's a disconnect between elected officials and the people themselves."

Connerly's confidence grows out of his success with the three previous initiatives. But this year, his attempts to get his proposal before voters in Arizona, Missouri and Oklahoma all failed because his campaign workers didn't gather enough valid signatures to get the initiatives on the ballot.

Connerly blames what he calls an overly restrictive initiative process in Oklahoma, as well as organized opposition by what he calls "blockers," who shadowed signature-gatherers and disputed their explanations of the amendments.

Opponents had a different name for themselves. "Our voter educators were simply that — voter educators," said Brandon Davis, political director of the Service Employees International Union in Missouri. "Ward Connerly should accept what Missourians said, and he should stop with the sore-loser talk."[42]

The opposition began deploying street activists to counter what they call the deliberately misleading wording of the proposed initiatives. In Colorado, Proposition 46 is officially described as a "prohibition against discrimination by the state" and goes on to ban "preferential treatment to any individual or group on the basis of race, sex, color, ethnicity or national origin."[43]

"We want an acknowledgement that disadvantage cannot be specifically determined based on looking at some race data or gender data," Corry says. But tutoring, counseling and other activities should be extended to all who need help because of their socioeconomic circumstances, she contends.

Likewise, a project to interest girls in science and math, for instance, would have to admit boys. "In a time when America is losing its scientific advantage by the second, why are you excluding potential Nobel prize winners because they're born with the wrong biology?" she asks rhetorically.

Hart says that many tutoring and similar programs tailored to low-income students in Colorado already welcome all comers, regardless of race or ethnicity. But she questions why a math and science program tailored for girls should have to change its orientation. Likewise, Denver's specialized public schools for American Indian students would have to change their orientation entirely. "Class-based equal opportunity programs are not substitutes for outreach, training and mentoring on the basis of race and gender," she says.

The issue of class comes up in personal terms as well. Corry portrays herself as the product of a troubled home who had to work her way through college and graduate school. Though her father was a lawyer, her mother abandoned the family and wound up living on the streets. And Corry depicts Hart as a member of the privileged

The Preference Program Nobody Talks About

How "legacies" get breaks at top colleges.

Many critics say race-based affirmative action gives minority college applicants an unfair advantage. But reporter Peter Schmidt found an even more favored population — rich, white kids who apply to top-tier schools.

"These institutions feel very dependent on these preferences," Schmidt writes in his 2007 book, *Color and Money: How Rich White Kids Are Winning the War Over College Affirmative Action.* "They throw up their hands and say, 'There's no other way we can raise the money we need.'"

Colleges admit these students — "legacies," in college-admission lingo — because their parents are donation-making graduates. Offspring of professors, administrators or (in the case of top state universities) politically influential figures get open-door treatment as well.

"Several public college lobbyists, working in both state capitals and with the federal government in and around Washington, have told me that they spend a significant portion of their time lobbying their own colleges' admissions offices to accept certain applicants at the behest of public officials," Schmidt writes.[1]

Especially in regard to legacies and the families' donations, Schmidt says, "There is a utilitarian argument that the money enables colleges to serve students in need. But there isn't a correlation between how much money they're bringing in and helping low-income students."

As deputy editor of the *Chronicle of Higher Education*, Schmidt has been covering affirmative action conflicts since his days as an Associated Press reporter writing about protests over racial tensions at the University of Michigan in the mid-1990s.

His book doesn't deal exclusively with applicants from privileged families — who, by the nature of American society, are almost all white and academically well-prepared. But Schmidt's examination of privileged applicants frames his reporting on the more familiar issues of preferences based on race, ethnicity and gender.

According to Schmidt, Harvard as of 2004 accepted about 40 percent of the legacies who applied, compared to about 11 percent of applicants overall. In the Ivy League in general, children of graduates made up 10-15 percent of the undergraduates.

Though the issue is sensitive for college administrators, Schmidt found some members of the higher-education establishment happy to see it aired.

"Admissions officers are the ones who are finding the promising kids — diamonds in the rough — and getting emotionally invested in getting them admitted, then sitting down with the development officer or the coach and finding that these kids are knocked out of the running," he says.

Some education experts dispute that conclusion. Abigail Thernstrom, a senior fellow at the conservative Manhattan Institute and vice-chair of the U.S. Commission on Civil Rights, opposes "class-based" affirmative action (as well as racial/ethnic preferences), calling it unnecessary. She says that when top-tier schools look at an applicant from a disadvantaged background "who is getting a poor education — a diamond in the rough but showing real academic progress — and compare that student to someone from Exeter born with a silver spoon in his mouth, there's no question that these schools are going to take that diamond in the rough, if they think he or she will be able to keep up."

But some of Schmidt's findings echo what affirmative action supporters have observed. James E. Coleman Jr., a law professor at Duke University, argues against the tendency to focus all affirmative action attention on blacks and Latinos. "The idea is that any white student who gets here deserves to be here. They're not questioned. This has always been true."

At the same time, Coleman, who is black, agrees with Schmidt that those who start out near the top of the socioeconomic ladder have access to first-class educations before they even get to college. Coleman himself, who graduated from Harvard and from Columbia Law School, says he never had a single white classmate in his Charlotte, N.C., schools until he got accepted to a post-high school preparatory program at Exeter, one of the nation's most prestigious prep schools. "I could tell that my educational background and preparation were woefully inadequate compared to students who had been there since ninth grade," he recalls. "I had to run faster."

Schmidt says the politics of affirmative action can give rise to tactical agreements between groups whose interests might seem to conflict. In one dispute, he says, "Civil rights groups and higher-education groups had a kind of uneasy alliance: The civil rights groups would not challenge the admissions process and go after legacies as long as affirmative action remained intact."

But, he adds, "There are people not at the table when a deal like that is struck. If you're not a beneficiary of one or the other side of preferences, you don't gain from that agreement."

---

[1] Peter Schmidt, *Color and Money: How Rich White Kids Are Winning the War Over College Affirmative Action* (2007), p. 32.

AT ISSUE

Would many black and Latino science and math majors be better off at lesser-ranked universities?

YES
Rogers Elliott
Professor emeritus, Department of Psychology and Brain Sciences, Dartmouth College

From testimony before U.S. Civil Rights Commission, Sept. 12, 2008

Race preferences in admissions in the service of affirmative action are harming the aspirations, particularly, of blacks seeking to be scientists.

The most elite universities have very high levels in their admission standards, levels which minorities — especially blacks — don't come close to meeting.

[Thus], affirmative action in elite schools, which they pursue vigorously and successfully, leaves a huge gap, probably bigger than it would be for affirmative action at an average school. That is what constitutes the problem.

At elite schools, 90 percent of science majors [got] 650 or above on the SAT math score. About 80 percent of the white/Asian group are 650 or above, but only 25 percent of the black group have that score or better. The gaps that are illustrated in these data have not gotten any better. They have, in fact, gotten a little bit worse: The gap in the SAT scores between blacks and whites, which got to its smallest extent in about 1991 — 194 points — is back to 209.

The higher the standard at the institution, the more science they tend to do. But the [lower-ranking schools] still do science, and your chances of becoming a scientist are better. Now, obviously, there are differences. The higher institutions have eliteness going for them. They have prestige going for them, and maybe getting a degree from Dartmouth when you want to be a doctor will leave you better off in this world even though you're not doing the thing you started with as your aspiration.

Seventeen of the top 20 PhD-granting institutions for blacks in this country, are HBCUs [historically black colleges and universities].

Elite institutions are very performance-oriented. They deliberately take people at a very high level to begin with — with a few exceptions — and then they make them perform, and they do a pretty good job of it. If you're not ready for the first science course, you might as well forget it. Some of these minority students had mostly A's . . . enough to get to Dartmouth or Brown or Cornell or Yale. They take their first course, let's say, in chemistry; at least 90 percent of the students in that course are bright, motivated, often pre-med, highly competitive whites and Asians. And these [minority] kids aren't as well-prepared. They may get their first C- or D in a course like that because the grading standards are rigorous, and you have to start getting it from day one.

NO
Prof. Richard A. Tapia
Director, Center on Excellence and Equity, Rice University

From testimony before U.S. Civil Rights Commission, Sept. 12, 2008

The nation selects leaders from graduates and faculty of U.S. universities with world-class science, technology, engineering and math (STEM) research programs. If we, the underrepresented minorities, are to be an effective component in STEM leadership, then we must have an equitable presence as students and faculty at the very top-level research universities.

Pedigree, unfortunately, is an incredible issue. Top research universities choose faculty from PhDs produced at top research universities. PhDs produced at minority-serving schools or less-prestigious schools will not become faculty at top research universities. Indeed, it's unlikely they'll become faculty at minority-serving institutions. A student from a research school with a lesser transcript is stronger than a student from a minority-serving institution with all A's.

So are the students who come from these minority-serving institutions incompetent? No. There's a level of them that are incredibly good and will succeed wherever they go. And usually Stanford and Berkeley and Cornell will get those. Then there's a level below that you can work with. I produced many PhDs who came from minority-serving institutions. Is there a gap in training? Absolutely.

We do not know how to measure what we really value: Creativity. Underrepresented minorities can be quite creative. For example, the Carl Hayden High School Robotic Team — five Mexican-American students from West Phoenix — beat MIT in the final in underwater robotics. They were not star students, but they were incredibly creative.

Treating everyone the same is not good enough. Sink or swim has not worked and will not work. It pays heed to privilege, not to talent. Isolation, not academics, is often the problem. We must promote success and retention with support programs. We must combat isolation through community-building and mentoring.

Ten percent of the students in public education in Texas are accepted into the University of Texas, automatically — the top 10 percent. They could have said look, these students are not prepared well. They're dumped at our doorstep, let's leave them. They didn't. The Math Department at the University of Texas at Austin built support programs where minorities are retained and succeed. It took a realization that here they are, let's do something with them.

Race and ethnicity should not dictate educational destiny. Our current path will lead to a permanent underclass that follows racial and ethnic lines.

AP Photo/Nati Harnik

TV cameramen in Lincoln, Neb., shoot boxes of signed voter petitions that qualified a proposed initiative to be put on the ballot in Nebraska this coming November calling for a ban on most types of affirmative action.

class, a granddaughter of former Solicitor General Cox and a graduate of Harvard University and Harvard Law School. "People like Melissa, I believe, are well-intentioned but misguided," Corry says. "The worst thing you can do to someone without connections is to suggest that they can't make it without preferences."

Hart, rapping Corry for bringing up personal history rather than debating ideas, adds that her father and his part of the family are potato farmers from Idaho.

"I am proudly the granddaughter of Archibald Cox, proud of the fact that he argued the *Bakke* case for the University of California, and proud to be continuing a tradition of standing up for opportunity in this country," she says.

The Nebraska campaign, taking place in a smaller state with little history of racial or ethnic tension and a university where competition for admission isn't an issue, has generated somewhat less heat. But as in Colorado, college-preparation and other programs of various kinds that target young women and American Indians would be threatened by the amendment, says Laurel Marsh, executive director of the Nebraska ACLU.

Over Their Heads?

The U.S. Civil Rights Commission is examining one of the most explosive issues in the affirmative action debate: whether students admitted to top universities due to

racial preferences are up to the academic demands they face at those institutions.

Math and the hard sciences present the most obvious case, affirmative action critics — and some supporters — say. Those fields are at the center of the commission's inquiry because students from high schools in low-income areas — typically minority students — tend to do poorly in science and math, in part because they require considerable math preparation in elementary and high school.

Sander of UCLA, who has been studying the topic, testified to the commission that for students of all races who had scored under 660 on the math SAT, only 5 percent of blacks and 3.5 percent of whites obtained science degrees. But of students who scored 820 or above on the SAT, 44 percent of blacks graduated with science or engineering degrees. Among whites, 35 percent graduated with those degrees — illustrating Sander's point that that issue is one of academic preparation, not race.

Abigail Thernstrom, the commission's vice-chair, says that most graduates of run-of-the-mill urban schools labor under a major handicap in pursuing math or science degrees. "By the time they get to college they're in bad shape in a discipline like math, where all knowledge is cumulative," she says. "The colleges are inheriting a problem that, in effect, we sweep under the rug."

Thernstrom, a longtime affirmative action critic, bases her views both on her 11 years of service on the Massachusetts state Board of Education and on data assembled by academics, including Sander. "Test scores do predict a lot, high school grades predict a lot," Sander says in an interview, disputing critics of his work who say students from deficient high schools can make up in college what they missed earlier.

Testifying to the commission on Sept. 12, Sander presented data showing that black and Hispanic high school graduates tend to be more interested than their white counterparts in pursuing science and math careers, but less successful in holding on to majors in those fields in college. Lower high school grades and test scores seem to account for as much as 75 percent of the tendency to drop out of those fields, he says.

Sander added that a student's possibilities can't be predicted from skin color and that the key factor associated with inadequate academic preparation is socioeconomic status. "We ought to view that as good news, because that means there's no intrinsic or genetic gap," he testified.

Rogers Elliott, an emeritus psychology and brain sciences professor at Dartmouth College, told the commission that the best option for many black and Hispanic students who want to pursue science or math careers is to attend lower-rated universities. Among institutions that grant the most PhDs to blacks, 17 of the top 20 are HBCUs, Elliott said, "and none of them is a prestige university."

Richard Tapia, a Rice University mathematician, countered that consigning minority-group students who aren't stars to lower-ranking universities would be disastrous. Only top-tier universities, he argued, provide their graduates with the credibility that allows them to assert leadership. "Research universities must be responsible for providing programs that promote success," he said, "rather than be let off the hook by saying that minority students should go to minority-serving institutions or less prestigious schools."

Tapia directs such a program — one of a handful around the country — that he says has helped Rice students overcome their inadequate earlier schooling. But he accepts Sander's and Elliott's data and says students with combined SAT scores below 800 would not be capable of pursuing math or science majors at Rice.

Tapia, the son of Mexican immigrants who didn't attend college, worked at a muffler factory after graduating from a low-achieving Los Angeles high school. Pushed by a co-worker to continue his education, he enrolled in community college and went on to UCLA, where he earned a doctorate. He attributes his success to a big dose of self-confidence — something that many people from his background might not have but that mentors can nurture.

A commission member sounded another practical note. Ashley L. Taylor Jr., a Republican lawyer from Richmond, Va., who is black, argued that colleges have a moral obligation to tell applicants if their SAT scores fall within the range of students who have a shot of completing their studies. "If I'm outside that range, no additional support is going to help me," he said.

Sander agrees. "African-American students and any other minority ought to know going into college the ultimate outcomes for students at that college who have their profile."

Tapia agreed as well. "I had a student that I was recruiting in San Antonio who had a 940 SAT and was going to Princeton. I said, 'Do you know what the average at Princeton is?' He said, "Well, my teachers told me

it was about 950.' I said, 'Well, I think you'd better check it out.' "

In fact, the average combined math and verbal SAT score of students admitted to Princeton is 1442.[44]

OUTLOOK

End of the Line?

Social programs don't come with an immortality guarantee. Some supporters as well as critics of affirmative action sense that affirmative action, as the term is generally understood, may be nearing the end of the line.

"I expect affirmative action to die," says Tapia. "People are tired of it. And if we had to depend on affirmative action forever, then there was something wrong. If you need a jump-start on your battery, and you get it jumped, fine. If you start needing it everywhere you go, you'd better get another battery."

Tapia's tone is not triumphant. He says the decline in public school quality is evidence that "it didn't work, and we didn't do a good job." But he adds that the disparities between the schooling for low-income and well-off students is what makes affirmative action necessary. "Sure, in an ideal world, you wouldn't have to do these things, but that's not the world we live in."

UCLA's Sander, who favors reorienting affirmative action — in part by determining an academic threshold below which students admitted by preference likely will fail — sees major change on the horizon. For one thing, he says, quantities of data are now accessible concerning admission standards, grades and other quantifiable effects of affirmative action programs.

In addition, he says, today's reconfigured Supreme Court likely would rule differently than it did on the 2003 University of Michigan cases that represent its most recent affirmative action rulings.

Justice O'Connor, who wrote the majority decision in the 5-4 ruling that upheld the use of race in law-school admissions, has retired, replaced by conservative Justice Samuel A. Alito. "The Supreme Court as it stands now has a majority that's probably ready to overrule" that decision, Sander says. A decision that turned on the newly available data "could lead to a major Supreme Court decision that could send shockwaves through the system."

For now, says Kahlenberg of The Century Foundation, affirmative action has already changed form in states that have restricted use of racial and ethnic preferences. "It's not as if universities and colleges have simply thrown up their hands," he says. "They now look more aggressively at economic disadvantages that students face. The bigger picture is that the American public likes the idea of diversity but doesn't want to use racial preferences to get there."

Anderson of Texas A&M agrees that a vocabulary development marks the shift. "We've been changing affirmative action and quotas to diversity," he says. "Diversity is seen as good, and has become part of our mainstream culture."

In effect, diversity has come to mean hiring and admissions policies that focus on bringing people of different races and cultures on board — people like Obama, for example. "Obama's talking about merit, and keeping the doors open for all Americans, and strengthening the middle class," Anderson says.

Obama, whose father was Kenyan and whose half-sister is half-Indonesian, also represents another facet of the changing face of affirmative action. "Our society is becoming a lot more demographically complicated," says Schmidt, of *The Chronicle of Higher Education* and author of a recent book on affirmative action in college admissions. "All of these racial groups that benefit from affirmative action as a result of immigration — they're not groups that have experienced oppression and discrimination in the United States. And people are marrying people of other races and ethnicities. How do you sort that out? Which parent counts the most?"

All in all, Schmidt says, the prospects for affirmative action look dim. "In the long term, the political trends are against it," he says. "I don't see a force out there that's going to force the pendulum to swing the other way."

At the same time, many intended beneficiaries — African-Americans whose history set affirmative action in motion — remain untouched by it because of the deficient schools they attend.

The catastrophic state of public schools in low-income America remains — and seems likely to remain — a point on which all sides agree. Whether anything will be done about it is another story.

Top schools will continue to seek diverse student bodies, says Coleman of Duke law school. But the public schools continue to deteriorate. "I haven't seen any effort by people who oppose affirmative action, or people who support it, to do anything to improve the public school system. We ought to improve the quality of education because it's in the national interest to do that."

NOTES

1. See "Transcript: Obama and Clinton Debate," ABC News, April 16, 2008, http://abcnews.go.com/Politics/DemocraticDebate/story?id=4670271&page=1.

2. *Ibid.*

3. *Ibid.*

4. See "Trends in Political Values and Core Attitudes: 1987-2007," Pew Research Center for People and the Press, March 22, 2007, pp. 40-41, http://people-press.org/reports/pdf/312.pdf.

5. See Alemayehu Bishaw and Jessica Semega, "Income, Earnings, and Poverty Data from the 2007 American Community Survey," U.S. Census Bureau, August 2008, p. 20, www.census.gov/prod/2008pubs/acs-09.pdf.

6. See Peter Schmidt, *Color and Money: How Rich White Kids Are Winning the War Over College Affirmative Action* (2007), p. 47.

7. See Anthony P. Carnevale and Stephen J. Rose, "Socioeconomic Status, Race/Ethnicity, and Selective College Admissions," The Century Foundation, March 2003, pp. 26, 79, www.tcf.org/Publications/Education/carnevale_rose.pdf.

8. See Christopher B. Swanson, "Who Graduates? Who Doesn't? A Statistical Portrait of High School Graduation, Class of 2001," The Urban Institute, 2004, pp. v-vi, www.urban.org/UploadedPDF/410934_WhoGraduates.pdf.

9. Quoted in Ira Katznelson, *When Affirmative Action Was White: An Untold History of Racial Inequality in Twentieth-Century America* (2005), p. 175.

10. See Roberta W. Francis, "Reconstituting the Equal Rights Amendment: Policy Implications for Sex Discrimination," 2001, www.equalrightsamendment.org/APSA2001.pdf.

11. See Clarence Thomas, *My Grandfather's Son: A Memoir* (2007), p. 126.

12. Quoted in "Justice Thomas Mocks Value of Yale Law Degree," The Associated Press, Oct. 22, 2007, www .foxnews.com/story/0,2933,303825,00.html. See also, Coleman profile in Berger&Montague, P.C., law firm Web site, www.bergermontague.com/attorneys .cfm?type=1.

13. See Linda Greenhouse, "Justices Back Affirmative Action by 5 to 4, But Wider Vote Bans a Racial Point System," *The New York Times*, June 24, 2003, p. A1.

14. "Barack Obama, July 27, 2008, Unity 08, High Def, Part II," www.youtube.com/watch?v=XIoRzNVTyH4& eurl= http://video.google.com/videosearch?q=obama %2UNITY&ie=UTF-8&oe=utf-8&rls=org.mozilla :enUS:official&c.UNITY is a coalition of the Asian-American Journalists Association, the National Association of Black Journalists, the National Association of Hispanic Journalists and the Native American Journalists Association, www.unity journalists.org.

15. See "Trends in Political Values . . .," *op. cit.*, pp. 40-41.

16. See http://abcnews.go.com/Politics/Democratic Debate/story?id=4670271.

17. See Jonathan Kaufman, "Fair Enough?" *The Wall Street Journal*, June 14, 2008.

18. See Christine MacDonald, "Ban lost in college counties," *Detroit News*, Nov. 9, 2006, p. A16; and "1996 General Election Returns for Proposition 209," California Secretary of State, Dec. 18, 1996, http:// vote96.sos.ca.gov/Vote96/html/vote/prop/prop-209 .961218083528.html.

19. See Sue Anne Pressley, "Florida Plan Aims to End Race-Based Preferences," *The Washington Post*, Nov. 11, 1999, p. A15.

20. See Walter Alarkon, "Affirmative action emerges as wedge issue in election," *The Hill*, March 11, 2008, http://thehill.com/campaign-2008/affirmative-action-emerges-as-wedge-issue-in-election-2008-03-11.html.

21. *Ibid.*, p. viii.

22. Catherine L. Horn and Stella M. Flores, "Percent Plans in College Admissions: A Comparative Analysis of Three States' Experiences," The Civil Rights Project, Harvard University, February 2003,

pp. 30-31, http://eric.ed.gov/ERICDocs/data/eric docs2sql/content_storage_01/0000019b/80/1a/ b7/9f.pdf.

23. See Carnevale and Rose, *op. cit.*, pp. 10-11.

24. *Ibid.*

25. Quoted in Terry H. Anderson, *The Pursuit of Fairness: A History of Affirmative Action* (2004), p. 76. Unless otherwise indicated, material in this subsection is drawn from this book.

26. For background, see the following *Editorial Research Reports:* Richard L. Worsnop, "Racism in America," May 13, 1964; Sandra Stencel, "Reverse Discrimination," Aug. 6, 1976; K. P. Maize and Sandra Stencel, "Affirmative Action Under Attack," March 30, 1979; and Marc Leepson, "Affirmative Action Reconsidered," July 31, 1981, all available in *CQ Researcher Plus Archive.*

27. Quoted in Anderson, *op. cit.*, p. 125. For more background, see Richard L. Worsnop, "Racial Discrimination in Craft Unions," *Editorial Research Reports*, Nov. 26, 1969, available in *CQ Researcher Plus Archive.*

28. *Griggs v. Duke Power*, 401 U.S. 424 (1971), http:// caselaw.lp.findlaw.com/scripts/get.casepl?court=US &vol=401&invol=424. For background, see Mary H. Cooper, "Racial Quotas," *CQ Researcher*, May 17, 1991, pp. 277-200; and Kenneth Jost, "Rethinking Affirmative Action," *CQ Researcher*, April 28, 1995, pp. 269-392.

29. *Ibid.*

30. See *University of California Regents v. Bakke*, 438 U.S. 265 (1978), http://caselaw.lp.findlaw.com/ scripts/getcase.pl?court=US&vol=438&invol=265.

31. See *United Steelworkers of America, AFL-CIO-CLC v. Weber, et al.*, 443 U.S. 193 (1979), http://caselaw.lp .findlaw.com/scripts/get.casepl?court=US&vol=443& invol=193; and *Fullilove v. Klutznick*, 448 U.S. 448 (1980), www.law.cornell.edu/supct/html/historics/ USSC_CR_0448_0448_ZS.html.

32. Unless otherwise indicated, this subsection is drawn from Anderson, *op. cit* and Jost, *op. cit.*

33. For background, see Cooper, *op. cit.*

34. Anderson, *op. cit.*, p. 186.

35. *Ibid.*, p. 206.

36. Quoted in *ibid.*, p. 233. Unless otherwise indicated this subsection is drawn from Anderson, *op. cit.*

37. *Ibid.*, p. 242.

38. *Ibid.*, p. 244.

39. For background, see Kenneth Jost, "Race in America," *CQ Researcher*, July 11, 2003, pp. 593-624.

40. Quoted in Greenhouse, *op. cit.*

41. "Controversial Bake Sale to Go On at CU, College Republicans Protesting Affirmative Action," 7 News, Feb. 10, 2004, www.thedenverchannel.com/news/ 2837956/detail.html.

42. Quoted in Kavita Kumar, "Affirmative action critic vows he'll try again," *St. Louis Post-Dispatch*, May 6, 2008, p. D1.

43. "Amendment 46: Formerly Proposed Initiative 2007-2008 #31," Colorado Secretary of State, undated, www.elections.colorado.gov/DDefault.aspx? tid=1036.

44. College data, undated, www.collegedata.com/cs/data/ college/college_pg01_tmpl.jhtml? schoolId=111.

BIBLIOGRAPHY

Books

Anderson, Terry H., *The Pursuit of Fairness: A History of Affirmative Action,* **Oxford University Press, 2004.**
A Texas A&M historian tells the complicated story of affirmative action and the struggles surrounding it.

Kahlenberg, Richard D., ed., *America's Untapped Resource: Low-Income Students in Higher Education,* **The Century Foundation Press, 2004.**
A liberal scholar compiles detailed studies that add up to a case for replacing race- and ethnic-based affirmative action with a system based on students' socioeconomic status.

Katznelson, Ira, *When Affirmative Action Was White: An Untold History of Racial Inequality in Twentieth-Century America,* **Norton, 2005.**
A Columbia University historian and political scientist argues that affirmative action — favoring whites — evolved as a way of excluding Southern blacks from federal social benefits.

Schmidt, Peter, *Color and Money: How Rich White Kids are Winning the War Over College Affirmative Action,* **Palgrave Macmillan, 2007.**
An editor at *The Chronicle of Higher Education* explores the realities of race, class and college admissions.

Sowell, Thomas, *Affirmative Action Around the World: An Empirical Study,* **Yale University Press, 2004.**
A prominent black conservative and critic of affirmative action dissects the doctrine and practice and its similarities to initiatives in the developing world, of which few Americans are aware.

Articles

Babington, Charles, "Might Obama's success undercut affirmative action," *The Associated Press,* **June 28, 2008, www.usatoday.com/news/politics/2008-06-28-3426171631_x.htm.**
In a piece that prompted a debate question to presidential candidate Barack Obama, a reporter examines a possibly paradoxical consequence of the 2008 presidential campaign.

Jacobs, Tom, "Affirmative Action: Shifting Attitudes, Surprising Results," *Miller-McCune,* **June 20, 2008, www.miller-mccune.com/article/447.**
A new magazine specializing in social issues surveys the long-running debate over university admissions. (*Miller-McCune* is published by SAGE Publications, parent company of CQ Press.)

Leonhardt, David, "The New Affirmative Action," *New York Times Magazine,* **Sept. 30, 2007, p. 76.**
A journalist specializing in economic and social policy explores UCLA's efforts to retool its admissions procedures.

Liptak, Adam, "Lawyers Debate Why Blacks Lag At Major Firms," *The New York Times,* **Nov. 29, 2006, p. A1.**
A law correspondent airs a tough debate over affirmative action's success, or lack of it, at big law firms.

Matthews, Adam, "The Fixer," *Good Magazine,* **Aug. 14, 2008, www.goodmagazine.com/section/Features/ the_fixer.**
A new Web-based publication for the hip and socially conscious examines the career of black businessman and affirmative-action critic Ward Connerly.

Mehta, Seema, "UCLA accused of illegal admissions practices," *Los Angeles Times*, Aug. 30, 2008, www .latimes.com/news/local/la-me-ucla30-2008aug30,0, 6489043.story.
Mehta examines the latest conflict surrounding the top-tier university's retailored admissions procedures.

Reports and Studies

Coleman, James E. Jr. and Mitu Gulati, "A Response to Professor Sander: Is It Really All About the Grades?" *North Carolina Law Review*, 2006, pp. 1823-1829.
Two lawyers, one of them a black who was a partner at a major firm, criticize Sander's conclusions, arguing he overemphasizes academic deficiencies.

Horn, Catherine L. and Stella M. Flores, "Percent Plans in College Admissions: A Comparative Analysis of Three States' Experiences," The Civil Rights Project, Harvard University, February 2003.
Educational policy experts with a pro-affirmative action perspective dig into the details of three states' alternatives to traditional affirmative action.

Prager, Devah, "The Mark of a Criminal Record," *American Journal of Sociology*, March 2003, pp. 937-975.
White people with criminal records have a better chance at entry-level jobs than black applicants with clean records, an academic's field research finds.

Sander, Richard H., "The Racial Paradox of the Corporate Law Firm," *North Carolina Law Review*, 2006, pp. 1755-1822.
A much-discussed article shows that a disproportionate number of black lawyers from top schools leave major law firms before becoming partners.

Swanson, Christopher B., "Who Graduates? Who Doesn't? A Statistical Portrait of Public High School Graduation, Class of 2001," *The Urban Institute*, 2004, www.urban.org/publications/410934.html.
A centrist think tank reveals in devastating detail the disparity in high schools between races and classes.

For More Information

American Association for Affirmative Action, 888 16th St., N.W., Suite 800, Washington, DC 20006; (202) 349-9855; www.affirmativeaction.org. Represents human resources professionals in the field.

American Civil Liberties Union, 125 Broad St., 18th Floor, New York, NY 10004; www.aclu.org/racialjustice/aa/index.html. The organization's Racial Justice Program organizes legal and voter support for affirmative action programs.

American Civil Rights Institute, P.O. Box 188350, Sacramento, CA 95819; (916) 444-2279; www.acri.org/index.html. Organizes ballot initiatives to prohibit affirmative action programs based on race and ethnicity preferences.

Diversity Web, Association of American Colleges and Universities, 1818 R St., N.W., Washington, DC 20009;

www.diversityweb.org. Publishes news and studies concerning affirmative action and related issues.

www.jessicacorry.com. A Web site featuring writings by Jessica Peck Corry, director of the Colorado campaign for a racial preferences ban.

Project SEAPHE (Scale and Effects of Admissions Preferences in Higher Education), UCLA School of Law, Box 951476, Los Angeles, CA 90095; (310) 206-7300; www.seaphe.org. Analyzes data on the effects of racial and other preferences.

U.S. Commission on Civil Rights, 624 Ninth St., N.W., Washington, DC 20425; (202) 376-7700; www.usccr.gov. Studies and reports on civil rights issues and implements civil rights laws.

3

Gender Pay Gap

Are Women Paid Fairly in the Workplace?

Thomas J. Billitteri

A suit filed by Betty Dukes, right, and other female Wal-Mart employees accuses the retail giant of sex discrimination in pay, promotions and job assignments in violation of the Civil Rights Act of 1964. The case, covering perhaps 1.6 million current and former Wal-Mart employees, is the biggest class-action lawsuit against a private employer in U.S. history.

From *CQ Researcher*, March 14, 2008.

"A n insult to my dignity" is the way Lilly Ledbetter described it.[1] For 19 years, she worked at the Goodyear Tire plant in Gadsden, Ala., one of a handful of women among the roughly 80 people who held the same supervisory position she did. Over the years, unbeknownst to her, the company's pay-raise decisions created a growing gap between her wages and those of her male colleagues. When she left Goodyear, she was earning $3,727 a month. The lowest-paid man doing the same work got $4,286. The highest-paid male made 40 percent more than she did.[2]

Ledbetter sued in 1998, and a jury awarded her back pay and more than $3 million in damages. But in the end, she lost her case in the U.S. Supreme Court.[3]

A conservative majority led by Justice Samuel A. Alito Jr. ruled that under the nation's main anti-discrimination law she should have filed a formal complaint with the federal government within 180 days of the first time Goodyear discriminated against her in pay. Never mind, the court said, that Ledbetter didn't learn about the pay disparity for years.

"The Supreme Court said that this didn't count as illegal discrimination," she said after the ruling, "but it sure feels like discrimination when you are on the receiving end of that smaller paycheck and trying to support your family with less money than the men are getting for doing the same job."[4]

The *Ledbetter* decision has added fuel to a long-burning debate over sex discrimination in women's wages and whether new laws are needed to narrow the disparity in men's and women's pay.

Women Closing the Pay Gap . . . Slowly

More than 40 years after women began demanding equal rights and opportunities, they still earn 77 percent of what men earn. The pay gap has been closing, however, because women's earnings have been rising faster than men's.

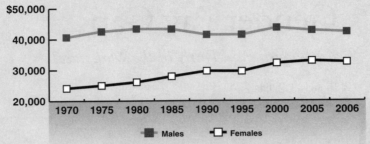

Median Annual Earnings of Full-time, Year-round Workers
(By gender, 1970-2006, in constant 2006 dollars)

Source: Carmen DeNavas-Walt, et. al., "Income, Poverty, and Health Insurance Coverage in the United States: 2006," U.S. Census Bureau, August 2007

"A significant wage gap is still with us, and that gap constitutes nothing less than an ongoing assault on women's economic freedom," declared U.S. Rep. Rosa L. DeLauro, D-Conn., at a congressional hearing on a pay-equity bill she is sponsoring, one of several proposed on Capitol Hill.

But that view is hardly universal. "Men and women generally have equal pay for equal work now — if they have the same jobs, responsibilities and skills," testified Diana Furchtgott-Roth, a senior fellow at the Hudson Institute, a conservative think tank, and former chief economist at the Labor Department in the George W. Bush administration.[5]

The wrangle over wages is playing out not just in Washington but in cities and towns across America. In the biggest sex-discrimination lawsuit in U.S. history, a group of female Wal-Mart employees has charged the retail giant with bias in pay and promotions. The case could affect perhaps 1.6 million women employees of Wal-Mart and result in billions of dollars in back pay and damages. (*See sidebar, p. 62.*)

The enormously complex gender-pay debate encompasses economics, demographics, law, social justice, culture, history and sometimes raw emotion. Few dispute that a wage gap exists between men and women. In 2006 full-time female workers earned 81 percent of men's weekly earnings, according to the latest U.S. Labor Department data, with the wage gap broader for older workers and narrower for younger ones. Separate U.S. Census Bureau data put the gap at about 77 percent of men's median full-time, year-round earnings.[6]

The fundamental issues are why the gap exists, how much of it stems from discrimination and what should be done about it.

Some contend the disparity can largely be explained by occupational differences between women and men, variations in work experience, number of hours worked each year and other such things.

June O'Neill, an economics professor at the City University of New York's Baruch College and former director of the Congressional Budget Office in the Clinton administration, says that the most important factors affecting the pay gap stem from differences in the roles of women and men in family life. When the wages of men and women who share similar work experience and life situations are measured, the wage gap largely disappears, she says. Reasons that the earnings disparity may appear bigger in some research, she says, include the fact that many studies do not control for differences in years of work experience, the extent of part-time work and differences in training and occupational choices. O'Neill notes that Labor Department data show median weekly earnings of female part-time workers exceed those of male part-timers. She also says the wage gap has been narrowing over time as women's work experience, education and other job-related skills have been converging with those of men.

"Large amounts of discrimination? No," she says. "Individual women may experience discrimination, and it's good to have laws that deal with it," she adds. "But those cases don't change the overall picture. The vast majority of employers don't harbor prejudice against women."

Yet others argue that beneath such factors as occupation and number of hours worked lies evidence of significant discrimination — covert if not overt.

"Women do not realize the enormous price that they pay for gender wage discrimination because they do not see big bites taken out of their paychecks at any one time," Evelyn F. Murphy, president of The Wage Project, a nonprofit organization that works on eliminating the gender wage gap and author of *Getting Even: Why Women Don't Get Paid Like Men and What To Do About It*, told a congressional panel last year.[7]

In her book, she told the hearing, she wrote of employers "who had to pay women employees or former employees to settle claims of gender discrimination, or judges and juries ordered them to pay up. The behavior of these employers vividly [illustrates] the commonplace forms of today's wage discrimination: barriers to hiring and promoting qualified women; arbitrary financial penalties imposed on pregnant women; sexual harassment by bosses and co-workers; failure to pay women and men the same amount of money for doing the same jobs," and "everyday discrimination" marked by "the biases and stereotypes which influence [managers'] decisions about women."

Women's advocates point to a 2003 General Accounting Office (GAO) study concluding that while "work patterns" were key in accounting for the wage gap, the GAO could not explain all the differences in earnings between men and women. "When we account for differences between male and female work patterns as well as other key factors, women earned, on average, 80 percent of what men earned in 2000. . . . We cannot determine whether this remaining difference is due to discrimination or other factors," the GAO report said.[8]

The study said that in the view of certain experts some women trade promotions or higher pay for job flexibility that allows them to balance work and family responsibilities.

Women's advocates point out that many women have little choice but to work in jobs that offer flexibility but pay less because they typically shoulder the bulk of family caregiving duties. And, they argue further, expectations within companies and society — typically subtle, but sometimes not — often channel women away from male-dominated jobs into female-dominated ones that pay less.

Gap Widens for College Graduates

College-educated women earn only 80 percent of what their male counterparts earn a year after graduation, when both male and female employees have the same level of work experience and (usually) no child-care obligations — factors often used to explain gender pay differences. The gap widens to 69 percent by 10 years after graduation.

Gap in Average Weekly Earnings for Bachelor's Degree Recipients
(For full-time workers)

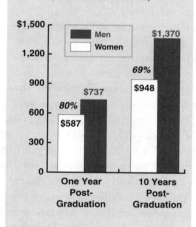

Source: "Beyond the Pay Gap," American Association of University Women, based on data from the "2003 Baccalaureate and Beyond Longitudinal Study," National Center for Education Statistics, U.S. Department of Education

"People who argue that [wage discrimination] is small will say a lot of it is due to women's choices," such as the choice to stay home with the children, work part time or enter lower-paying fields, says Reeve Vanneman, a sociology professor at the University of Maryland, College Park, who studies gender inequality. But, he says, it's misleading to explain most of the wage gap in that way, especially when mid-career and older female workers are concerned.

"Why do women make those choices? Part of the reason is because they are discriminated against in the job. They see men getting rewarded more and promoted more than they are."

Women face unequal work not just on the job but at home, too, Vanneman says, with husbands not picking up their share.

Part of the wage gap stems from weak government enforcement, some argue. A U.S. inspector general's report stated last fall that the Equal Employment Opportunity Commission, which enforces federal employment-discrimination laws, is "challenged in accomplishing its mission" because of "a reduced workforce and an increasing backlog of pending cases." The agency has experienced a "significant loss of its workforce, mostly to attrition and buyouts...offered to free up resources," the report said.[9]

The news on gender discrimination in pay is not all bad. The wage gap has narrowed considerably in recent decades. For example, Labor Department data show that for 35- to 44-year-olds, the earnings ratio of women to men rose from 58 percent in 1979 to 77 percent in 2006. For 45- to 54-year-olds, it went from 57 percent to 74 percent.[10] Among the youngest workers, ages 16 to 24, only about 5 percentage points separated median weekly wages of men and women in 2006.[11]

Still, many experts say the progress of the 1980s and early '90s has slowed or stalled in recent years, with the wage gap stuck in the range of 20 to 24 percent, although it is not entirely clear why. Some argue that entrenched wage discrimination remains a major culprit.

In a study of college graduates last year, the American Association of University Women Educational Foundation found that one year out of college, women working full time earn only 80 percent as much as their male colleagues, and 10 years after graduation the gap widens to 69 percent. Even after controlling for hours worked, training and education and other factors, the portion of the pay gap that remains unexplained is 5 percent one year after graduation and 12 percent a decade afterward, the study found.[12] (See graph, p. 51.)

"These unexplained gaps are evidence of discrimination," the study concluded.

Employer advocates challenge such conclusions, though. Michael Eastman, executive director of labor policy at the U.S. Chamber of Commerce, questions the assumption "that whatever gap is not explained must be due to discrimination. An unexplained gap is simply that — it's unexplained."

Election-year politics and the recent shift toward Democratic control of Congress — along with the Supreme Court's decision in the Ledbetter case — have helped to reinvigorate the pay debate. Proposed gender-pay bills have strong support from women's-rights groups and some economists, who argue that the Equal Pay Act and Title VII of the Civil Rights Act of 1964 — the main avenues for attacking wage discrimination — fall short.

Presidential contender Sen. Hillary Rodham Clinton, D-N.Y., is sponsoring the Senate version of the DeLauro bill; another presidential hopeful, Sen. Barack Obama, D-Ill., is one of the 22 co-sponsors, although he didn't sign on to it until more than a month after she introduced it. Among other things, the measure would raise penalties under the Equal Pay Act, which bars paying men and women differently for doing the same job.[13]

Obama is co-sponsoring a more controversial bill, introduced in the Senate by Sen. Tom Harkin, D-Iowa, that advocates the notion of comparable worth; the idea, generally speaking, suggests that a female-dominated occupation such as social work may merit wages that are comparable to those of a male-dominated job such as a probation officer.[14] The Harkin measure would bar wage discrimination in certain cases where the work is deemed comparable in skill, effort, responsibility and working conditions, even if the job titles or duties are different. (See sidebar, p. 60.)

A third effort would undo the Supreme Court's ruling in the Ledbetter case.[15] A bill passed the House last summer, and advocates are hoping the Senate version — sponsored by Sen. Edward M. Kennedy, D-Mass., and co-sponsored by Clinton and Obama — moves forward soon. But the Bush administration has threatened a veto, and business interests are vehemently opposed.

As the debate over wage disparities continues, these are some of the questions being discussed:

Is discrimination a major cause of the wage gap?

When economist David Neumark studied sex discrimination in restaurant hiring in the mid-1990s, he discovered something intriguing: In expensive restaurants, where waiters and waitresses can earn more than they can at low-price places, the chances of a woman getting a wait-staff job offer were 40 percentage points lower than those of a man with similar experience.[16]

The study is a telling bit of evidence that the wage gap is real and that discrimination plays a significant part in it, says Vicky Lovell, director of employment and work/life programs at the Institute for Women's Policy Research, an advocacy group in Washington. She estimates that perhaps a third of the wage gap stems from discrimination — mostly "covert" bias that occurs when people make false assumptions about the ability or career commitment of working women.

Lovell has little patience with those who say the wage gap stems from non-discriminatory reasons that simply haven't yet been identified. "That's just specious," she says. "If we can't explain why women on average get paid less, what is the alternative explanation?"

The role of discrimination lies at the heart of the pay-gap debate. Researchers fall into different camps.

Some see little evidence that bias plays a big part in the gap. When adjusted for work experience, education, time in the labor force and other variables, wages of men and women are largely comparable, they contend.

"This so-called wage gap is not necessarily due to discrimination," the Hudson Institute's Furchtgott-Roth said in congressional testimony. "Decisions about field of study, occupation and time in the work force can lead to lower compensation, both for men and women."[17]

What's more, "some jobs command more than others because people are willing to pay more for them," she said. "Many jobs are dirty and dangerous.... Other highly paid occupations have long, inflexible hours.... Women are not excluded from these or other jobs but often select professions with a more pleasant environment and potentially more flexible schedules, such as teaching and office work. Many of these jobs pay less."

Pay Gap Exists Despite Women's Choices

Those who discount the seriousness of gender pay bias often blame differences in men's and women's salaries on women's choices to study "softer" sciences or to have children. But a recent study shows that the pay gap persists even when women choose not to have children and when they choose male-dominated fields of study and occupation — such as business, engineering, mathematics and medicine. The pay gap is greatest in the biology, health and mathematics fields. Women out-earn men only in the history professions.

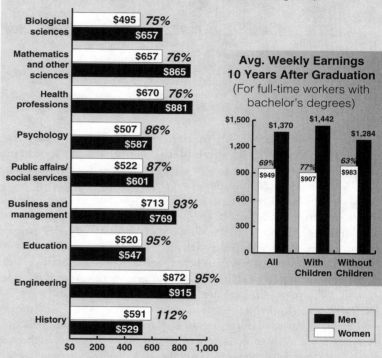

Avg. Weekly Earnings One Year After Graduation
(For full-time workers with bachelor's degrees)

	Women	%	Men
Biological sciences	$495	75%	$657
Mathematics and other sciences	$657	76%	$865
Health professions	$670	76%	$881
Psychology	$507	86%	$587
Public affairs/ social services	$522	87%	$601
Business and management	$713	93%	$769
Education	$520	95%	$547
Engineering	$872	95%	$915
History	$591	112%	$529

Avg. Weekly Earnings 10 Years After Graduation
(For full-time workers with bachelor's degrees)

	Women	Men
All	$949 (69%)	$1,370
With Children	$907 (77%)	$1,442
Without Children	$983 (63%)	$1,284

■ Men □ Women

Source: "Behind the Pay Gap," American Association of University Women, 2007

Wage Disparities Highest Among Asians

The median weekly earnings for women are lower than men's across all ethnic groups. The largest disparity is among Asians, where men earn $183 more on average per week than their female counterparts. The average difference for all groups is $143.

Median Weekly Earnings of Full-Time Workers
(by gender and ethnicity, 2006)

Source: "Highlights of Women's Earnings in 2006," Bureau of Labor Statistics, September 2007

occupation earn less than their male counterparts. A typical college-educated woman working full time earns $46,000 a year compared to $62,000 for college-educated male workers — a difference of $16,000.

"The pay gap between female and male college graduates cannot be fully accounted for by factors known to affect wages, such as experience (including work hours), training, education and personal characteristics," the AAUW study says. "In this analysis the portion of the pay gap that remains unexplained after all other factors are taken into account is 5 percent one year after graduation and 12 percent 10 years after graduation. These unexplained gaps are evidence of discrimination, which remains a serious problem for women in the work force."[18]

Warren Farrell, who in the 1970s served on the board of the New York City chapter of the National Organization for Women, argues in his 2005 book — *Why Men Earn More: The Startling Truth Behind the Pay Gap — and What Women Can Do About It* — that women pay an economic price by seeking careers that are more fulfilling, flexible and safe. With a stated goal of helping women gain higher pay, Farrell offers 25 "differences in the way women and men behave in the workplace." Those who earn more, he says, work longer hours, are more willing to relocate, require less security and produce more, among other things.

O'Neill, of Baruch College, points out that women are much more likely to go into occupations that will allow them to work part time, and typically "that doesn't pay as well."

She studies data that track the work histories of women and men over a long period of time. "Women have just not worked as many weeks and hours over their lives as men," she says. "When you adjust for that, you explain most of the [pay] difference.... You're still left with a difference, but then there are other things that become harder to measure."

The AAUW study found that even women who make the same choices as men in terms of fields of study and

"This research asked a basic but important question: If a woman made the same choices as a man, would she earn the same pay? The answer is no," Catherine Hill, director of research at the AAUW, told a House Committee on Education and Labor hearing last year.

Speaking more generally about pay inequity, Linda Meric, national director of 9to5, National Association of Working Women, a Milwaukee-based advocacy group, says that "when you control for all the other so-called factors" that might explain the wage gap, "there is still a gap."

"And many of those so-called factors are not independent of discrimination and stereotypes of women. One is time in the work force. If there aren't policies that allow women to get jobs and maintain and advance in employment at the same time they are meeting their responsibility in terms of family caregiving, that's not an independent factor. It's something that influences the pay gap significantly."

Heather Boushey, senior economist at the Center for Economic and Policy Research, a Washington think tank, noted that time away from the workforce strongly affects lifetime earnings. She said it is a myth that women choose lower-paying occupations because they provide the flexibility to better manage work and family. "The empirical evidence shows that mothers are actually less likely to be employed in jobs that provide them with greater flexibility."[19]

Echoing that sentiment, Beth Shulman, co-director of the Fairness Initiative on Low Wage Work, a public policy advocacy group also in Washington, says, "We have kind of an Ozzie and Harriet workplace, with a full-time worker and the wife at home," but "70 percent of women with children are in the workplace." She adds, "Our structures haven't kept up with that. So women who are primary caregivers get punished."

Shulman, author of *The Betrayal of Work: How Low-Wage Jobs Fail 30 Million Americans*, says that while overt gender discrimination exists in the job market, an equally important contributor to the wage gap is the lack of flexibility for low-income working women with families. For example, she says, female factory employees with family responsibilities often find it difficult to accept better-paying manufacturing jobs because such jobs often require mandatory overtime.

Shulman also says that three-fourths of women in low-wage jobs don't have paid sick days. So when a child is sick or an elderly parent needs help, women may be forced to leave the workforce and then re-enter it — something that has a huge effect on wages over time.

"Low-wage workers get kind of ghettoized into these part-time jobs that have poor wages, poor benefits and less government protection," Shulman says.

In a 1998 study, Cornell University economists Francine Blau and Lawrence Kahn found that 40 percent of the pay gap is unexplained after adjusting for gender differences in experience, education, occupation and industry. Blau cautions that such an estimate is conservative, because variables such as women's choices of occupation or industry and even their education and work experience can themselves be affected by discrimination. On the other hand, she acknowledges that some of the unexplained differences may be due to unmeasured productivity characteristics that increase men's earnings relative to women's earnings.

Applying that 40 percent figure to current government wage-gap data would suggest that 8 to 9 cents of each dollar in wage disparity is unexplained, with an unknown portion of that amount caused by discrimination.

Martha Burk, who directs the Corporate Accountability Project for the National Council of Women's

AP Photo/Ron Edmonds

U.S. Rep. Rosa L. DeLauro, D-Conn., is sponsoring one of several pay-equity bills in Congress. Presidential contender Sen. Hillary Rodham Clinton, D-N.Y., is sponsoring the Senate version of DeLauro's bill; Sen. Barack Obama, D-Ill., is one of the 22 co-sponsors. "A significant wage gap is still with us, and that gap constitutes nothing less than an ongoing assault on women's economic freedom," DeLauro says.

Organizations, a coalition of more than 200 women's groups, says some of the pay gap stems from "historical discrimination" rooted in a time when employers could legally exclude women from certain jobs and pay them less for the kinds of jobs they typically did hold, such as teaching and clerical work.

Burk, who led the fight to open the Augusta (Ga.) National Golf Club to women, says those female-dominated jobs "were systematically devalued, and that has carried through to modern times."

Are new laws needed to close the gender pay gap?

When President John F. Kennedy signed the Equal Pay Act in 1963, he called it "a first step."[20]

Over the decades, the pay gap has narrowed significantly, but the push for new laws to curb gender-pay inequity goes on, fueled in part by the view among women's advocates that progress toward wage equity has slowed or stalled in recent years.

"The best way is for corporations to behave as socially responsible corporate citizens [and] examine their wage practices," says Lovell of Women's Policy Research. "But that is not going to happen. I don't see any reason to think the private sector is going to address this issue on its own. A few will to the extent they can within their own workforces. But if corporations individually or within industry groups aren't going to make this a priority, then that's why we have a government."

Opponents of new laws have sharply different views, though.

Roger Clegg, president and general counsel of the Center for Equal Opportunity, a conservative think tank in Falls Church, Va., says some gender discrimination will always exist but that existing laws can address it. Besides, Clegg says, the amount of gender discrimination that remains in the American work force "is greatly exaggerated by the groups pushing for legislation."

Much of the support for new laws rests on the view that some jobs pay poorly because females historically have dominated them. Jocelyn Samuels, vice president for Education and Employment at the National Women's Law Center, a Washington advocacy group, told a congressional hearing last year that 95 percent of child-care workers are female while the same proportion of mechanical engineers are male.

Moreover, she said, wages in fields dominated by women "have traditionally been depressed and continue to reflect the artificially suppressed pay scales that were historically applied to so-called 'women's work.' " Maids and housecleaners — 87 percent of whom are women — make roughly $3,000 per year less than janitors and building cleaners, 72 percent of whom are men, she said. "Current law simply does not provide the tools to address this continuing devaluation of traditionally female fields."[21]

To attack that situation, some advocates back the comparable-worth theory, arguing that women should be paid commensurate with men for jobs of equivalent value to a company, even if the work is different. But critics argue that such an approach violates the free-market principles of supply and demand for labor and that it could hurt both the economy and the cause of women.

"The comparable-worth approach has the government setting wages rather than the free market, and a great lesson of the 20th century is that centrally planned economies and centrally planned wage and price systems do not work," Clegg says.

Carrie Lukas, vice president for policy and economics at the Independent Women's Forum, a conservative group in Washington that backs limited government, contends that "government attempts to 'solve' the problem of the wage gap may in fact exacerbate some of the challenges women face, particularly in balancing work and family."

In an opinion column last year, she criticized the Clinton/DeLauro bill, which calls for guidelines to help companies voluntarily "compare wages paid for different jobs...with the goal of eliminating unfair pay disparities between occupations traditionally dominated by men or women." Lukas wrote that the bill would "give Washington bureaucrats more power to oversee how wages are determined, which might prompt businesses to make employment options more rigid." Flexible job structures would become less common, she argued. Why, Lukas wondered, "would companies offer employees a variety of work situations and compensation packages if doing so puts them at risk of being sued?"[22]

Not only might women suffer from new laws, but so would employers, some argue. Washington lawyer Barbara Berish Brown, vice-chair of the American Bar Association's Labor and Employment Law Section, said in a hearing on the Clinton/DeLauro bill that she is "unequivocally committed" to erasing gender-pay bias, but that existing laws suffice.

"All that the proposed changes will do is encourage more employment-related litigation, which is already drowning the federal court docket, and make it much more difficult, if not impossible, for employers, particularly small businesses, to prove the legitimate, nondiscriminatory reasons that explain differences between the salaries of male and female employees," she said.[23]

But longtime activists such as Burk, author of *Cult of Power: Sex Discrimination in Corporate America and What Can Be Done About It*, say existing laws are not effective enough to stamp out wage bias. "It has always been the view of conservatives that if you pay women equally, it's going to destroy capitalism," she says. "So far capitalism has survived quite well."

Is equity possible after the Supreme Court's *Ledbetter* ruling?

After the Supreme Court ruled in the Goodyear pay-discrimination case, Eleanor Smeal, president of the Feminist Majority, urged congressional action to reverse the decision. "We cannot stand by and watch a Bush-stacked court destroy in less than a year Title VII — the bedrock of women's rights and civil rights protection in wage-discrimination cases," she said.[24]

Yet, such outrage at the Supreme Court is matched by praise from business advocates. "We think the court got it exactly right," says Eastman, the U.S. Chamber of Commerce labor policy official.

In the 5-4 ruling, the court said workers can't sue under Title VII of the Civil Rights Act, the main federal anti-discrimination law, unless they file a formal complaint with the EEOC within 180 days of a discriminatory act. And in Ledbetter's case, the clock didn't start each time a new paycheck was issued. The 180-day timeline applies whether or not the employee immediately spots the discrimination.

Critics argue that because pay decisions are seldom broadcast throughout a company, the ruling makes it difficult — if not impossible — for an employee to detect bias until it may have gone on for years. "The ruling essentially says 'tough luck' to employees who don't immediately challenge their employer's discriminatory acts, even if the discrimination continues to the present time," said Marcia Greenberg, co-president of the National Women's Law Center.[25]

"With this misguided decision, the court ignores the realities of the 21st-century workplace," Margot Dorfman, chief executive officer of the U.S. Women's Chamber of Commerce, told a congressional panel this year. "The confidential nature of employee salary information complicates workers' abilities to recognize and report discriminatory treatment."[26]

Lovell, of the Institute for Women's Policy Research, says the *Ledbetter* ruling "seems to reflect a complete lack of understanding of the labor market and a complete lack of concern for individuals who are at any kind of disadvantage in the labor market." Workers wouldn't necessarily know right away that they were being discriminated against, she says. When Congress passed Title VII, it "was trying to establish an avenue for people who are discriminated against to pursue their claims…, not trying to make it impossible."

In a strongly worded dissent to the *Ledbetter* ruling, Justice Ruth Bader Ginsburg noted that pay disparities often occur in small increments, evidence of bias may develop over time, and wage information is typically hidden from employees. At the end of her dissent she wrote that "the ball is in Congress' court" to correct the Supreme Court's "parsimonious reading of Title VII" in the *Ledbetter* decision, just as Congress dealt with a spate of earlier Supreme Court decisions with passage of the 1991 Civil Rights Act.

Business groups have stood firm in the face of such impassioned views, though.

An exchange between Eastman of the U.S. Chamber of Commerce and law professor Deborah Brake last fall on the National Public Radio show "Justice Talking" helped underscore how polarizing the *Ledbetter* decision has been between advocates for women and for employers.[27]

Brake, a professor at the University of Pittsburgh School of Law who once litigated sex-discrimination cases for the National Women's Law Center, said she thought it was questionable whether the ruling was even good for employers.

"What an employee is supposed to do, let's say from the moment in time that they are hired, is search around the workplace and make sure that they're not being paid less if it's a woman than her male colleagues," she said on the radio program.

"If she has the slightest inkling or suspicion that she might be paid less than her male colleagues, she'd better immediately file a pay-discrimination claim. At every raise decision she better be sniffing around to make sure that her raise wasn't less than that of her male colleagues. And if she hears that someone got a higher raise than her who was a male, to preserve her rights under [the *Ledbetter* ruling] she'd better immediately file an EEOC claim. I don't think that is in the best interest, long-term, of employer or employees."

Eastman, though, said Title VII "has a strong incentive for employees to file claims quickly so that matters are resolved while all the facts and evidence are fresh and in people's minds. And it is very difficult for employers to defend themselves from allegations made many, many years down the line."

Brake said it wasn't the 180-day limit that bothered her. "What I'm objecting to is a ruling that starts the clock running before any employee has enough reason or incentive to even think about filing a discrimination claim," she said.

BACKGROUND

Early Wage Gap

From the republic's beginning, women have played an integral role in American economic growth and prosperity, yet a wage gap has always been present.

During the Industrial Revolution of the 19th century, as the nation's productivity and wealth exploded, young, single women moved from farm to city and took jobs as mill workers, teachers and domestic servants.

The factory work wasn't easy, and owners exploited women and girls as cheap sources of labor. In 1830, females often worked 12 hours a day in "boarding-house mills" — factories with housing provided by mill owners. They earned perhaps $2.50 a week. "Minor infractions such as a few minutes' lateness were punished severely," historian Richard B. Morris noted, and "one-sided contracts gave them no power over conditions and no rewards for work."[28]

Still, young women flocked to manufacturing jobs in the cities. In Massachusetts, among the earliest states to industrialize, a third of all women ages 10 to 29 worked in industry in 1850, according to Harvard University economist Claudia Goldin.[29]

As demand for goods grew along with the nation's population, the wages of women working full time in manufacturing rose slowly as a percentage of men's pay. The wage gap narrowed from about 30 percent of men's earnings in 1820 to 56 percent nationwide in 1885, according to Goldin.[30]

But progress came more slowly, if at all, in ensuing years and decades.

In manufacturing, Goldin noted in a 1990 book on the economic history of American women, "The ratio of female to male wages . . . continued to rise slowly across most of the nineteenth century but reached a plateau before 1900."[31]

As the 20th century dawned, some women's advocates pushed for equal pay for equal work between the sexes. But others questioned the equal-pay idea. In 1891, the British economist Sidney Webb pointed to "the impossibility of discovering any but a very few instances in which men and women do precisely similar work, in the same place and at the same epoch."[32]

By the turn of the 20th century, women's jobs had started growing more diverse. Women found work not only in domestic service and manufacturing but also in teaching, sales and clerical positions. Still, only 21 percent of American women worked outside the home in 1900, and most left the labor force upon or right after marriage.[33]

Women seeking to move up in the business world faced huge cultural hurdles. In 1900 *Ladies' Home Journal* told its readers: "Although the statement may seem a hard one, and will unquestionably be controverted, it nevertheless is a plain, simple fact that women have shown themselves naturally incompetent to fill a great many of the business positions which they have sought to occupy. . . . The fact is that no one woman in a hundred can stand the physical strain of the keen pace which competition has forced upon every line of business today."[34]

Women's labor participation gradually rose in the early decades of the 20th century, fueled in part by World War I, which ended in 1918. By 1920, almost a quarter of all U.S. women were in the labor force, and 46 percent of single women worked.[35]

World War I advanced women's status, historian Michael McGerr noted. "Although the number of employed women grew only modestly during the 1910s, the wartime departure of men for military service opened up jobs traditionally denied to women in offices, transportation and industry. Leaving jobs as domestic servants, seamstresses and laundresses, women became clerks, telephone operators, streetcar conductors, drill press operators and munitions makers. Women's new prominence in the work force led in turn to the creation of a Women's Bureau in the Department of Labor."[36]

In 1920 women gained the right to vote with adoption of the 19th Amendment. Soon afterward, Quaker activist Alice Paul introduced the first version of today's Equal Rights Amendment. In 1982 the amendment fell three states short of ratification, and its passage remains controversial today.[37] (*See "At Issue," p. 66.*)

CHRONOLOGY

1900-1940 *Women make economic gains but face discrimination.*

1914 Start of World War I marks a period of advancement in the status of women, who go to work in traditionally male jobs.

1919 Women gain the right to vote through the 19th Amendment.

1923 The Equal Rights Amendment is introduced, but it falls three states short of ratification.

1930 Half of single women are in the labor force, and the labor-participation rate among married women approaches 12 percent.

1938 Fair Labor Standards Act establishes rules for a minimum wage, overtime pay and child labor.

1940-1960 *Women make major contribution to wartime manufacturing efforts but don't gain wage equality with men.*

1942 National War Labor Board urges employers to equalize pay between men and women in defense jobs.

1945 Congress fails to approve Women's Equal Pay Act.

1955 Census Bureau begins calculating female-to-male earnings ratio.

1960-1980 *Major anti-discrimination laws helps women to fight pay bias.*

1963 Equal Pay Act bans gender pay discrimination in equal jobs.

1963 *The Feminine Mystique* by Betty Friedan challenges idea that women can find happiness only through marriage.

1964 Title VII of the Civil Rights Act bans job discrimination on the basis of race, color, religion, national origin and sex.

1965 Equal Employment Opportunity Commission founded.

1966 National Organization For Women is formed.

1973 Supreme Court's *Roe v. Wade* ruling overturns laws barring abortion, energizes the women's movement.

1979 National Committee on Pay Equity is formed.

1980-2000 *Gender pay gap continues to narrow, but progress toward wage equality shows signs of slowing in the 1990s.*

1981 Supreme Court ruling in *County of Washington v. Gunther* allows female jail guards to sue for sex discrimination but declines to authorize suits based on theory of comparable worth.

1993 Family and Medical Leave Act requires employers to grant unpaid leave for medical emergencies, birth and care of newborns and other family-related circumstances.

2001-Present *States expand laws to help working families, while several major corporations face gender-bias accusations.*

2001 Wal-Mart employees file for sex-discrimination claim against the retailer, to become the largest class-action lawsuit against a private employer in U.S. history.

2004 California grants up to six weeks partial pay for new parents.

2004 Equal Employment Opportunity Commission and Morgan Stanley announce $54 million settlement of sex-discrimination suit....Wachovia Corp. agrees to pay $5.5 million in a pay-discrimination case involving more than 2,000 current and former female employees.

2007 San Francisco requires employers to provide paid sick leave to all employees, including temporary and part-time workers.

2007 In *Ledbetter v. Goodyear*, Supreme Court rules that a female worker's pay-discrimination claim was invalid because it was filed after a 180-day deadline.

Debating the Comparable-Worth Doctrine

Would the approach help close the gender gap?

Imagine a company whose employees include a man who supervises telephone linemen and a woman who supervises clerical employees. They oversee the same number of workers, report to the same number of bosses, work the same hours and their jobs have been deemed of equal value to the company. Should their paychecks be the same?

Should the man get extra points for having to work outside in the cold? Should the woman get extra points for having a college degree or more years of experience?

Or, as some argue, should competitive market forces and the laws of supply and demand determine how much the man and woman earn?

Such questions lie at the heart of the debate over "comparable worth." The doctrine argues that when jobs require similar levels of skill, effort, responsibility and working conditions, the pay should be the same — even if the duties are entirely different.

Advocates of comparable worth say the market historically has undervalued jobs traditionally held by women — such as social work, secretarial work and teaching — and that such inequity has been a major contributor to the gender pay gap. If comparable worth were taken into account, they argue, it would even out wage inequality between those working in jobs dominated by women and those traditionally held by men when an impartial evaluation deems the jobs are of equal value to an employer.

Advocates also say neither the Equal Pay Act of 1963 — which bars unequal pay for the same job — nor Title VII of the Civil Rights Act of 1964, which bans discrimination based on race, color, gender, religion and national origin in hiring and promotion, do what the comparable-worth doctrine would do: Root out bias against entire occupations traditionally dominated by females.[1]

Although women began entering non-traditional fields decades ago, Labor Department data show that certain occupations still are filled mostly by females. For example, in 2006, 89 percent of paralegals and legal assistants were women, while only 33 percent of lawyers were women. And only 7 percent of machinists were women, while 84 percent of special-education teachers were female.[2]

"There's a lot of [job] segregation, and the closer you look, the more segregation you find," says Philip Cohen, a sociologist at the University of North Carolina who studies gender inequality. "Under current law, it's very difficult to bring legal action successfully and say the pay gap between men and women is discrimination, because the employer can say 'they're doing different jobs.' "

But critics say comparable worth would disrupt the traditional market-based system of determining wages based on the laws of supply and demand. "You would have people moving into occupations where there was really no shortage" of workers, says June O'Neill, an economist at the City University of New York's Baruch College. "You would have gluts in some [job categories] and shortages in others."

In 2000 testimony before a congressional panel, O'Neill outlined what she saw as the dangers of adopting a comparable-worth approach. Because there is no uniform way to rank occupations by worth, she says, such a policy would "lead to politically administered wages that would depart from a market system of wage determination." Pay in traditionally female occupations would likely rise — appointing people

During the Great Depression of the 1930s, the proportion of single women who were working stayed more or less flat. But the percentage of married women who worked rose to almost 14 percent by 1940 — a jump of more than 50 percent over the 1920 rate.[38] World War II brought millions more women into the labor force, as females — characterized by the iconic image of Rosie the Riveter — took jobs in defense plants doing work traditionally performed by men.

Equal-Pay Initiatives

As women proved their mettle behind the drill press and rivet gun, advocates continued to push for equal pay. In 1942 President Franklin D. Roosevelt had the National War Labor Board urge employers to equalize wage rates between men and women "for comparable quality and quantity of work on the same or similar operations."[39]

favorable to the comparable-worth idea "would all but guarantee that result," she said. But that higher pay would raise costs for employers, leading them to put many women out of work, she suggested. "The ironic result is that fewer workers would be employed in traditionally female jobs."

Not only that, but some employers would respond to the higher wage levels by providing fewer non-monetary benefits, such as favorable working hours, that help accommodate women with responsibilities at home, O'Neill said. "Apart from the inefficiency and inequality it would breed," she concluded, "I find comparable worth to be a truly demeaning policy for women. It conveys the message that some cannot compete in non-traditional jobs and can only be helped through the patronage of a job evaluator."

Critics also say that comparable worth would put the government into the role of setting wages for private business, an idea that is anathema to business interests.

"Who determines what is equal value?" asks Michael Eastman, executive director of labor policy at the U.S. Chamber of Commerce. "Equal value to society? Who's setting wages then? Is the government coming up with guidelines? For example, are truckers equal to nurses, and who's making that comparison? We've never had the government setting private-sector wage rates like that."

Martha Burk directs the Corporate Accountability Project for the National Council of Women's Organizations.

Supporters of comparable worth brush off such concerns. Martha Burk, a longtime women's activist, notes that a bill proposed by Sen. Tom Harkin, D-Iowa, would require companies to disclose how they pay women and men by job categories, a practice that alone would lead to more equitable wages. "What you have is a government solution that is not telling anybody what to pay their employees," she says. It would only "increase the transparency so the company can solve its own problem if it has one."

As to the notion that comparable worth amounts to government intrusion in the private market, Burk says, "Free marketers think anything short of totally unregulated capitalism is interfering in the free market."

"It may be that markets are efficient from the point of view of employers," adds Vicky Lovell, director of employment and work/life programs at the Institute for Women's Policy Research in Washington. "But I don't think they're efficient from the point of view of workers."

[1] For background, see June O'Neill, "Comparable Worth," *The Concise Encyclopedia of Economics*, The Library of Economics and Liberty, www.econlib.org.

[2] "Women in the Labor Force: A Databook," U.S. Department of Labor, Report 1002, September 2007, Table 11, pp. 28-34.

In the closing months of the war, the first bill aimed at barring gender pay discrimination came to the floor of Congress. The Women's Equal Pay Act of 1945 went nowhere, though.[40]

By 1960, more than a third of women were working, and among single, white women ages 25 to 34, the labor participation rate was a then-record 82 percent.[41] But most women continued to work in low-paying clerical, service and manufacturing jobs, and the wage gap between males and females was wide. By 1963, women made only 59 cents for every dollar in median year-round earnings paid to men.[42] Women who tried to break into so-called "men's" occupations faced huge resistance.

That year, after decades of struggle by women's advocates for federal legislation on gender pay equity, Congress passed the Equal Pay Act as an amendment to the Fair Labor Standards Act of 1938. In signing the act, President Kennedy said the law "affirms our determination that

Did Wal-Mart Favor Male Workers?

Women's suit seeks billions in damages.

Dedra Farmer, the daughter of an auto mechanic, worked in the Tire Lube Express Division of Wal-Mart Stores, the only female in her district who held a salaried manager position in that division. During her 13 years with the retail giant, she told a congressional panel last year, she saw evidence that women — herself among them — earned less than men holding the same jobs.

Farmer said she complained to Wal-Mart's CEO through e-mails, expressed her concern at a store meeting and was assured by the store manager that she'd get a response. "The response I received was a pink slip," she said.[1]

Farmer has joined a class-action lawsuit accusing Wal-Mart of sex discrimination in pay and promotions. The case, which could cover perhaps 1.6 million current and former female employees and result in billions of dollars in damages, is the biggest workplace discrimination lawsuit in the nation's history.

Filed in 2001 by Betty Dukes and five other Wal-Mart employees, the case has gone through a series of legal maneuverings, most recently in December, when a three-judge panel of the U.S. 9th Circuit Court of Appeals reaffirmed its certification as a class-action lawsuit but left the door open for Wal-Mart to ask for a rehearing on that status. If the appeals court does not reconsider the class-action designation, the company reportedly will petition the Supreme Court.[2]

The stakes in the case are high. Goldman Sachs Group last year estimated potential damages at between $1.5 billion and $3.5 billion if the retailer loses, and punitive damages could raise the figure to between $13.5 billion and $31.5 billion.[3]

The company's lawyers have asserted that a class-action suit is an inappropriate vehicle to use because Wal-Mart's employment policies are decentralized, and individual store managers and district managers make pay and promotion decisions.[4]

Theodore J. Boutrous Jr., a lawyer for Wal-Mart, has said that decisions by thousands of managers at 3,400 Wal-Mart stores during six years were "highly individualized and cannot be tried in one fell swoop in a nationwide class action."[5] He has also said the company has a "strong diversity policy and anti-discrimination policy."[6]

But Brad Seligman, executive director of the Impact Fund, a nonprofit group in Berkeley, Calif., representing the plaintiffs, said, "No amount of PR or spin is going to allow Wal-Mart to avoid facing its legacy of discrimination."[7]

A statistician hired by the plaintiffs said it took women an average of 4.38 years from the date of hire to be promoted to assistant manager, while it took men 2.86 years. Moreover, it took an average of 10.12 years for women to become managers compared with 8.64 for men.[8]

The statistician, Richard Drogin, of California State University at East Bay, also found that female managers made an average annual salary of $89,280, while men in the same position earned an average of $105,682. Female hourly workers earned 6.7 percent less than men in comparable positions.[9]

when women enter the labor force they will find equality in their pay envelopes."[43]

The measure, as finally adopted, stopped short of ensuring the elusive comparable-worth standard that women's advocates had so long sought. Instead, the bill made it illegal to discriminate in pay and benefits on the basis of sex when men and women performed the same job at the same employer.

Under the law, for example, a company couldn't pay a full-time female store clerk less per hour than a male one for doing the same job in stores located in the same city. But the law was silent on situations in which, say, the work of a female secretarial supervisor was deemed to be of comparable worth to that of a male who supervised the same company's truck drivers.

While the Equal Pay Act marked progress, it was far from an airtight guarantee of "equality in . . . pay envelopes." For example, the law initially did not cover executive, administrative or professional jobs; that exemption was lifted in 1972. Yet, one study argues that courts have interpreted the act so narrowly that white-collar female workers have had trouble winning claims through its provisions.[44]

Appellate Judge Andrew J. Kleinfeld has dissented in the case, arguing that certifying the suit as a class action deprived the retailer of its right to defend against individual cases alleging bias. In addition, he argued that female employees who were discriminated against would be hurt by class-action status, because women "who were fired or not promoted for good reasons" would also share in any award if Wal-Mart lost the case.[10]

Business lobbies also have urged that the class-action certification be reversed. An official of the U.S. Chamber of Commerce, which filed a "friend of the court" (*amicus curiae*) brief in the case, warned of "potentially limitless claims" against companies "with limited ability to defend against them." He added: "The potential financial exposure to an employer facing a class action of this size creates tremendous pressure to settle regardless of the case's merit."[11]

But women's advocates argue that a class-action approach is appropriate. It "provides the only practical means for most women in low-wage jobs to redress discrimination in pay because of such workers' often tenuous economic status," stated an *amicus* letter written to the appeals court on behalf of the U.S. Women's Chamber of Commerce.[12]

A Wal-Mart store manager reads the store's weekly sales results to other workers. Male hourly workers at Wal-Mart earn 6.7 percent more than women in comparable positions, a pay-equity study contends.

Added Margot Dorfman, chief executive officer of the group: "A woman with family responsibilities often isn't in a position to quit her job or risk antagonizing her employer with a challenge to a bad workplace practice."[13]

[1] Statement of Dedra Farmer before House Committee on Education and Labor, April 24, 2007.

[2] Amy Joyce, "Wal-Mart Loses Bid to Block Group Bias Suit," *The Washington Post*, Feb. 7, 2007, p. 1D.

[3] Details of the Goldman Sachs analysis are from Steve Painter, "Judges modify sex-bias decision; Wal-Mart appeal likely to see delay," *Arkansas Democrat-Gazette*, Dec. 12, 2007.

[4] Steven Greenhouse and Constance L. Hays, "Wal-Mart Sex-Bias Suit Given Class-Action Status," *The New York Times*, June 23, 2004.

[5] Joyce, *op. cit.*

[6] Quoted in Bob Egelko, "Wal-Mart sex discrimination suit advances; Appeals court OKs class action status for 2 million women," *San Francisco Chronicle*, Feb. 7, 2007, p. B1.

[7] Joyce, *op. cit.*

[8] *Ibid.*

[9] *Ibid.*

[10] Painter, *op. cit.*

[11] "U.S. Chamber Files Brief in Wal-Mart Class Action," press release, U.S. Chamber of Commerce, Dec. 13, 2004, www.uschamber.com/press/releases/2004/december/04-159.htm.

[12] Mark E. Burton Jr., Hersh & Hersh, San Francisco, et al., letter submitted to 9th U.S. Circuit Court of Appeals, March 27, 2007, www.uswcc.org/amicus.pdf.

[13] PR Newswire, "U.S. Women's Chamber of Commerce Joins Fight in Landmark Women's Class Action Suit Against Wal-Mart," March 28, 2007.

Perhaps more significantly, the law gives companies several defenses for pay disparities: when wage differences stem from seniority or merit systems, are based on quantity or quality of production, or stem from "any other factor other than sex."

That last provision, critics say, can sometimes allow business practices that may seem gender-neutral on the surface but discriminate nonetheless.

The Equal Pay Act took effect in 1964, and that same year Congress passed Title VII of the Civil Rights Act of 1964, a broad measure that prohibits employment discrimination on the basis of race, color, religion, national origin and sex, and covers hiring, firing and promotion as well as pay. A measure called the Bennett Amendment, sponsored by Rep. Wallace F. Bennett, a Utah Republican, sought to bring Title VII and the Equal Pay Act in line with each other.

In ensuing years, the overlap of the Equal Pay Act and Title VII created confusion but also helped to animate the battle against wage discrimination. Part of the conflict over pay equity played out in the courts in the 1970s and '80s.

Key Court Rulings

In a case that initially raised hopes for the theory of comparable worth, the U.S. Supreme Court ruled 5-4 to allow female jail guards to sue for sex discrimination. The women, called "matrons," earned 30 percent less than male guards, called "deputy sheriffs."[45] The women argued that while they had fewer prisoners to guard and more clerical duties than the male guards, their work was comparable. An outside job evaluation showed that the women did 95 percent of what the men were doing, but received $200 less a month than the men.[46]

Prior to the Supreme Court's ruling, *The Washington Post* noted at the time, "the only sure grounds for a pay discrimination claim by a woman under federal law was 'unequal pay for equal work' — an allegation that she was paid less than a man holding an identical job. The jail matrons and women's rights lawyers said that lower pay for a comparable, if not equal, job could also be the basis for a sex-discrimination charge."

Justice William J. Brennan wrote that a claim of wage discrimination under Title VII did not have to meet the equal work standards of the Equal Pay Act. Thus, noted Clare Cushman, director of publications at the Supreme Court Historical Society, "a woman employee could sue her employer for gender-based pay discrimination even if her company did not employ a man to work the same job for higher pay."[47]

Still, Cushman wrote, while the court "opened the door slightly for women working in jobs not strictly equal to their male counterparts, it also specifically declined to authorize suits based on the theory of comparable worth."

In 1985 that theory suffered a blow that continues to resonate today, partly because of the personalities who were involved. In *AFSCME v. the State of Washington*, the 9th U.S. Circuit Court of Appeals overturned a lower court's ruling ordering Washington to pay more than $800 million in back wages to some 15,000 state workers, most of them women.[48]

The case turned on the question of whether employers were required to pay men and women the same amounts for jobs of comparable worth, rather than equal wages for the same jobs. It eventually ended in a draw when the state negotiated a settlement with AFSCME (American Federation of State, County and Municipal Employees union).[49]

Judge Anthony M. Kennedy, who now sits on the U.S. Supreme Court and presumably could help decide a comparable-worth case should one arise before the justices, wrote the appellate court's decision. Kennedy wrote: "Neither law nor logic deems the free-market system a suspect enterprise." During this same period, two other personalities who now sit on the high court also expressed negative views on comparable worth. As a lawyer in the Reagan administration, John Roberts, now chief justice, described it as "a radical redistributive concept."[50] And the EEOC, then under Chairman Clarence Thomas, rejected comparable worth as a means of determining job discrimination. "We found that sole reliance on a comparison of the intrinsic value of dissimilar jobs — which command different wages in the market — does not prove a violation of Tile VII," Thomas stated.[51]

The views of Thomas and Roberts reflected the conservative policies of the Reagan administration during the 1980s. Yet despite the political tenor of that era, women made major strides toward workplace equality. From 1980 to 1992, the wage gap in median weekly earnings of full-time female wage and salary workers narrowed from 64 percent to 76 percent after adjusting for inflation. But it shrank only from 77 percent to 81 percent from 1993 — the year that Democratic President Bill Clinton took office and the Family and Medical Leave Act was enacted — to 2006.[52]

Measuring Progress

Experts debate whether and to what degree women's gains may have slowed or stopped in recent years. Some point to huge political gains in this decade, including Sen. Clinton's role in the presidential race and the rise of Rep. Nancy Pelosi, D-Calif., to speaker of the House. Others cite such evidence as a recent study showing that female corporate directors, though a small minority in boardrooms, out-earn male directors.[53]

But many scholars believe women's gains have indeed slowed.

Vanneman, the University of Maryland sociologist, has carefully charted a number of trends linked to the so-called gender revolution, and on his Web site he notes that he and several colleagues are studying the pace of women's progress.

"For much of the last quarter of the 20th century, women gradually reduced gender inequalities on many fronts," he wrote, citing such trends as women entering the labor force in growing numbers, the opening of previously male-dominated jobs to women, the narrowing wage gap, women's role in politics and a growing openness in public opinion about the participation of women in public and community life.

But, he added, "all this changed in the early to mid-1990s." A "flattening of the gender trend lines" is seen in nearly all parts of society, he added: working-class and middle-class, black, white, Asian and Hispanic, mothers with young children and those with older ones, and so on. "All groups experienced major gender setbacks during the 1990s. The breadth of this reversal suggests something fundamental has happened to the U.S. gender structure."

In an interview, Vanneman says he has no theories as to what accounts for that reversal — only hunches — as he continues to study the phenomenon. One hunch is that the flattening started happening in the 1980s but didn't show up in a big way until the 1990s. He also says he suspects the reversal in women's progress gathered momentum in the 1990s as the "culture of parenting" changed. Americans, he says, became less accepting of women trying to balance busy careers with the pressures of motherhood, a shift that has put women in more of a bind than they felt in previous periods. As a result, many women have backed away from high-paying careers and devoted more time to family, he says.

"There's been tremendous growth in expectations of what it means to be a good parent," Vanneman says.

Cornell University economist Blau agrees that progress in women's wages slowed in recent years, though she sees some evidence that the picture has brightened a bit.

One reason for the slowdown in the 1990s, she says, may have been that the increase in demand for white-collar and service workers shifted into a lower gear compared to the 1980s, when many women benefited from a surge in hiring for white-collar jobs, including ones that required computer skills, while blue-collar jobs dominated by men began to wane.

In addition, Blau says that during the eighties, as many women began to stay in the workforce even after marriage and childbirth, employers' view of the value of female workers improved. That, she says, helped narrow the wage gap at a faster pace than in earlier decades.

Blau also sees evidence that men were doing more at home in the 1980s than ever before. That trend didn't go away in the past decade, she says, but it hasn't grown much either.

CURRENT SITUATION
Prospects in Congress

As concerns over the progress of gender equity grow, women's advocates are hoping that the Democrat-controlled Congress will pass new laws this year. But proposed legislation is likely to face stiff opposition.

Reversing the Supreme Court's *Ledbetter* decision seems to have the best chance of making it through Congress. The House passed the Ledbetter Fair Pay Act last July 31 by a 225-199 vote, largely along party lines.[54] A companion bill in the Senate, called the Lilly Ledbetter Fair Pay Restoration Act, had garnered 37 co-sponsors as of early March. Momentum continued this year with a Senate hearing.

In introducing the Senate version of the bill last July, Sen. Kennedy said it "simply restores the status quo" that existed before the *Ledbetter* decision "so that victims of ongoing pay discrimination have a reasonable time to file their claims."[55]

But employer advocates such as the U.S. Chamber of Commerce dispute such descriptions. Pointing to the House version that passed last summer, chamber officials said it would broaden existing law to apply to unintentional as well as intentional discrimination and would lead to an "explosion of litigation second-guessing legitimate employment and personnel decisions."[56]

The Bush administration has threatened a veto, saying last year that if the House bill came to the president, "his senior advisers would recommend that he veto" it.[57] The measure would "impede justice" by allowing employees to sue over pay or other employment-related discrimination "years or even decades after the alleged discrimination occurred," the administration said. Moreover, the House bill "far exceeds the stated purpose of undoing the court's decision" by "extending the expanded statute of limitations to any 'other practice' that remotely affects an individual's wages, benefits, or other compensation in the future."

Is the Equal Rights Amendment to the Constitution still needed?

YES
Idella Moore
Executive Officer, 4ERA

Written for *CQ Researcher*, February 2008

We still need the Equal Rights Amendment (ERA) because sex discrimination is still a problem in our country. Like race or religious discrimination, gender discrimination is intended to render its victims economically, socially, legally and politically disadvantaged. But unlike racism and religious intolerance — whose practice against certain groups is localized within countries or regions — sex discrimination is universal. Why, then, in our court system are race and religious discrimination considered more serious offenses?

Today, American women — of all races and religions — are still fighting to achieve equal opportunity, pay, status and recognition in all realms of our society. At this moment, the largest class-action lawsuit in the history of this country is being argued on behalf of 1.6 million women who were discriminated against purely because of their gender. If the ERA had been ratified back in the 1970s, by now these types of lawsuits would be extinct.

We still need the ERA because ratification of the amendment will elevate "sex" to, in legal terms, a so-called suspect class. A suspect class has the advantage in discrimination cases. Gender, as yet, is not afforded that advantage. As we've seen with race, suspect class status increased the chance of favorable outcomes in discrimination cases. This, in turn, served as a deterrent. Consequently, in our society racism is now socially unacceptable. Sex discrimination, however, is not.

We still need the ERA because the continuing struggle for legal equality for women should be seen as a shameful and embarrassing condition of our society. Yet today lawmakers — sworn to represent all their constituents — proudly voice their objections to granting legal equality to women and without any fear of consequences to their political careers. How different our reactions would be if they were espousing racism.

The Equal Rights Amendment will perfect our Constitution by explicitly guaranteeing that the privileges, laws and responsibilities it contains apply equally to men and women. As it stands today the Constitution is sometimes interpreted that way, but women, as a universally and historically disadvantaged group, cannot rely on such interpretations. We have seen these "interpretations" vary and change, often due to the whims of the political climate. Therefore, without the ERA any gains women make will always be tenuous.

I see the Equal Rights Amendment, too, as a pledge to ourselves and posterity that we recognize that sexism exists and that we as a country are determined to continue perfecting our democracy by proudly and unequivocally guaranteeing that one's gender will no longer be a detriment to achieving the American dream.

NO
Phyllis Schlafly
President, Eagle Forum

Written for *CQ Researcher*, February 2008

The Equal Rights Amendment (ERA) was fiercely debated across America for 10 years (1972-1982) and was rejected. ERA has been reintroduced into the current Congress under a slightly different name, but it's the same old amendment with the same bad effects.

The principal reason ERA failed is that although it was marketed as a benefit to women, its advocates were never able to prove it would provide any benefit whatsoever to women. ERA would put "sex" (not women) in the Constitution and just make all our laws sex-neutral.

ERA advocates used their massive access to a friendly media to suggest that ERA would raise women's wages. But ERA would have no effect on wages because our employment laws are already sex-neutral. The equal-pay-for-equal-work law was passed in 1963, and the Equal Employment Opportunity Act — with all its enforcement mechanisms — was passed in 1972.

Supreme Court Justice Ruth Bader Ginsburg's book *Sex Bias in the U.S. Code* spells out the changes ERA would require, and it proves ERA would take away benefits from women. For example, the book states that the "equality principle" would eliminate the concept of "dependent women." This would deprive wives and widows of their Social Security dependent-wife benefits, on which millions of mothers and grandmothers depend.

Looking at the experience of states that have put ERA language into their constitutions, we see that ERA would most probably require taxpayer funding of abortions. The feminists aggressively litigate this issue. Their most prominent victory was in the New Mexico Supreme Court, which accepted the notion that since only women undergo abortions, the denial of taxpayer funding is sex discrimination.

ERA would also give the courts the power to legalize same-sex marriages. Courts in four states have ruled that the ERA's ban on gender discrimination requires marriage licenses to be given to same-sex couples. In Maryland and Washington, those decisions were overturned by a higher court by only a one-vote margin. The ERA would empower the judges to rule either way.

If all laws are made sex-neutral, the military draft-registration law would have to include women. We don't have a draft today, but we do have registration, and those who fail to register immediately lose their college grants and loans and will never be able to get a federal job.

Eric Dreiband, a former EEOC general counsel in the Bush administration, told this year's hearing on the Senate bill that the measure would subject state and local governments, unions, employers and others to potentially unlimited penalties and could expose pension funds to "potentially staggering liability."[58]

Still, women's advocates remain sanguine about the measure's prospects. "My hope is that the bill will move expeditiously [this] spring" in the Senate and that "the president will reconsider and recognize how important this fix to the law is," says Samuels of the National Women's Law Center.

The other two main bills on gender pay equity could have rougher sledding.

Sen. Clinton's Paycheck Fairness Act is similar to a bill by the same name proposed during her husband's presidential administration. As of early March, the bill had garnered 22 co-sponsors in the Senate and 226 in the House.

Among other things, it would strengthen penalties on employers who violate the Equal Pay Act, make it harder for companies to use the law's defense for wage differences based on factors "other than sex," and bar employers from retaliating against workers who share wage information with each other. It also calls for the Labor Department to draw up guidelines aimed at helping employers voluntarily evaluate job categories and compare wages paid for different jobs with the aim of eliminating unfair wage differences between male- and female-dominated occupations.

The bill has drawn enthusiastic support from some women's advocates, but it also has opponents. Washington lawyer Brown said the goal of the provision on voluntary guidelines was "nothing more than the discredited 'comparable-worth' theory in new clothing."[59]

The Fair Pay Act, proposed by Sen. Harkin and Del. Eleanor Holmes Norton, D-D.C., a former EEOC chair, steps even closer to embracing the comparable-worth theory and thus, many observers believe, is likely to face stiff headwinds. The main ideas have circulated in Congress for years.

As Harkin describes it, the bill requires employers to provide equal pay for jobs that are comparable in skill, effort, responsibility and working conditions, regardless of sex, race or national origin, and it bars companies from reducing other employees' wages to achieve pay equity.[60]

Again, advocates such as Samuels are hopeful Congress will pass both the Paycheck Fairness and Fair Pay Act and that the president won't veto them if they do make it to his desk. "The hope would be that the level of support for these bills both in Congress and among the public is so substantial, and they so clearly are a necessary step toward ensuring true equality of wages, that the president would understand the necessity for them and sign them," she says.

But business opposition is likely to be strong. Eastman at the U.S. Chamber of Commerce lists a variety of complaints about both bills, such as their provisions for punitive damages and their allowances for class-action suits against employers. "The case has not been made that these bills are justified," he says.

State Action

While women's advocates hold out hope for congressional action, they also are turning their attention to the states in hopes of pressing legislatures to stiffen laws on pay equity and make local economies friendlier to gender issues. As of April 2007, all but 11 states and the District of Columbia had laws on equal pay.[61]

Minnesota has had a system of comparable worth, or "pay equity," for public employees since the 1980s, and last year proposals were made to expand the system to private employers that do business with the state. The Minnesota program gave smaller raises to public workers in male-dominated jobs and bigger raises to those in female-dominated ones, according to a former staff member of the Minnesota Commission on the Economic Status of Women. The system shrank the pay gap from 72 percent to nearly equal pay.[62]

A report by the Institute for Women's Policy Research said in 2006 that while women's wages had risen in all states in inflation-adjusted terms since 1989, "in no state does the typical full-time woman worker earn as much as the typical man." It would take 50 years "at the present rate of progress" for women to achieve wage parity with men nationwide, it said.[63]

Some advocates are unwilling to wait that long. In Colorado, for example, a Pay Equity Commission appointed by Donald J. Mares, executive director of the state Department of Labor and Employment, worked since last June to formulate policy recommendations to curb gender and racial pay inequities in the private and public sectors. The 12-member commission includes policy analysts, business and labor union representatives, academics and advocates for women and minorities.[64]

Meric, the 9to5 director and a Colorado resident, said her group was instrumental in getting the state to appoint the commission. Although the panel has no authority to force employers to alter pay practices, Meric hopes the commission's work leads to change. One key recommendation, she says, is that employers do more to create flexible policies so that workers — especially women with caregiving responsibilities — aren't penalized for meeting both work and family responsibilities.

Mares told the Colorado Women's Legislative Breakfast in February that another recommendation calls for making the commission permanent, so it can continue to monitor gender pay equity in the state and help educate businesses on good practices.

In Colorado, he said, the average woman makes 79 cents for every dollar earned by the average man. "Every day you as a community walk in the door," he told the gathering of women, "your pay is being discounted. That's not good."[65]

Better negotiating skills could help narrow the gender wage gap, in the view of women's advocates. The Clinton/DeLauro bill calls for grants to help women and girls "strengthen their negotiation skills to allow the girls and women to obtain higher salaries and the best compensation packages possible for themselves."

It's a talent that many women don't exercise, says Linda Babcock, an economist at Carnegie Mellon University in Pittsburgh and co-author of the recent book *Women Don't Ask: Negotiation and the Gender Divide.* Babcock found in a study of Carnegie Mellon students graduating with master's degrees in public policy that only 12.5 percent of females tried to negotiate for better pay when they received a job offer, while 51.5 percent of males did. Afterward, the females earned 8.5 percent less than the males.

Babcock sees several reasons why women are not inclined to negotiate more, including that they have been socialized by American culture to be less assertive than men. And, she says, women who do try to bargain for better wages often are subjected to "backlash" by employers and peers.

Not that women are incapable of negotiating, Babcock stresses. While they may not always stand up for themselves in seeking higher wages, women outperform men when negotiating on behalf of somebody else, she has found.

"It's really striking," she says. "If we were missing some gene, we wouldn't really be able to turn it on on behalf of somebody else."

OUTLOOK

Pressure for Change

Some women's advocates are not especially sanguine about the possibility of big strides on the gender-wage front, at least in the near future.

"I don't think five years is long enough [for there] to be much change, particularly if we don't see much concerted effort among employers," says Lovell of the Institute for Women's Policy Research.

Big change would require a "push from the federal government" or "some dramatic effort on the part of socially conscious employers," she says. "That hasn't happened before, and I don't think it will in the next few years."

Still, observers believe that social and political shifts will produce new pressure for changes in the way employers deal with wage equity.

Meric says 9to5's "long-term agenda" is to have the theory of comparable worth enshrined in law as well as to have "guaranteed minimum labor standards" for all workers that include paid sick leave and expanded coverage under the Family and Medical Leave Act. In Colorado, she hopes the recommendations outlined by the Pay Equity Commission will serve as a model for other states and "move us closer" to that long-term goal. "Basic protections should apply to workers wherever they live in the United States."

"In the last five or 10 years we have seen progress stall in [achieving] gender equality," says Philip Cohen, a sociologist at the University of North Carolina at Chapel Hill who studies gender inequity. But in coming years, he says he is inclined to think that college-educated women will exert increasing pressure on federal and state lawmakers and employers to make policy changes that can narrow the wage gap.

"If you look back to feminism in the '60s," Cohen says, "a lot of women had college degrees but weren't able to take advantage of their skills in the marketplace, and that became the 'feminine mystique' " explored in Betty Friedan's groundbreaking 1963 book.

Today, "Women are outnumbering men in college graduation rates, and I think we are going to see more and more women looking around for better opportunities. If they don't see gender equality resulting, they're going to be very dissatisfied."

And that dissatisfaction, Cohen says, could well show up in the political arena.

Samuels of the National Women's Law Center hopes the debate in Congress and fallout from the Supreme Court's *Ledbetter* decision will spur further gains in wage equity for women.

"Unfortunately, over the course of the last several years things have pretty much stagnated," she says. "I do hope that the recent public attention paid to wage disparity will cause employers to take a look at their pay scales and try to do the right thing."

NOTES

1. Testimony of Lilly Ledbetter before the Committee on Education and Labor, U.S. House of Representatives, on the Amendment of Title VII, June 12, 2007.

2. Testimony of Lilly Ledbetter before Senate Committee on Health, Education, Labor and Pensions, Jan. 24, 2008.

3. *Ledbetter v. Goodyear Tire & Rubber Co. Inc.,* 550 U.S. __ (May 29, 2007).

4. Ledbetter testimony, *op. cit.,* June 12, 2007.

5. Diana Furchtgott-Roth, testimony on the Paycheck Fairness Act before House Committee on Education and Labor, April 24, 2007.

6. "Highlights of Women's Earnings in 2006," U.S. Department of Labor, September 2007, Table 1, p. 7. Data are for median usual weekly earnings of full-time wage and salary workers ages 16 and older. For the Census Bureau data, see www.census.gov/compendia/statab/tables/08s0628.pdf. The Census Bureau data represent median full-time, year-round earnings for male and female workers 15 years old and older as of March 2006.

7. Testimony before Senate Committee on Health, Education, Labor and Pensions, April 12, 2007.

8. "Women's Earnings: Work Patterns Partially Explain Difference between Men's and Women's Earnings," U.S. General Accounting Office, October 2003.

9. U.S. Equal Employment Opportunity Commission, Office of Inspector General, "Semiannual Report to Congress," April 1, 2007-Sept. 30, 2007, Oct. 30, 2007, p. 7.

10. "Highlights of Women's Earnings," *op. cit.,* p. 1.

11. *Ibid.,* Table 1, p. 7.

12. Judy Goldberg Dey and Catherine Hill, "Behind the Pay Gap," American Association of University Women Educational Foundation, 2007.

13. Paycheck Fairness Act, HR 1338, S 766.

14. Fair Pay Act, S 1087 and HR 2019, sponsored in the House of Representatives by Del. Eleanor Holmes Norton, D-D.C.

15. Lilly Ledbetter Fair Pay Act, HR 2831 and Fair Pay Restoration Act, S 1843.

16. David Neumark, with the assistance of Roy J. Bank and Kyle D. Van Nort, "Sex Discrimination in Restaurant Hiring: An Audit Study," *The Quarterly Journal of Economics,* August 1996.

17. Furchtgott-Roth testimony, *op. cit.*

18. Dey and Hill, *op. cit.*

19. Testimony of Heather Boushey before House Committee on Education and Labor, April 24, 2007, p. 4.

20. John F. Kennedy, Remarks Upon Signing the Equal Pay Act, June 10, 1963, quoted in John T. Woolley and Gerhard Peters, *The American Presidency Project* [online], Santa Barbara, Calif., University of California (hosted), Gerhard Peters (database), www.presidency.ucsb.edu/ws/?pid=9267.

21. Testimony of Jocelyn Samuels before Senate Committee on Health, Education, Labor and Pensions, "Closing the Gap: Equal Pay for Women Workers," April 12, 2007, p. 6.

22. Carrie Lukas, "A Bargain At 77 Cents To a Dollar," *The Washington Post,* April 3, 2007, p. 23A.

23. Testimony of Barbara Berish Brown before Senate Committee on Health, Education, Labor and Pensions, April 12, 2007.

24. Quoted in Justine Andronici, "Court Gives OK To Unequal Pay," *Ms. Magazine,* summer 2007, accessed at www.msmagazine.com/summer2007/ledbetter.asp.

25. Quoted in Michael Doyle, "Justices Put Bias Lawsuits on Tight Schedule," *Kansas City Star,* May 30, 2007, p. 1A.

26. Testimony of Margot Dorfman before Senate Committee on Health, Education, Labor and Pensions on the "The Fair Pay Restoration Act: Ensuring Reasonable Rules in Pay Discrimination Cases," Jan. 24, 2008, pp. 2-3.

27. "Employment Discrimination: Post-Ledbetter Discrimination," "Justice Talking," National Public Radio, Oct. 22, 2007, accessed at www.justicetalking.org/transcripts/071022_EqualPay_transcript.pdf.

28. Richard B. Morris, ed., "The U.S. Department of Labor Bicentennial History of the American Work," U.S. Department of Labor, 1976, p. 67.

29. Claudia Goldin, *Understanding the Gender Gap: An Economic History of American Women* (1990), p. 50.

30. *Ibid.,* Figure 3.1, p. 62, and text pp. 63, 66.

31. *Ibid.,* p. 67.

32. Quoted in *ibid.,* p. 209.

33. *Ibid.,* Table 2.1, p. 17, citing U.S. Census data.

34. "Setting a New Course," *CQ Researcher,* May 10, 1985, citing Julie A. Matthaei, *An Economic History of Women in America* (1982), p. 222.

35. Goldin, *op. cit.,* p. 17.

36. Michael McGerr, *A Fierce Discontent: The Rise and Fall of the Progressive Movement in America, 1870-1920* (2003), pp. 295-296.

37. For background, see Richard Boeckel, "Sex Equality and Protective Laws," *Editorial Research Reports,* July 13, 1926; and Richard Boeckel, "The Woman's Vote in National Elections," *Editorial Research Reports,* May 31, 1927, both available at *CQ Researcher Plus Archive,* www.cqpress.com.

38. *Ibid.*

39. American Association of University Women, "A Brief History of the Wage Gap, Pay Inequity, and the Equal Pay Act," www.aauw.org/advocacy/laf/lafnetwork/library/payequity_hist.cfm. For background, see K. R. Lee, "Women in War Work," *Editorial Research Reports,* Jan. 26, 1942, available at *CQ Researcher Plus Archive,* www.cqpress.com.

40. *Ibid.*

41. Goldin, *op. cit.,* Table 2.2, p. 18.

42. *Ibid.,* Table 3.1, p. 60.

43. John F. Kennedy, *op. cit.*

44. Juliene James, "The Equal Pay Act in the Courts: A De Facto White-Collar Exemption," *New York University Law Review,* Vol. 79, November 2004, p. 1875.

45. Clare Cushman, *Supreme Court Decisions and Women's Rights,* CQ Press (2000), p. 146. The case is *County of Washington v. Gunther* (1981). For background, see Sandra Stencel, "Equal Pay Fight," *Editorial Research Reports,* March 20, 1981, and R. Thompson, "Women's Economic Equity," *Editorial Research Reports,* May 10, 1985, both available at *CQ Researcher Plus Archive,* www.cqpress.com.

46. Deborah Churchman, "Comparable Worth: The Equal-Pay Issue of the '80s," *The Christian Science Monitor,* July 22, 1982, p. 15.

47. Cushman, *op. cit.*

48. James Warren, "Fight for Pay Equity Produces Results, But Not Parity," *Chicago Tribune,* Sept. 8, 1985, p. 13.

49. Judy Mann, "New Victory in Women's Pay," *The Washington Post,* Aug. 27, 1986, p. 3B.

50. Linda Greenhouse, "Judge Roberts, the Committee Is Interested in Your View On...," *The New York Times,* Sept. 11, 2005, p. 1A.

51. "Women Dealt Setback on 'Comparable Worth,'" *Chicago Tribune,* June 18, 1985, p. 1.

52. "Highlights of Women's Earnings in 2006," *op. cit.,* Table 13, p. 28.

53. Martha Graybow, "Female U.S. corporate directors out-earn men: study," Reuters, Nov. 7, 2007. The study of more than 25,000 directors at more than

3,200 U.S. companies was done by the Corporate Library. It found that female directors earned median compensation of $120,000 compared with $104,375 for male board members.

54. Libby George, "House Democrats Prevail in Effort to Clarify Law on Wage Discrimination," *CQ Weekly*, Aug. 6, 2007, p. 2381.

55. Sen. Edward Kennedy, statement on S 1843, "Statements on Introduced Bills and Joint Resolutions," Senate, July 20, 2007, accessed at www.thomas.gov.

56. U.S. Chamber of Commerce, "Letter Opposing HR 2831, the Ledbetter Fair Pay Act," July 27, 2007, accessed at www.uschamber.com/issues/letters/2007/070727_ledbetter.htm.

57. "Statement of Administration Policy: HR 2831, Lilly Ledbetter Fair Pay Act of 2007," Executive Office of the President, Office of Management and Budget, July 27, 2007, accessed at www.whitehouse.gov/omb/legislative/sap/110-1/hr2831sap-r.pdf.

58. Statement of Eric S. Dreiband before Senate Committee on Health, Education, Labor and Pensions, Jan. 24, 2008, pp. 11-13.

59. Barbara Berish Brown testimony, *op. cit.*

60. Statement of Sen. Tom Harkin at the Health, Education, Labor and Pensions Committee Hearing on Equal Pay for Women Workers, April 12, 2007, accessed at www.harkin.senate.gov/pr/p.cfm?i=272330.

61. National Conference of State Legislatures, "State Laws on Equal Pay," April 2007.

62. H.J. Cummins, "Legislature will look at closing the gender gap," *Star Tribune*, April 23, 2007, p. 1D.

63. Heidi Hartmann, Olga Sorokina and Erica Williams, "The Best and Worst State Economies for Women," Institute for Women's Policy Research, December 2006.

64. "Pay Equity Commission holds first meeting," *Denver Business Journal*, June 26, 2007.

65. Remarks of Donald Mares, Colorado Women's Legislative Breakfast, Feb. 12, 2008, accessed at www.youtube.com/watch?v=UIO0mlHb6b8&feature=related.

BIBLIOGRAPHY

Books

Cushman, Clare, *Supreme Court Decisions and Women's Rights, CQ Press*, 2000.
In clear prose, the director of publications for the Supreme Court Historical Society covers the waterfront of Supreme Court cases and issues involving women's rights, including those related to pay equity and discrimination in the workplace.

Farrell, Warren, *Why Men Earn More*, AMACOM, 2005.
The only man elected three times to the board of directors of the National Organization for Women's New York chapter argues that the pay gap can no longer be ascribed to discrimination, and he seeks "to give women ways of earning more rather than suing more."

Goldin, Claudia, *Understanding the Gender Gap: An Economic History of American Women*, Oxford University Press, 1990.
A Harvard University economics professor traces the evolution of female workers and gender differences in occupations and earnings from the early days of the republic to the modern era.

Murphy, Evelyn, with E.J. Graff, *Getting Even: Why Women Don't Get Paid Like Men — and What to Do About It*, Touchstone, 2005.
The former Massachusetts lieutenant governor writes in this anecdote-filled book that the "gender wage gap is unfair" and "it's not going away on its own."

Articles

Hymowitz, Carol, "On Diversity, America Isn't Putting Its Money Where Its Mouth Is," *The Wall Street Journal*, Feb. 25, 2008.
Progress for women and minorities in business has stalled or moved backward at many of the nation's largest companies, and the inequality shapes perceptions about who can or should fill leadership roles.

Murphy, Cait, "Obama flunks Econ 101," *Fortune*, CNNMoney.com, June 6, 2007, http://money.cnn.com/2007/06/04/magazines/fortune/muphy_payact.fortune/index.htm.

The presidential candidate is "flirting with a very bad idea" by co-sponsoring the Fair Pay Act, "a bill that would bureaucratize most of the labor market," Murphy argues.

Parloff, Roger, and Susan M. Kaufman, "The War Over Unconscious Bias," *Fortune*, Oct. 15, 2007.
Wal-Mart and other companies are facing accusations of gender pay bias and other forms of job discrimination, "but the biggest problem isn't their policies, it's their managers' unwitting preferences."

Reports and Studies

Dey, Judy Goldberg, and Catherine Hill, "Behind the Pay Gap," *American Association of University Women Educational Foundation*, April 2007, www.aauw.org/ research/upload/behindPayGap.pdf.
A study of college graduates concludes that one year out of college women working full time earn only 80 percent as much as their male colleagues and that a decade after graduation the proportion falls to 69 percent.

Foust-Cummings, Heather, Laura Sabattini and Nancy Carter, "Women in Technology: Maximizing Talent, Minimizing Barriers," *Catalyst*, 2008, www .catalyst.org/files/full/2008%20Women%20in%20 High%20Tech.pdf.
Technology companies are making progress at creating more diverse work environments, but women in the high-technology field still face barriers to advancement, such as a lack of role models, mentors and access to networks.

Hartmann, Heidi, Olga Sorokina and Erica Williams, *et al.*, "The Best and Worst State Economies for Women," *Institute for Women's Policy Research*, IWPR No. R334, December 2006, www.iwpr.org/ pdf/R334_BWStateEconomies2006.pdf.
The advocacy group concludes that women's wages have risen in all states since 1989 after adjusting for inflation, but that in "no state does the typical full-time woman worker earn as much as the typical man."

For More Information

Eagle Forum, PO Box 618, Alton, IL 62002; (618) 462-5415; www.eagleforum.org. Conservative social-policy organization opposed to ratification of the Equal Rights Amendment.

4ERA, 4355J Cobb Parkway, #233, Atlanta, GA 30339; (678) 793-6965; www.4era.org. Single-issue organization advocating ratification of the Equal Rights Amendment.

Institute for Women's Policy Research, 1707 L St., N.W., Suite 750, Washington, DC 20036; (202) 785-5100; www .iwpr.org. Research organization that focuses on gender pay as well as other issues affecting women, including poverty and education.

National Committee on Pay Equity, c/o AFT, 555 New Jersey Ave., N.W., Washington, DC 20001-2029; (703) 920-2010; www.pay-equity.org. Coalition of women's and civil rights organizations, labor unions, religious, professional,

legal and educational associations and others focused on pay-equity issues.

National Women's Law Center, 11 Dupont Circle, N.W., Suite 800, Washington, DC 20036; (202) 588-5180; www.nwlc.org. Advocacy group that focuses on employment, health, education and economic-security issues affecting women and girls.

9to5, National Association of Working Women, 207 E. Buffalo St., #211, Milwaukee, WI 53202; (414) 274-0925; www.9to5.org. Grassroots organization focusing on economic-justice issues for women.

U.S. Chamber of Commerce, 1615 H St., N.W., Washington, DC 20062-2000;.(202) 659-6000; www .uschamber.com. Represents business interests before Congress, government agencies and the courts.

Gay Marriage Showdowns

Will Voters Bar Marriage for Same-Sex Couples?

Kenneth Jost

Kate Sheppard and Kory O'Rourke celebrate with their children after obtaining a marriage license at San Francisco City Hall on June 17, 2008. A California Supreme Court ruling in May made California the second state, after Massachusetts, to legalize same-sex marriage. Opponents quickly gained approval to put a state constitutional amendment on the Nov. 4 ballot that would allow marriage in California only "between a man and a woman."

From *CQ Researcher*, September 26, 2008.

Jennifer Pizer and Doreena Wong met on their first day at New York University Law School in 1984. They graduated in 1987 and moved to California together three years later.

Jenny and Doreena were still together on May 15, 2008, when the California Supreme Court issued its stunning, 4-3 decision establishing a constitutional right to marriage for same-sex couples in the state. As one of the Lambda Legal Defense and Education Fund lawyers in the case, Pizer spoke at a press conference in San Francisco after the decision was released and then flew home to Los Angeles for a rally in the heart of gay West Hollywood.

"You're not going to do anything funny, are you?" Doreena asked Jenny in the car as they drove to the rally. Pizer feigned ignorance even as she was thinking that the event was the perfect time to pop "the question."

So, as she finished her remarks, Pizer looked down toward her partner's face in the crowd and said, "Now, I'd like to ask a question I've waited 24 years to ask: Doreena Wong, will you marry me?"

"Yes, of course," Wong replied. Standing at the microphone, Pizer relayed the answer to the cheering crowd: "She said yes!"

Television cameras recorded the moment, but Pizer admits months later that she has yet to see the full video clip. For even as gay rights advocates are celebrating the victory — and Jenny and Doreena are planning their Oct. 5 wedding in Marin County — opponents of gay marriage are working hard to reverse the state court's decision.

Less than three weeks after the decision, opponents won legal approval to put a state constitutional amendment on the Nov. 4

Most States Ban Gay Marriage

Voters in 26 states have approved constitutional amendments banning marriage for same-sex couples.* Most of the measures may also ban other forms of recognition, such as domestic partnership or civil unions; some may ban legal recognition for unmarried opposite-sex couples as well. Arizona, California and Florida will be voting on Nov. 4 on constitutional amendments to define marriage as the union of one man and one woman.

Seventeen other states have enacted statutory bans on same-sex marriage since 1995. Iowa's ban was ruled unconstitutional by a state trial court; the state's appeal is pending. In addition, pre-existing laws in Maryland, New York, Wyoming and the District of Columbia have been interpreted to limit marriage to opposite-sex couples.

California and Massachusetts are the only states recognizing same-sex marriage. Some states with gay-marriage bans recognize civil unions or domestic partnerships. Two states — New Mexico and Rhode Island — appear not to have addressed the issue.

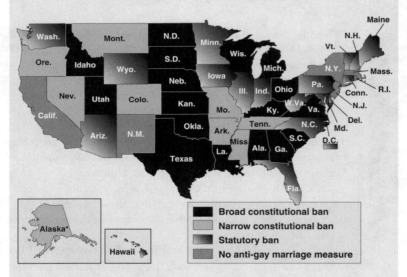

Broad constitutional ban
Narrow constitutional ban
Statutory ban
No anti-gay marriage measure

* *CQ Researcher* follows the National Gay and Lesbian Task Force in not counting Hawaii as a constitutional ban state, unlike Stateline.org and the Williams Institute. Hawaii adopted a constitutional amendment in 1998 authorizing the legislature to define marriage in opposite-sex terms, and the legislature did so. However, the point of the constitutional bans is to preclude change, and Hawaii's amendment does not.

Sources: National Gay and Lesbian Task Force; Williams Institute, UCLA School of Law; Stateline.org

recognition of same-sex marriages from other states as well.

"Marriage has always been understood as the union of one man and one woman by California citizens and by other people in the country," says Mathew Staver, founder and chairman of Liberty Counsel, a Christian public-interest law firm, and one of the lawyers who argued against gay marriage before the California Supreme Court. "That provides the best environment for society."

"We absolutely agree that marriage is a special word for a special institution," Pizer responds. "We disagree that the social institution should be available only in a discriminatory manner and that it serves any social purpose to exclude gay and lesbian couples."

The debate over the ballot measure has not deterred but in fact has encouraged gay and lesbian couples in California to get to the altar — or to city hall. By one estimate, some 5,000 same-sex couples got married in California within the first week after the court ruling became effective on June 17. The first-week spike receded, but the weddings are continuing — spurred by the widespread assumption that marriages performed before Nov. 4 will remain valid even if Proposition 8 is approved.

Hollywood celebrities have been among those tying the knot, including TV talk show host Ellen de Generes and ex-"Star Trek" actor George Takei. De Generes wed Portia de Rossi, her girlfriend of the past four years, in an intimate, picture-book ceremony at their Beverly Hills home on Aug. 16. Takei and his longtime partner Brad Altman exchanged self-written vows in a more lavish ceremony

ballot that would allow marriage in California only "between a man and a woman." If accepted by a simple majority of the state's voters, Proposition 8 would prohibit marriage for gay and lesbian couples in California and bar

at the Japanese American National Museum in downtown Los Angeles on Sept. 14. "May equality long live and prosper," Takei said as he left the ceremony amid a horde of photographers and well-wishers.[1]

Most of the newlyweds, however, are non-celebrities, many of them in long-term relationships that had already been registered under a 2003 California law as domestic partnerships with nearly complete marriage-like rights and responsibilities. "There's almost no change" over domestic partnership status, explains David Steinberg, news desk copy chief at the *San Francisco Chronicle*, who married his longtime partner Gregory Foley in July. Steinberg says he and Foley, a nurse at Kaiser Permanente, decided to get married anyway "because they might take it away."

The state high court decision made California the second state, after Massachusetts, to allow marriage for same-sex couples. The Supreme Judicial Court of Massachusetts issued a 4-3 decision in November 2003, holding that the state had "no constitutionally adequate reason" for denying same-sex couples the legal benefits of marriage. The court gave the legislature 180 days to respond but later issued an advisory opinion saying that civil union status would not be an adequate substitute for marriage. When the legislature failed to act by the deadline, the high court decision took effect, and same-sex marriages began in Massachusetts on May 17, 2004.[2]

The California Supreme Court ruled similarly but more directly that the state's constitution guarantees a "fundamental right to marry" to "all Californians, whether gay or heterosexual, and to same-sex couples as well as opposite-sex couples." The majority opinion — written by the Republican-appointed chief justice, Ronald George — specifically rejected civil union or domestic partnership status.[3]

Ten States, D.C. Recognize Same-Sex Unions

Ten states and the District of Columbia grant some legal recognition to same-sex couples, ranging from limited spousal rights in Hawaii to fully recognized marriages in California and Massachusetts. Hawaii was the first to recognize same-sex couples' rights in 1997. An estimated 85,000 same-sex couples have gained legal recognition under the various laws.

State	Date	Provisions
California	2008	Marriage approved by California Supreme Court; Proposition 8 on Nov. 4 ballot would overturn ruling
Connecticut	2005	Civil unions approved by legislature; marriage suit pending before Connecticut Supreme Court
District of Columbia	2002	Limited domestic partnership law enacted by D.C. Council in 1992; delayed by Congress until 2002
Hawaii	1997	"Reciprocal beneficiaries" (limited spousal rights)
Maine	2004	Limited domestic partnership law approved by legislature
Massachusetts	2004	Marriage legalized as required by November 2003 ruling by Supreme Judicial Court; constitutional amendment to overturn ruling failed to qualify for ballot
New Hampshire	2007	Civil unions approved by legislature
New Jersey	2006	Civil unions approved by legislature to comply with October 2006 ruling by New Jersey Supreme Court
Oregon	2007	Domestic partnership law approved by legislature
Vermont	2000	Civil unions approved by legislature following ruling by Vermont Supreme Court in December 1999
Washington	2007	Limited domestic partnership law approved by legislature; marriage suit rejected by state Supreme Court, July 2006

Sources: National Gay and Lesbian Task Force; Williams Institute, UCLA Law School

Getty Images/David McNew

An anti-gay protester demonstrates against same-sex marriage during the 38th annual LA Pride Parade on June 8, 2008, in West Hollywood, Calif. Constitutional amendments that would deny marriage rights to same-sex couples are on the ballot in Arizona and Florida, as well as California.

The ruling invalidated a statutory initiative to define marriage as between one man and one woman approved by slightly over 61 percent of the state's voters as Proposition 22 in March 2000. Gay marriage opponents had already begun circulating an initiative to write the "one-man, one-woman" definition of marriage into the state constitution. By June 2, they had submitted petitions with approximately 1.1 million signatures — sufficient for the secretary of state to certify the proposed constitutional amendment for the Nov. 4 ballot.

The state Supreme Court added to the urgency of the opposition by declining to stay its decision pending the Nov. 4 vote. Same-sex marriages began in California on June 17. The first marriage license in San Francisco went to two longtime lesbian activists, Del Martin and Phyllis Lyons, who had been together for more than 50 years. San Francisco Mayor Gavin Newsom officiated at the ceremony. Martin died 10 weeks later — at age 87.

Besides Massachusetts and California, eight other states and the District of Columbia permit some legal recognition for same-sex couples, including four that permit civil unions with virtually the same rights and responsibilities as marriage. (*See chart, p. 75.*) On the opposite side, 26 states have constitutional amendments that prohibit marriage for same-sex couples, and another 17 have similar statutory bans. In addition, the federal Defense of Marriage Act — known as DOMA — prohibits federal recognition for same-sex marriages. The 1996 law also provides that states need not recognize same-sex marriages from other states. (*See map, p. 74.*)

Massachusetts recorded approximately 11,000 same-sex marriages in the three years after the state high court ruling, according to demographer Gary Gates, a senior research fellow at the Williams Institute, UCLA School of Law. He says an exact count is not possible in California because marriage licenses are no longer recording the parties' sex, but a projection based on the increased number of marriages in the months after the state high court ruling indicates more than 5,000 same-sex couples married in the first week after the decision.

All told, Gates and his colleagues at the institute — which studies sexual-orientation policy and law, primarily funded by a gay philanthropist — estimate that 85,000 same-sex couples have taken advantage of recognition provisions in those states permitting that status. But a higher percentage of same-sex couples are opting to marry than are registering for civil union or domestic partnership.[4]

Supporters of marriage equality say the growing number of same-sex couples in legally protected relationships is eroding opposition to gay marriage. "We're seeing a growing public understanding that ending gay couples' exclusion from marriage helps families and harms no one," says Evan Wolfson, executive director of Freedom to Marry, self-described as a gay and non-gay partnership advocating marriage rights for same-sex couples.

Opponents disagree. They point to the gay marriage bans already enacted as the better gauge of public attitudes on the issue. "Supporters of same-sex marriage have a real uphill climb if they hope to undo what has been accomplished in the past 10 years by supporters of traditional marriage," says Peter Sprigg, vice president for policy at the Family Research Council, a Christian organization based in Washington, D.C., promoting traditional marriage.

An initial poll in California indicated the ballot measure was ahead, but statewide surveys in August and September showed the proposition trailing by at least 14 percentage points.[5] Two other states — Arizona and Florida — will be voting on similar constitutional amendments on Nov. 4. Arizona's measure needs a majority vote; Florida requires a 60 percent vote for a state constitutional amendment (*see p. 90*).

In addition to those three ballot measures, Arkansans will be voting on a statutory initiative to prohibit unmarried couples — whether same-sex or opposite-sex — to adopt or take foster children. The initiative was proposed after a regulation barring adoption or placement with same-sex couples was overturned in court.[6]

As the debates over same-sex marriage continue, here are some of the major questions being discussed:

Should same-sex couples be allowed to marry?

George Gates and Brian Albert met in 1991 when they both worked at the Human Rights Campaign, a gay rights advocacy organization. They held a big commitment ceremony at the posh Jefferson Hotel in Washington, D.C., in 1996 and, five years later, a small wedding on Cape Cod in Massachusetts.

With both of them still in Washington — now working for other nonprofit groups — Gates and Albert get no tangible benefits from their Massachusetts marriage license. But Gates says he and Albert viewed it as a matter of equal rights to take advantage of the Bay State's welcoming attitude toward gay couples. "We did feel that our relationship was no different from an opposite-sex couple," Gates says, "and we felt we were entitled to the same benefits and responsibilities as they are."

Opponents counter that gay marriage amounts to a redefinition of an institution created by God and universally understood in opposite-sex terms until the recent gay marriage movement. "Never before have we had such a serious effort to make such a profound change to the institution of marriage," says Lynn Wardle, a professor at Brigham Young University Law School, which is

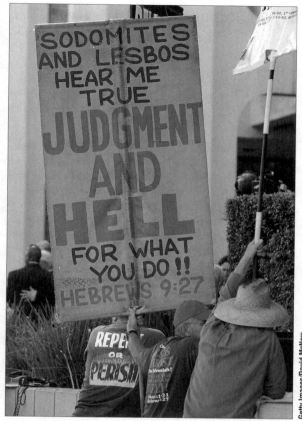

Getty Images/David McNew

Anti-gay religious protesters picket at the marriage ceremony of Robin Tyler and Diane Olson, in Beverly Hills, Calif., on June 17, 2008. The two women were plaintiffs in one of the lawsuits that led to the overturning of California's gay marriage ban in May.

operated by the Church of Jesus Christ of Latter-day Saints — the Mormons.

The opposing sides disagree more concretely about the effects that legal recognition of same-sex relationships has or would have on heterosexual marriage, on children and on gay men and lesbians themselves. Opponents say gay marriage will harm the institution of marriage, hurt children and have no significant effect for same-sex couples. Supporters of gay marriage say legal recognition will promote stable relationships for same-sex couples, benefit children in same-sex families and have no effect whatsoever on opposite-sex marriages.

Liberty Counsel's Staver, who is also dean of the School of Law at Liberty University, the Lynchburg, Va., school

Census Won't Recognize Same-Sex Marriages

"It really is something out of Orwell," a critic says.

Even though Massachusetts and California recognize same-sex marriages, the U.S. Census Bureau will not count gay or lesbian spouses as married in the 2010 census.

Census Bureau officials say the policy of treating married same-sex couples as unmarried partners is dictated by the federal Defense of Marriage Act, which bars the federal government from recognizing same-sex marriages for any purpose under federal law. The law "requires all federal agencies to recognize only opposite-sex marriages for the purpose of administering federal programs," Census spokesman Stephen Bruckner explained shortly after the policy was disclosed in July.[1]

The policy has drawn criticism from same-sex couples in both states and from gay rights advocacy groups. "To have the federal government disappear your marriage I'm sure will be painful and upsetting," Shannon Minter, legal director for the San Francisco-based National Center for Lesbian Rights, told the *San Jose Mercury News*, which first disclosed the policy. "It really is something out of [*1984* author George] Orwell."[2]

Demographer Gary Gates at the pro-gay marriage Williams Institute, UCLA School of Law, says the decision amounts to deliberately producing inaccurate population data. "Bureau officials should acknowledge the reality that same-sex couples can legally marry in this country," he says, "and stop altering the accurate responses of same-sex couples who describe themselves as married."[3]

Anti-gay marriage groups, however, defend the bureau's decision. "We're dealing with a government entity that is given certain charters and mandates, and they have to subscribe to public law," says Kris Mineau, president of the Massachusetts Family Institute.[4]

The bureau's decision will not affect the overall population count, which the Constitution requires every 10 years in order to apportion seats in the House of Representatives among the 50 states. But detailed information from the household questionnaires is used by the bureau and by independent researchers to provide demographic analyses in such areas as family structure and size, income and education.

The bureau says the questionnaires used in the 2010 census will not be destroyed, so the data will theoretically be available to independent researchers later. But the bureau's decision will slow a count of married same-sex couples in California, where marriage licenses now identify spouses only as "Party A" and "Party B."

---

[1] Quoted in Eric Moskowitz, "Federal rules mean thousands of same-sex marriages in Massachusetts will be ignored in the U.S. 2010 Census," *The Boston Globe*, July 27, 2008, p. B1.

[2] See Mike Swift, "U.S. Census Bureau won't count same-sex marriages," *San Jose Mercury News*, July 12, 2008.

[3] See Gary J. Gates, "Making same-sex marriages count," *Los Angeles Times*, July 18, 2008, p. A25.

[4] Quoted in Moskowitz, *op. cit.*

---

founded by the late televangelist Jerry Falwell, calls same-sex marriage "a huge, unknown sociological experiment done . . . with no understanding of the implications on our children and our society." Wardle and other opponents say recognition of same-sex marriages will contribute to a further decrease in the percentage of people who are married.

Wolfson of Freedom to Marry bluntly disagrees. Same-sex marriage "is not going to change anything" for heterosexual marriages, he says. Gay author Jonathan Rauch goes further to argue that recognizing gay marriage would strengthen the institution overall. "America needs more marriages, not fewer," Rauch wrote in a recent op-ed article, "and the best way to encourage marriage is to

encourage marriage, which is what society does by bringing gay couples inside the tent."[7]

Opponents also argue that heterosexual marriages are the best environment for raising children. "Same-sex marriage says as a matter of policy moms and dads are irrelevant to the raising of children," Staver says. In like vein, the pro-Proposition 8 Campaign for Families and Children says on its Web site, "From the commitment of a man and woman in marriage comes the best opportunity for children to thrive."

Gay and lesbian advocacy groups cite studies by, among others, the American Academy of Pediatrics to argue that children raised in families with gay or lesbian parents fare as well overall as children raised in opposite-sex households.[8] "All the evidence and common sense arguments indicate that this will help children who are being raised by gay families without hurting other children at all," says Wolfson.

Opponents also say, some more bluntly than others, that same-sex couples are not entitled to marriage because so many couples — particularly gay men — have short-lived, sometimes non-monogamous relationships. "Whether we like it or not, a big part of the gay agenda for decades has been to repudiate what are regarded as overly restrictive expectations of monogamy and sexual fidelity," University of Pennsylvania Law School Professor Amy Wax wrote in an online debate sponsored by the Washington-based Federalist Society, a prominent conservative organization for lawyers.[9]

Gay marriage advocates counter that allowing marriage for same-sex couples would actually help stabilize their relationships. "Marriage advocates argue that marriage provides a mechanism and incentive to form more stable unions," Williams Institute researcher Gates says. "If that's true, then you would expect the same effect among gays and lesbians."

Apart from the individual points of disagreement, Wardle insists that supporters of traditional marriage should not be forced to prove the case against gay marriage. "When you have a proposal to redefine a basic social institution, the burden of proof is on those who advocate a change," Wardle says.

"Both sides agree that marriage is a powerful institution," Gates rejoins. "Opponents make the argument that we have to be so cautious. Proponents say that's exactly why this is important. This is an important social institution, and you're leaving gay people out of it."

Should state constitutions prohibit marriage for same-sex couples?

The anti-gay organization Focus on the Family dispatched its vice president for public affairs, Ron Prentice, to California in 2003 to launch and become executive director of the affiliated California Family Council. Now, as chairman of the Protect Marriage/Yes on Proposition 8 campaign, Prentice is helping lead the effort to overturn the California Supreme Court's gay marriage ruling by amending the state constitution to define marriage as only "the union of one man and one woman." "We are going to change the constitution and say on Nov. 4, 'Judges, you can't touch this,'" Prentice says.[10]

Gay marriage opponents have enjoyed great success with the strategy. Constitutional amendments limiting marriage to opposite-sex couples have been approved by voters in 26 states, which together represent about 43 percent of the U.S. population. Hawaii voters approved an amendment in 1998 authorizing the legislature to limit marriage to opposite-sex couples. Only once — in Arizona in 2006 — have voters rejected an anti-gay marriage amendment.

Supporters say Proposition 22 represents a legitimate political response to the state high court ruling. "A victory in California will not only protect marriage," the Alliance for Marriage, in Merrifield, Va., says on its Web site, "but will send a strong democratic rebuke by voters to radical, activist groups who've used the courts" to try to gain recognition for same-sex couples.

Gay marriage supporters, however, say the tactic is antithetical to American democracy. "The whole idea of amending constitutions to fence out groups of people is yet another debasement of American fundamentals," says Freedom to Marry's Wolfson. "That is a radical idea: the idea that you amend constitutions to carve out a group of people, shove them outside, and say they can't go to the legislature, that they are permanently treated as second class by the constitution where they live."

Overall, state courts have been responsible for the most dramatic gains realized so far by advocates of legal recognition of same-sex relationships. The Vermont Supreme Court in 1999 became the first state high court to require marriage-like rights for same-sex couples; the state legislature enacted a civil union law five months later. The Supreme Judicial Court of Massachusetts effectively required recognition of gay marriage with a November 2003 ruling that took effect six months later. Opponents of the Massachusetts ruling have tried but failed to get the state legislature to put a constitutional amendment before the voters to overturn the decision.

Gay marriage opponents say the California Supreme Court invited retaliation with a decision that not only nullified the 2000 ballot measure but also used the state constitution's equal-protection clause to require the highest level of scrutiny for any laws discriminating on the basis of sexual orientation. Wardle calls the ruling "a very clear example" of judges "openly using their position to promote their political preference."

Gay marriage supporters had failed to match their victory in Massachusetts until the California ruling — suffering defeats in closely watched cases in New York and Washington. "To their credit, a number of state supreme courts are behaving more judiciously," Wardle says. But gay marriage supporters are hoping the California ruling may influence supreme court justices in two other states — Connecticut and Iowa — with pending marriage cases. "The California Supreme Court is recognized as by far the most influential state high court in the country," says Lambda Legal's Pizer.

In California, opponents of Proposition 8 won a significant tactical victory with the decision by Attorney General Jerry Brown to list the measure's title on the ballot as, "Eliminates Right of Same-sex Couples to Marry." Prop. 8 supporters tried but failed to get a state court judge to order a change in what they called "an inherently argumentative" title.

Prop. 8 opponents are using the title to frame their campaign message. "We think it's always wrong to be voting on taking away people's rights," says Dale Kelly Bankhead, statewide campaign manager for Equality California/No on Proposition 8.

In a later skirmish, Prop. 8 opponents tried but failed to block the initiative from the ballot. In a petition to the state Supreme Court, they argued that the measure amounted to a "revision" of the state constitution that — under the constitution — could not be put on the ballot without a two-thirds vote of the legislature. Prop. 8 supporters called the lawsuit a "desperate" effort to avoid a vote. The court unanimously declined to hear the request, but the issue could be revived if the measure passes in November.

Should states recognize same-sex marriages from other states?

A one-page legal memorandum that New York Gov. David A. Paterson's legal counsel David Nocenti sent to state agency directors on May 14 quietly handed gay marriage supporters a major victory. Following up a ruling by a state appellate court in February, Nocenti directed state agencies to recognize same-sex marriages from other jurisdictions — a list that then included Massachusetts and five countries: Belgium, Canada, the Netherlands, South Africa and Spain.*

---

* The Norwegian Parliament completed approval of a gay marriage law on June 17; the law will take effect on Jan. 1, 2009.

Nocenti made no announcement of the directive — issued, by coincidence one day before the California Supreme Court's ruling. But three days later Paterson disclosed the move in a videotaped message to the annual dinner of the Empire State Pride Agenda, a gay rights advocacy group. Paterson, who had supported unsuccessful bills in the state legislature to legalize same-sex marriage, called the directive "a strong step to marriage equality."[11]

The possibility that states would either choose or be required to recognize same-sex marriages from other states has been a major concern of gay marriage opponents ever since a Hawaii court's preliminary approval of an ultimately unsuccessful gay marriage suit in 1993. Gay marriage opponents included a provision in the federal Defense of Marriage Act (DOMA) in 1996 strengthening states' discretion to refuse to recognize same-sex marriages. At the same time, they began building a firewall against same-sex marriage by pushing for gay marriage bans in individual states.

The United States' federal system leaves marriage laws generally to states — with the inevitable consequence of differences from state to state. For example, some states permit and others prohibit marriages between first cousins.

Northwestern University law Professor Andrew Koppelman, an expert in an area known as "conflict of laws," says state courts over time have developed some general rules for when to recognize out-of-state marriages that would not be valid within their own state. In general, Koppelman says, states recognize marriages for people who travel through or move to a state with laws otherwise precluding legal status for the union. But states will not recognize a marriage for residents who go to another state to circumvent the state's law — especially if the law reflects a strong public policy.

Gay marriage opponents say the gay marriage bans fit that situation. "You don't have to recognize that status," Brigham Young Professor Wardle says, referring to a same-sex marriage from another state. "It's up to the state to choose for itself what domestic status it will recognize."

Koppelman — who supports same-sex marriage — says, however, that the state bans are "badly drafted" and ignore the real-world situations that will inevitably arise as same-sex couples travel or move from the state where their marriage was performed. "A blanket non-recognition rule is absolutely loony," Koppelman says.

Courts in two non-gay marriage states have already bowed to states that grant legal recognition to same-sex couples. In June, the Virginia Supreme Court ruled that Vermont rather than Virginia courts have jurisdiction over a custody dispute between two former lesbian partners following the dissolution of their Vermont civil union. Lisa Miller, who gave birth to a daughter during the civil union and moved to Virginia after the dissolution, had sought to block visitation rights that a Vermont court had granted to her former partner, Janet Jenkins. In an earlier decision, the federal appeals court for Oklahoma invalidated a state law refusing to recognize an out-of-state, court-approved adoption by a same-sex couple.[12]

Both courts said that the Constitution's "Full Faith and Credit Clause" required the state court to recognize court judgments from other states. Koppelman notes that despite widespread misunderstanding, the constitutional provision does not apply to the more common instances that do not involve litigation already in progress.

The California gay marriage ruling heightened the stakes for both sides because the state has no residency requirement to be married. Massachusetts had been enforcing a 1913 law that barred marriages for out-of-state residents if the union would not be recognized in their home states. But the state repealed the law in August. As a result, businesses in Massachusetts and California are now actively encouraging same-sex couples to come to their states to be wed. (*See sidebar, p. 86*.)

Gay marriage opponents still maintain that states can enforce bans against recognizing same-sex unions from other states. "It's contrary to the strong public policy in those states," says the Family Research Council's Sprigg.

Koppelman disagrees. "Same-sex marriage ought to be, as a general matter, recognized," he says. But gay marriage supporters are sufficiently concerned about their prospects that they are urging same-sex couples not to initiate legal challenges to the state bans at this time.

BACKGROUND

Coming Out

The history of same-sex relationships is long, but the issue of legally recognizing those relationships is of recent origin.

Up until the mid-20th century, gay and lesbian couples in the United States generally kept a low profile politically and even socially. An outbreak of repressive laws and policies dating from the 1920s helped give rise to a gay rights movement and by the 1970s to a self-identified gay and lesbian community. Marriage, however, was not a priority or even a widely agreed on goal until the AIDS epidemic and the so-called lesbian baby boom of the 1980s prompted many gay men and lesbians to view legal recognition of their relationships as a practical necessity.[13]

Male couples and female couples can be found in history and literature from ancient times to the present. In the United States, same-sex couples formed part of the gay subcultures present but only somewhat visible in many major cities from the turn of the 20th century. Gay and lesbian couples generally drew as little attention to themselves as possible. As one example, the 1993 book *Jeb and Dash* recounts through posthumously published diaries the secret love affair between two government employees in Washington from 1918 to 1945.[14]

The federal government and many state and local governments began cracking down on homosexuals during the period between the two world wars. "Sexual perverts" were barred from entering the country and were made subject to exclusion from the military. Disorderly conduct and anti-sodomy laws were used to break up gay organizations and arrest individuals looking for or engaging in gay sex.

After the repeal of Prohibition in 1933, gay bars could still be shut down through license suspensions or revocations. The repressive atmosphere increased after World War II as homosexuals came to face the same kind of persecution as suspected communists. The historian David K. Johnson suggests that more federal employees lost jobs because of suspected homosexuality during what he terms "the lavender scare" than were dismissed because of suspected communist leanings.[15]

Threatened in their workplaces and gathering places, gay men and lesbians in the 1950s formed the forerunners of the present-day gay rights movement. The gay Mattachine Society and the Daughters of Bilitis both adopted assimilationist stances: no garish costumes, no lavish parades. In 1965, however, a fired government astronomer, Franklin Kameny, staged the first "gay rights" picketing outside the White House, aimed at reversing policies generally barring homosexual from federal employment. Then in

CHRONOLOGY

Before 1970 *Gay rights movement begins to form; same-sex marriage low on agenda.*

1968, 1969 Metropolitan Community Church in Los Angeles performs first public weddings for same-sex couples.

1970s-1980s *Gay rights measures enacted in some states, localities; AIDS epidemic, "lesbian baby boom" spur interest in legal recognition for relationships.*

1971 Marriage rights suits filed by gay Minnesota couple and lesbian couple in Kentucky are rejected; over next six years, 15 states pass laws defining marriage as opposite-sex union.

1984 Berkeley, Calif., becomes first city to provide domestic-partner benefits to employees.

1986 U.S. Supreme Court upholds state anti-sodomy laws.

1990s *Gay marriage rulings spur backlash in Congress, states; Vermont court is first to require state to give legal recognition for same-sex couples.*

1993 Hawaii Supreme Court requires state to justify ban on same-sex marriage; trial court rules ban unconstitutional in 1996, but ruling is nullified by state constitutional amendment approved in 1998.

1996 Congress passes and President Bill Clinton signs Defense of Marriage Act, which bars federal benefits for same-sex couples and buttresses states' authority not to recognize same-sex marriages from other states.

1998 Alaska trial court rules ban on same-sex marriage unconstitutional; ruling nullified by constitutional amendment approved by voters in November.

1999 California passes limited domestic-partnership law for same-sex couples; rights under law expanded in 2001, 2003. . . . Vermont Supreme Court rules state must allow same-sex couples to enjoy legal benefits accorded to heterosexuals; state legislature implements ruling by passing civil-unions law in April 2000.

2000-Present *Massachusetts recognizes marriage for same-sex couples; after setbacks in several states, gay rights supporters win second pro-marriage ruling in California; opponents qualify ballot measure to overturn decision.*

2003 U.S. Supreme Court rules anti-sodomy laws unconstitutional; majority opinion does not address gay marriage. . . . Supreme Judicial Court of Massachusetts rules same-sex couples entitled to same rights as opposite-sex couples; gives legislature 180 days to act.

2004 Voters in 13 states pass gay marriage bans, all by substantial margins. . . . Federal Marriage Amendment fails in Senate (and again in 2006).

2005 Connecticut legislature passes civil-union law — first state to act without court mandate.

2006 State high courts in New York and Washington uphold laws limiting marriage to opposite-sex couples. . . . New Jersey Supreme Court rules same-sex couples entitled to same benefits, protections as opposite-sex couples, with legislature to choose between "marriage" or "civil unions"; legislature approves civil-union bill two months later.

2007 Connecticut Supreme Court hears arguments in gay-marriage case; decision still pending in fall 2008. . . . Judge in Iowa rules gay-marriage ban unconstitutional; same-sex couple weds before decision is stayed pending appeal to state Supreme Court.

2008 California Supreme Court says same-sex couples entitled to marriage, anti-gay laws presumptively unconstitutional (May 15); ruling goes into effect one month later after justices decline request for stay. . . . Gay-marriage opponents qualify Proposition 8 for Nov. 4 ballot; measure would define marriage as union of one man, one woman; bar recognition of same-sex marriages from other states. . . . Same-sex marriage bans also on ballot in Arizona, Florida; Arkansas to vote on banning adoptions, foster-child placements with unmarried couples.

1969 the gay patrons of the Stonewall Inn in New York City rose up in protest after a police raid on the Greenwich Village bar. The disturbance attracted little attention in the straight world but quickly became a rallying point for a newly assertive gay and lesbian community.

Marriage was not high on the community's agenda, however.[16] Many other issues were more pressing: pushing for gay rights ordinances, fighting employment bans and seeking to repeal anti-sodomy laws. In any event, many gay and lesbian activists actively opposed marriage, as Yale University historian George Chauncey recounts. Gay liberation celebrated sexual freedom, not committed relationships. And many lesbians viewed marriage as an inherently patriarchal institution to be reformed (or even abolished), but certainly not to be imitated.

The activists' views should not be overemphasized, Chauncey cautions. "Most lesbians and gay men across the country looked for a steady relationship," he writes. Indeed, the Metropolitan Community Church, a gay congregation formed in 1968 in Los Angeles, began blessing same-sex unions at its creation and performed 150 marriages in its first four years. In addition, same-sex couples in Minneapolis and Louisville, Ky., filed lawsuits in 1971 seeking to win the right to marry. Courts in both cases said marriage was only for opposite-sex couples, even though the state laws did not say so. To fill in the gap, 15 states passed laws from 1973 to 1977 limiting marriage to heterosexual couples.

The AIDS epidemic brought gay men face to face with the consequences of legally unrecognized relationships. The illness or death of a "longtime companion" became even more painful when hospitals, funeral homes or government agencies refused to give any regard to the relationship. Medical costs and medical decision-making were difficult issues as long as the patient lived; at death, many survivors had bitter conflicts with their deceased lover's "real" family over funeral arrangements and disposition of property.[17]

Meanwhile, the growing interest in childrearing also focused attention on the disadvantages of legally unrecognized relationships. Gay men and lesbians who had children from previous opposite-sex marriages typically faced difficulties in winning custody or sometimes even visitation rights. As historian Chauncey explains, the lesbian baby boom of the 1980s "represented something new: a generation of women who . . . no longer felt obliged

to marry a man in order to have a child." A biological mother's relationship to her child was not legally difficult, but her partner could gain a legal relationship only through a cumbersome second-parent adoption. Moreover, couples who split up had no assurance that courts would respect or enforce agreed-on custody and visitation rights.

Debating Marriage

Marriage gradually moved toward the top of the gay rights agenda in the 1990s as dissenting views within the GLBT (gay, lesbian, bisexual, transgender) community were either transformed or suppressed. An initial victory in Hawaii, however, resulted in a major setback with congressional passage of the federal Defense of Marriage Act in 1996. DOMA limited federal status to opposite-sex couples and buttressed states' prerogatives to refuse to recognize same-sex marriages from other states. Gay rights advocates' later successes in winning civil unions in Vermont in 1999 and marriage in Massachusetts in 2003 were offset by losses in other state courts and a new flurry of so-called mini-DOMAs approved by voters in the 2004 election cycle.[18]

For gay rights advocates, Hawaii ended as a ballot-box defeat after a potential judicial victory. In 1993, the Hawaii Supreme Court held the state's ban on same-sex marriage presumptively unconstitutional and ordered a trial for the state to try to show a compelling interest to justify the restriction. The trial opened in Honolulu just as the Senate was about to complete action on DOMA in Washington. The judge ruled for the gay couples who brought the suit, but the state high court kept the appeal under advisement long enough for voters to approve a state constitutional amendment in 1998 that authorized the legislature to limit marriage to opposite-sex couples. The next year, the state Supreme Court dismissed the suit.

In Washington, Republican lawmakers cited the Hawaii Supreme Court's 1993 ruling as the motivation for the bills introduced in May 1996 that led to DOMA's enactment four months later. The bills provided that no state was obligated to recognize a same-sex marriage from another state. In a second section, the measures defined "marriage" and "spouse" in opposite-sex terms for federal law, thus precluding same-sex couples from filing joint tax returns or qualifying for any federal marital or spousal benefits. Opponents said the federal provision was discriminatory, the state law provision either unconstitutional or unnecessary. But the Republican-controlled

McCain and Obama Diverge Over Legal Recognition

But both oppose same-sex marriage.

Democrat Barack Obama and Republican John McCain both oppose marriage for same-sex couples. But the two presidential nominees diverge significantly on a secondary question about gay and lesbian relationships: Should they receive legal recognition?

Obama favors civil unions that would give same-sex couples all of the legal rights of marriage. The Illinois senator also wants to repeal the federal Defense of Marriage Act (DOMA), which defines marriage as the union of a man and a woman. And he displays both positions along with other gay-rights stances on a full page devoted to GLBT (gay, lesbian, bisexual and transgender) issues on his campaign Web site (www.obama.com).

Sen. John McCain voted for the Defense of Marriage Act in 1996.

announcing their positions in late June.[1]

Unsurprisingly, major GLBT advocacy organizations are supporting Obama in the Nov. 4 presidential balloting. "Sen. Obama has consistently shown that he understands, as we do, that, GLBT rights are civil rights and human rights," Human Rights Campaign President Joe Solmonese said in formally endorsing the Democratic ticket on June 6.

But the Log Cabin Republicans, a gay GOP group, is backing McCain — four years after withholding its endorsement from President Bush. In announcing the endorsement on Sept. 2, President Patrick Sammon pointed to McCain's two Senate votes in 2004 and 2006 oppos-

McCain says same-sex couples should be allowed to establish some rights through "legal agreements," but he appears to oppose civil unions. The Arizona senator voted for DOMA in 1996; the Republican Party platform opposes repealing it. McCain's campaign Web site has no GLBT page; the only tacit references to GLBT issues endorse the one-man, one-woman definition of marriage and oppose "activist" judges (www.mccain.com).

The two candidates also differ on the Nov. 4 ballot proposition in California to overturn the state Supreme Court's decision granting full marriage rights to same-sex couples. McCain favors the measure; Obama opposes it. Both candidates called little attention to statements

ing the proposed constitutional amendment to bar recognition of same-sex marriages by the federal or state governments.

Obama also voted against the amendment and restates his opposition on his Web site. McCain does not mention the amendment on his site.

McCain says on his Web site that only the definition of marriage as the union of one man and one woman "sufficiently recognizes the vital and unique role played by mothers and fathers in the raising of children, and the role of the family in shaping, stabilizing and strengthening communities and our nation." He has been less clear in his position on civil unions.

Congress approved the measure by wide margins: 342-67 in the House, 85-14 in the Senate. President Bill Clinton endorsed the bill as it moved through Congress and then quietly signed it on Sept. 21.[19]

State legislatures followed suit by approving statutes or submitting for voter approval constitutional amendments similarly aimed at precluding legal recognition for same-sex couples. As in Hawaii, a state constitutional amendment

Campaigning in New Hampshire in March 2007, McCain said he opposed the civil union legislation recently enacted in the state. "Anything that impinges or impacts the sanctity of the marriage between men and women, I'm opposed to it," McCain was quoted as saying in a conference call with several political bloggers.[2] Appearing on the "Ellen de Generes Show" in May 2008, however, McCain said "people should be able to enter into legal agreements" and should be encouraged to do so.[3]

Obama professes strong support for civil unions for same-sex couples. "Barack Obama supports full civil unions that give same-sex couples equal legal rights and privileges as married couples," the campaign Web site states. The entry goes on to call for repealing DOMA and providing federal rights and benefits to same-sex couples in civil unions or other legally recognized unions.

Obama has been somewhat reticent, however, on same-sex marriage. "My religious beliefs say that marriage is something sanctified between a man and a woman," Obama was quoted as saying during his 2004 Senate campaign in Illinois.[4] In the presidential campaign, however, he has generally answered questions about gay marriage only indirectly by explaining his support for civil unions — as can be seen in an undated CNN video clip from a campaign town hall meeting in Durham, N.H.[5]

On his Web site, Obama also calls for adoption rights for "all couples and individuals, regardless of sexual orientation." On his site, McCain — father of an adoptive child — calls for promoting adoption as "a first option" for crisis pregnancies. But the site makes no reference to McCain's statement in a newspaper interview opposing adoption by gay couples or individuals. Under criticism, McCain modified the statement the next day to say that adoption is a state issue.[6]

Sen. Barack Obama wants to repeal the Defense of Marriage Act.

Longtime gay-rights advocate Winnie Stachelberg says the contrast between the two candidates on GLBT issues "could not be more clear." Stachelberg, a senior vice president at the Center for American Progress, a Democratic think tank, says Obama would promote gay and lesbian equality if elected president. She complains that McCain has "studiously" avoided reaching out to the GLBT community.

From the opposite side, Family Research Council policy Vice President Peter Sprigg voices satisfaction with the Republican platform's support for "traditional marriage," while acknowledging ambivalence about McCain's votes against the Federal Marriage Amendment, a proposed constitutional amendment that would define marriage as a union of one man and one woman. But he complains that Obama "is unwilling to support any kind of actions that would defend the traditional definition of marriage. I kind of think he's playing word games in saying that he does not support same-sex marriage."

---

[1] See Michael Finnegan and Cathleen Decker, "Quiet stands on gay marriage," *Los Angeles Times*, July 2, 2008, p. A12.

[2] Ryan Sanger, "Exclusive: John McCain Comes Out Against NH Civil Unions," *New York Sun*, April 27, 2007, www.nysun.com/blogs.

[3] Quoted in Jim Brown, "Is McCain for civil unions?" *OneNewsNow*, May 28, 2008, www.onenewsnow.com/elections. OneNewsNow is a service of the American Family News Network, a Christian news service.

[4] Quoted in Eric Zorn, "Change of subject," *Chicago Tribune*, March 25, 2007, p. C2.

[5] CNN, "Election 2008: GLBT Issues," www.cnn.com/ELECTION/2008/issues/issues.samesexmarriage.html.

[6] See Michael Cooper, "Facing Criticism, McCain Clarifies His Statement on Gay Adoption," *The New York Times*, July 16, 2008, p. A15.

---

approved by Alaska voters in 1998 wiped out a trial court's ruling tentatively backing gay marriage. Gay rights lawyers, however, scored significant victories with cases in two New England states: Vermont and Massachusetts.

The Vermont Supreme Court's ruling in December 1999 held the denial of marital benefits to same-sex couples to violate the state constitution's equal protection provisions and ordered the state legislature to remedy the

Will Gay Weddings Bring Economic Boom?

California and Massachusetts are not cashing in yet.

Same-sex couples from other states who travel to California and Massachusetts to get marriage licenses may spark a modest economic boom in the two states.

The pro-gay marriage Williams Institute at UCLA School of Law forecasts $64 million in added revenue for state and local governments in California from out-of-state couples coming to wed. A similar study for Massachusetts — prepared this summer as the state was about to repeal a law limiting marriage for out-of-state couples — projects a $5 million revenue gain over three years.[1]

Gay-marriage opponents discount the studies. "Those claims are highly suspect, particularly since the only study was done by a blatantly pro-homosexual think tank," says Kris Menau, president of the Massachusetts Family Institute, the major advocacy group working against gay marriage in the Bay State. "We see no evidence of a great migration by out-of-state homosexual couples to come here to marry."

Anecdotal evidence is ambiguous. A justice of the peace in the gay mecca of Provincetown, Mass., told *The Washington Post* she had to use scheduled vacation days in August to perform weddings for out-of-state couples. "I have a full-time job, and this has become a full-time job," Rachel Peters said.[2]

In California, however, the head of a nationwide trade association for wedding professionals says the predicted boomlet has yet to materialize. "The goal was to pull several hundred thousand from other states to come here and get married," says Richard Markel, president of the Sacramento-based Association for Wedding Professionals International. "I haven't seen it totally yet."

Five months after Massachusetts became the first state to extend marriage to same-sex couples, the respected business magazine *Forbes* forecast that gay weddings could mean an additional $16.8 billion for the nation's $70-billion-a-year wedding industry. A wedding industry newsletter cited in the Williams Institute studies puts the average cost of a wedding in the United States today at $30,000.[3]

For its calculations, the Williams Institute used a more conservative figure of about $3,000 per wedding — assuming that out-of-state couples would be somewhat budget-strapped on their celebrations. But after adding in anticipated tourist spending, the institute predicted $111 million in added spending in Massachusetts from more than 30,000 out-of-state couples over the next three years. In California, the institute predicted that 51,000 California couples and 67,000 out-of-staters would spend $638.6 million.

Whatever the exact figure proves to be, gay marriage has been a definite boon for the lesbian couple who founded the Rainbow Wedding Network in 2002. Co-founder Cindy Sproul says she returned to her home state of North Carolina in August with a good-sized boost for the $4.5-million-a-year business after hosting

inequality. The legislature responded in April 2000 with a law creating the marriage-like "civil union" status for same-sex couples. The law took effect July 1, 2000, and by 2008 an estimated 1,485 same-sex civil unions had been registered in the state.

In Massachusetts, lawyers from the Boston-based Gay and Lesbian Advocates and Defenders filed suit in April 2001 on behalf of seven same-sex couples who had been together for periods ranging from three to 30 years. The trial judge rejected the suit the next year, but in November 2003 the Supreme Judicial Court of Massachusetts issued its epochal, 4-3 decision mandating legal recognition for same-sex couples. The ruling gave the state legislature a 180-day deadline to comply. The first marriages were performed on May 17, 2004.

The victory in Massachusetts, however, proved costly for gay rights advocates by re-energizing opponents of gay marriage, who qualified ballot measures in 13 states in 2004 aimed at banning marriage for same-sex couples. Voters approved all 13: two in early voting in September and 11 more in November. With more than 20 million voters casting ballots, the measures triumphed overall by a better than 2-1 margin. Gay rights advocates had looked to Oregon as their only realistic chance of stemming the tide, but the gay marriage ban prevailed there with 57 percent of the vote.

four wedding expos in California in July.

Sproul and Markel both say gay weddings are similar to opposite-sex weddings. Most gay weddings are performed in places of worship and officiated by clergy, Sproul says, though same-sex couples are somewhat more likely to write their own vows than opposite-sex couples. Same-sex couples — typically older than opposite-sex couples — are also more likely to be paying for weddings themselves rather than their parents.

Sproul says most of the companies that advertise through the network are straight-owned, and the owners have no problems with serving gay ceremonies. Markel agrees. "A majority of the people in the business got into the business because they enjoy celebrations, and they enjoy helping people," he says.

Both Sproul and Markel point to some exceptions, however. "We've had some very hostile e-mails and death threats," says Sproul. Markel quotes one photographer as saying he would shoot a lesbian wedding but — using an epithet — questioned whether he would photograph two men getting married.[4]

The Williams Institute predicted that New York would be the major source of out-of-state couples for Massachusetts.

Same-sex wedding cake figurines will be in demand if the gay-wedding business grows as expected in California.

For California, the institute forecast influxes from New York, Texas, North Carolina and the nearby Pacific Coast and South western states. But among those who already traveled to California to tie the knot was Sproul and her partner Marianne Puechl, who got married on July 22 on a Malibu beach.

The newlyweds flew home the next day to North Carolina. The state enacted a statutory ban on recognizing same-sex marriages in 1996. "We hope [the marriage] will be recognized some day," Sproul says.

[1] See Brad Sears and M.V. Lee Badgett, "The Impact of Extending Marriage to Same-Sex Couples on the California Budget," Williams Institute, June 2008, www.law.ucla.edu/WilliamsInstitute/publications/EconImpactCAMarriage.pdf. The Massachusetts study is referenced in Keith B. Richburg, "A Milestone for Gays, A Boon for Massachusetts," *The Washington Post*, Sept. 3, 2008, p. A3. The Charles R. Williams Project on Sexual Orientation and the Law was established at UCLA in 2001 after a $2.5 million contribution from Williams, a gay businessman and philanthropist.

[2] Quoted in Richburg, *op. cit.*

[3] Aude Lagorce, "The Gay-Marriage Windfall: $16.8 Million," *Forbes*, April 5, 2004. See Sears and Badgett, *op. cit.*, p. 7.

[4] See My Thuan Tran, "Gay Weddings Not Quite a Piece of Cake," *Los Angeles Times*, June 21, 2008, p. B1.

The battles continued. Gay marriage advocates suffered two big disappointments in July 2006 when the highest state courts in New York and Washington both narrowly rejected suits seeking to require marriage equality for same-sex couples. The 4-2 ruling in New York and the 4-3 decision in Washington both said the issue was for state legislatures to decide. In the same month, the Georgia Supreme Court and the federal appeals court for Nebraska reinstated constitutional amendments banning same-sex marriages that lower courts had ruled invalid.

Meanwhile, however, gay rights supporters had scored legislative victories in some states. The California legislature passed a domestic partnership law in 2003 giving same-sex couples virtually all the rights of marriage. Connecticut passed a civil union law in 2005 — the first state to do so without a court mandate. By the end of 2007, same-sex couples had marriage-like status available in three other states: civil unions in New Jersey and New Hampshire and domestic partnerships in Oregon.

California Showdown

Supporters and opponents of legal recognition for same-sex couples have waged virtually nonstop battles against each other in California for nearly a decade. Opponents won the first round in 2000 with the voter-approved Proposition 22 defining marriage in opposite-sex terms.

Supporters won the next round in 2003 with enactment of a domestic partnership law granting all marriage rights allowed under state law. Two gay marriage bills were passed by the legislature but vetoed by Republican Gov. Arnold Schwarzenegger, while the landmark gay marriage case moved toward the state Supreme Court. Instead of settling the issue, the court's May 15 ruling only set the stage for another ballot-box showdown.

California voters' approval of the state Defense of Marriage Act in the March 7, 2000, election followed a fractious campaign that cost both sides together more than $16 million. The late state Sen. William "Pete" Knight, a Republican from Los Angeles County's high desert and father of an estranged gay son, drafted the 14-word initiative. Roman Catholic and Mormon churches did much of the legwork supporting the initiative, which carried with 61.4 percent of the vote. "California is not ready for a marriage between a man and a man," Knight told supporters on election night. Gay rights advocates vowed to regroup. "We're stronger and more galvanized than ever before," said Gwen Baldwin, executive director of the Los Angeles Gay & Lesbian Center.[20]

The setback came six months after California had become the first state to provide domestic partner status for same-sex couples without court intervention. The bill that Democratic Gov. Gray Davis signed into law in September 1999 was limited; it provided hospital visitation rights and, for public employees, health insurance coverage for partners. With Davis in office, the Democratic-controlled legislature significantly expanded the rights of domestic partners in 2001 and again two years later. The California Domestic Partner Rights and Responsibilities Act of 2003 essentially gave state-registered domestic partners all of the rights, benefits and duties of marital spouses recognized by state law. Davis signed the bill on Sept. 22, 2003, before a huge and appreciative crowd at San Francisco's GLBT center in the Castro district. Knight said the law circumvented Proposition 22.

Barely six weeks after signing the bill, Davis was recalled by California voters — who blamed him for a variety of economic problems — and replaced by Republican Schwarzenegger. The change left gay rights groups with a gay-friendly governor from a gay-unfriendly party. Twice — in 2005 and again in 2007 — the Democratic-controlled legislature passed same-sex marriage bills, each time by bare majorities on party-line votes. Schwarzenegger vetoed both bills, saying they amounted to end-runs around the 2000 ballot initiative. "The governor believes the matter should be determined not by legislative action — which would be unconstitutional — but by court decision or another vote of the people of our state," Schwarzenegger's press secretary, Margita Thompson, explained after the first veto in September 2005.[21]

In the meantime, Democratic San Francisco Mayor Newsom had tried to take matters into his own hands in February 2004 by directing the county clerk's office to issue marriage licenses to same-sex couples on request. About 4,000 such licenses were issued over the next month before the California Supreme Court ordered a halt. On Aug. 12, the court voided the same-sex marriages that had been performed. The city-county government then joined with half a dozen same-sex couples in seeking to invalidate Proposition 22 and win a court ruling to permit gay marriage.

The gay marriage plaintiffs won an initial ruling from a San Francisco Superior Court judge in March 2005, but a state appeals court reversed the decision by a split 2-1 vote in July 2006. The seven-justice California Supreme Court scheduled an extraordinary four hours of arguments in the case for March 4, 2008. Attorneys on the plaintiffs' side took some encouragement from some of the questions that Chief Justice George posed.

Still, neither side was completely prepared for the strongly written opinion that George authored for the 4-3 majority on May 15. Shannon Minter, legal director for the National Center for Lesbian Rights, who argued the case for the plaintiffs, called the decision "a powerful affirmation of love, family and commitment."

Liberty Counsel's Staver, one of the lawyers on the other side, said the court had "abandoned the rule of law and common sense."

CURRENT SITUATION
Gay Marriage Ban Trailing

Californians appear to be closely divided on whether to permit gay marriage, but a ballot measure to overturn the state Supreme Court decision granting marriage rights to same-sex couples is trailing in the most recent public opinion surveys. Both sides in the statewide contest, however, expect the election to be close and are planning to

Should the Defense of Marriage Act (DOMA) be repealed?

YES

Evan Wolfson
*Executive Director, Freedom to Marry;
author,* Why Marriage Matters:
America, Equality, and Gay People's
Right to Marry

Written for *CQ Researcher*, September 2008

Congress should repeal the federal anti-marriage law. Couples who are legally married by a state such as Massachusetts or California should not be treated as legal strangers or denied rights by the federal government.

DOMA says that no matter what the need or purpose for any given program, the government will categorically deny all federal protections and responsibilities to married couples it doesn't like, i.e., those who are gay. This is an intrusive departure from more than 200 years in which couples properly married under state law then qualified for the more than 1,138 federal benefits of marriage such as Social Security, tax treatment as a family unit, family unification under immigration law and access to a spouse's health coverage. Through DOMA, Congress for the first time ever gave itself the power to say who is married, a power that under the Constitution belongs to the states.

Even worse, by denying rights such as family leave, child support and survivor benefits to one set of married couples, DOMA penalizes not only the couples themselves but their children. If the government wants to promote strong families, it should treat all married couples, and their children, equally.

There are far better reasons to treat marriages with respect than there are for destabilizing them — for all couples, gay and non-gay alike. And there are many constitutional and legal reasons why DOMA should be repealed: It denies one group of families an important and meaningful safety net; it violates the right of equal protection; it upends the traditional ways in which our country has treated married couples; and it's a power-grab by the federal government at the expense of the states.

Most important, however, Congress should reverse DOMA's radical wrong turn because it leaves no one better off, but it harms some people severely.

When DOMA was stampeded into law back in 1996, no gay couples were married anywhere in the world; Congress was voting on a hypothetical. But today real-life married couples are cruelly affected by DOMA's double standard, and Americans better understand the unfairness of depriving these families of the federal rights and responsibilities that will help them protect their loved ones. Even conservative former Georgia Republican Rep. Bob Barr, the original sponsor, has acknowledged DOMA to be abusive and now calls for its repeal.

In the United States, we don't have second-class citizens, and we shouldn't have second-class marriages. Couples who have made a personal commitment in life deserve an equal commitment under the law.

NO

Peter Sprigg
*Vice president for policy, Family
Research Council; author,* Outrage: How Gay
Activists and Liberal Judges Are Trashing
Democracy to Redefine Marriage

Written for *CQ Researcher*, September 2008

Cases asserting a "right" to same-sex marriage were heard in both Hawaii and Alaska in the early 1990s. Both states responded with constitutional amendments to forestall such judicial activism, but the cases triggered a national response as well. Fearing that if even one state legalized homosexual marriages, those marriages might then have to be recognized in every state and by the federal government, a bill was introduced in Congress to accomplish two things. First, it declared that for every purpose under federal law (such as taxation, Social Security, immigration and federal employee benefits), marriage would be defined only as the union of one man and one woman.

Second, it declared that no state would be required to recognize a same-sex marriage or other same-sex union that was legally contracted in another state. The Defense of Marriage Act (DOMA) passed both houses of Congress by large, bipartisan majorities and was signed into law by President Bill Clinton in 1996.

Many states followed with statewide DOMAs declaring homosexual marriage contrary to the public policy of that state. The 45 state DOMAs show a strong national consensus in favor of defining marriage as the union of one man and one woman.

Two state courts (Massachusetts and California) have succeeded in forcing homosexual marriage upon their unwilling populations. But the federal DOMA has been effective in preventing the imposition of this radical social experiment on the rest of the country. Unfortunately, some members of Congress (including Sen. Barack Obama, D-Ill.) are now proposing to repeal DOMA. This would open the door for federal taxpayers to subsidize homosexual relationships through domestic partner benefits and pave the way for lawsuits demanding recognition of same-sex unions from Massachusetts and California in every state.

Family Research Council opposes giving formal recognition or benefits to homosexual relationships under any circumstances, for numerous reasons. Even those who support same-sex marriage, however, should acknowledge that such a radical redefinition of our most fundamental social institution should not be imposed by the federal government in the face of a strong consensus among the states against it. Still less should we allow unelected judges from one or two states to force such a policy upon every other state.

The federal Defense of Marriage Act has served us well in the 12 years since its enactment, and it should not be tampered with.

spend about $20 million each on advertising and voter mobilization before the Nov. 4 balloting.

The two most recent polls find Proposition 8 trailing by 17 or 14 percentage points: 40 percent to 54 percent in a late August poll by the Public Policy Institute of California (PPIC); 38 percent to 55 percent in a September survey by the long-established Field Poll. The margins approximate the gap for Prop. 8 supporters recorded by the Field Poll in late May, shortly after the California Supreme Court's gay marriage ruling.[22]

A poll by the *Los Angeles Times* and the Los Angeles TV station KTLA one week earlier in May found 54 percent in favor of and 35 percent opposed to the ballot measure. At the time, Prop. 8 campaign officials described the Field Poll as "an outlier" and called the *Times*/KTLA poll a more accurate gauge of public opinion.[23]

After the most recent surveys, Prop. 8 campaign spokeswoman Jennifer Kerns blamed the gap on the ballot title that Attorney General Brown gave to the measure. Still, Kerns is predicting a "close" race that will turn on the level of enthusiasm among voters on both sides.

"There's a great deal of passion in support of this, which bodes well for Election Day," Kerns said. "People who feel most passionate are the people who go to the polls."[24]

The PPIC poll, in fact, found greater interest in the ballot measure among supporters than among opponents. More than half of those in favor of the measure — 57 percent — called the outcome "very important," compared to 44 percent of those opposed.

For their part, gay marriage supporters are also describing the race as close. "It's a dead heat," says Equality California Campaign Director Bankhead.

The PPIC poll found Californians evenly divided — 47 percent to 47 percent — on letting gay and lesbian couples marry. The earlier Field Poll had found a majority in favor: 54 percent to 39 percent. That was the first time in more than a decade of polling that a survey had found a majority in favor of same-sex marriage in the state.

With more than a month before the election, both campaigns are still in low gear. Political observers in the state report few visible signs of the campaign. The Yes on Prop. 8 campaign is reporting having raised around $17.8 million — much of it from religious or socially conservative groups from outside the state. Equality California has raised $12.4 million, also much of it from out of state.[25]

Despite the current edge in fund-raising, Prop. 8 supporters face some daunting obstacles in winning approval for the measure. In a state where Democrats hold an 11-percentage-point edge over Republicans in voter registration, the PPIC poll found Democrats opposing the measure by better than a 2-to-1 margin (66 percent to 29 percent). Republicans favor the measure — 60 percent to 34 percent. But the state's leading Republican, Schwarzenegger, opposes it.

Prop. 8 supporters also cannot rely on the kind of conservative religious constituencies that helped win passage of same-sex marriage bans in other states. "You don't have nearly the same presence of religious conservatives in California as you do in other states," says Jack Pitney, a political science professor at Claremont-McKenna College in Pomona. The Field Poll found Prop. 8 trailing — 44 percent to 48 percent — in inland counties, where religious conservatives are strongest.

To offset the disadvantages, Prop. 8 supporters are making special efforts to target Latino voters — the state's biggest ethnic minority and thought to be socially conservative. But the Public Policy Institute found Latinos opposed to the measure — 54 percent to 41 percent — by only a slightly smaller margin than whites (55 percent to 39 percent). PPIC President Mark Baldassare said the poll did not have a sufficient number of African- or Asian-American respondents for a valid measure of those groups.

Despite the wide margin, Baldassare says the campaign is "early" and the vote "hard to predict." Pitney, however, says Prop. 8 supporters are unlikely to overcome the gap. "It loses," he says. "The pattern in California ballot initiatives is that once a measure starts losing by a large margin in the polls, it almost never passes."

Gay marriage opponents are also lagging in one of the two other states with ballot measures to forestall recognition of same-sex unions. The measures — Amendment 2 in Florida and Proposition 102 in Arizona — would amend the states' constitutions to define marriage as a union of one man and one woman.

In Florida, the most recent poll shows 55 percent in favor of and 41 percent opposed to Amendment 2 — short of the 60 percent majority required for a constitutional amendment.[26]

In Arizona, Proposition 102, a measure submitted by the Republican leaders of the state Senate and House, would fortify a statutory gay marriage ban adopted in

1996. A broader measure that would have blocked any legal recognition for same-sex or unmarried straight couples failed, 48 percent to 52 percent, in 2006.[27]

Marriage Cases Waiting

Gay rights advocates hope — and gay marriage opponents worry — that the California Supreme Court's decision recognizing same-sex marriages could influence justices considering similar suits already pending in two other states: Connecticut and Iowa.

The California ruling is legally significant because it is the only state high court ruling to date holding that gays are a constitutionally protected class and that laws discriminating against gays are subject to "strict scrutiny" — the highest level of constitutional review. "It was just a matter of time before courts would acknowledge that," says Lambda Legal attorney Pizer.

Judges in the Connecticut and Iowa marriage cases had already signaled their interest in reconsidering the legal standard for judging laws adversely affecting gay men and lesbians. Three of the Connecticut Supreme Court's seven justices asked about treating gays as a specially protected class when the gay marriage case was argued in May 2007.[28] In Iowa, Polk County District Court Judge Robert Hanson applied strict scrutiny in striking down the state's gay marriage ban in a 63-page decision in August 2007.[29]

Gay marriage opponents acknowledge the California Supreme Court to be one of the most influential of state tribunals. "What happens in California is noticed not only around the country but around the world," says Brigham Young law Professor Wardle.

Lawyers and advocates on both sides are waiting impatiently for the Connecticut high court to rule on the case, *Kerrigan v. Department of Health*, after deliberating for well over a year. The plaintiffs — eight same-sex couples — filed their suit in state court in New Haven in September 2004. Connecticut enacted a civil union statute the following year.

In July 2006, Superior Court Judge Patty Jenkins Pittman ruled in a 25-page decision that in light of the legislature's "courageous and historic step," the plaintiffs had "failed to prove that they have suffered any legal harm that rises to constitutional magnitude." The couples appealed the ruling, represented by lawyers from the Boston-based Gay & Lesbian Advocates & Defenders.

The first marriage license in San Francisco went to lesbian activists Del Martin, left, and Phyllis Lyons, who had been together for more than 50 years. Martin, 87, died 10 weeks after San Francisco Mayor Gavin Newsom performed the ceremony on June 17.

The Connecticut Supreme Court includes four Republican-appointed justices and three appointed by Gov. Lowell P. Weicker Jr., a one-time Republican elected to the statehouse under auspices of his self-styled Connecticut Party. The Republican-appointed chief justice, Chase Rogers, recused herself from the marriage case because her husband's law firm filed a brief on behalf of a gay rights organization. A retired Democratic-appointed justice, David Borden, sat on the case in her place. Borden was one of the three justices to question lawyers during arguments about applying strict scrutiny to laws discriminating against homosexuals.

The Iowa case, *Varnum v. Brien*, began with a suit filed by six same-sex couples in 2005 after they were denied marriage licenses by the office of the then-Polk County recorder, Tim Brien, in Des Moines. Lawyers from Lambda Legal's regional office in Chicago represented the plaintiffs.

In his decision, Judge Hanson applied strict scrutiny in ruling that the gay marriage ban violated the state constitution's due process and equal-protection clauses. Hanson found that the county attorney's office had failed to prove that the state's ban would "promote procreation," "encourage child rearing by mothers and fathers," "promote stability for opposite-sex marriages," "conserve resources" or "promote heterosexual marriage."

Lawyers completed filing briefs with the Iowa Supreme Court in June; the court has yet to schedule arguments.

The seven justices on the court include two Republican and five Democratic appointees.

In a unique twist, one gay couple managed to get married the day after the ruling before Hanson agreed to the county attorney's motion to stay the decision pending appeal. Sean Fritz and Tim McQuillan, both in their early 20s, heard of the ruling on Aug. 30 and drove to Des Moines the next day to be wed. They found a judge who was willing to waive the normal three-day waiting period and got their marriage license at 10:45 a.m. Hanson issued his stay 45 minutes later.[30]

OUTLOOK

'It's About Marriage'

The Massachusetts Supreme Court's decision in 2003 granting marriage rights to same-sex couples produced a strong backlash. Public opinion polls registered a sharp drop in support for same-sex marriage, and gay marriage opponents won enactment of gay marriage bans in 13 states in the 2004 election cycle.

Five years later, no comparable backlash has emerged in the wake of the California Supreme Court's decision in favor of marriage equality — either nationally or in California itself. Nationwide surveys generally indicate a majority of Americans continue to oppose same-sex marriage, but surveys registered only a slight increase in opposition after the May 15 decision. And in August a poll by *Time* magazine actually found an even split between supporters and opponents: 47 percent to 47 percent.[31]

In addition, polls over the past five years indicate a stable majority of between 55 percent and 60 percent in favor of allowing either marriage or civil union status for same-sex couples. As one indication of the popular acceptance of some legal recognition of same-sex couples, supporters of California's Proposition 8 to overturn the state high court decision are arguing that the ruling was unnecessary. The state's domestic partnership law, they say, already gives same-sex couples all the rights of marriage.

With Prop. 8 trailing in the polls, gay marriage opponents are still professing optimism about the outcomes in California and the two other states — Arizona and Florida — with proposed constitutional amendments to ban marriage for same-sex couples. Liberty Counsel's Staver, who is helping to organize support for the measures

in Florida and California, envisions three victories to bring the total number of constitutional gay marriage bans to 30. That number, Staver says, is "getting very close to enough to ratify a federal constitutional amendment" to ban same-sex marriages nationwide.

However, the Federal Marriage Amendment, a proposed constitutional amendment that would redefine marriage as a union of one man and one woman, appears to face all but insurmountable obstacles at present. After failing in the Senate twice in 2004 and 2006, the amendment now seems all the more unlikely to win approval in a Democratic-controlled Congress. Public opinion polls over the past several years also show a majority of Americans opposed.

Gay marriage opponents may actually find themselves on the defensive in Congress if Democratic Sen. Barack Obama is elected president. Both Obama and the Democratic Party platform call for repealing the federal Defense of Marriage Act. Repeal would not be a foregone conclusion, however, since Congress passed the law in 1996 with overwhelming bipartisan majorities.

Whether or not DOMA is repealed, states will remain free to decide whether to allow marriage for same-sex couples for their own residents. The existing state constitutional bans mean that for the foreseeable future there will be a patchwork of state laws on the issue — barring the currently remote likelihood of a U.S. Supreme Court decision on the subject.

"We're going to be getting increasingly disparate treatment of same-sex unions," says Mark Strasser, a professor at Capital University Law School in Columbus, Ohio.

Gay marriage proponents think that time is on their side as more Americans come to see or know legally recognized same-sex couples in their communities or workplaces. "Most Americans don't have to fully love the idea of ending discrimination," Freedom to Marry's Wolfson adds. "They just have to realize that they can live with it. And, overall, it benefits the country."

Some opponents also expect eventual recognition for same-sex marriage. "I actually think that it will come within a generation," University of Pennsylvania Professor Wax said in the Federalist Society online debate.

Opponents continue, however, to mount their arguments against recognizing same-sex marriage. In a newly published volume, *What's the Harm?*, Brigham Young's Wardle organizes opposing essays around four perceived harms from legalizing same-sex marriage: to families and child rearing, to responsible sexual behavior and procreation, to the meaning of marriage and to "basic human freedoms," including religious liberty.[32]

"The issue is about marriage," says Wardle. "It's about protecting a basic social institution."

"It's about marriage," echoes Wolfson. "We're not looking to create something new. We're talking about allowing every American to exercise the same freedom to marry, to have the same responsibilities, the same respect as every other American."

NOTES

1. For photo coverage, see "George Takei Beams Up Marriage," *E News*, Sept. 15, 2008; "Ellen & Portia Share the Wedding-Day Love," *ibid.*, Sept. 10, 2008, www.eonline.com.

2. The decision is *Goodridge v. Massachusetts*, 798 N.E.2d 941 (Mass. 2003). For background, see Kenneth Jost, "Gay Marriage," *CQ Researcher*, Sept. 5, 2003, pp. 721-748.

3. The decision is *In re Marriage Cases*, 43 Cal. 4th 757 (2008). For next-day coverage, see Maura Dolan, "Gay Marriage Ban Overturned," *Los Angeles Times*, May 16, 2008, p. A1; Bob Egelko, "California Supreme Court, in 4-3 decision, strikes down law that bans marriage of same-sex couples," *San Francisco Chronicle*, May 16, 2008, p. A1.

4. See Gary J. Gates, M.V. Lee Badgett and Deborah Ho, "Marriage, Recognition and Dissolution by Same-Sex Couples in the U.S.," Williams Institute, July 2008, www.law.ucla.edu/williamsinstitute/publications/Couples%20 Marr%20Regis%20Diss.pdf.

5. Mark Baldassare, *et al.*, "Californians and Their Government: Statewide Survey," Public Policy Institute of California, August 2008, www.ppic.org/content/pubs/survey/S_808MBS.pdf; Field Poll, September 2008, www.field.com/fieldpoll.

6. See Charlie Frago, "Petitions to restrict adoption hit mark," *Arkansas Democrat-Gazette*, Aug. 26, 2008.

7. Jonathan Rauch, "Gay Marriage Is Good for America," *The Wall Street Journal*, June 21, 2008, p. A9. Rauch is a senior writer at *National Journal*, a guest scholar at the Brookings Institution and author of *Gay Marriage: Why It Is Good for Gays, Good for Straights, and Good for America* (2004).

8. See Jost, "Disputed Studies Give Gay Parents Good Marks," *op. cit.*, pp. 732-733.

9. "Same-Sex Marriage," Aug. 6, 2008, www.fed-soc.org/debates/dbtid.24/default.asp.

10. Quoted in Tracie Cone and Lisa Leff, "Gay marriage foes mobilize for ban in California," The Associated Press, Aug. 24, 2008.

11. The memo is posted on the Web site of the New York County Bar Association, www.nycbar.org/pdf/memo.pdf. For coverage, see Jeremy W. Peters, "New York Backs Same-Sex Unions From Elsewhere," *The New York Times*, May 29, 2008, p. A1.

12. The cases are *Miller-Jenkins v. Miller-Jenkins*, Virginia Supreme Court (June 6, 2008); *Finstuen v. Crutcher*, 496 F.3d 1139 (10th Cir. 2007). For coverage, see Frank Green, "Ruling comes in same-sex custody case," *Richmond Times-Dispatch*, June 7, 2008, p. A1; Robert E. Boczkiewicz, "Victory for gay adoptive parents," *The Oklahoman*, Aug. 8, 2008, p. 1A.

13. Background relies heavily on George Chauncey, *Why Marriage? The History Shaping Today's Debate Over Gay Equality* (paperback ed. 2005). See also William N. Eskridge Jr., *The Case for Same-Sex Marriage: From Sexual Liberty to Civilized Commitment* (1996); Allene Phy-Olsen, *Same-Sex Marriage* (2006), pp. 63-72.

14. Ina Russell (ed.), *Jeb and Dash: A Diary of Gay Life 1918-1945* (1993).

15. See David K. Johnson, *The Lavender Scare: The Cold War Persecution of Gays and Lesbians in the Federal Government* (2004).

16. See Chauncey, *op. cit.*, pp. 89-96.

17. *Ibid.*, pp. 96-104.

18. See *ibid.*, pp. 123-136; Jost, *op. cit.* (2003).

19. See "New Law Discourages Gay Marriages," *CQ Almanac 1996*.

20. See these stories by Jennifer Warren in the *Los Angeles Times*: "Ban on Gay Marriages Wins in All Regions but Bay Area," March 8, 2000, p. A23; "Gays Differ Sharply on Their Next Steps," March 9, 2000,

p. A3. The vote on the measure was 4,618,673 yes (61.4 percent) to 2,909,370 no (38.6 percent).

21. Quoted in Nancy Vogel and Jordan Rau, "Gov. Vetoes Same-Sex Marriage Bill," *Los Angeles Times*, Sept. 30, 2005, p. B3. The second veto was on Oct. 12, 2007.

22. See Denis C. Theriault, "Opposition growing to Prop. 8," *San Jose Mercury News*, Sept. 18, 2008; Jessica Garrison, "Bid to ban gay marriage trailing," *Los Angeles Times*, Aug. 28, 2008, p. B1.

23. See "California Poll: Same-Sex Marriage Is OK," The Associated Press, May 28, 2008.

24. Quoted in "Weak support for gay marriage ban," *Monterey County Herald*, Aug. 28, 2008.

25. See Dan Morain and Jessica Garrison, "Backers of California same-sex marriage ban are out-fundraising opponents," *Los Angeles Times*, Sept. 23, 2008; Aurelio Rojas, "Pitt's just another big giver in gay marriage showdown," *Sacramento Bee*, Sept. 23, 2008. Campaign finance filings can be found on the California secretary of state's Web site: http://cal-access.sos.ca.gov/Campaign/Measures/Detail.aspx?id=1302602&session=2007.

26. Mary Ellen Klas, "Same-sex marriage ban falling short," *The Miami Herald*, Sept. 9, 2008, p. B5.

27. See Amanda J. Crawford, "Consistent Message Doomed Prop. 107," *The Arizona Republic,* Nov. 9, 2006, p. 21.

28. The Connecticut case is *Kerrigan v. Department of Health*; for documents, see the Web site of Gay & Lesbian Advocates & Defenders: www.glad.org/marriage/Kerrigan-Mock/kerrigan_documents.html. For coverage, see Lynne Tuohy, "Supreme Court Justices Hear Arguments on Whether State Must Allow Marriage for Same-Sex Couples, Not Just Civil Unions," *Hartford Courant*, May 15, 2007, p. A1; Thomas B. Scheffey, "Following In California's Footsteps?" *Connecticut Law Tribune*, June 30, 2008, p. 1.

29. The Iowa case is *Varnum v. Brien*; for the lower court decision, see Freedom to Marry Web site: www.freedomtomarry.org/pdfs/iowa_ruling.pdf. For coverage, see Jeff Eckhoff and Jason Clayworth, "Judge: ban on gay marriage invalid," *Des Moines Register*, Aug. 31, 2007, p. 1A.

30. See Cara Hall, "Gay couple eyes court rulings," *Des Moines Register*, June 12, 2008, p. 1E.

31. For a compilation, see PollingReport.com, www.pollingreport.com/civil.htm, visited Sept. 19, 2008.

32. Lynn D. Wardle (ed.), *What's the Harm? Does Legalizing Same-Sex Marriage Really Harm, Individuals, Families or Society?* (2008).

BIBLIOGRAPHY

Books

Chauncey, George, *Why Marriage? The History Shaping Today's Debate Over Gay Equality, Basic Books*, 2004.
A Yale University historian compactly links the increased visibility of the gay and lesbian community and changes in the institution of marriage to the gay rights movement's effort to attain marriage equality. Includes chapter notes.

Koppelman, Andrew, *Same Sex, Different States: When Same-Sex Marriages Cross State Lines, Yale University Press*, 2006.
A professor at Northwestern University Law School argues that, based on established legal principles, courts should recognize marriages between same-sex couples that are recognized in their home state. Includes detailed notes.

Phy-Olsen, Allene, *Same-Sex Marriage, Greenwood Press*, 2006.
A professor emeritus of English at Austin Peay State University provides a thorough and balanced account of the background and current debate over same-sex marriage. Includes chapter notes, 24-page annotated bibliography.

Rauch, Jonathan, *Gay Marriage: Why It Is Good for Gays, Good for Straights, and Good for America, Times Books*, 2004.
A writer for *The Atlantic* and *National Journal* and guest scholar at the Brookings Institution argues that gay marriage would be beneficial by establishing marriage as the norm for gay men and lesbians, reversing the trend toward alternatives to marriage and making the country "better unified and truer to its ideals." The book bears this dedication: "For Michael. Marry me when we can."

Rimmerman, Craig A., and Clyde Wilcox, *The Politics of Same-Sex Marriage, University of Chicago Press,* 2007.
Fourteen essays examine various aspects of the politics of same-sex marriage, including litigation and public opinion on the issue. Notes with each essay. Rimmerman is a professor of public policy studies and political science at Hobart and William Smith Colleges; Wilcox, a professor of government at Georgetown University.

Savage, Dan, *The Commitment: Love, Sex, Marriage, and My Family, Dutton,* 2005.
An author and sex-advice columnist relates with humor and poignancy how his mother goaded him into marrying his boyfriend of 10 years — and how their young son picked out skull rings to symbolize the union.

Stanton, Glenn, and Dr. Bill Maier, *Marriage on Trial: The Case Against Same-Sex Marriage and Parenting, InterVarsity Press,* 2004.
The authors, both with the anti-gay organization Focus on the Family, use a question-and-answer format to argue against recognizing same-sex marriage or encouraging child-rearing in same-sex households.

Wardle, Lynn D. (ed.), *What's the Harm? Does Legalizing Same-Sex Marriage Really Harm Individuals, Families or Society? University Press of America,* 2008.
Twenty contributors on both sides of the issue debate the potential impact of further legalizing same-sex marriage. Each essay includes notes. Wardle is a professor at Brigham Young University School of Law.

Articles

Denizet-Lewis, Benoit, "Young Gay Rites," *The New York Times Magazine,* April 8, 2008.
The writer examines the impact of the Massachusetts ruling legalizing gay marriage through the lives of several same-sex couples in the state.

Reports and Studies

Gates, Gary J., M. V. Lee Badgett and Deborah Ho, "Marriage, Registration and Dissolution by Same-Sex Couples in the U.S.," *Williams Institute, UCLA School of Law,* July 2008, www.law.ucla.edu/WilliamsInstitute/publications/Couples%20Marr%20Regis%20Diss.pdf.
The study provides data on same-sex couples who have taken advantage of legal recognition — marriage, civil union or domestic partnership — allowed in 10 states and the District of Columbia. Gates is the co-author with Jason Ost of *The Gay and Lesbian Atlas* (Urban Institute, 2004).

On the Web

Federalist Society, "Same Sex Marriage," Aug. 6, 2008, www.fed-soc.org/debates/dbtid.24/default.asp.
Four law professors — two on each side — debate marriage rights for same-sex couples in an online forum sponsored by the conservative lawyers' organization.

Vestal, Christine, "Calif. gay marriage ruling sparks new debate," *Stateline.org,* June 12, 2008, www.stateline.org/live/printable/story?contentId=310206.
The online state news service provides an overview of the same-sex marriage debate following the California Supreme Court ruling, along with a national map and state-by-state chart.

Note: For additional earlier titles, see bibliography in Kenneth Jost, "Gay Marriage," CQ Researcher, Sept. 5, 2003, p. 746.

For More Information

Alliance Defense Fund, 5100 N. 90th St., Scottsdale, AZ 85260; (800) 835-5233; www.alliancedefensefund.org. Legal alliance defending the right to speak and hear biblical beliefs.

Alliance for Marriage, P.O. Box 2490, Merrifield, VA 22116; (703) 934-1212; www.allianceformarriage.org. Research and education organization promoting traditional marriage.

Equality California, 2370 Market St., Suite 200, San Francisco, CA 94114; (415) 581-0005; www.eqca.org. Supports GLBT civil rights protections in California.

Family Research Council, 801 G St., N.W., Washington, DC 20001; (202) 393-2100; www.frc.org. Promotes traditional marriage in national policy debates.

Freedom to Marry, 116 W. 23rd St., Suite 500, New York, NY 10011; (212) 851-8418; www.freedomtomarry.org. Gay and non-gay partnership working to win marriage equality.

Gay and Lesbian Advocates and Defenders, 30 Winter St., Suite 800, Boston, MA 02108; (617) 426-1350; www.glad.org. Opposes discrimination based on sexual orientation, gender identity and HIV status.

Human Rights Campaign, 1640 Rhode Island Ave., N.W., Washington, DC 20036; (202) 628-4160; www.hrc.org. America's largest civil rights organization supporting GLBT equality.

Lambda Legal Defense and Education Fund, 3325 Wilshire Blvd., Suite 1300, Los Angeles, CA 90010; (213) 382-7600; www.lambdalegal.org. GLBT civil rights litigation group.

Liberty Counsel, P.O. Box 540774, Orlando, FL 32854; (800) 671-1776; www.lc.org. Nonprofit litigation group supporting traditional family values.

National Gay and Lesbian Task Force, 1325 Massachusetts Ave., N.W., Suite 600, Washington, DC 20005; (202) 393-5177; www.thetaskforce.org. Promotes equality for gays and lesbians.

Protect Marriage; www.protectmarriage.com. California group opposed to gay marriage initiatives in upcoming ballot.

5

Aging Baby Boomers

Will the 'Youth Generation' Redefine Old Age?

Alan Greenblatt

Former mortgage broker Jamie Sims checks crawfish traps at her family's fish farm in Harrisburg, Ark., Sims, 44, is among the baby boomers who have left the "rat race" for less stressful jobs. Beginning in 2008, the oldest boomers will turn 62 — old enough to collect Social Security. In 2030, when the number of Americans over 65 hits 72 million, some experts predict dangerous strains on entitlement programs and the federal budget.

From *CQ Researcher*,
October 19, 2007.

Five years ago, Honda introduced a boxy SUV — the Element — with an ad campaign billing it as a combined "base camp" and "dorm room on wheels" and featuring images of 20-somethings cruising down to the beach.

But Honda's appeal to the youth market missed its target, triggering the interest of baby boomers instead. During its first year on the market, the average age of Element buyers was 42.[1]

Several other cars initially targeted at younger consumers have been "hijacked" by baby boomers — the generation of 78 million Americans born between 1946 and 1964 — including the Toyota Matrix, the Pontiac Vibe and the Dodge Neon.[2]

It's not surprising that the notably nostalgic boomer generation seeks out youthful, environmentally friendly products. Throughout their middle-age years, they have remained loyal to the music and culture of their youth as well as youthful in their habits, priorities and pursuits. "In their eagerness to live up to their label as the youth generation, boomers — with the help of marketers, to be sure — have created a youth-oriented consumer mindset that is proving difficult to shift," writes Diane Crispell, a consumer-behavior consultant associated with Cornell University.[3]

Since baby boomers apparently "never devised an exit strategy from their youth," as one wag put it, marketers and others wonder what they are going to be like as old people. Beginning in January 2008, the oldest boomers will turn 62 — old enough to start collecting Social Security. The nation's first baby boomer, in fact, just applied for Social Security benefits on Oct. 15, to great media fanfare. Kathleen Casey-Kirschling, 62, a retired schoolteacher in Cherry Hill, N.J., was born a second after the stroke of midnight

Number of Seniors Is Rising Rapidly

One in five Americans will be over age 65 by 2050. Such a profound demographic change raises fundamental questions about the federal government's ability to pay for all the aging boomers who will be depending on Social Security, Medicare and other entitlements.

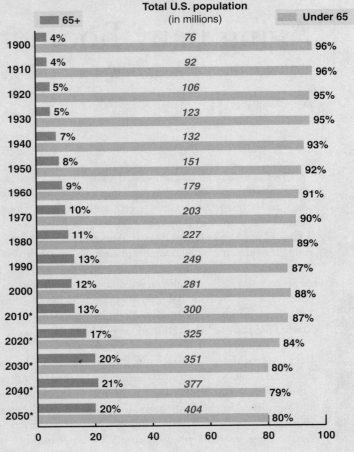

Percentage of People Age 65 and Older, 1900-2050

Total U.S. population (in millions)

Year	65+	Total U.S. population (in millions)	Under 65
1900	4%	76	96%
1910	4%	92	96%
1920	5%	106	95%
1930	5%	123	95%
1940	7%	132	93%
1950	8%	151	92%
1960	9%	179	91%
1970	10%	203	90%
1980	11%	227	89%
1990	13%	249	87%
2000	12%	281	88%
2010*	13%	300	87%
2020*	17%	325	84%
2030*	20%	351	80%
2040*	21%	377	79%
2050*	20%	404	80%

* projected

Percentages may not total 100 due to rounding.

Source: Robert B. Friedland and Laura Summer, "Demography Is Not Destiny, Revisited," Center on an Aging Society, Georgetown University, March 2005

on entitlement programs and the federal budget to a decline in the stock and housing markets as boomers start liquidating their assets.

But boomers are growing old at an opportune time. The nature of aging is changing. Many boomers feel younger than their parents did at the same age, and that's not just in their imaginations — or their consumer patterns. Life expectancy has gone up significantly during boomers' lifetimes, as has "health expectancy." Fewer seniors suffer from chronic disabilities than was the case 25 years ago, so millions of today's senior citizens — and the boomers who will follow — can lead active and productive lives until a later age.

Frieda Birnbaum, the 60-year-old New Jersey psychologist who gave birth to twins in May, is just an extreme example of how life milestones are shifting to later ages. The percentage of Americans working in their late 60s has shot up by more than half since 1985. And the percentage of those over 55 who say they are exercising 100 or more days per year has jumped 33 percent.[4]

"Compared to older people in the past, boomers will actually have a greater opportunity and ability to live a youthful old age," write J. Walker Smith and Anne Clurman, senior executives at the Yankelovich consumer research firm, which coined the term "baby boomer."[5]

"Four decades of Yankelovich research," they write in their new book, *Generation Ageless*, "has found one thing about boomers over and over again — an unwavering determination to not get old."[6]

Surveys conducted by AARP, the main advocacy organization for older Americans, show that up to 80 percent of boomers intend to work past 65. Many have expressed

on Jan. 1, 1946. By 2030, according to the Census Bureau, the number of Americans over 65 is expected to double — to 72 million. That demographic leap has led to a variety of dire predictions, from unbearable strains

the desire to pursue entirely new second careers, such as social work or teaching.

"Most previous generations thought about whether they're wealthy and able to leave trust funds," says Andrew Achenbaum, a historian of aging at the University of Houston. "Boomers think, 'I have to feel that I'm accomplishing something; I really am concerned about legacy.' Legacy has always been a minor motif, but it's going to be major with the boomers."

Less altruistic boomers will work simply to pay the bills. Taken together, boomers are affluent, but there are vast asset disparities within this huge cohort. Contrary to concerns about hostility between boomers and younger generations as Social Security and Medicare expenditures shoot up, New York University political scientist Paul C. Light noted, "It is far more likely that the baby boomers will divide against themselves in an intra-generational war between the haves and have-nots."[7]

Still, it remains an open question as to how much longer boomers will actually work — and whether those who want to work will be able to find work that's more "meaningful" or lucrative than, say, being a greeter at Wal-Mart.

Numerous nonprofits and job placement services have sprung up to help seniors interested in gainful employment, but some observers doubt boomers will be able to work much more than previous generations did during their retirement years. At the same time, some sectors — such as nursing and government — realize that boomers comprise a disproportionate share of their workforces so they are offering accommodations to keep seniors on longer. By 2010, an estimated 26 million workers will be 55 or over — a 46 percent increase since 2000.[8]

But surveys indicate that not all employers are eager to provide the flexibility senior workers say they want as they begin to dial back the number of hours they're willing to work. A few million boomers have already chosen to leave

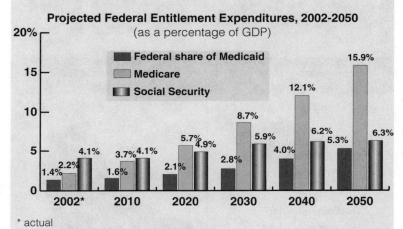

Entitlement Spending Will Skyrocket

The cost of the three key federal entitlement programs for older Americans is expected to increase dramatically in coming decades, reaching 28 percent of gross domestic product (GDP) by 2050. Comptroller General David M. Walker says the expected "tsunami of spending" threatens the nation's economic stability.

Projected Federal Entitlement Expenditures, 2002-2050
(as a percentage of GDP)

* actual

Source: Robert B. Friedland and Laura Summer, "Demography Is Not Destiny, Revisited," Center on an Aging Society, Georgetown University, March 2005

the workplace, and about 40 percent of retirees left their jobs involuntarily — due to layoffs or health issues.

However, when it comes to actually retiring, "There's a difference between saying it and doing it, and a difference between saying it and being able to do it," says Eric Kingson, a professor of social work at Syracuse University.

Still, it's clear boomers intend to age differently from their parents. They won't all be exercise demons or work full time. But because there are so many boomers, trends that take root among even a small percentage of them can have an outsized influence on society. "When we thought hula hoops were in, hula hoops were in," says Jerry Abramson, a boomer himself who sees the continuing influence of his generation at work in his capacity as mayor of Louisville, Ky. "When we thought bell bottoms were to be worn, they were worn." Today, he notes, "Boomers are changing housing patterns, recreation facilities and community-center programming."

"They've changed everything all along, haven't they," says Sara E. Rix, interim director of economic studies at AARP's Public Policy Institute. "I think they are going to

Seniors' Assets Total More Than $5 Trillion

Americans over age 65 hold one-fifth of the stocks and bonds owned by Americans — more than $5.25 trillion (left). According to a Government Accountability Office (GAO) study of current retirees, overall spending down of assets is slowing, and many retirees are actually continuing to purchase stocks. If baby boomers behave like current retirees, the GAO says, "a rapid and mass sell-off of financial assets seems unlikely." One-third of all baby boomers have no financial assets (right).

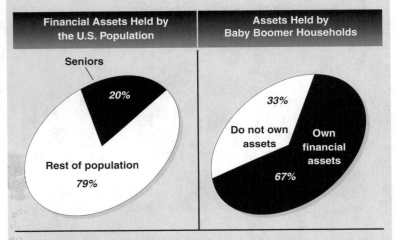

Percentages may not total 100 due to rounding.

Source: "Baby Boom Generation: Retirement of Baby Boomers Is Unlikely to Precipitate Dramatic Decline in Market Returns, but Broader Risks Threaten Retirement Security," Government Accountability Office, July 2006

and complete college than their parents, and few clouds darkened their economic horizon until the oil shocks and high inflation of the 1970s. As a result, boomers have been "one of the most prosperous generations in U.S. history," concludes a Congressional Budget Office study.[10]

But projections are mixed on whether boomers can afford to retire comfortably. Wealth in households headed by Americans 55 or over has doubled since 1989, reaching nearly $250,000 in 2004. During the same period, however, accumulated assets among those in their late 30s had dropped by more than 25 percent, to less than $50,000.[11]

Some boomers, however, are clearly struggling. "The top one-third of the boomers will have lots of choices, and the bottom one-third will be working until they drop just to keep food on the table," said Paul Hodge, chairman of the Global Generations Policy Institute at Harvard University.[12]

Waiting lists for affordable and subsidized senior housing have grown, while bankruptcy filings are rising faster for those 55 and older than for any other age group, due primarily to mortgage debt and health costs.[13]

And while boomers may be wealthy as a collective group, there is clearly a divide among them in terms of savings and wealth. The wealthiest 10 percent of boomers hold more than two-thirds of their generation's accumulated financial assets, says the Government Accountability Office (GAO).[14] And a substantial difference in wealth accumulation exists between older boomers (born between 1946 and 1954) and younger boomers (those born between 1955 and 1964), according to the Bureau of Labor Statistics.[15]

"Boomers have a hugely disproportionate amount of assets compared to the other generations," says Frederick R. Lynch, a sociologist at Claremont McKenna College

see themselves as young for a much longer period of time and act accordingly."

Or as Birnbaum put it after her twins were born: "I don't feel like I'm 60. I don't know what 60 is meant to be."[9]

As boomers prepare to change the nature of senior citizenship — and as the nation prepares for them — here are the questions being debated:

Can boomers afford to retire?

Baby boomers owe their lives to affluence. The economic boom that followed World War II — when America dominated the global economy and wages skyrocketed — convinced millions of parents they could afford to raise multiple children.

As the resulting boomers grew up, the future continued to look promising. They were much more likely to attend

in California and author of a forthcoming book about boomers. But about half of the boomers are vulnerable, in terms of retirement savings and retiree health benefits, he points out. "There's a whole working class — Joe Six-Pack America — even among the boomers."

Boomers should be better off in retirement than their parents because they have enjoyed higher per capita incomes and accumulated wealth at about the same rate as their parents. But some studies indicate that boomers have saved slightly less and spent a bit more than their forebears. One insurance ad noted, "The generation that 'wouldn't trust anyone over 30' never planned on a 30-year retirement."[16]

"Charge cards like VISA and MasterCard came into the market just when the boomers came out of college and started making money, so they've been in debt ever since," says Charles F. Longino Jr., director of the Reynolda Gerontology Program at Wake Forest University.

Boomers also tend to have more of their assets tied up in financial instruments such as stocks, which leaves them more vulnerable to economic cycles than previous generations. If there's a badly timed recession or stock market decline, Lynch says, "It's going to be really tough. They're betting on the market."

Some of the younger boomers may be vulnerable to the widespread switch in private pension plans from "defined-benefit" programs, which offered fixed, guaranteed payments throughout retirement, to "defined-contribution" plans (such as 401(k)s), which are stock-market-based accounts that can decline precipitously in value.[17]

"The risks of aging have shifted more to the individual," says Michael A. Smyer, codirector of the Center on Aging and Work at Boston College. "With the shift from defined-benefit to defined-contribution plans, the risk of planning for your future is more on your shoulders."

Rix, of AARP, says older Americans are also nervous about cuts in retiree health benefits, which may be why more seniors are working at least until 65, when they become eligible for Medicare. "Surveys overwhelmingly reveal that older workers expect to work in retirement, both for financial as well as non-financial reasons," she says. "But when pressed for the main reason, the financial reason rises to the top."

The proportion of Americans still working in their late 60s has been rising. In California, a study found that the proportion of people 55 to 69 who were working rose by about 10 percent between 1995 and 2006.[18]

The fact that boomers are expected to live longer than prior generations also prompts many to consider delaying retirement. Boomers "aren't saving and are really strapped," says Susan Krauss Whitbourne, a psychologist at the University of Massachusetts, Amherst who has studied boomers. "Now they're living longer, and the chickens have to come home to roost at some point."

But Yankelovich's Smith doubts claims that boomers haven't saved enough. "Compared to prior generations, boomers are in better shape financially," he says. "And, since they're not going to retire [at the traditional retirement age], those that predict doom and gloom are wrong."

Will boomers bankrupt America?

Like earlier generations, boomer seniors will depend heavily on government programs such as Social Security and Medicare, which have brought poverty rates among the over-65 population down from more than one-third in 1960 to just over 10 percent.[19] But with the number of seniors rising rapidly, can the federal government afford to pay for all the aging boomers who will be signing up?

Over the past two years, Comptroller General David M. Walker — the nation's top accountant — has been conducting his own "fiscal wake-up tour," crisscrossing the nation with a doomsday message for policy makers and the public about the government's long-term financial liabilities. "The most serious threat to the United States is not someone hiding in a cave in Afghanistan or Pakistan, but our own fiscal irresponsibility," he told CBS' "60 Minutes" in July.

His biggest worry: the federal entitlement programs designed to benefit seniors. "The first baby boomer will reach 62 and be eligible for early retirement or Social Security on Jan. 1, 2008," he said. The boomers will be eligible for Medicare three years later, and when they start retiring en masse it will create a potential "tsunami of spending that could swamp our ship of state."[20]

For years, economists have warned that the enormous boomer bulge in the senior population could impose enormous burdens on both the economy and government finances. Social Security and Medicare expenditures could rise from their current 8.5 percent of national economic output to 15 percent, according to the Congressional Budget Office (CBO). The federal debt

Getty Images/Chip Somodevilla

Many baby boomers are doing the things they couldn't afford in their youth, or during their working years. For Dean and Beverly Avery, of San Leandro, Calif., owning a 1966 Ford Mustang fulfilled a longtime dream.

could nearly double, rising from its current 37 percent of the economy to about 100 percent — a level previously reached only during World War II.[21]

But longstanding fears about the impact of this demographic upheaval on the stock market may not materialize, according to some experts. A GAO report concluded that retiring boomers will not suddenly sell off enormous amounts of stock market assets all at once. Because the boomer generation spans such a broad range of time — 1946 to 1964 — they will not all be retiring and liquidating at the same time. While the oldest boomers are on the cusp of Social Security eligibility, the youngest boomers will not reach 60 for another 15 years yet.[22]

"There will be no market meltdown — people don't all sell at once," says Barbara D. Bovbjerg, director of Education, Workforce and Income Security Issues at the GAO.

For similar reasons, some economists doubt that aging boomers will cause chaos in the housing market, since they won't be selling their homes all at once.

Similarly, there are fewer worries that millions of boomers reaching retirement age will cripple the nation's workforce, since many will continue working for some time, and replacements can be found for them in most fields. Fear of a coming labor shortage due to boomer retirements is "way overstated," says Peter Capelli, director of the human resources program at the University of Pennsylvania's Wharton School and author of an influential paper on the topic. "The biggest high-school class ever to graduate in the U.S. is the class that graduated this year, 2007."

But what about the impact of the boomer tsunami on entitlement programs for seniors? Like Comptroller General Walker, many observers worry that the number of seniors will increase exponentially as the number of workers paying taxes to support the entitlement programs dwindles. The ratio of seniors to working-age population (ages 25 to 64) will rise by 30 percent over each of the next two decades, says Dowell Myers, a demographer at the University of Southern California.

"I think it's going to be pretty bad," says Laurence J. Kotlikoff, an MIT economist and coauthor of *The Coming Generational Storm*, a 2004 book about the threat retired boomers represent to the entitlement programs. "The county is basically insolvent. We can't afford the policies we've got in place, let alone the projected growth."

Peter Diamond, a colleague of Kotlikoff at MIT who has written extensively about Social Security, is more sanguine. The baby boomers are already factored into Social Security projections, he notes. And, although the program's trust fund is expected to run out of money by 2040, Diamond and many other economists believe modest tweaking can keep the program in the black much longer.

But he does share widespread concerns that Medicare costs are rising at unsustainable rates, especially since the addition of a prescription-drug benefit that went into effect in 2006. Medicare and Medicaid, says GAO's Bovbjerg, are "the primary threat to the fiscal stability of the federal government."

Even those most pessimistic about Medicare costs believe the issue will have to be confronted, but in the broader context of the U.S. health system as a whole, not as a problem triggered just by boomer retirements.

Lynch, at Claremont McKenna College, agrees Medicare will be tougher to fix than Social Security but says both are more fixable than the "doomsday" economists believe. "The solution is that boomers will work longer and won't retire as early as prior generations," Lynch says. "The proportion of people retiring early is starting to decline."

Laura L. Carstensen, director of Stanford University's Center on Longevity, agrees that boomers will break the nation's bank — but only if the programs don't change and if boomers act like previous generations in terms of their retirement patterns — and she doubts either premise will come to pass.

"Boomers could [bankrupt the country] if . . . Social Security and Medicare don't change and people continue to retire at 65," Carstensen says. "But relatively modest changes could turn that around."

Will boomers change the nature of aging?

An old joke says old age is always 15 years away. Boomers seem to have taken that to heart. According to a 1996 survey, boomers believed old age began at age 79 — and at that time life expectancy was just over 76 years. As Smith and Clurman point out in *Generation Ageless*, boomers literally thought they'd die before they got old.[23]

"Boomers are not going to give up their aspirations for youthfulness," Smith says. "It is the defining characteristic of the boomer sensibility."

When approaching his own 60th birthday last year, President Bush, said he "used to think 60 was really old. Now I think it's young, don't you? It's not that old, it really isn't."[24]

Is Bush right? The fact that the boomers are "coming of old age" at a time when life expectancy is lengthening may pose some demographic and economic challenges. But many sociologists and gerontologists believe the generation that refuses to grow up can change — in healthy ways — how Americans think about aging.

"There's no question that we're going to change the meaning of late life," says Achenbaum, the University of Houston historian. "Chronological age, per se, is going to be a miserable predictor of what contingencies and opportunities might arise."

Today, the average man reaching 65 can expect to live for 17 more years, while women will live for 20.[25] People are not only living longer but in many cases are staying healthy longer. The rate of chronic disabilities is down to just 19 percent among those over 65, compared with 26 percent in 1982.[26]

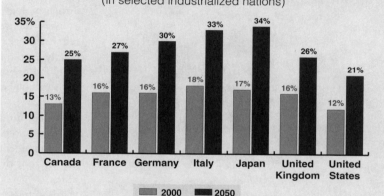

U.S. Is 'Younger' Than Other Countries

The proportion of today's U.S. population 65 and older is smaller than in other industrialized nations and is projected to be even smaller in 2050. The elderly will be 21 percent of the U.S. population in 2050 compared with 30 percent or more in Japan, Italy and Germany. Population aging is a worldwide phenomenon, mostly due to declining fertility rates and increasing life expectancies.

Percentage of Population Age 65 and Older, 2000 and 2050*
(in selected industrialized nations)

	Canada	France	Germany	Italy	Japan	United Kingdom	United States
2000	13%	16%	16%	18%	17%	16%	12%
2050	25%	27%	30%	33%	34%	26%	21%

* projected

Source: Robert B. Friedland and Laura Summer, "Demography Is Not Destiny, Revisited," Center on an Aging Society, Georgetown University, March 2005

"We get so many [article] pitches about people over 60 riding their bikes across the country, or running in 100-mile marathons, we have to tell them, 'Great, but this is not really news any more,' " said Margaret Guroff, health editor of *AARP* magazine.[27]

Not everyone is a marathoner, of course. But boomers, who created the jogging fad in the 1970s and turned aerobics into a multibillion-dollar industry in the 1980s, are now returning to fitness centers in record numbers.

They are also beginning to impose their preferences on a variety of services for seniors, demanding, for instance, that meals be less about gravy and more about fresh, healthy foods — organic, if possible, thank you very much.

Stanford psychologist Carstensen says boomers will update old-age behaviors and consumer demands — demanding, for instance, Starbucks coffee in nursing homes and cell phones designed so that "aging eyes can make out the numbers."

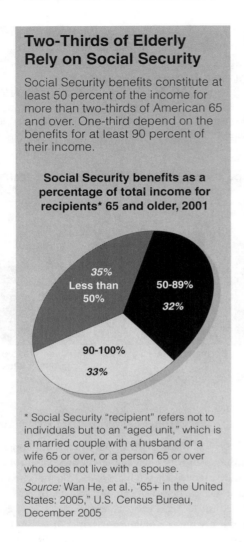

Two-Thirds of Elderly Rely on Social Security

Social Security benefits constitute at least 50 percent of the income for more than two-thirds of American 65 and over. One-third depend on the benefits for at least 90 percent of their income.

Social Security benefits as a percentage of total income for recipients* 65 and older, 2001

- 35% Less than 50%
- 50-89% 32%
- 90-100% 33%

* Social Security "recipient" refers not to individuals but to an "aged unit," which is a married couple with a husband or a wife 65 or over, or a person 65 or over who does not live with a spouse.

Source: Wan He, et al., "65+ in the United States: 2005," U.S. Census Bureau, December 2005

"We are offering more and more activities to keep the mind alert," said Becky Batta, director of a senior center in Annapolis, Md. "The baby boomers are coming and they demand it. They are completely different from other generations of seniors."[28]

But some research indicates boomers are less physically active than their parents and grandparents were at the same age, more likely to be sitting at a desk at work or in a car than actually working out. They also are more likely to suffer chronic problems such as high cholesterol and diabetes. Half of those 55 to 64 have high blood pressure, while 40 percent are obese, according to the Centers for Disease Control and Prevention.

"People are not as healthy as they approach retirement as they were in older generations," said Mark D. Hayward, a University of Texas sociologist. "It's very disturbing."[29]

"They talk to their fellow boomers and realize, I'm not the only one getting bad mammograms or having brittle bones," says Lynch, the Claremont McKenna sociologist. "They will realize there are some obstacles you can't overcome. Boomers are going to discover that if their biology says no, it's going to be no."

Nevertheless, Lynch and others believe boomers have redefined every stage of life they've entered, so being elderly should be no exception. And even if many inevitably fall prey to disease or age-related afflictions, millions will be healthier and more productive than society expects.

"Sheer numbers alone are going to cause a shift in attitudes and perception," says Rix of AARP. "When you see more older people who are active, vibrant and staying on the job, that will change perceptions about being older."

"Boomers are going to redefine what it means to be old and what it means to retire," says *Generation Ageless* coauthor Smith.

But that doesn't mean boomers can stay young forever. They may listen to cooler music than their parents did, but their ears are still getting older. About one in six boomers already suffers from hearing loss. In fact, the AARP reports, 10 million people between 45 and 64 suffer from hearing loss, compared to 9 million people over 65.[30] All those years of 115-decibel rock concerts and wearing a Walkman have taken their toll, experts say.

And while it's possible to retard the effects of aging through diet and exercise — which is why life expectancy keeps rising — it's not possible to turn back the clock. But that doesn't mean that being 60 will be the same experience that it was, say, in the 1960s.

"We hear 60 is the new 40," says Marc Freedman, CEO of Civic Ventures, a group that seeks to link seniors with job opportunities. "I'm convinced that 60 is the new 60 — that we're actually inventing a new stage of life now between the end of midlife careers and true old age and retirement."

BACKGROUND

'Fertility Splurge'

Following a long period of decline in birth rates — dating back to the Industrial Revolution — demographers in the 1930s predicted that the U.S. population would stagnate and was unlikely to rise above 150 million by century's end. But birth rates shot up immediately after World War II, quickly rising to more than 4 million births per year.

All told, about 76 million children were born in the United States between 1946 and 1964. (Several million have died, but immigrants have more than made up for those numbers.)

"Simply put, the baby boom was a 'disturbance' which emanated from a decade-and-a-half-long fertility splurge on the part of American couples," concluded the Population Research Bureau in 1980.[31]

Childbearing long delayed — first by the Depression and then by war — was put off no longer. Women married younger and had their first babies at an earlier age than at any time in modern history.[32] The fertility rate, which averaged 2.1 children per woman during the 1930s, peaked at 3.7 in the late 1950s. [33] The number of babies being born certainly surprised the General Electric Co. in January 1953. It promised five shares of stock to any employee who had a baby on Oct. 15, the company's 75th anniversary. GE expected maybe eight employees would qualify. Instead they had to hand over stock to 189 workers.[34]

But the baby boom of 1946 to 1964 was not simply triggered by the return of young soldiers from war. In fact, the boom accelerated through the 1950s. As the economy flourished and families moved to expansive new homes in suburbs, children became not just affordable but expected. As sociologist William Simon put it, those who didn't want children "were an embarrassed and embattled minority. It was almost evidence of a physical or mental deficiency."[35]

The time was ripe, economically, for many more people to have children than had done so during the Great Depression of the 1930s. The gross domestic product (GDP) expanded rapidly, growing from $227 billion in 1940 to $488 billion in 1960. Median family income and real wages climbed steadily due to tight labor markets, while inflation remained low. The Servicemembers'

Readjustment Act of 1944, commonly known as the GI Bill of Rights, helped more of the middle class buy their first homes and get college educations, significantly increasing their lifetime earnings.[36]

"Never had so many people, anywhere, been so well off," observed *U.S. News & World Report* in 1957.[37]

The increases in personal income and home ownership led to comfortable childhoods for millions of boomers in a largely peaceful and prosperous time, notable for its advances in medicine, such as the Salk polio vaccine of 1955.[38]

"Many American children, regardless of their family background, grew up during the baby boom with an expectation of nearly limitless growth and opportunity," wrote two University of Michigan social scientists in a recent essay.[39]

Those expectations were fueled by rising rates of education. Between 1955 and 1975, enrollment in elementary and high schools increased 41 percent. Not only were more students attending school but they were also spending more time in school than their parents or grandparents — 40 more days per school year, on average. Less than 20 percent of the school-age population had graduated from high school during the 1920s; by the 1970s more than 75 percent did.[40]

Near-universal high school — and the G.I. Bill's liberal higher-education benefits — led to a boom in college enrollments. From 1965 to 1980, college enrollment more than doubled, from 5.9 million to 12 million, making boomers the first generation in which vast numbers took college nearly for granted.[41]

Boomer Culture

Away from the schoolbooks, boomers were important and influential consumers. "They were the first generation of children to be isolated by Madison Avenue as an identifiable target," noted journalist Landon Y. Jones in his 1980 history of the boomers, *Great Expectations*. "From the cradle, the baby boomers had been surrounded by products created especially for them, from Silly Putty to Slinkys to skateboards. New products, new toys, new commercials, new fads [were] integral to the baby boomer experience."[42]

Boomers identified strongly with their favorite toys and trends. They grew up with TV — which evolved

CHRONOLOGY

1940s-1950s *High postwar birth rates fuel suburban growth.*

1946 First baby boomers are born.

1954 Bill Haley's "Rock Around the Clock" inaugurates the rock 'n' roll era that became symbolic of the baby boom generation.

1959 More than 50 million Americans are under age 14, representing 30 percent of the population.

1960s-1970s *Boomers continue to affect trends, increasing rates of college education and drug use.*

1960 A record 66.5 million Americans are employed. . . . Real wages are up almost 30 percent from 1940 levels. . . . Nine of 10 U.S. households have TVs. . . . Sun City opens in Arizona, pioneering the idea of a retirement community.

1963 In a defining moment for the generation, President John F. Kennedy is assassinated in Dallas.

1964 The Beatles appear on "The Ed Sullivan Show," attracting a record 70 million viewers. . . . Free Speech Movement at University of California, Berkeley launches era of student protests.

1965 Forty-one percent of all Americans are under age 20.

1968 Martin Luther King Jr. and Robert F. Kennedy are assassinated. . . . Antiwar protests spread worldwide.

1970 College enrollment reaches nearly 8 million, vs. 3.6 million in 1960. . . . First Earth Day ushers in modern environmental activism.

1973 Arab oil embargo triggers inflation.

1976 Writer Tom Wolfe names the "Me Decade."

1980s-1990s *Boomers set aside youthful rebelliousness to take a leading role in wealth creation and politics.*

1980 College enrollment reaches 12 million students.

1983 Congress raises age for full Social Security payments from 65 to 67.

1984 *Newsweek* declares it the "year of the yuppie," arguing that young, urban professionals are reshaping the economic and social landscapes.

1986 The Age Discrimination Employment Act is amended to eliminate mandatory retirement ages.

1990 One in two Americans lives in suburbs — double the 1950 ratio.

1992 Bill Clinton becomes first boomer president.

2000s *Oldest boomers enter their 60s, raising concerns about funds for their retirements.*

2000 For every American 65 or over there are 3.4 workers paying into Social Security — a ratio that will shrink to 2.0 by 2030.

2003 Congress passes prescription-drug benefit for seniors.

2005 President George W. Bush calls for privatizing Social Security, gets little support.

2006 President Bush, part of the baby boom generation's leading edge, turns 60. . . . Pension Protection Act allows workers to dip into their pensions while working past 62.

2007 Federal Reserve Chairman Ben S. Bernanke says Social Security and Medicare will swallow 15 percent of annual economic output by 2030. . . . Federal Aviation Administration proposes increasing retirement age for pilots from 60 to 65. . . . Survey finds older Americans enjoy active sex lives. . . . The nation's first baby boomer, Kathleen Casey-Kirschling, 62, a retired schoolteacher in Cherry Hill, N.J., applies for Social Security benefits.

Jan. 1, 2008 Oldest boomers turn 62, become eligible for Social Security.

2010 Number of workers 55 and over to hit 26 million — a 46 percent increase since 2000; number to reach 33 million by 2025.

from being a novelty at the start of the baby boom to a near-universal household appliance by its end. And the first rock records had a secret, defiant meaning for kids whose parents were listening to softer sounds. The success of Bill Haley's 1954 "Rock Around the Clock" was "the first inkling teenagers had that they might be a force to be reckoned with in numbers alone," wrote music critic Lillian Dixon.[43]

But not everything was sunshine and light. Part of the push toward education was fueled by competition in science and engineering with the Soviet Union, which challenged U.S. scientific superiority when it launched the world's first satellite, *Sputnik I*, into space on Oct. 4, 1957. And in a world in which youngsters practiced "duck and cover" drills at school and the government encouraged citizens to build bomb shelters in their back yards, the Cold War was ever-present in the boomer subconscious.

"The baby boomers never forgot the lesson that their world could someday end in a flash of light and heat while they were crouched helplessly in gyms and basements among heating ducts and spare blackboards," Jones writes.[44]

For many, the lesson of formative events such as the Cuban missile crisis of 1962 and the assassination of President John F. Kennedy the following year, was *carpe diem* — live for today. Perhaps partly as a result, the boomers developed an ethos that stressed the pursuit of personal fulfillment rather than focusing on mundane concerns, such as making a living.

Many rejected the "plastic" values of their parents. During the Free Speech Movement that began in 1964 at the University of California, Berkeley, protester Jack Weinberg told a reporter, "We have a saying in the movement, never trust anyone over 30," encapsulating an attitude that rejected the values of the past.[45]

The Free Speech Movement inaugurated an era of student protests, largely focusing on the war in Vietnam. Of the 27 million men who came of draft age from 1964 to 1973, only 11 million served in the military.[46] But protests seemed near-universal during the 1960s, especially in 1968 — the year anti-war demonstrators engaged in bloody clashes with police outside the Democratic National Convention in Chicago.

"After 1968, the antiwar demonstration was the standard adolescent rite of passage," recalled author Annie Gottlieb.[47]

The era's frequent protests — which included demands for equal rights for women and minorities — along with events such as the massive Woodstock rock concert in 1969 — gave visible representation to the enormous size of the baby boom generation. "If the baby boomers could not be heard as individuals, they were delivering a testimony of bodies that was deafening," Jones writes.[48]

From Me to Yuppies

While boomers may have been born due to social pressures and norms, as they grew up they continued to reject traditional social roles. Boomer women were more likely to work outside the home, and the percentage of women in the workforce skyrocketed. As adults, boomers were less likely than their parents to belong to a church, club or traditional nuclear family.

In previous generations, according to University of Massachusetts professor of psychology Whitbourne, the years from ages 30 to 65 were typically quiescent — a time to prepare for retirement and to anticipate and enjoy grandchildren. But as boomers entered adulthood in the 1970s, they remained disproportionately focused on themselves — a phenomenon dubbed by author Tom Wolfe the "me decade" in a celebrated 1976 *New York* article.

"The old alchemical dream was changing base metals into gold. The new alchemical dream is: changing one's personality — remaking, remodeling, elevating and polishing one's very self," Wolfe wrote.[49]

Not everyone fit that description, but many middle-aged boomers remained active, youth-oriented and in search of new challenges (and cosmetic surgery). The bestselling self-help book of the 1970s was *Looking Out for #1*.

"Boomers are America's self-absorbed generation," according to Leslie M. Harris, managing partner of Mature Marketing and Research in Teaneck, N.J. "For them, it wasn't enough to grow up, get married, become parents, be responsible and get off the stage as their parents did."[50] Instead, they wanted to stay in shape, protest a war, run a marathon or join a cult.

In 1985, *Time* magazine wished the boomers a happy early 40th by writing that "trendiness became a generational hallmark: From pot to yoga to jogging, they embraced the In thing of the moment and then quickly changed it for another."[51]

There's No Place Like Home

Cities are providing enhanced services to help seniors stay put.

Beacon Hill in Boston is home to the Massachusetts state capitol and once was home to the 19th-century novelists Henry James and Louisa May Alcott. Over the last few years, however, the swanky area has become known for something new — a concierge service that helps older residents continue to live independently in their own homes.

For up to $780 a year, Beacon Hill Village connects subscribers to carefully screened vendors who can provide them with nursing care, carpentry — or help chasing down a stray cat. The program has become a model for more than 100 such cooperative community efforts across the country.

Help with small tasks — getting to a grocery store or finding a tradesman who would otherwise be unwilling to come out for a small job — can help seniors stay in their own homes much longer. "A few neighborhood-based, relatively inexpensive strategies can have an enormous effect," said Philip McCallion, director of the Center for Excellence in Aging Services at the State University of New York in Albany. "If people don't feel so overwhelmed, they don't feel pushed into precipitous decisions that can't always be reversed."[1]

Beacon Hill residents are among the nation's most affluent — Sen. John F. Kerry, D-Mass., secured a $6 million mortgage on his townhouse there to keep his 2004 presidential campaign afloat. Most of the other areas where residents are setting up copycat programs are also well-to-do. But social-service agencies are launching similar ventures in lower-income areas as well.

With so many aging baby boomers, the nation is experiencing a new phenomenon: what sociologists call NORCs, or naturally occurring retirement communities. Although many seniors do retire to Sun Belt states — including Florida, Arizona and, increasingly, Nevada — they represent a tiny fraction of the senior population as a whole. As the number of senior citizens continues to rise, therefore, many more communities across the nation are having to adjust to offering the types of services older residents require.

"This whole idea of aging in place is important," says William H. Frey, a demographer at the Brookings Institution. "The fact is, most seniors don't move very far. If they do move, it's likely to be locally."[2]

Fewer than 2 percent of those 55 to 64 move across state lines in any given year — a number that grows smaller for those older than 65.[3] The senior population has doubled in recent years in Nevada, as it's become a magnet for seniors who want to follow the sun. But even in New York, which

This much-celebrated move toward self-centeredness may have been encouraged by the decade's economic downturns — among the first to confront the boomers. The 1973 embargo by oil-producing nations led to a period of high unemployment and rising inflation — and boomers faced stiff job competition due to their numbers. Indeed, the rising number of college graduates both lifted expectations and led to disappointment for those who found that a degree was no longer the automatic ticket to prosperity it had once been.

In 1975, some 2 million additional jobs were needed to keep up with population growth. Instead, the nation shed 1 million jobs in a recession, leading to an 8.5 percent unemployment rate.[52] In his history of the boomers, Steve Gillon writes, "A generation raised on the expectations of the good life would confront the cold, hard reality that their quality of life could actually decline."[53]

Things were never quite as rosy for the younger boomers who came of age in a less prosperous time. But by the time the older boomers began reaching their peak earning years in the 1980s — the era when they were dubbed "yuppies," for young, urban professionals — they had triggered a housing boom and were collecting half of all U.S. personal income.[54]

Boomer Critiques

Perhaps thanks to the advent of the birth control pill in 1960 and the fact that more women had careers of their own, boomers were slower to become parents than their parents had been. Between 1965 and 1976 — the era of

has the slowest projected growth among states in the 55-to-64 set, their numbers will still increase by 33 percent from 2000 to 2010.

Frey's research indicates that areas with strong job markets in recent decades now have plenty of people aging into the 55-to-64-year-old group. "In Georgia, for instance, the senior population will increase by more than 40 percent from 2010 to 2020 due to the aging of existing residents, vs. less than 3 percent due to migration," he writes.[4]

Many communities recognize they will have to gear up to offer enhanced services to an aging population. Staying at home is not only what seniors want but also something that can save governments money. The fastest-growing expense within Medicaid is nursing home care.

"The real future in the mayor business is changing demographics," says Louisville Mayor Jerry Abramson. "Aging boomers are changing housing patterns, public transportation and recreational facilities and community-center programming."

State and local governments have initiated many scattered programs to accommodate the rising population of seniors, from enlarged road signs on highways to "senior housing" zoning designations that allow small rental units. Many communities are expanding transportation programs to offer rides to those needing medical care — although few are as ambitious as the Northeastern Colorado Association of Local Governments, which provides heavily subsidized transportation to the sole dialysis center within a 9,500-square-mile region.

A 2005 survey by the National League of Cities found that more municipal officials were concerned about the increase in seniors (72 percent) than about other demographic issues such as growth and immigration.[5] In most communities, though, increased awareness of the need to step up services for seniors has not yet translated into much action. To some extent, that's because Congress approved aging-in-place pilot programs in its 2006 reauthorization of the Older Americans Act, but did not fund them.

Kate Sarosy, the mayor of Casper, Wyo., has been conducting a survey of local agencies to find out what they're doing for seniors and what they need to do to prepare for the coming wave.

"What we're finding is that they're all in a panic," Sarosy says. "They haven't begun to plan for baby boomers. They're having a hard enough time keeping up with their current seniors."

---

[1] Jane Gross, "A Grass-Roots Effort to Grow Old at Home," *The New York Times*, Aug. 14, 2007, p. A1.

[2] For more information, see Marcia Clemmitt, "Caring for the Elderly," *CQ Researcher*, Oct. 13, 2006, pp. 841-864.

[3] Jane Adler, "There's No Place Like Home for Aging Boomers," *Chicago Tribune*, Dec. 24, 2006, p. 4.

[4] William H. Frey, "Mapping the Growth of Older America: Seniors and Boomers in the Early 21st Century," The Brookings Institution, May 2007, p. 1.

[5] Haya El Nasser, "Cities Gird for Getting Grayer," *USA Today*, May 14, 2007, p. 1A.

---

the so-called baby bust — fertility dropped below replacement levels among whites.[55]

After just two decades, Americans were back to marrying later and producing fewer children. In 1990, only 32 percent of women 20-24 were married, compared to 70 percent in 1960.

But the baby bust was followed by the uptick known as the "echo boom," when many boomers became parents themselves, racking up 64 million live births between 1977 and 1993.[56]

The boomer's sense of individualism was not purely selfish. Last year, Leonard Steinhorn, a communications professor at American University, argued that the boomers made enormous strides in areas such as gender equality and environmental protection. "Boomers will never go

down in history as a generation that fought a great war to protect liberty," he wrote, "but boomers should go down in history as a generation that fought a great cultural war to expand and advance liberty."[57]

Steinhorn was writing in response to the 1989 book *Destructive Generation*, the 2000 *Esquire* article "The Worst Generation" and other critiques of boomers that blamed their personal habits and quest for self-fulfillment for every social ill from climbing divorce rates to teen drug use.

The debate about boomers' sex-drugs-and-rock-'n'-roll values became a recurring motif in politics — especially after Bill Clinton, who would become the first boomer president, emerged on the national stage in 1992. Political scientists have noted that the boomers failed to coalesce behind a single political party, with many growing more

Many Older Americans Continue Working

Traditional definition of 'retirement' is outmoded.

When Don Davidson was in his 50s, he decided it was time for a change. The longtime publishing executive's plan: turn his hobby — carpentry — into a business. Davidson pulled it off and now employs his two sons full time installing cabinets and refinishing furniture.[1]

Davidson is part of a growing trend of older workers who reach the traditional retirement age and decide to strike out in a new direction, rather than just withdraw from the workforce. "We now know that baby boomers are going to work longer than their parents did, whether they want to or not," writes Marc Freedman, founder of Civic Ventures, which promotes employment among older Americans. "Four out of five boomers consistently tell researchers that they expect to work well into what used to be known as the retirement years."[2]

The percentage of people in their late 60s who are working has increased from 18 percent in 1985 to nearly 30 percent in 2006, according to the Employee Benefit Research Institute, and it's a trend that's likely to continue. "The old, traditional definition of retirement doesn't work any more," says Sandra Timmermann, director of the MetLife Mature Market Institute, in Westport, Conn. "It's really no longer a fixed date."

A bipartisan group of senators has introduced legislation to provide a tax credit for employers who allow workers 62 and above to work flexible hours while retaining full pension and health-care benefits. The bill has not yet seen action, however.

"We're watching a very, very pleasant change occur, where industries that have traditionally relied on bright-eyed and bushy-tailed young people are now relying on older workers to do these jobs," said Bob Skladany, vice president of RetirementJobs.com., an online service that matches companies most-suited to older workers with seniors seeking a job or project that matches their lifestyle. "I think we're seeing the first wave of a fairly substantial shift."[3]

People are working past 65 for a number of reasons. Some simply need the money, recognizing that longer life expectancy leaves them with greater financial needs, as well as time on their hands. Meanwhile, some of the financial disincentives under Social Security and private pension laws that once penalized elderly people for working have been dropped. And many potential retirees relish the chance to do something different and, perhaps, more meaningful.

"Baby boomers grew up in the JFK era," says Frank Benest, city manager of Palo Alto, Calif., referring to John F. Kennedy, who served as president from 1961 to 1963 and challenged young Americans to give back to their country by volunteering. "They value the idea of contributing to their community. All of a sudden, they have the opportunity to do that."

Some economists are predicting that not only will boomers want to continue working but employers will need them. Within the next seven years, the number of workers age 55 and older will increase at four times the rate

fiscally conservative during the 1980s but remaining socially liberal, with views on race, AIDS, drugs and women's rights distinctly different from their parents' generation.

Yet, there were divisions — particularly over women's role in society. While most boomers celebrated the fact that a majority of women were working by the 1980s, in her address against the "counterculture" at the 1992 Republican National Convention Marilyn Quayle, wife of Vice President Dan Quayle, dismissed women who "wish to be liberated from their essential natures as women." Quayle noted that "Dan and I are members of the baby

boom generation, too," but "not everyone demonstrated, dropped out, took drugs, joined in the social revolution or dodged the draft."[58]

Former Kansas Sen. Bob Dole, the Republican presidential nominee in 1996, returned to the generational theme during his acceptance speech at the GOP convention, charging that the Clinton administration was made up of a soft "corps of the elite who never grew up, never did anything real, never suffered and never learned." That night, *Newsweek* reported, Clinton was celebrating his 50th birthday at a "plush summer home" in Jackson Hole, Wyo., singing Beatles songs.[59]

of the overall labor force. The Conference Board, a business research group, estimates that by 2010, 64 million workers — 40 percent of the nation's public and private workforce — will have reached retirement age.[4]

"Employers are going to need to find ways to engage older workers longer and more fully than has been the case in the past," says Michael A. Smyer, codirector of Boston College's Center on Aging and Work.

But others are more skeptical. A study released in May by a rival center at Boston College found that employers are "lukewarm" about retaining older workers.[5] Relatively few employers are interested in providing the flexible hours older workers desire — or paying for their health coverage, which can be more expensive than covering the young.

Some economists say the labor shortage predicted by demographers — created by the smaller generation that followed the boomers — should be manageable, especially given the global labor pool. Many retirees will be replaced in the workforce by immigrants — an advantage the U.S. has over other aging industrialized nations with less-open borders — while other jobs will be shipped abroad. And not all the boomers are going to hit retirement age at the same time.

In addition, a 2006 study by the consulting firm McKinsey & Co. found that 40 percent of current retirees left their jobs earlier than they had planned, either because of layoffs, downsizing or health reasons.[6] Given that several other studies have found that people often end up retiring earlier than they expected, from the employer point of view the need to keep aging boomers on the payroll is proving less pressing than some had expected.

Companies that have looked at the aging workforce have come to the conclusion that it isn't as big a problem as they thought," says Mary Young, a senior researcher at the Conference Board. "It's much more manageable."

Young concedes the aging baby boomers will exacerbate existing shortages in professions such as nursing and teaching and create new ones in fields such as engineering and the utilities industry. Governments also employ a disproportionate share of older workers.

But what about the rest of the labor market? Aging workers are attractive to industries with high turnover, such as retail. Both CVS and Borders have programs that allow people to follow the seasons, working summers in the North and winters in the South. Not all employers are likely to be that flexible, however.

"Endless surveys have found contradictory conclusions," says Ron Manheimer, executive director of the Center for Creative Retirement at the University of North Carolina, Asheville. "Human-resources directors say seniors are loyal and reliable and have better customer-service attitudes — but they're not planning to hire them."

[1] Emily Brandon, "You Can Use This Time of Your Life for a Whole New Beginning," *U.S. News & World Report*, June 12, 2006, p. 55.

[2] Marc Freedman, *Encore: Finding Work That Matters in the Second Half of Life* (2007), p. 9.

[3] Jonathan Peterson, "Older Workers Becoming Valued Prize for Firms," *Chicago Tribune*, Sept. 10, 2007, p. 1.

[4] Christopher Conte, "Expert Exodus," *Governing*, February 2006, p. 22.

[5] Andrew D. Eschtruth, *et al.*, "Employers Lukewarm About Retaining Older Workers," Boston College Center for Retirement Research, May 2007.

[6] Sandra Block, "Off to Work They Go, Even After Retirement Age," *USA Today*, Aug. 31, 2007, p. 1B.

At his own nominating convention in 2000, George W. Bush echoed the points made by Quayle and Dole, seeking to castigate Democrats for the purported failings of the baby boom generation. "Our current president embodied the potential of a generation," Bush said. "So many talents, so much charm, such great skill. But in the end, to what end? So much promise, to no great purpose."[60]

But President Bush, of course, is a boomer — only five weeks older than Bill Clinton. "This year, the first of about 78 million baby boomers turn 60, including two of my dad's favorite people, me and President Clinton," Bush said during his 2006 State of the Union address. "This milestone is more than a personal crisis. It is a national challenge. The retirement of the baby boom generation will put unprecedented strains on the federal government."[61]

Combined spending for Social Security, Medicare and Medicaid will consume 60 percent of the federal budget by 2030, Bush said, presenting future Congresses with "impossible choices — staggering tax increases, immense deficits or deep cuts in every category of spending."

Bush had spent a good chunk of 2005 touting his plan to revamp Social Security, meant to be the signature domestic achievement of his second term. But the plan — which would have allowed workers born after

1950 to put part of their payroll taxes into private investment accounts in exchange for cuts to traditional benefits — went nowhere. A Washington Post/ABC News Poll found that 58 percent of those surveyed said the more they heard about Bush's plan the less they liked it.[62]

CURRENT SITUATION

Entitlement Bills

Perhaps not coincidentally, the nation's budget-busting entitlement programs for seniors have barely been mentioned during the current run-up to the 2008 presidential campaign. Candidates are generally mum about how to deal with the fact that Social Security is expected to run through its surplus accounts by 2040 and Medicare and Medicaid — the federal government's two main federal health-care programs — pose an even more dire fiscal threat.[63]

President Bush has proposed a "means test" for the prescription-drug benefit under Medicare that he created in 2003. His idea is to charge higher premiums and deductibles for upper-income seniors. But the proposal died as part of his budget earlier this year, and the Senate voted down a similar idea in March.[64]

A pair of proposals are pending in Congress to create a commission to craft an entitlement-reform package that Congress and the next president would have to act upon. Bush had created a commission to address Social Security, but the current proposals would go further, tackling other entitlements as well. If approved, a new commission would be jointly appointed by the president and Congress and would likely craft a proposal that Congress would have to accept or reject, without amendment.

Some expect a solution for Medicare and Medicaid will be found within a broader discussion of reforming the U.S. health system. The three leading Democratic presidential candidates — Sens. Hillary Rodham Clinton and Barack Obama and former Sen. John Edwards — have all unveiled proposals for universal or near-universal health-coverage laws.

Clinton launched a plan in September that involved fewer government mandates than the universal-coverage package she designed as first lady in 1993. "I learned that people who are satisfied with their current coverage want assurances that they can keep it," she said. "Part of our

health-care system is the best in the world, and we should build on it; part of the system is broken, and we should fix it."[65]

Clinton's new plan, in fact, resembles a law passed in Massachusetts in 2006 that requires nearly all residents to buy private health insurance if they don't qualify for government-run coverage such as Medicaid. That bill was signed into law by Mitt Romney — then Massachusetts governor and now a leading GOP presidential candidate. The bill, along with other health-care expansion laws in Vermont and Maine, triggered a wave of activity among nearly two-dozen states that considered (but did not pass) universal health legislation this year.[66]

But Romney has not talked about his state's law much on the campaign trail this year, saying that he would prefer to leave it to states to create insurance mandates without prodding from Washington. Romney and the other Republican candidates have talked about problems with the U.S. health-care system but in general favor reforms within the insurance markets, states or the use of tax credits to resolve them, rather than any sort of new federal program.

Clearly, the fact that millions of boomers will soon reach retirement age has many economists worried about entitlement expenses and recommending cutting back coverage — or at least raising the age of eligibility. Federal Reserve Chairman Ben S. Bernanke told the Senate Budget Committee in January that by 2030 Social Security and Medicare will cost the equivalent of 15 percent of U.S. annual economic output — up from 8.5 percent today.

"The longer we wait [to make changes], the more severe, the more draconian, the more difficult the adjustment is going to be," Bernanke said. "I think the right time to start is about 10 years ago."[67]

Many of these financing problems, particularly in Social Security, are rooted in the fact that the ratio of workers to retirees is shrinking. There were 16.5 workers paying into the system for every Social Security beneficiary in 1950, but now there are only 3.3 workers for every retiree — a number that continues to decline.

One solution, AARP and other advocates for the elderly say, would be for boomers to work longer than their parents did. While the percentage of people in their late 60s still working has risen, they remain a minority. People working just a few years longer — and collecting fewer years worth of Social Security checks — could erase its

Do aging boomers pose a threat to fiscal solvency?

YES

David M. Walker
Comptroller General,
Government Accountability Office

From a speech delivered in Washington, D.C., Nov. 7, 2006

The United States government is on an imprudent and unsustainable fiscal path. We do not face an immediate crisis, but we face large and growing structural imbalances that are growing every second of every minute of every day due to continuing deficits, known demographic trends — the demographic tidal wave, the demographic tsunami which is represented by the retirement of the baby boom generation — and rising healthcare costs.

Let me give you some numbers. In 1965, 43 percent of the federal government's budget was for defense. Fast-forward 40 years to 2005: It was down to 20 percent. Where did the money go? Social Security, Medicare and Medicaid. In 1965, we spent zero money on Medicare and Medicaid because they didn't exist. In 2005, 19 percent of the federal budget was for Medicare and Medicaid, and growing rapidly. In 1965, [Congress] got to decide how two-thirds of the money was going to be spent. In 2005, it was down to 39 percent; stated differently, 61 percent of the budget is on autopilot, and that number is going up every year.

You project it out for 75 years, which is what the Social Security and Medicare trustees are required to do every year. If you take the difference between what we promised and what's funded for Social Security and Medicare alone, you'll find that the total liabilities and unfunded commitments of the United States in the last five years have gone up from a little over $20 trillion — and you got to add 12 zeroes to the right of that 20 to get a feel for that number — to over $46 trillion in five years. It's going up every second of every minute of every day.

How much is $46 trillion? It's over 90 percent of the entire net worth of every American in the United States. It's $156,000 for every man, woman and child in the United States. People talk about eliminating the death tax. How about eliminating the birth burden, that $156,000? No wonder newborn babies cry. Somebody's giving them the bill.

Let's tell it like it is. This is mortgaging the future of our kids and grandkids, big time. And for the first time in the history of the United States, the baby boom generation, of which I am a member, may be the first generation in the history of this country to leave this country in a situation where their kids and grandkids will not have a higher quality of life. That is not acceptable, and we need to start doing something about it.

NO

Robert B. Friedland and Laura Summer
Center on an Aging Society,
Georgetown University

From *Demography Is Not Destiny, Revisited* (2005)

That our society is aging is well known. Media stories and political rhetoric abound concerning the impending demographic challenges as the population age 65 and older is anticipated to more than double by the year 2030. Much of the hand wringing concerns an expectation of dire fiscal consequences for publicly financed programs, such as Medicare and Social Security, of which older people tend to be the principal beneficiaries.

What is not said is that planning for the future on the basis of demographic projections alone is a fool's game. Population projections can be wrong, but even if they turn out to be correct, other factors, particularly those related to the economy and public policies, can have a decidedly greater impact on the future than simply the growing number of older people.

At any point in the past century, one could have easily anticipated a dramatic increase in the size and proportion of the population age 65 and older. Since 1900, the number of Americans age 65 and older has doubled three times. Since 1960, the population age 65 and older has doubled while the overall population has only grown 57 percent. However, since 1960 the nation's income (as measured by real gross domestic product) has nearly quadrupled.

Economic growth has made the nation more prosperous and has enabled many to enjoy a higher standard of living than would have been possible a generation earlier. . . .

There are legitimate reasons to be concerned about growth in federal entitlement spending, but there is more reason to be concerned about economic growth. Small differences in sustained economic growth will have a dramatic impact on the fiscal future of society. If real economic growth averages about 2 percent per year between now and 2050, then, depending on the policy choices we make, government expenditures as a proportion of the economy in 2050 might not be substantially larger than today, and we will still be able to meet the promises made to future beneficiaries.

It would be foolish to assume society will simply grow its way out of the difficult choices that the aging of the population will require. It would be equally foolish to assume that the future will be completely dismal if there is no radical restructuring of government programs. If public policies support the market transitions necessary for economic growth during demographic transitions, then we can afford to meet the challenges of the retirement of the baby boomers.

deficits for the next 75 years, say some economists.[68] This would also help solve any workforce shortages posed by the aging baby boom generation.

"The old model was people worked for 40 or 45 years and then they slipped off into retirement and focused on leisure and recreation," says Freedman of Civic Ventures. "We've said to older people as they moved into their 50s and 60s, please leave the labor market."

The proposed Older Worker Opportunity Act, sponsored by Sen. Herb Kohl, D-Wis., and a bipartisan group of cosponsors, would provide a tax credit of up to 25 percent of a worker's wages to employers who allow workers age 62 and above to participate in a flexible program that allows them to work full or part time while retaining full pension and health-care benefits. But the measure has seen no action, and not everyone is convinced there will be jobs for millions of seniors to continue working.

Ron Manheimer, executive director of the Center for Creative Retirement at the University of North Carolina, Asheville, says that while human resources directors say seniors are loyal and reliable, many employers are not planning to hire them.

There are a number of reasons why employers are not embracing the idea of workers staying on the job longer. Some are concerned that their younger employees will leave if they have less hope of advancement. According to Mary Young, a senior researcher at the Conference Board, 65 percent of the employees who leave IBM do so because they see little hope of promotional opportunities. "That's another reason why holding onto boomers isn't a cure-all," she says.

Other employers are worried that older workers don't always keep their skill sets up to date in an age of rapid technological change, while still others worry about the amount of hours they can count on older workers logging. Workers with long experience have also had years of salary increases and their health-insurance premiums can be more costly.

Finally, says Lynch of Claremont McKenna College, "There isn't a whole lot of evidence for it yet, but I think boomers are going to face age discrimination."

Fiscal Cancer

In Christopher Buckley's satirical novel *Boomsday*, which opens with a mob of young people rioting in front of a Florida gated community "known to harbor early retiring boomers" in protest of a Social Security payroll tax hike, a 29-year-old character suggests a solution to the entitlement funding problem: pay retirees to commit suicide. As an incentive, volunteers could take one last, lavish vacation at government expense, and their children would be exempt from estate taxes. Even so, the government would come out ahead. If only 20 percent of boomers committed suicide, Social Security and Medicare would remain solvent.

In the world of reality, rather than satire, some analysts criticize today's seniors and boomers for what the critics perceive as a selfish insistence on expensive benefits that will have to be paid for by younger workers.

"At nearly every critical juncture, they have preferred the present to the future," wrote former Clinton adviser Paul Begala. "They've put themselves ahead of their parents, ahead of their country, ahead of their children — ahead of our future."[69]

The central question of *Immigrants and Boomers*, a recent book by USC demographer Myers, is whether aging boomers will support programs that benefit younger generations — dominated by other ethnic groups — at the short-term expense of programs that benefit themselves.

"We actually transfer resources from workers up to the elderly, essentially wasting resources," says Myers. "Younger people would have more years of life and could recoup the investment."

The question of whether older, mostly white seniors will support education and other domestic programs used by mostly brown- and black-skinned younger people is becoming increasingly pressing. During the recent congressional debate over the State Children's Health Insurance Program, the House version of the bill would have provided $15 billion more to cover more children who lack insurance. It would have been funded by cutting subsidies to Medicare managed-care plans and providers. The House voted to support the legislation, but the idea proved unpalatable in the Senate, which passed a more modest expansion of the program. Bush vetoed the final bill, however.

MIT economist Diamond notes, "Surveys suggest that young people are very supportive of paying taxes for Medicare and Social Security for their parents," because it removes a potential burden for them. "On the flip side, older people are supportive of education for their grandchildren."

But Medicare and Social Security are entitlement programs, Diamond points out, so spending increases for programs benefiting older people are automatic. Yet Congress must allocate funds annually for programs that primarily benefit younger Americans.

Asked why no one in Congress is taking the lead on reforming entitlement programs, Sen. Kent Conrad, D-N.D., told CBS's "60 Minutes" in July, "It's always easier to defer, to kick the can down the road to avoid making choices. You know, you get in trouble in politics when you make choices."[70]

But Comptroller General Walker warns of the dire consequences of ignoring the entitlement problem: "We suffer from a fiscal cancer," he says. "It is growing within us, and if we do not treat it, it could have catastrophic consequences for our country."[71]

'Sandwich' Generation

Boomers will add to the rising number of seniors — but their parents, in many cases, will still be around. Those 85 and over now make up the fastest-growing segment of the U.S. population, according to the National Institute on Aging. That means that even as the boomers enter what has traditionally been considered old age, they are "sandwiched" between still-living parents and their own children and grandchildren.

A 2005 survey found that 13 million boomers were "deeply involved" in the care of their aging parents, with 25 percent living with their parents.[72] "The children who are caring for the elderly are elderly themselves," says Stanford's Carstensen.

Many boomers already wonder who, in turn, will take care of them when they are frail themselves. That may be particularly true for boomer women, since far more of them are entering old age single — divorced or having never been married — than has traditionally been the case.[73]

"Boomers are quite different from earlier generations as they're approaching this age," says William H. Frey, a demographer at the Brookings Institution. For example, boomer women "are much more likely to have lived independent lives, been head of households and worked."

But there's a great deal of economic inequality within the baby boom generation, he notes, which means many retirees will have a hard time making ends meet. In

Older baby boomers like Marcos Zavala, 59, a locomotive engineer for The Indiana Rail Road Co., pose a dilemma for demographic experts. While many older boomers are expected to retire soon, up to 80 percent of boomers intend to work past 65 to help make ends meet. Two-thirds of the nation's retirees rely on Social Security for at least 50 percent of their income.

addition, Frey says, boomers didn't have as many children as their parents' generation, so they "can't rely on them for support."

As a result, while federal policy makers puzzle over how to pay for entitlement programs, state and local governments are gearing up to provide more services to aging boomers. For example, by 2030 the number of drivers 65 or older could nearly double, to 65 million, according to the GAO.[74] So states and cities are scrambling to enlarge and brighten road signs.

In California, where an elderly driver slammed into an open-air market in 2003, killing 10 people, the Department of Motor Vehicles has launched a pilot program that requires drivers to pass more intensive vision, memory and reflex tests when renewing their licenses. "What we can do is try to identify drivers who can't drive safely," said David Hennessy, a former DMV research program specialist. "This is something we've become especially sensitive to because of the aging of the baby boomers."[75]

Transportation is a key concern for the elderly. Millions of seniors find themselves stranded in the suburbs when they can no longer drive safely. The Denver-area Seniors' Resource Center has responded by offering different transportation modes for seniors, ranging from paratransit — unscheduled rides in vans or cars — to volunteer drivers and taxis.[76]

Other local governments have similar ride-dispatch programs or even training programs to familiarize seniors with public transportation. They are also widening sidewalks to accommodate wheelchairs and encourage walking and, in a few cases, changing zoning laws to allow "granny flats" and other small, one-floor housing units in dense areas (particularly near grocery stores).

An aging population, says Syracuse social work professor Kingson, is a sign that society has successfully fostered an economy that helps people lead long, prosperous lives. "Population aging is not just about the old," he says. "It's about how all of our institutions are going to change."

"The question is, How do we adapt as we get older, and how does society adapt to our needs?" says Achenbaum, the University of Houston historian of aging. "There's no question that we're going to change the meaning of late life."

OUTLOOK

Late Boomers

A decade from now, the oldest boomers will be in their 70s, while the youngest will be well into their 50s. It's almost impossible to predict what their lives will be like and how they will have changed the nature of senior citizenship.

By the time the youngest boomers are in their 60s, their older brothers and sisters may have changed senior housing patterns, workforce participation rates and leisure and recreational activities. Patterns of health and disability may also have changed as medical science continues to make new breakthroughs.

In a few years, Smith of Yankelovich predicts, "We'll probably be talking about one thing especially — that is, a world in which there's an older generation that can expect to live actively past the age of 90."

Even as boomers get older, Smith says, "They have this psychological view of themselves, this youthful mindset, which helps motivate health-service providers to pioneer new solutions to help them stay young a little bit and beat the odds."

Signs already point in that direction. Not only are boomers buying cars aimed at younger drivers, but they also are changing the marketing and packaging of products and services aimed at older people. For instance, many boomers are too vain to wear hearing aids. "To appeal to boomer vanity, many companies are making hearing aids that look more like cousins of Bluetooth or iPod earbuds," said Gordon Wilson, vice president for marketing at Oticon, which manufactures hearing aids.[77]

"Boomers are different from their predecessors and are also different from the generation beyond them," said Washington dermatologist Tina Alster. "They know they can get [cosmetic surgery]. What they're not realistic about is they think they can just come in and do one thing and take away five decades of sin."[78]

Indeed, age does have its consequences. Even optimists like Smith concede there is a downside to living longer. Boomers who had expected to be retired for only a decade or so may start to run out of money as they continue to age. And, although disabilities and frailties are getting pushed farther back in life, they do occur.

"One of the biggest questions for baby boomers is 'Who is going to take care of me?' " says AARP's Rix. "Old age, in the sense of increasing physical problems, is being pushed back, but ultimately more and more people are going to be facing the types of problems that require assistance."

As the senior population grows — both in sheer numbers and as a percentage of the population as a whole — new strains will be put on social services, government budgets and, perhaps, the economy as a whole. Myers, the USC demographer, believes a labor shortage will force employers to adapt their policies to accommodate and retain older workers. He also thinks there will be a side benefit: Society will start to offer better-quality educations to minority groups who traditionally have languished in schools. "We'll be scouting for workers among minority youth," he says. "We haven't worried too much about them before, but suddenly they're going to be precious." Like many observers, Myers worries about the growth in the cost of senior entitlement programs and recommends moving back eligibility ages.

"It's very heard to make changes when the changes you're going to make are seen as painful by the voting public," says MIT economist Diamond. "But the longer you wait, the harder it is to deal with it."

Social scientists can get frustrated at policy makers, who have known for decades that the baby boomers would eventually reach old age. Yet, even though they are on the cusp of retirement age, there appears to be little movement in addressing the core issues.

Employers are not much better prepared, says the Wharton School's Capelli. Even though he doubts that aging boomers will create a labor shortage, he's surprised that employers aren't preparing for the fact that many of their older workers with legacy skills are going to leave.

"The surprise is not that these people are going to retire," he says. "The surprise is that nobody was planning for it."

Whether boomers resort to rocking chairs and playing with their grandchildren or end up changing the nature of old age by living longer, staying healthier and continuing to work past 65, there are so many of them that American society must change in profound ways to this new senior population.

"The question is, 'How do we adapt as we get older and how does society adapt to our needs?'" says Achenbaum, of the University of Houston. "There's no question that we're going to change the meaning of late life."

NOTES

1. Leonard Steinhorn, *The Greatest Generation: In Defense of the Baby Boom* (2006), p. 82.

2. J. Walker Smith and Ann Clurman, *Generation Ageless: How Baby Boomers Are Changing the Way We Live Today . . . and They're Just Getting Started* (2007), p. 97.

3. Leslie M. Harris, ed., *After Fifty: How the Baby Boom Will Redefine the Mature Market* (2003), p. vii.

4. Smith and Clurman, *op. cit.*, p. 89.

5. *Ibid.*, p. 28.

6. *Ibid.*, p. 24.

7. Paul C. Light, *Baby Boomers* (1988), p. 10.

8. Susan Krauss Whitbourne and Sherry L. Willis, eds., *The Baby Boomers Grow Up: Contemporary Perspectives on Midlife* (2006), p. 283.

9. "60-Year-Old Woman Delivers Twin Boys," Fox News, May 24, 2007, www.foxnews.com/story/0,2933,274726,00.html.

10. "Baby Boomers' Retirement Prospects: An Overview," Congressional Budget Office, November 2003.

11. Dennis Cauchon, "Generation Gap? About $200,000," *USA Today*, May 21, 2007, p. 1A.

12. Mindy Fetterman, "Retirement Unfolds in Five Stages for Hearty Boomers," *USA Today*, June 26, 2006, p. 1B.

13. Kathleen Day, "Bankruptcies Rise Fastest for Over-55 Group," *The Washington Post*, April 27, 2007, p. D3.

14. "Baby Boom Generation," Government Accountability Office, July 2006, p. 8.

15. Sharon A. DeVaney and Sophia T. Chiremba, "Comparing the Retirement Savings of the Baby Boomers and Other Cohorts," Bureau of Labor Statistics, March 15, 2005.

16. Sonia Arrison, "80 Is the New 65," *Los Angeles Times*, March 13, 2007, p. A17.

17. For background, see Alan Greenblatt, "Pension Crisis," *CQ Researcher*, Feb 17, 2006, pp. 145-168.

18. Maria L. La Ganga, "More People Over 55 Are Working 9 to 5," *Los Angeles Times*, April 3, 2007, p. B1.

19. Robert B. Friedland and Laura Summer, "Demography Is Not Destiny, Revisited," Georgetown University Center on an Aging Society, March 2005, p. 43.

20. See "Wake-Up Call; David Walker, America's top accountant, going on tour to sound the alarm that America cannot sustain current level of spending," transcript from "60 Minutes," CBS News, July 8, 2007.

21. Steven R. Weisman, "Fed Chief Warns That Entitlement Growth Could Harm Economy," *The New York Times*, Jan. 19, 2007, p. C1.

22. "Baby Boom Generation," *op. cit.,* p. 16.

23. Smith and Clurman, *op. cit.*, p. 35.

24. Sheryl Gay Stolberg, "A Touchy Topic," *The New York Times*, July 6, 2006, p. A1.

25. Nell Henderson, "Aging Is Inevitable, But Boomers Put 'Old' on Hold," *The Washington Post*, Sept. 12, 2007, p. H1.

26. Kim Painter, "Boomers Leap the 60 Hurdle," *USA Today*, Dec. 18, 2006, p. 4D.

27. Joel Achenbach, "The Rise of the Alpha Geezer," *The Washington Post*, Sept. 9, 2007, p. B3.

28. Leslie Walker, "Cross-Training Your Brain to Maintain Its Strength," *The Washington Post*, Sept. 12, 2007, p. H2.

29. Rob Stein, "Baby Boomers Appear to Be Less Healthy Than Parents," *The Washington Post*, April 20, 2007, p. A1.

30. Stephanie Rosenbloom, "The Day the Music Died," *The New York Times*, July 12, 2007, p. G1.

31. Light, p. 23.

32. Herbert S. Klein, "The U.S. Baby Bust in Historical Perspective," in Fred R. Harris, ed., *The Baby Bust: Who Will Do the Work? Who Will Pay the Taxes?* (2006), p. 115.

33. Light, *op. cit.*, p. 23.

34. Steve Gillon, *Boomer Nation: The Largest and Richest Generation Ever and How It Changed America* (2004), p. 1.

35. Light, *op. cit.*, p. 24.

36. For background on the G.I. Bill, see "Record of the 78th Congress (Second Session)," in *Editorial Research Reports*, Dec. 20, 1944, available from *CQ Researcher Plus Archive*, http://cqpress.com.

37. Gillon, *op. cit.*, p. 6.

38. For background, see H. B. Shaffer, "Progress Against Polio," in *Editorial Research Reports*, March 14, 1956, available from *CQ Researcher Plus Archive*, http://cqpress.com.

39. Abigail J. Stewart and Cynthia M. Torges, "Social, Historical and Developmental Influences on Psychology of the Baby Boom at Midlife," in Whitbourne and Willis, *op. cit.*, p. 31.

40. Light, *op. cit.*, p. 121.

41. Whitbourne and Willis, *op. cit.*, p. 12.

42. Landon Y. Jones, *Great Expectations: America and the Baby Boom Generation* (1980), p. 1.

43. Gillon, *op. cit.*, p. 7.

44. Jones, *op. cit.*, p. 52.

45. Steinhorn, *op. cit.*, p. 82.

46. Gillon, *op. cit.*, p. 51.

47. Annie Gottlieb, *Do You Believe in Magic?* (1987), p. 47.

48. Jones, *op. cit.*, p. 99.

49. Tom Wolfe, "The 'Me' Decade and the Third Great Awakening," *New York*, Aug. 23, 1976, p. 26.

50. Harris, *op. cit.*, p. 2.

51. Evan Thomas, "Growing Pains at 40," *Time*, May 19, 1986, p. 22.

52. Jones, *op. cit.*, p. 152.

53. Gillon, *op. cit.*, p. 22.

54. *Ibid.*, p. 117. Also see Roger Thompson, "Baby Boom's Mid-life Crisis," in *Editorial Research Reports*, Jan. 8, 1988, available in *CQ Researcher Plus Archive*, http://library.cqpress.com.

55. Klein, *op. cit.*, p. 173.

56. William Sterling and Stephen Waite, *Boomernomics: The Future of Money in the Upcoming Generational Warfare* (1998), p. 3.

57. Steinhorn, *op. cit.*, p. xiii.

58. Ronald Brownstein, "GOP Takes Politics of '92 Race Personally," *Los Angeles Times*, Aug. 21, 1992.

59. Howard Fineman, "Bring on the Baby Boomers," *Newsweek*, Aug. 26, 1996, p. 18.

60. John Diamond, "Bush: 'They Have Not Led, We Will,' " *Chicago Tribune*, Aug. 4, 2000, p. 1.

61. President Bush, State of the Union address, Jan. 31, 2006, http://www.whitehouse.gov/news/releases/2006/01/20060131-10.html.

62. Jonathan Weisman, "Skepticism of Bush's Social Security Plan Is Growing," *The Washington Post*, March 15, 2005, p. A1.

63. "Social Security's Future — FAQs," Social Security Administration, www.ssa.gov/qa.htm.

64. Jonathan Weisman, "Means Test Sought for Medicare Drug Plan," *The Washington Post*, Oct. 5, 2007, p. A1.

65. Patrick Healy and Robin Toner, "Wary of Past, Clinton Unveils a Health Plan," Sept. 18, 2007, p. A1.

66. Alan Greenblatt, "Gimme Coverage," *Governing*, June 2007, p. 40.

67. Steven R. Weisman, "Fed Chief Warns That Entitlement Growth Could Harm Economy," *The New York Times*, Jan. 19, 2007, p. C1.

68. Walker and Clurman, *op. cit.*, p. 51.

69. Paul Begala, "The Worst Generation," *Esquire*, April 2000.

70. "60 Minutes," *op. cit.*

71. *Ibid.*

72. "Thirteen Million Baby Boomers Care for Ailing Parents, 25% Live With Parents," *Senior-Journal.com*; http://seniorjournal.com/NEWS/Boomers/5-10-19BoomersCare4Parents.htm.

73. Mary Elizabeth Hughes and Angela M. O'Rand, "The Lives and Times of the Baby Boomers," Russell Sage Foundation/Population Research Bureau, 2004, p. 19.

74. William Neikirk, "States Told to Prep for Gray Driver Boom," *Chicago Tribune*, April 12, 2007, p. 3.

75. Rong-Gong Lin II, "DMV Tests a Tough New Test," *Los Angeles Times*, Sept. 30, 2007, p. B1.

76. Christopher Swope, "Stranded Seniors," *Governing*, June 2005, p. 40.

77. Rosenbloom, *op. cit.*, p. G1.

78. Tina Alster, "As Boomers Hit Their 60s, the Clock Takes on a New Face," *The Washington Post*, Sept. 12, 2007, p. H5. Also see David Masci, "Baby Boomers at Midlife," *CQ Researcher*, July 31, 1998, pp. 649-672.

BIBLIOGRAPHY

Books

Freedman, Marc, *Encore: Finding Work That Matters in the Second Half of Life*, PublicAffairs, 2007.
The founder of Civic Ventures, which aims to match up seniors with meaningful work, offers specific examples of workers who have changed the course of their careers.

Gillon, Steve, *Boomer Nation: The Largest and Richest Generation Ever and How It Changed America*, Free Press, 2004.
The History Channel's Gillon writes a sympathetic history of the boomers whose birth, he says, is the "single greatest demographic event in American history."

Jones, Landon Y., *Great Expectations: America and the Baby Boom Generation*, Coward, McCann & Geoghegan, 1980.
A journalist's pioneering history of the boomers is still a reliable portrait of the generation's first 35 years.

Smith, J. Walker, and Ann Clurman, *Generation Ageless: How Baby Boomers Are Changing the Way We Live Today . . . and They're Just Getting Started*, Collins, 2007.
As their subtitle suggests, these two executives with the consumer research firm Yankelovich, Inc., believe boomers remain influential and that sectors other than travel and financial services need to recalibrate their offerings to them as they age.

Whitbourne, Susan Krauss, and Sherry L. Willis, eds., *The Baby Boomers Grow Up: Contemporary Perspectives on Midlife* (2006), p. 283.
Fresh research from social scientists about boomer psychology, employment patterns, health and other issues.

Articles

Adler, Jerry, "Hitting 60," *Newsweek*, Nov. 14, 2005, p. 50.
Newsweek kicks off a long series of "Boomer Files" articles about the boomers, their habits and cultural impact.

Begala, Paul, "The Worst Generation," *Esquire*, April 2000.
The political adviser makes the case that boomers have been the most self-indulgent generation in American history.

Conte, Christopher, "Expert Exodus," *Governing*, February 2006, p. 22.
Government workers were hired in abundance in the 1960s and '70s, leaving the public sector with a much older workforce than private companies.

Gross, Jane, "A Grass-Roots Effort to Grow Old at Home," *The New York Times*, Aug. 14, 2007, p. A1.
Seniors are banding together to create neighborhood-based associations that provide everything from medical care to carpentry in an effort to stay longer in their own homes.

Hulbert, Mark, "Baby Boomers Are Cashing In. So What?" *The New York Times*, May 27, 2007, p. 5.
Economists show that although boomers are beginning to liquidate assets, the total size of their investments, such as 401(k) plans, will continue to grow.

Peterson, Jonathan, "At Some Companies, Older Skilled Workers Are Golden," *Los Angeles Times*, Sept. 3, 2007, p. 1.
Since the generation following the boomers is 16 percent smaller, many firms and the government are realizing they'll have to offer incentives to keep boomers on the payroll longer.

Studies and Reports

"Baby Boom Generation: Retirement of Baby Boomers Is Unlikely to Precipitate Dramatic Declines in Market

Returns, but Broader Risks Threaten Retirement Security," *Government Accountability Office*, July 2006, www.gao.gov/new.items/d06718.pdf.
Boomers will have to sell off assets to fund retirement but will not sell all at once, so financial markets won't tank.

DeVaney, Sharon A., and Sophia T. Chiremba, "Comparing the Retirement Savings of the Baby Boomers and Other Cohorts," *Bureau of Labor Statistics*, March 16, 2005, www.bls.gov/opub/cwc/cm20050114ar01p1.htm.
Older baby boomers are more likely than other generational cohorts to hold a retirement account.

Eschtruth, Andrew D., Steven A. Sass and Jean-Pierre Aubry, "Employers Lukewarm About Retaining Older Workers," *Boston College Center for Retirement Research*, May 2007, www.bc.edu/centers/crr/issues/wob_10.pdf.
A quarter of workers now in their 50s will probably want to work after traditional retirement age.

Feinsod, Roselyn, et al., "The Business Case for Workers 50+," *AARP*, Dec. 2005, http://assets.aarp.org/rgcenter/econ/workers_fifty_plus.pdf.
Many Americans plan to work past the retirement age and replacing them would be expensive. Older employees want a mix of benefits and flexibility that a small but growing number of employers are willing to offer.

Frey, William H., "Mapping the Growth of Older America: Seniors and Boomers in the Early 21st Century," *The Brookings Institution*, May 2007, www3.brookings.edu/views/articles/200705frey.pdf.
A demographer finds that "pre-seniors" (55-to-64-year-olds) are now the fastest-growing age group.

Hughes, Mary Elizabeth, and Angela M. O'Rand, "The Lives and Times of the Baby Boomers," *Russell Sage Foundation/Population Research Bureau*, 2004.
Two Duke University sociologists offer an overview of boomers — their educational attainment levels, family lives, income levels, racial divides and their likely futures.

For More Information

Administration on Aging, One Massachussetts Ave., N.W., Suites 4100 and 5100, Washington, DC 20201; (202) 619-0724; www.aoa.gov. Facilitates communication between other federal agencies and older Americans.

AARP, 601 E St., N.W., Washington, DC 20049; (888) 687-2277; www.aarp.org. The leading advocacy organization for older Americans.

Civic Ventures, 139 Townsend St., Suite 505, San Francisco, CA 94107; (415) 430-0141; www.civicventures.org/index.cfm. A think tank and employment incubator that advises older adults and employers about putting experienced Americans to work.

Government Accountability Office, 441 G St., N.W., Washington, DC 20548; (202) 518-3000; http://gao.gov. Congress' investigative arm; publishes studies about the baby boom generation's impact on federal entitlement programs.

National Academy on an Aging Society, 1220 L St., N.W., Suite 901, Washington, DC 20005; (202) 408-3375; www.agingsociety.org/agingsociety/index.html. Conducts research on public policy issues concerning the aging of America.

National Association of Area Aging Agencies, 1730 Rhode Island Ave., N.W., Suite 1200, Washington, DC 20036; (202) 872-0888; www.n4a.org. Umbrella organization for local aging agencies.

National Council on Aging, 1901 L St., N.W., 4th Floor, Washington, DC 20036; (202) 479-1200; www.ncoa.org. Provides a network for sharing information and ideas between more than 14,000 groups providing services to seniors.

National Institute on Aging, Building 31, Room 5C27, 31 Center Dr., MSC 2292, Bethesda, MD 20892; (301) 496-1752; www.nia.nih.gov. Leads the government's scientific effort to study the nature of aging.

6

Future of Marriage

Is Traditional Matrimony Going Out of Style?

David Masci

Businesswoman Dimitra Hengen of Alexandria, Va., has no desire to remarry after divorcing her husband of 18 years. She is among a growing number of Americans who no longer see marriage as necessary to their happiness.

Courtesy Dimitra Hengen

From *CQ Researcher*, May 7, 2004.

For Washington-area lawyers Melissa Jurgens and Jim Reed, their four years of marriage has meant greater happiness and stability than they have ever known.

"After I married Jim, I had someone I could talk to all the time and who could support me in ways that my friends, as wonderful as they are, just can't," Jurgens says. "He's more than a friend: He's committed to me, like I am to him, and that makes all the difference. We're going to be living together for the next 50 years, and so he needs to ensure that I'm OK and happy."

Marriage has been equally transforming for Reed. "I was 35 at the time and had already been a lawyer for 10 years," he says. "But I didn't really feel established until I married my wife. Marrying Melissa committed me to certain things — like my career path and staying here in Washington . . . [in part] because we want to have children."

But Dimitra Hengen, a successful businesswoman in Alexandria, Va., wants no part of the institution. "A lot of people do much better living on their own rather than in a marriage," says Hengen, who divorced her husband in 2001 after 18 years of marriage. "I'm not saying that we don't need companionship, but you need to assess this need against the things you have to give up when you marry, like your independence — and I don't want to give those things up."

Instead, she sees herself having long-term relationships that may be permanent, but will never lead to matrimony. "I have a wonderful boyfriend right now, and he wants to marry me, but I don't even want to live with him," she says. "I want to be able to come and go as I please, to travel, to see friends and family, all without the

More Americans Remaining Unmarried

The percentage of American men and women who remain unmarried jumped dramatically in the past 30 years. In the 25-29 age group, the percentage of unmarried men rose from just 19 percent in 1970 to 54 percent in 2002; among women it nearly quadrupled, to 40 percent. Marriage experts say increased wealth and new freedoms are largely behind the trend.

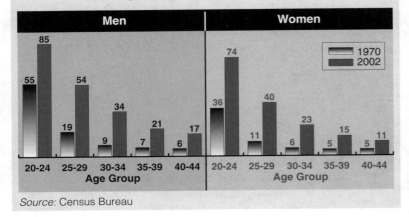

Percentage of Americans Who Never Married

Source: Census Bureau

Nonetheless, many marriage scholars argue that despite current trends, marriage will never really go out of style.

"We're a pair-bonding species, and we have a deep need at the species level to love and be loved by another and a need to pass on a part of ourselves to the next generation," says David Blankenhorn, founder and president of the Institute for American Values, a marriage advocacy group. "Marriage is the institution that encompasses these two great needs."

The case for marriage is further bolstered by research showing that married men and women are healthier, wealthier and happier than their single or divorced counterparts, Blankenhorn says. Children, too, are more likely to do better in school and less likely to have disciplinary trouble if they live in homes with married parents, he says.[2]

compromises that come with marriage. I don't need marriage."

Hengen is among a growing number of Americans who see marriage as more of an option than a necessity, according to Laura Kipnis, a professor of media studies at Northwestern University and author of *Against Love: A Polemic.* "As the economic necessity of it has become less pressing, people have discovered that they no longer need marriage," she says. "It restricts our choices and is too confining, which is why fewer people are marrying."

Indeed, in the last 50 years, the percentage of American households headed by married couples has fallen from nearly 80 percent to an all-time low of 50.7 percent, according to the Census Bureau. Meanwhile, the percentage of marriages that end in divorce jumped from roughly 25 percent to 45 percent.

As a result of the changing marriage and divorce statistics, married couples with children now comprise only 25 percent of all American households. The number is expected to fall to 20 percent by 2010. By that time, the bureau predicts, single adults will make up 30 percent of all households.[1]

Marriage advocates note there is already some evidence to support the institution's resiliency, pointing out that divorce rates, for instance, have leveled off and even declined slightly in recent years.

And compared with Northern Europeans, Americans are still marriage fanatics. In Denmark, for instance, 60 percent of all children are born out of wedlock, compared to 34 percent in the United States.[3] (*See sidebar, p. 128.*)

But many experts worry that the United States eventually will become more like Europe, with cohabitation and single parenthood replacing marriage as the dominant social institution. They point out that between 1996 and 2002, the number of cohabiting couples rose from 2.8 million to nearly 4.3 million, a trend that is expected to continue in the coming years.

Marriage advocates say that increasing rates of cohabitation can and should be stopped with education and other measures. They support a recent $1.5 billion Bush administration proposal to promote marriage among the poor, who are more likely to have children outside wedlock than middle-class Americans.

Fewer Couples Are Marrying . . .

The percentage of U.S. households headed by married couples has declined steadily from more than three-quarters of all homes in 1950 to barely more than half today.

Percentage of Married-Couple Households

. . . And More Couples Are Cohabiting

More than 4 million unmarried American couples were living together in 2002, a 50 percent increase over the number just six years earlier.

Households Headed by Opposite-Sex Unmarried Couples

Source: U.S. Census Bureau, Current Population Survey, March 2002

But critics of the initiative, which is part of a planned reauthorization of the nation's welfare law, argue that marriage-promotion schemes are unlikely to work. And even if they did prove successful, they say, the money could better be spent on education or job training, which help women become more self-sufficient and less in need of a husband.

The Bush administration also has promoted marriage by rewriting the tax code to eliminate the so-called marriage penalty — which required some married couples to pay higher taxes together than if they had remained single. The penalty was eliminated when Congress passed the first Bush tax cuts in 2001, but the provisions expire at the end of the year. On April 28, the House passed legislation making the elimination of the penalty permanent. However, prospects in the Senate are uncertain.[4]

The drive to promote marriage, especially through taxes or benefits, has angered many singles. Unmarried workers, for instance, complain that pension and health benefits favor those with spouses and children. If the percentage of single employees in America were to surpass the percentage of married employees, singles could begin demanding equal treatment with regard to employee benefits and other government policies that currently favor married workers. Already, 40 percent of the nation's largest 500 companies have re-examined their "marriage-centric" benefit policies. For instance, Bank of America has redefined "family" to include non-traditional household members — such as domestic partners or adult children living at home.[5]

At the same time, many widows and widowers feel that they can't afford to remarry because they will lose health, pension and other benefits tied to their deceased spouses.

Ironically, concerns over the state of heterosexual marriage come as more gay couples are forming families by adopting children, and national debate flares over

Single mothers would be encouraged to marry under President Bush's Healthy Marriage Initiative, which would provide $1.5 billion to establish marriage-education programs, advertise pro-marriage messages and teach marriage skills.

whether same-sex partners should be allowed to wed. While most polls show that roughly two-thirds of Americans oppose same-sex unions, several state courts in recent years have expanded marriage rights for homosexuals — ranging from civil unions now allowed in Vermont to matrimony approved by the Massachusetts Supreme Court in November. In addition, mayors and other public officials in several cities — notably San Francisco and Portland, Ore. — have issued thousands of marriage licenses to gay couples.[6]

Social conservatives have responded by proposing an amendment to the U.S. Constitution defining marriage as the union of a man and a woman — effectively banning same-sex marriages. Supporters of the amendment, including President Bush, argue a constitutional amendment is needed to prevent liberal judges and officials from watering down the millennia-old definition of marriage until it loses all meaning and significance.

But the amendment's critics, including gay-rights groups and even some conservatives, note that the Constitution is usually changed to expand rights rather than take them away.

As the experts debate the future of one of the most important social institutions in human history, here are some of the questions they are asking:

Is the future bleak for traditional marriage?

Nearly half of all American marriages end in divorce, according to the U.S. Census Bureau. Meanwhile, the nation's marriage rate has been steadily dropping, from an annual rate of 9.9 marriages per thousand population in 1987 to 8.4 per thousand in 2001.[7] Today, just over 50 percent of all households are headed by married couples, compared with three-quarters of the same group 40 years ago.

At the same time, the number of households with unmarried couples has risen dramatically. In 1977, only 1 million Americans were cohabiting; today, it's 5 million. The percentage of children raised in single-parent households also has jumped, tripling in the last 40 years — from 9.4 percent in 1960 to 28.5 percent in 2002.

Some scholars say that while traditional marriage will not disappear, it will never again be the country's preeminent social arrangement. "Ozzie and Harriet are moving out of town, and they're not coming back," says Stephanie Coontz, a professor of history at The Evergreen State College in Olympia, Wash., and author of the upcoming *A History of Marriage*. "Americans now have too many choices — due to new technologies and economic and social opportunities — and it would take a level of repression unacceptable to nearly everyone to force us to begin marrying and stay married at the same levels we once did."

Coontz argues that marriage thrived during more socially and economically restricted times. "For thousands of years, marriage has been humanity's most important economic and social institution," she says. "It gave women economic security and helped men financially, through dowry payments and socially by connecting them to another family."

However, Coontz continues, the recent expansion of individual wealth and freedom — especially among women — makes the economic argument for marriage much less compelling. "We no longer need a spouse for economic security or to [financially] take care of us when we get old," she says. "We can do these things for ourselves now."

Northwestern's Kipnis agrees. "Look, marriage has essentially been an economic institution with some romantic aspects tacked onto it," she says. "Once you take away the economic need, marriage becomes different, and for many people it becomes confining."

Indeed, only 38 percent of Americans in first marriages say that they are happy, Kipnis says, citing a 1999 survey by the National Marriage Project at Rutgers University.[8] "What does that say about everyone else?" she asks.

Some of the unhappiness may be stoked by a consumer culture that emphasizes choice and happiness, raising unrealistic expectations, she says. "Our emotional and physical needs have expanded a lot, and people now expect that the person they are with is going to meet those needs," Kipnis says. "That makes it much harder to find someone else they feel they can marry."

It also increases the likelihood of divorce, says Diane Sollee, founder and director of the Coalition for Marriage, Family and Couples Education, in Washington. "There are a lot of unnecessary divorces because when married people feel unhappy, they assume that they married the wrong person. So they find someone new, and when that person doesn't make them happy, they move on to the next one."

But supporters of marriage believe that the institution is beginning to make a comeback, in part because of the very changes that the pessimists cite. "All of this mobility and freedom also means that we're living in an increasingly impersonal mass society," says William Doherty, director of the Marriage and Family Therapy Program at the University of Minnesota in St. Paul. "Marriage will continue to be important. We will continue to need someone who is permanently and unquestionably in our corner."

Others argue that marriage will survive and thrive because it is still the best way to organize society on a personal level. "People are beginning to see how much we need marriage, because it is the only effective way to raise children," says Tom Minnery, vice president of public policy at Focus on the Family, a Christian-oriented advocacy group in Colorado Springs. "And this isn't something that just religious people are saying; it's accepted by the scientific community too." (*For debate over research on the impact of marriage, see "Current Situation," p. 135.*)

There are already signs of a reverse in the trend away from marriage, Minnery says. He points out, for instance,

that the annual divorce rate has dropped, from five per 1,000 people in 1982 to four in 2002.[9]

In addition, the upward trend toward working motherhood has halted, Minnery says, noting that the percentage of working mothers (72 percent) has held steady since 1997, after rising dramatically during the previous 20 years.[10]

Optimists also contend that young people today take marriage more seriously than people did 20 or 30 years ago. "I actually think the institution of marriage is going to become more stable in the future," Doherty says. "People are becoming more sober and serious about marriage and doing more to prepare themselves for it, like taking marriage-education classes before they wed."

He attributes the trend, in part, to painful memories. "The children of divorce are keen to not make the same mistakes as their parents," he says. "Even those whose parents stayed together feel this way, because they were able to witness the divorce revolution in other families."

"When you look at young people, they're more conservative and religious than their parents," Minnery agrees. "Just like the Baby Boomers rebelled against what their parents did, the children of Baby Boomers are rebelling against their parents and their lifestyle choices."

Finally, marriage optimists say the institution will endure because it fulfills basic human needs. "There's still a great hunger for stable, loving, intimate relationships," Doherty says, "and marriage is still the best way to have them."

Would President Bush's initiative to promote marriage improve poor people's lives?

In 2001, President Bush proposed a new initiative to promote marriage for lower-income Americans as part of the reauthorization of the federal government's welfare program. Recently, the president expanded his proposal.

Bush's new Healthy Marriage Initiative would provide $1.5 billion over five years in grants to states, local governments and private charities for a variety of activities, including establishing marriage-education programs in schools and community centers and advertising pro-marriage messages. Funds could also be spent to teach marriage skills to people preparing to tie the knot or to mentor troubled married couples.[11]

The proposal was included in the massive Welfare Reform Reauthorization bill passed by the House last

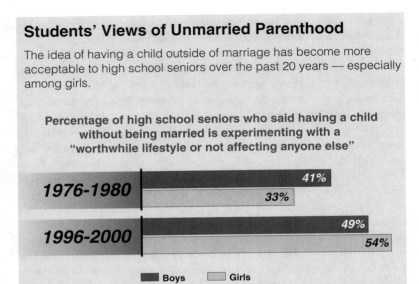

Students' Views of Unmarried Parenthood

The idea of having a child outside of marriage has become more acceptable to high school seniors over the past 20 years — especially among girls.

Percentage of high school seniors who said having a child without being married is experimenting with a "worthwhile lifestyle or not affecting anyone else"

1976-1980
41%
33%

1996-2000
49%
54%

■ Boys □ Girls

Source: "Monitoring the Future," Survey Research Center, University of Michigan

year. But the measure stalled in the Senate in March after legislators could not agree on several issues unrelated to the marriage proposal.[12] No new date has been set for the Senate to revisit the bill.

The University of Minnesota's Doherty says Bush's initiative is "well worth the effort" because research shows that marriage can dramatically improve the physical and emotional health of adults and especially children. "We know that the best way to raise children is in a healthy marriage, and yet our welfare policy hasn't reflected that," he says.

Blankenhorn agrees. "We know that in most cases children do better in a home where their biological parents are married," he says. "So we have a real interest in seeing that people with children marry and making those marriages work."

But opponents of the initiative contend it will neither be effective nor do much social good.

"You can't really push people into a decision this big, absent fraud or coercion," Evergreen State's Coontz says. "This will change a few minds, but not many."

Moreover, says Barbara Risman, a sociology professor at North Carolina State University in Raleigh, changing minds in one direction should not be the focus of federal efforts. Instead, she says, the government should better equip people to make their own decisions.

"The focus on marriage, as opposed to self-sufficiency, is a negative issue for women," she says. "The government shouldn't promote marriage. It should promote the ability of anyone to live with the kind of families they want to have."

There are better ways to spend welfare funds, critics like Risman and Coontz add. "No matter what [supporters] say, this is going to divert funds, because at the end of the day you only have so much welfare money to spend," Coontz says. "And we have such pressing needs in child care, health and in so many other areas."

In fact, opponents say, marriage promotion could do more harm than good, in part because the pool of good husbands is likely to be much smaller among the group the administration would target — poor women. "Poor people have a lot of barriers in their lives, like economic instability, substance abuse and a history of being on the giving or receiving end of domestic violence," says Lisalyn Jacobs, vice president for government relations at the National Organization for Women's (NOW) Legal Defense and Education Fund, a women's-rights group. "It may not be such a great idea to get or keep some of these people together, because they're not very stable."

Marriage also might not be such a good idea if the potential marriage partner is abusive, she adds. "Sixty percent of all women on welfare have been victims of domestic violence at some time in their lives," Jacobs says, "so encouraging people to stay together might put many of them at greater risk of injury."

But Sollee disputes that argument. "Women generally get beat up when they're single or in a cohabiting relationship," she says, pointing to the National Crime Victimization Survey, which shows that two-thirds of the violence against women is not committed by husbands but by casual dating partners.[13] "Marriage stabilizes relationships and makes domestic violence much less likely."

Indeed, the idea that those on welfare are different with regard to marriage is "insulting to the poor," she says. "Poor people pair up for the same reason everyone else does: Because they're human, and they want to form love relationships. Given that, we need to give them the skills needed to make the right choices."

Should the Constitution be amended to define marriage as the union of a man and a woman?

After months of pressure from religious and conservative groups alarmed at what they saw as a rising tide of same-sex marriages, President Bush on Feb. 24 publicly endorsed amending the U.S. Constitution to define marriage as the union of a man and a woman. Bush said the decision had been forced upon him by "activist judges and local officials" in Massachusetts, California and elsewhere, who had made "an aggressive attempt" to redefine marriage.

"On a matter of such importance, the voice of the people must be heard," the president said at a White House press conference announcing his support for the amendment. "Activist courts have left the people with one recourse. If we are to prevent the meaning of marriage from being changed forever, our nation must enact a constitutional amendment to protect marriage in America." (*See "At Issue," p. 136.*)

The federal government had addressed the issue in 1996, when Congress passed, and President Bill Clinton signed, the Defense of Marriage Act, which defined marriage as being between a man and a woman for purposes of federal law. It explicitly prevents any jurisdiction from being forced to accept another's definition of marriage. In other words, if Massachusetts legalizes same-sex unions and marries two men, New York is not required to acknowledge the marriage if the couple subsequently moves there.

But because the Constitution's Full Faith and Credit Clause requires each state to recognize the lawful actions of other states, Bush and gay marriage opponents say federal courts might overrule the law and require other states to recognize gay marriages performed outside their jurisdiction.

The amendment, proposed by Colorado Republicans Sen. Wayne Allard and Rep. Marilyn M. Musgrave, would still allow states to pass civil-union or domestic-partnership laws that could grant same-sex partners and others the same rights as married couples. But marriage would be strictly limited to heterosexual couples.

Gay-rights activists, civil-liberties organizations and even some conservatives and libertarians oppose the amendment, albeit for different reasons.

"It's a perversion to use our founding document to discriminate against a group of people when it has traditionally been used to expand liberties and rights," says Kevin M. Cathcart, executive director of the Lambda Legal Defense and Education Fund, a gay-rights group in New York. "It's ironic that on the 50th anniversary of the *Brown v. Board of Education* decision, which struck down the doctrine of 'separate but equal,' we're on the verge of writing it back into the Constitution."[14]

Other opponents of gay marriage say a Constitutional amendment is heavy-handed and unnecessary. Former Rep. Bob Barr, who as a conservative Republican from Georgia helped write the 1996 Defense of Marriage Act, calls the amendment an unwarranted intrusion into an area traditionally left to the states.

"Changing the Constitution is just unnecessary — even after the Massachusetts decision, the San Francisco circus and the Oregon licenses," Barr told the House Judiciary Subcommittee on the Constitution on March 30. "We have a perfectly good law on the books that defends marriage on the federal level and protects states from having to dilute their definitions of marriage by recognizing other states' same-sex marriage licenses."[15]

Opponents also call the administration's claim that the amendment process was "forced" on them by activist judges and mayors a cynical election-year ploy. "I don't think this is really about gay marriage at all, but is a distraction meant to focus attention away from the [Iraq] war and the deficit and all of the other problems this administration is dealing with," Cathcart says.

Moreover, by trying to amend the Constitution, they say, conservatives are trying to cut off the emerging national debate on same-sex marriage. "You know, we're really just beginning this debate all over the country, and already they want to amend the Constitution," Cathcart says. "They accuse liberals of trying to use the courts and local officials to circumvent debate, but that's actually what they're doing."

"Amending the Constitution is something that you traditionally do when you've run out of remedies," agrees NOW's Jacobs. "It seems to me that we've only just begun to try to work this one out."

But supporters point out that to become law any constitutional amendment must first win the support of two-thirds of Congress and three-quarters of the nation's state legislatures. "This would be a wonderful way to

Will U.S. Follow Europe's Cohabitation Trend?

In their 21 years together, Stig Skovlind and Malene Breining Nielsen of Denmark have dated, lived together and raised three children — but they never got married. "We trust each other. We don't need a document," Malene said.[1]

More and more Europeans — particularly from the Nordic countries — are cohabiting. Nineteen percent fewer Europeans got married in 2002 than in 1980, compared to a 5.7 percent drop in the United States during the same period. Meanwhile, almost 20 percent of young Europeans — and 40 percent of Swedes — are cohabiting, compared to 7.7 percent in the United States.[2]

Northern European demographics could be headed toward a point at which "marriage and cohabitation have become indistinguishable," says Kathleen Kiernan, a professor of social policy and demography at the London School of Economics. And many cohabiting Nordic couples are having children. More than half of Swedish mothers ages 25 to 29 give birth to their first child out of wedlock, and more than a quarter of Norwegian mothers.

About 80 percent of those surveyed in Sweden, Finland and Denmark consider cohabiting couples with children a "family." But in mostly Catholic Southern European countries like Italy, Spain and Portugal, attitudes about cohabitation are more conservative; only 44 percent of Italians, for instance, view unwed couples with children as a family.

Still, many European courts now accommodate the emerging class of cohabiting partners and parents. In Sweden, Finland and Denmark, "family law has come to be applied to married and cohabiting couples in the same way," writes Kiernan. And in 1998 the Netherlands began recognizing both homosexual and heterosexual partnerships as if they were "functionally equivalent to marriage," she notes.

In France, so-called PACS (pacte civil de solidarite) offer unwed heterosexual and homosexual couples some of the same rights accorded to marriage; more than 130,000 couples have signed PACS. Even in Italy, the government is considering granting some legal rights to unmarried couples.

Some experts worry the United States may be headed in the same direction as Europe. "Our marriage rate continues to drop, our divorce rate is high and our cohabitation rate continues to climb," says David Popenoe, professor of sociology and co-director of the National Marriage Project at Rutgers University.

And recent U.S. demographic surveys support his concerns: The number of American couples cohabiting rose by 72 percent in the 1990s, with nearly half of them raising children.[3]

debate the issue, given all of the hurdles that have to be jumped before it became part of the Constitution," says Minnery of Focus on the Family. "There would be a debate in Congress and then in every state legislature in the country. That seems pretty thorough to me."

And while proponents admit the amendment process has traditionally been about expanding rights, they argue that gay marriage presents a unique challenge to American society that calls for a unique solution.

"Our founding documents, like the Declaration of Independence, tell us that our rights have come from our creator or, to put it another way, they are part of natural law," says Ed Vitagliano, pastor of Harvester Church in Pontotoc, Miss., and a spokesman for the conservative American Family Association. "When you talk about redefining marriage, you're really talking about an overthrow of this natural order or natural law, because

marriage is something that predates government. So this is a big deal, a once-in-a-lifetime debate about whether to overturn the natural order upon which our rights are based. That requires a big response."

BACKGROUND

Origins of Marriage

Marriage has meant very different things in different places at different times. "Marriage has been continually evolving through the centuries, and it's still doing so," the University of Minnesota's Doherty says.

For instance, the idea of choosing a mate or "marrying for love" became commonplace only in the 18th and 19th centuries and only in some cultures. Marriages are still arranged by families throughout much of Asia, Africa and

The rising numbers worry experts because cohabiting couples tend to break up more than married couples, Popenoe says, and there is no safety net in the United States for kids who slip through the financial cracks when parents separate.

"We can't agree that . . . welfare provisions are proper, and we don't [want] to give up our hard-earned taxes in times of need," says Popenoe.

But in Northern Europe, expansive welfare measures provide a safety net for children when relationships break down, Kiernan says. Unwed European mothers — cohabiting or not — have the same rights as married mothers and, although the law is less clear-cut for men, unwed fathers generally have a financial duty to their children once paternity is established.

Critics of cohabitation note that children of single parents have a higher incidence of psychiatric problems, Popenoe says. "Kids are much better off when raised by two married biological parents than . . . by a single parent or a broken cohabiting couple and are then thrust on the welfare state," he says.

Protesters in Paris oppose the Civil Solidarity Pacts (PACS) being considered by the French National Assembly, which would give traditional rights to homosexual and unwed couples.

But Stephanie Coontz, national co-chair of the Council on Contemporary Families, notes that while "transitions can be hard on kids," the effects of a parental breakup on kids can be exaggerated.

Americans will have to get used to a broader definition for family, she concludes. "There's been a worldwide transformation of marriage — it will never again have a monopoly on organized child care or on the caring for dependents," she says.

— Benton Ives-Halperin

---

[1] Jennie James, *et al.*, "All In The Family . . . or Not," *Time*, Sept. 17, 2001, p. 54.

[2] Marriage statistics are from "Demography: EU Population Up by 0.3% in 2002," *European Report*, Sept. 3, 2003, and National Center for Health Statistics; cohabiting statistics — which are for those ages 25 to 34 — come from the U.S. Census Bureau and Kathleen Kiernan, "Unmarried Cohabitation and Parenthood: Here to Stay?" Conference on Public Policy and the Future of the Family, Oct. 25, 2002. Unless otherwise noted, other data are from Kiernan.

[3] Laurent Belsie, "More Couples Living Together, Roiling Debate on Family," *The Christian Science Monitor*, March 13, 2003, p. 1.

---

the Middle East today. Moreover polygamy — long rejected by Western cultures — is common in many Muslim countries and among various ethnic groups.

Still, there have been some constants, especially in the West. For instance, until recently, most people saw marriage as a necessary right of passage into adulthood, rather than a choice. Moreover, definitions of marriage were — and to some degree still are — largely dictated by the Judeo-Christian ethic, which sees the institution as a permanent, unbreakable union between a man and a woman.

While polygamy was allowed in early Jewish life, ancient Hebrew laws on marriage eventually came to stress monogamy. Laws strictly forbade adultery (the prohibition is one of the Ten Commandments) and incest. Restrictions against divorce also were enforced, making it almost impossible for Jewish couples to legally separate.

Christian thinkers built on this tradition. St. Mark, in his New Testament gospel, echoed the Old Testament when he said "from the beginning of the creation God made them male and female. For this cause shall a man leave his father and mother, and cleave to his wife; and the twain shall be one flesh: so then they are no more twain, but one flesh. What therefore God hath joined together, let no man put asunder."[16]

But the Christian emphasis on monogamy and fidelity was more than a reaffirmation of ancient Jewish traditions or the teachings of the new church's founders; it also was a reaction to what Christians viewed as the weak marriage laws of Rome, which allowed couples to separate and gave women an unusual amount of personal freedom.

While many of today's matrimonial traditions — such as the wearing of bridal veils and the exchange of wedding rings — date back to ancient Rome, the Christian church

CHRONOLOGY

17th-19th Centuries *Less-restrictive ideas and laws concerning marriage develop among colonies and later states in the New World.*

1620 Puritans arrive in America and establish more liberal marriage and divorce laws than those in England.

1770s The struggle for U.S. independence spurs debate on the rights of women and the obligations of marriage.

1800s Western territories pass liberal divorce laws in the transition to statehood to attract settlers.

1867 All but three states have abolished the most restrictive divorce laws.

1870 Only 3 percent of all U.S. marriages end in divorce.

1900-1960 *War and social changes bring women more freedom.*

1900 U.S. divorce rate stands at 8 percent.

August 1920 American women get the right to vote after Tennessee becomes the 36th state to ratify the 19th Amendment.

1925 The divorce rate is 25 percent.

1941 U.S. entry into World War II brings millions of American women into the work force.

1947 California Supreme Court rules that the state's miscegenation law violates the state Constitution, making California the first state to abolish limits on interracial marriage.

1960-Present *The civil and women's rights movements and the sexual revolution dramatically change the institution of marriage.*

1960 Nation's divorce rate is 26 percent.

1964 Civil Rights Act prohibits discrimination based on gender.

1967 U.S. Supreme Court overturns state miscegenation laws.

1969 Gov. Ronald Reagan, R-Calif., signs the first no-fault divorce law.

1972 The launch of *Ms.* magazine heralds the arrival of the woman's movement. Jessie Barnard's book *The Future of Marriage* argues that marriage is often detrimental to women.

1974 The number of American children whose parents divorce in a year reaches 1 million.

1980 Divorce rate hits 50 percent.

1989 Psychologist Judith Wallerstein argues in her book *Second Chances* that the impact of divorce on kids is worse than previously thought.

1992 Vice President Dan Quayle criticizes decision by TV sitcom character "Murphy Brown" not to wed her child's father.

1996 Hawaiian Supreme Court rules the state cannot ban gay marriages.

1998 Hawaiians amend Constitution to permit ban on gay marriage.

1999 Vermont Supreme Court rules that gay couples are entitled to the same benefits as married people.

2000 *The Case for Marriage* attempts to counter earlier arguments that many people don't need marriage.

Nov. 18, 2003 Massachusetts Supreme Court rules the state's law prohibiting gay marriage violates the state Constitution.

April 29, 2004 Massachusetts legislature adopts constitutional amendment banning gay marriage but allowing civil unions.

May 17, 2004 Massachusetts must begin issuing marriage licenses to same-sex couples, according to a state Supreme Court order; governor is seeking an emergency stay of the deadline until action is completed on new constitutional amendment.

2006 Massachusetts gay-marriage amendment would take effect, if approved again by the legislature and by a statewide voter referendum.

rejected Rome's lax marriage laws and instead transformed marriage into a divinely ordained sacrament. Separation or divorce were strictly forbidden, although widows could remarry after a spouse's death. Jesus himself condemned divorce, calling men who leave their wives for others — even if legally sanctioned — adulterers.[17]

The only option for irreconcilable couples was to petition the church for an annulment, which did not dissolve the marriage but declared that it had been invalid from the start and hence had never actually taken place. Annulments were usually employed in cases of bigamy or when a husband and wife were closely related. Otherwise, annulments were difficult to obtain. Petitions, even from kings — like England's Henry VIII — were routinely denied.

Besides fidelity, the church emphasized the dominant role of the husband, continuing the tradition of the Jews and of most other ancient cultures at the time. Ironically, Christian teachings held that men and women were equal in God's eyes but, nonetheless, women were to be "in submission" to their husbands.[18]

Sweeping Changes

Sweeping changes in the state of marriage did not begin to occur until the 16th-century Protestant Reformation, which rejected much of the institutionalization of religion and stressed individual choice. Many Protestants, including the movement's founder, Martin Luther, cast off the notion that marriage was a holy sacrament to be regulated entirely by the church. Instead, Luther wrote, marriage was "a secular and outward thing having to do with wife and children, house and home and with other matters that belong to the realm of the government, all of which have been completely subjected to reason." According to Luther, the laws of marriage and divorce "should be left to the lawyers and made by secular government."[19]

Luther's new attitudes set the stage for ensuing changes. "The Reformation set out a bunch of new ingredients on the table . . . but it took the Enlightenment and the spread of wage labor to bring these ingredients together," says Coontz, who is writing a book on the history of marriage.

The 17th- and 18th-century Enlightenment "brought to dominance the notion that people have the right to organize their lives as they see fit," says Coontz, leading to the belief that "marriage should be a love match."

Indeed, during subsequent centuries, more and more people, especially among the educated classes, eschewed family obligation and chose their own mates.

If the Enlightenment gave people the intellectual justification for choosing their spouses, wage labor — brought on by the Industrial Revolution — gave them the means. Having a job with a steady wage disconnected people from rural life and its familial and other controls, giving them both geographic and social mobility.

"If you didn't want to marry the person your parents had chosen for you, you could leave and find someone else," Coontz says.

New World Flexibility

The first European settlers in the New World took with them many of their old laws and customs, including rules on matrimony. But as the late historian Daniel Boorstin has pointed out, views on marriage in England's American colonies and eventually the United States were always more flexible than those in Europe, in large part to fit the needs of a less socially rigid and more mobile society.

"The rights of married women and their powers to carry on business and to secure divorce were much enlarged," he wrote about matrimony laws in early North America. "The law protected women in ways unprecedented in the English common law."[20]

In colonial Massachusetts and Connecticut, Puritan-influenced law even allowed for divorce if a spouse could prove that the other had neglected a fundamental duty of the marriage, such as providing food and shelter. However, by modern standards, divorce was difficult to obtain in colonial America, so it was rare. In many jurisdictions, divorce could only be granted by legislative action (the passage of a private bill), which meant that usually only the rich had the resources to legally separate.

Between the American Revolution and the Civil War, most states greatly liberalized their divorce laws as part a trend to expand individual freedoms, such as voting and other rights and eventually to abolish slavery. For instance, by 1867, 34 of the 37 states had abolished legislative divorce, giving the courts authority to grant divorces. Still, the divorce rate remained relatively low — about 3 percent in 1870.

During the great westward migration just before and after the Civil War, many of the new states carved out of the Western territories, such as Nevada, passed liberal divorce laws in an effort to attract more settlers.

Can You Click to Find Your Soul Mate?

"**A**ren't you just a little curious?" purrs an ad on Match.com's Web site. "With 8 million profiles to choose from, imagine the possibilities."

Match.com and other large Internet dating services do offer singles many choices, but most online services are not concerned with whether a potential customer is seeking a friend, a date or a spouse.

But a relatively new online service, eHarmony.com, actively plays cupid, limiting its clientele to those looking to get married. Its Web site is full of pictures of happily married or engaged couples who met through the service. Potential subscribers are urged to join "when you're ready to find the love of your life."

Americans have been seeking love online for more than a decade, but in recent years Internet dating has become much more widespread and socially acceptable.

"The traditional institutionalized means for getting people together are not working as well as they did previously," says Norville Glenn, a sociology professor at the University of Texas. "There's a need for something new, and the Internet is filling it."

Last year an estimated 21 million Americans spent $313 million on Internet dating, a figure likely to more than double by 2008, according to the Internet market research firm Jupiter Communications.[1]

Most of the biggest services, like Match.com and Yahoo Personals, are largely search engines with millions of profiles, each usually containing a photo and personal information ranging from height and weight and likes and dislikes to "latest book read." Users scroll through the results, e-mailing anyone who catches their fancy.

But eHarmony works differently. Founded in August 2000 by clinical psychologist Neil Clark Warren, it does not allow users to choose whom they're going to contact. Instead, it matches people based on an exhaustive personality survey completed by each subscriber.

"I saw more than 7,000 patients over the years, and so many of these troubled people were in bad marriages," he says of his 35 years as a therapist. "It struck me that the most important need we have is to get marriage right."

Warren conducted more than 500 "divorce autopsies," usually interviewing both spouses and sometimes even the children of the divorced parents. He found that most people in failed marriages chose their spouse for the wrong reasons, such as physical appearance, sense of humor or financial status, while neglecting more important concerns. "The true things people need for a happy marriage are on the inside, like character and intellect, rather than the shape of their nose," he says.

Moreover, he found, those who succeed at marriage are usually paired with someone who shares most of their basic values and beliefs. "We're told that opposites attract, but that's not so," he says. "When people have a lot in common, they have much less to negotiate, fewer things to compromise on."

For instance, different work ethics or attitudes about how to raise children might not be a problem on a first or second date, but they can breed resentment when a couple is living together.

eHarmony tries to match clients by requiring all new subscribers to fill out an extensive 436-item questionnaire based on what Warren says are "29 dimensions for compatibility," such as spirituality, education, sexual desire and kindness.

After computers match candidates with similar traits, early communication is limited to e-mail, but eventually, matched singles can move on to calling and then meeting for dates.

Around the same time, some Mormon settlers in the Western territory began practicing polygamy. The practice, which only involved men marrying multiple wives, began after the religion's leader, Brigham Young, declared it acceptable in 1852.

While most Mormon marriages involved only two people, a substantial minority of Mormon men had two or three wives. Still, the practice was relatively short-lived due to pressure from the rest of the country. As a condition for Utah joining the Union, the Mormon Church banned new polygamous marriages in 1890.*

Meanwhile, many states, new and old, were also passing so-called miscegenation laws, prohibiting marriage between members of different races. Support for miscegenation goes back to the earliest English settlements and, while largely prompted by a desire to prevent whites

---

* Polygamy is still practiced illegally in parts of Utah.

Although eHarmony is less than four years old, it has attracted 4 million users and is adding roughly 10,000 new customers per day, Warren says. It has also spawned imitators, such as TrueBeginnings. Even Match.com, the nation's largest online service, is offering a short personality test to its registered users.

Warren claims his company has already connected at least 2,500 couples who have gotten married, and who, for the most part, are "doing very well so far."

Paul Consbruck, 42, of Jacksonville Fla., says he is now happily married

Neil Clark Warren, founder of eHarmony.com, says most marriages fail because people choose mates for the wrong reasons.

to someone he met through eHarmony. "A lot of dating takes place on a very superficial level," he says. "eHarmony tells you to step back and look at what's really important to you before you get attracted to someone physically."[2]

About 15 percent of all eHarmony applicants are turned away for a variety of reasons including emotional difficulties, substance abuse or concerns about truthfulness.

"A lot of people are not ready for marriage, and so we encourage them to get better and then reapply," Warren says. "It's painful, but the alternative is to match them when

they're not ready and bring other, healthier people down with them. That's not fair."

Because eHarmony takes such an active interest in who and even how its subscribers meet, it has been likened to an old-fashioned matchmaker, a comparison Warren does not reject. In today's increasingly urban and mobile society, he says, a service like eHarmony can provide "the kind of wisdom that people might have found with their family or community in the past," when most people lived in small towns.

Indeed, Warren believes that it is "extremely hard" for a single person to find a suitable partner without a service like eHarmony. "You need to tap into a large pool of people so that you can increase your odds of finding your soul mate," he says. "That's why I think that in 10 or 15 years, virtually every person will find their husband or wife on the Internet."

---

[1] Figures cited in Adrienne Mand, "Dr. Love is In," ABCNEWS.com, March 26, 2004.

[2] Quoted in Anna Kuchment, "The Internet: Battle of the Sexes," *Newsweek International*, Dec. 2, 2003.

---

and African-Americans from marrying, was not confined to the South or to black-white couplings. Indeed, the first miscegenation law in North America was passed by the Maryland Assembly in 1664. Later, Western states prohibited Asians and whites from marrying.

Miscegenation laws were still common until the second half of the 20th century. In 1948, the California Supreme Court became the first state court to strike down a law banning racial mixing, arguing that it violated the

Constitution's 14th Amendment guaranteeing equal protection under the law.[21] Nearly 20 years later, in *Loving v. Virginia*, the Supreme Court followed suit, repealing miscegenation laws nationwide.

Shifting Views

The last 50 years have witnessed dramatic, indeed unprecedented, changes in perceptions of marriage in the United States. During the 1960s and '70s, the general purposes and

Female shipyard workers in California were among the millions of women brought into the work force by World War II, helping to plant the seeds for social changes in the 1960s that experts say have contributed to the decline in marriage nationwide.

characteristics of marriage changed, reflecting new attitudes about freedom and self-realization, especially for women.

"In the middle of the 20th century, we began to shift away from an institutional view of marriage — that it is based on economic viability, child rearing and a sense that this is something that all adults should do," Doherty says. "And we moved into what is called psychological marriage — the notion that you marry by choice and that you do so primarily for romantic reasons."

This change was prompted by several important social and cultural changes, beginning with the entry of married women into the workplace during World War II. In 1930, only 12 percent of all married women worked outside the home.[22] By 1944, with the labor demands of rising war production and 12 million young men taken out of the work force for military service, that figure had nearly tripled, to 35 percent.[23]

Work brought many women, regardless of their marital status, more than financial independence. "In the absence of men, women found doors suddenly open to them in higher education and in the professions; the Army and Navy admitted women for the first time," writes Nancy F. Cott, a professor of history at Yale University and author of *Public Vows: A History of Marriage and the Nation.*[24]

After the war, as men returned home to resume their lives and jobs, many women — both married and unmarried — left work to focus on starting and raising families. While many women wanted to leave their war jobs, some were pressured or even forced to do so in order to make room for returning soldiers.

But during the 1960s, social changes erupted that significantly altered the marital landscape. The civil rights and anti-war movements, the rise of youth culture and the sexual revolution wrought profound changes in society's mores. Old ideas, such as female premarital chastity or the notion that a woman went to college only to find a suitable husband, began to collapse.

By the early 1970s, the women's-liberation movement was demanding equality in the workplace and elsewhere and advising American women that they had other choices in life besides getting married and having children. Indeed, many feminists declared housework demeaning and encouraged women to "find themselves."

By the 1980s, a majority of married women with children at home were working, driven not only by new freedoms and a desire for a career but also by financial need, as the cost of housing, education and health care rose faster than wages. Today, about 72 percent of married women with children are in the labor force, a figure that has held steady since the mid-1990s.

A dramatic rise in the rate of divorce accompanied these societal changes. From 1955 to 1975, the divorce rate more than doubled, from 23 percent of all marriages to 48 percent, where it has remained, give or take a few points, ever since.[25]

Many say that the advent of no-fault divorce laws across the country in the late 1960s and early '70s helped spur the divorce trend. No-fault laws made it significantly easier to untie the knot. "These new laws did a terrible disservice to couples, but they especially hurt the weakest people in our society — children," says Richard Land, a spokesman for the Southern Baptist Convention. "No-fault divorce sent people the message that it was easy and alright to breakup their families."

But others disagree. "When you look at almost all the reputable research, you find that no-fault divorce was not the cause of the divorce rate rising," says Coontz of Evergreen State. "It was the rising demand for divorce that caused the rising divorce rate. No-fault only speeded up what was going to happen anyway."

CURRENT SITUATION

Gay-Marriage Push

Although the debate over same-sex marriage has only recently risen to national attention, the issue is by no means new. Gay-rights advocates have been working for decades to secure matrimonial and related rights for same-sex couples. And they have had some successes, as well as setbacks.

In 1996, Hawaii's Supreme Court ruled that the state government's ban on same-sex marriages violated the state Constitution. Then in 1998, Hawaiian voters amended their state constitution giving the legislature authority to ban same-sex marriage. In 1999, Vermont's highest court ruled that gay couples were entitled to the same benefits as married people, although it stopped short of giving same-sex couples the right to marry.

But developments in the last six months have clearly pushed gay marriage to the top of the national domestic agenda, making it one of the most hotly debated topics of the year and an important issue in the 2004 presidential election.

New interest in same-sex unions was sparked by a Nov. 18 Massachusetts Supreme Court decision holding that denying gay couples the right to marry violated the state constitution's equal protection clause. The court gave the state until May 17 to start allowing same-sex couples to wed.

For gay advocates and some civil-rights proponents, the decision was long overdue. "A court finally had the courage to say that this really is an issue about human equality and human dignity, and it's time that the government treats these people fairly," said Mary Bonauto, the lawyer for Gay & Lesbian Advocates & Defenders who argued the case.

But the decision also prompted a substantial backlash and not just in Massachusetts, where both Republicans

Married life improves the physical and psychological health of both adults and children, according to many social researchers. But skeptics contend that bad marriages aren't counted in the statistics.

and Democrats proposed amending the state Constitution to reverse the court's ruling. President Bush also weighed in, arguing that "activist judges" had no right to rewrite the rules of marriage.

After several months of negotiation, Massachusetts Gov. Mitt Romney, a Republican, and legislative leaders proposed amending the state Constitution defining marriage as a union of a man and a woman, but allowing for gay civil unions. It was approved by the legislature on April 29, but it must be passed again next year and then approved by voters before it would take effect — in 2006 at the earliest.

Meanwhile, Romney has asked the state legislature to pass emergency legislation allowing him to seek a stay of the Massachusetts Supreme Court's May 17 marriage order. The governor contends the state should not be required to issue marriage licenses to same-sex couples while the legislature and citizens are debating the issue.

"This is a decision that is so important it should be made by the people," Romney said on April 15, the day the legislation was filed.[26]

But the Massachusetts court ruling has emboldened gay-marriage supporters elsewhere. On Feb. 12, newly elected San Francisco Mayor Gavin Newsom authorized the city government to begin issuing marriage licenses to gay and lesbian couples. The response surprised even supporters: In the first five days 2,500 couples came from

Will same-sex marriage hurt traditional marriage?

YES
Maggie Gallagher
*President, Institute for
Marriage and Public Policy*

Written for *The CQ Researcher*, April 2004

Same-sex marriage divides people into two camps: Those who say that gay marriage will affect only gays and those who believe that court-ordered same-sex marriage will dramatically alter the legal, shared public understanding of marriage.

If the Massachusetts court had decided the "right to marry" includes the right to polygamy, would that affect only those who want a polygamous marriage? Of course not. The entire marriage culture would shift if polygamy were to become a "normal" marriage variant. Monogamy would no longer be a core part of our definition of marriage.

I'm not saying that same-sex marriage will lead to polygamy. I'm pointing out that legally changing the definition of marriage affects everyone and would radically transform what marriage is. It would hurt the traditional form of marriage by:

- Sending a terrible message to the next generation: The law will say that two men or two women raising children are just the same as a mom and dad; thus, social institutions would be bending to adult sexual desires, regardless of who gets hurt.
- Creating an abyss between "civil" and "religious" marriage. Civil marriage would be divorced from religious traditions that gave rise to marriage and which continue to sustain marriage as a social institution. Government should be more modest about redefining marriage to make it unrecognizable to most religious traditions in this country.
- Neutering our shared language about parenting. You won't be able to say "children need mothers and fathers, and marriage has something important to do with getting this for children" because it will no longer be true, and because the government will be committed to the idea that two mothers or two fathers are just as good as a mom and a dad in raising children.
- Marginalizing or silencing traditional advocates of marriage. Marriage is a public act. Faith-based organizations that fail to endorse and accept same-sex marriage may find themselves driven from the public square: their broadcasting licenses, tax-exempt statuses and school accreditation at risk.

If gay marriage is a civil right, then people who believe that children need moms and dads will be treated like bigots. How will we raise young men to become reliable husbands and fathers in a society that officially promotes the idea that fathers don't matter?

NO
Kevin Cathcart
*Executive Director, Lambda Legal
Defense and Education Fund*

Written for *The CQ Researcher*, April 2004

The current, unprecedented dialogue about marriage for same-sex couples isn't an abstract discussion about politics or religion — it's about real people's lives and the human cost of denying basic equality to an entire group of Americans.

Denying marriage to same-sex couples blocks hundreds of thousands of families nationwide from the critical rights and protections that others take for granted. Same-sex couples are left vulnerable and scrambling to cobble together a patchwork of legal documents that still don't provide them with the security and protections they want and need.

To understand why marriage is so important for so many same-sex couples, look no further than Lydia Ramos. Lydia's partner of 14 years died in a car accident, triggering a legal and emotional nightmare. The coroner refused to turn the body over to Lydia, and the daughter they raised together was taken away by her partner's relatives after the funeral. Mother and daughter were kept apart for months — at a time when they most needed each other.

Mother and daughter were finally reunited after Lambda Legal fought a long and gut-wrenching legal battle on their behalf. But if Lydia and her late partner had been able to marry, their daughter would never have been put through such a nightmare.

If they had been able to marry, it would not have changed the marriages of their heterosexual neighbors and co-workers. Heterosexual marriages are not on such shaky ground that they will fall apart simply because loving, committed same-sex couples are given equal access to the rights and protections provided by marriage.

The nation is about to see that in Massachusetts, where lesbian and gay couples will soon begin getting married. That state's highest court — with six of seven justices appointed by Republican governors — ruled that only marriage can fix the inequalities in how the state treats same-sex couples.

Within the next year, our lawsuit on behalf of seven New Jersey couples is expected to reach that state's high court. Our cases in New York, Washington state and California will ask the same fundamental questions addressed by the Massachusetts court.

The courts and political leaders are beginning to recognize that anything less than marriage treats same-sex couples differently, and that separate is never equal. Loving couples are being kept from our nation's promise of fairness, and we'll fight for them for as long as it takes to win equality.

around the country to marry, often waiting for hours in long lines for their chance to get hitched.

In 2000, California voters had approved a ballot measure — Proposition 22 — which defined marriage as a union between a man and a woman, leading many, including the state's popular new governor, Arnold Schwarzenegger, to criticize Newsom for ignoring the law. "If the people change their minds and they want to overrule [Proposition 22], that's fine with me," Schwarzenegger said on March 2. "But right now, that's the law, and I think that every mayor and everyone should abide by the law."[27]

But Newsom justified his decision as an attempt to abide by the U.S. Constitution's requirement to treat all people equally. "I've got an obligation that I took seriously to defend the Constitution," he said. "There is simply no provision that allows me to discriminate."[28]

However, as a result of a court order, San Francisco stopped granting licenses on March 12. The state Supreme Court is currently deciding whether the 4,000 same-sex marriage licenses eventually issued by the city are valid, and a decision is expected in the next month or so.[29]

Even if the marriage licenses are ultimately invalidated, supporters of same-sex unions see developments in San Francisco as the event that most fully energized the gay-marriage movement. "Sometimes when you're in a civil rights struggle, you reach a tipping point," Lambda Legal's Cathcart says. "We reached it in San Francisco when you saw thousands of couples lining up to pay for their license and legally get married. Mayor Newsom lit the spark, and what happened afterwards inspired others around the country to act."

Indeed, mayors in a handful of other cities — including New Paltz, N.Y., and Asbury Park, N.J. — and the commissioners of Multnomah County, Ore., (which includes Portland) followed Newsom's lead.

But conservatives and marriage traditionalists view events in San Francisco differently. "The mayor of San Francisco has done more than anyone else to solidify support behind a federal amendment," the Southern Baptist Convention's Land says. "When a public official defies not only public opinion but [also] the law he's sworn to uphold, average people get outraged."

As evidence, Land points to a March CBS poll showing that 59 percent of Americans favor a constitutional amendment allowing marriage only between a man and woman. In December, only 35 percent favored such an amendment.[30]

AFP Photo/Hector Mata

Two newly married women ignore demonstrators protesting same-sex marriages at San Francisco City Hall on Feb. 20. San Francisco Mayor Gavin Newsom authorized the city government to begin issuing marriage licenses to gay and lesbian couples on Feb. 12, helping put the controversial issue at the top of the national agenda.

But gay-marriage advocates remain optimistic, in part because polls show that resistance to same-sex unions increases with age. "I'm very confident that we're going to win on this issue over the long term, because most young people just don't think this is a big deal," Cathcart says. "The young are already largely with us, and the coming generations will be even more supportive."

Impact of Marriage

In the last decade, a consensus has begun to emerge that marriage can provide tangible benefits, both for adults and their children. Indeed, most sociologists now agree that a good marriage generally improves the lives of all involved.

"Even the skeptics have come around on this," the University of Minnesota's Doherty says. "Marriage, and by that we mean a good marriage, is good for people."

Linda J. Waite, a sociology professor at the University of Chicago and co-author of *The Case for Marriage*, agrees, claiming the amount of evidence available on the benefits of matrimony is "fairly overwhelming."

Waite and Doherty point to a host of studies by sociologists and others, which found that men and women in happy marriages enjoy better mental and physical health and more financial security than similarly situated unmarried counterparts.

For instance, a 1990 study published in the *Journal of Marriage and Family* found that singles had higher mortality rates than their married counterparts. The mortality rate for single men was a whopping 250 percent higher, while unmarried women had a 50 percent higher rate. The study found the unwed were particularly at greater risk for those diseases that hinged on behavior, like lung cancer and cirrhosis of the liver.[31]

"It helps to have someone around who is a stakeholder in your health," Doherty says. "People do better when there is someone who can push you or nag you to keep to your diet, exercise regularly, take your medicine and just take proper care of yourself."

Likewise, married people are happier — almost twice as happy as singles and roughly three times happier than those who are widowed or divorced, according to studies.[32]

Children growing up in homes where both biological parents are present also do better than those living with a single parent. For instance, they are more likely to graduate from high school and have fewer discipline problems if raised by a married biological mother and father.

But some experts question at least parts of the consensus, arguing that much of the research paints too rosy a picture of the institution. "We need to remember that when we're talking about the benefits of marriage, we're really only talking about good marriages," Evergreen State's Coontz says. "When researchers tell you that marriage in general is good even after they average together good and bad marriages, it's very skewed," she continues. "Good marriages tend to last, while many bad marriages aren't being counted because they end in divorce. So good marriages tend to get overcounted, and the bad ones are undercounted."

Skeptics also contend that while marriage may provide real benefits for men, it is not necessarily as beneficial for women. "Men are used to being taken care of by women, first their mothers and then their wives," North Carolina State's Risman says. "So, sure, marriage is a good deal for men because without a wife to replace their mother they usually don't take good care of themselves."

But women marry for very different reasons, she says. "In a world where women still earn 75 cents or less for every dollar earned by men, they still have to marry for economic and other practical reasons."

According to Risman, a woman might be mentally or physically healthier in marriage because her husband's income provides health insurance or better food and shelter. "So, yes, maybe she's better off, but only because society's inequalities force her to marry," she says. "We need to construct a society where women don't need to marry in order to have these things."

But researchers who tout the benefits of marriage dispute both arguments. Coontz's contention of research bias is incorrect, they say, because the studies account for both divorce and bad marriages. "The best research looks at transitions into and out of marriage and includes all marriages experienced by respondents, including those prior bad marriages that ended in divorce," Waite says.

Others question the contention that marriage favors men more than women. "There are some areas where men benefit more and some where women are better off," says Maggie Gallagher, president of the Institute for Marriage and Public Policy. "But in all cases, in every area, both benefit."

OUTLOOK
Privatizing Marriage

The 20th century witnessed unparalleled changes in married life, at least in rich countries. In the United States, Europe and elsewhere, a trend away from obligations to parents, community and other forms of collective responsibility gave way to more of an emphasis on individual choice. People had more freedom to decide whom to marry and whether to stay in that marriage.

Many scholars predict even more dramatic changes for married life in the coming decades. "People will view their marriages more and more as a private relationship, rather than as a public institution as they did in the past," Blankenhorn of the Institute for American Values says. "You already see this happening, as, for instance, many couples no longer speak the traditional vows but write their own."

Blankenhorn contends that the "privatization of marriage," as he calls it, will make marriages less stable, because the built-in guideposts and expectations that once accompanied married life are disappearing. "Without a set of public expectations — like permanence or total commitment to each other — people are more on their own, and that's more risky."

> "Men will look more and more at a woman's earning potential when they decide on a mate and be less interested in women as homemakers."
>
> — **William Doherty,**
> **Director, Marriage and Family Therapy Program,**
> **University of Minnesota**

The University of Minnesota's Doherty agrees that marriage will continue shifting from a public to a private institution. But he sees a positive side to such a change: Future marriages, to some degree, will be on a more equal footing. "Men will look more and more at a woman's earning potential when they decide on a mate and be less interested in women as homemakers," Doherty says. "They're going to want someone who can bring in as much as they do or just a little less. This will bring a new level of gender equality to marriage."

Still, Doherty adds, husbands will continue to want to make at least as much as their wives. "Men are hardwired by evolution to be 'providers,' and this is still backed up by our culture," he says. "So a man who makes less than his wife will still feel 'inadequate.' "

However, some scholars see few, if any, additional changes occurring in married life — at least in the United States. "We've had so much happen in the last 100 years that I think we're entering a period of stability," North Carolina State's Risman says. "We're going to spend the next 50 years absorbing the changes of the last 50."

Risman says several trends bolster her belief. "The divorce rate has stabilized over the last few decades, which to me is a sign that the rate of change has slowed significantly." Moreover, things don't seem to be changing all that much for women at home, she points out.

"Housework remains women's work, and there is no sign that that is changing." The fact that the number of women staying at home with their children and putting off a career is holding steady is a sign that "women still don't think they can have it all, because their husbands still haven't been told that they can't have it all."

Evergreen State's Coontz agrees that marriage won't change too much in the immediate future, but neither will it remain static. "We're going to spend the next few decades sorting through the enormous changes we've seen in marriage."

In particular, she says, couples will have to work through the consequences of gender equality in marriage. "Men can no longer count on being the boss anymore," she says. "So, we're going to see more efforts to develop new habits, new emotional expectations, new time schedules and new negotiating skills as we sort out the details of this new reality."

NOTES

1. Figures cited at www.census.gov/population/www/socdemo/ms-la.html.

2. For background, see David Masci, "Children and Divorce," *The CQ Researcher*, Jan. 19, 2001, pp. 25-40.

3. Figures cited in *National Vital Statistics Reports*, Centers for Disease Control and Prevention, Vol. 52, No. 10, Dec. 17, 2003, p. 1.

4. Amy Fagen, "Permanent Tax Cut OK'd," *The Washington Times*, April 29, 2004, p. A1.

5. Michelle Conlin, "Unmarried America," *Business Week*, Oct. 20, 2003, p. 106.

6. For background, see Kenneth Jost, "Gay Marriage," *The CQ Researcher*, Sept. 5, 2003, 721-748.

7. Figures cited in *National Vital Statistics Reports*, Centers for Disease Control and Prevention, Vol. 50, No. 14, Sept. 11, 2002, p. 1.

8. Barbara Defoe Whitehead and David Popenoe, "The State of Our Unions: The Social Health of Marriage in America," National Marriage Project, Rutgers University, 1999.

9. Figures cited in *National Vital Statistics Reports*, *op. cit.*, and "U.S. Per Capita Divorce Rates Every Year: 1940-1990," Centers for Disease Control and Prevention, www.cdc.gov/nchs/fastats/pdf/43-9s-t1.pdf.

10. Figures cited in Claudia Wallis, "The Case for Staying at Home," *Time*, March 22, 2004, p. 51. See also, Sarah Glazer, "Mothers' Movement," *The CQ Researcher*, April 4, 2003, pp. 297-320.

11. Amy Fagen, "Senate Mulls Pro-Marriage Funds," *The Washington Times*, April 1, 2004, p. A5.

12. Bill Swindell, "Welfare Reauthorization Becomes Another Casualty in Congress' Partisan Crossfire," *CQ Weekly*, April 3, 2004, p. 805.

13. Figures cited in Ronet Bachman and Linda E. Saltzman, "Violence Against Women: Estimates from the Redesigned Survey," *National Crime Victimization Survey Special Report*, August 1996, p. 4.

14. For background, see Kenneth Jost, "School Desegregation," *The CQ Researcher*, April 23, 2004, pp. 345-372.

15. Barr's testimony available at www.house.gov/judiciary/barr033004.pdf.

16. *The Gospel According to St. Mark*, 10:6-9.

17. *The Gospel According to St. Mark*, 10:11-12.

18. *1 Corinthians,* 14:34-35.

19. Quoted in Daniel J. Boorstin, *The Americans: The Colonial Experience* (1958), p. 67.

20. *Ibid.*, p. 187.

21. Nancy Cott, *Public Vows: A History of Marriage and the Nation* (2000), p. 184.

22. *Ibid.*, p. 167.

23. *Ibid.*, p. 187.

24. Quoted in *ibid.*, p. 185.

25. Figures available from the Bureau of the Census at www.census.gov/prod/2004pubs/03statab/vitstat.pdf.

26. Cheryl Wetzstein, "Romney Moves to Get Vote on Same-Sex 'Marriage,'" *The Washington Times*, April 16, 2004, p. A3.

27. Quoted in Dean E. Murphy, "Scwharzenegger Backs Off His Stance Against Gay Marriage," *The New York Times*, March 2, 2004, p. A11.

28. Quoted in Dean E. Murphy, "San Francisco Mayor Exults in Move on Gay Marriage," Feb. 18, 2004, p. A18.

29. Maura Dolan, "State High Court Seeks Briefs on Validity of Gay Marriage," *Los Angeles Times*, April 15, 2004, p. B6.

30. CBS News Poll, March 15, 2004, at www.cbsnews.com/stories/2004/03/15/opinion/polls/main606453.shtml.

31. Catherine E. Ross, John Mirowsky and Karen Goodsteen, "The Impact of the Family on Health: Decade in Review," *Journal of Marriage and the Family 52* (1990), p. 1061.

32. Cited in Linda J. Waite and Maggie Gallagher, *The Case for Marriage: Why Married People are Happier, Healthier and Better Off Financially* (2000), p. 67.

BIBLIOGRAPHY

Books

Cott, Nancy F., *Public Vows: A History of Marriage and the Nation, Harvard University Press*, 2000.
A professor of history and American studies at Yale University examines marriage in the United States from European settlement in the New World through social revolutions following World War II.

Waite, Linda J., and Maggie Gallagher, *The Case for Marriage: Why Married People Are Happier, Healthier and Better Off Financially*, Doubleday, 2000.
A sociology professor of at the University of Chicago (Waite) and the director of the Marriage Program at the Institute for American Values (Gallagher) review recent research showing the benefits of marriage.

Articles

"The Case for Gay Marriage," *The Economist*, Feb. 28, 2004, p. 9.
The venerable English news weekly argues that same-sex couples should be allowed to marry, asking: "Why should one set of loving, consenting adults be denied a right that other such adults have and which, if exercised, will do no damage to anyone else?"

Belsie, Laurent, "More Couples Living Together, Roiling Debate on Family," *The Christian Science Monitor*, March 13, 2003, p. A1.
The author explores the social impact of the growth of cohabiting couples in the United States.

Conlin, Michelle, "UnMarried America," *Business Week*, Oct. 20, 2003, pp. 106-116.
Conlin's cover story reports that the dramatic decline in traditional families has significant implications for American business and society.

Crary, David, "Will the Institution of Marriage Continue? It's Debatable," *Los Angeles Times*, Feb. 10, 2002, p. A20.
The article explores the marriage movement, which seeks to promote the social benefits of matrimony.

Gallagher, Maggie, "Massachusetts vs. Marriage," *The Weekly Standard,* Dec. 1, 2003.

A pro-marriage writer argues that expanding the definition of marriage to include same-sex couples will make the institution largely meaningless.

Hubler, Shawn, "Nothing But 'I Do' Will Do Now for Many Gays," *Los Angeles Times,* March 21, 2004, p. A1.

Hubler explores the evolution of attitudes about gay marriage within the homosexual community.

Jost, Kenneth, "Gay Marriage," *The CQ Researcher,* Sept. 5, 2003, pp. 721-748.

The author provides a broad overview of the debate over gay marriage.

Kmiec, Douglas R., "Marriage is Based on Procreation, a Fact No Claim of Gay 'Equality' Can Avoid," *Los Angeles Times,* March 14, 2004, p. M1.

A professor of constitutional law at Pepperdine University argues that marriage is not a matter of rights but a question of public policy.

Lyall, Sarah, "In Europe, Lovers Now Propose: Marry Me a Little," *The New York Times,* Feb. 15, 2004, p. A3.

Lyall looks at legal arrangements short of marriage that are gaining popularity in Europe.

Munro, Neil, "Supporting Marriage, But for What Goal?" *National Journal,* Jan. 3, 2004.

An overview of the debate over the benefits of marriage.

Reich, Robert B., "Marriage Aid That Misses the Point," *The Washington Post,* Jan. 22, 2004, p. A25.

The former secretary of Labor argues that the money President Bush would like to spend on marriage promotion would be better used on job training and education.

Wallis, Claudia, "The Case for Staying At Home," *Time,* March 22, 2004, p. 50.

Wallis examines the trend among professional women who put their careers on hold to care for their children.

Reports and Studies

Coontz, Stephanie, and Nancy Folbre, *Marriage, Poverty and Public Policy,* Council on Contemporary Families, April 2002.

A professor of history and family studies at The Evergreen State College (Coontz) and an economics professor at the University of Massachusetts (Folbre) argue that money devoted to promoting marriage among welfare recipients would be better spent on education and other social services.

Why Marriage Matters: Twenty-One Conclusions from the Social Sciences, Center for the American Experiment, Coalition for Marriage, Family and Couples Education and Institute for American Values, 2002.

Three marriage-advocacy groups catalog many of the arguments traditionally given in favor of marriage.

For More Information

Coalition for Marriage, Family and Couples Education, 5310 Belt Rd., N.W., Washington, DC 20015-1961; (202) 362-3332; www.smartmarriages.com. Nonpartisan group that promotes marriage education.

Council on Contemporary Families, 208 E. 51st St., Suite 315, New York, NY 10022; www.contemporaryfamilies .org. Left-leaning think tank that researches marriage and other family issues.

Focus on the Family, 8685 Explorer Dr., Colorado Springs, CO, 80995; (719) 531-3400; www.family.org.

Christian advocacy group that opposes gay marriage.

Institute for American Values, 1841 Broadway, Suite 211, New York, NY 10023; (212) 246-3942; www .americanvalues.org. Promotes the renewal of marriage and family life.

Lambda Legal Defense and Education Fund, 120 Wall St., Suite 1500, New York, NY 10005-3904; (212) 809-8585; www.lambdalegal.org. National gay-rights organization that represents plaintiffs in a number of gay-marriage cases.

7

Student Aid

Will Many Low-Income Students Be Left Out?

Marcia Clemmitt

Harvard graduates celebrate at commencement on June 7, 2007. In December the university cut education costs by up to 50 percent for families that earn $120,000 to $180,000; tuition is waived for students from families earning under $60,000. Meanwhile, as merit-based scholarships replace need-based aid across the country, many low-income and minority students are finding it hard to finance their college dreams.

From *CQ Researcher*,
January 25, 2008.

E ach month, Lucia DiPoi, a 24-year-old graduate of Tufts University in Boston, pays $900 toward her college debt — $65,000 in private loans, $19,000 in federal loans.

The first in her family to attend college, DiPoi blames some of her plight on her lack of financial knowledge. When grants and federal loans didn't cover enough of her fees, she took out private loans with interest rates of more than 13 percent. "How bad could it be?" she figured.[1]

Now she knows. DiPoi's loan burden is well above the average $20,000 in debt that American college graduates face, but it's not unusual. Many students from low-income families — and graduate students in particular — also face large burdens.

As the cost of higher education rises, grants for needy students have lagged behind, and more students are dependent on loans to finance their education. At the same time, worries about college costs have been reaching higher up the socioeconomic scale. In response, states and private colleges have launched new merit-based scholarships that shift some aid from the neediest students to middle- and even upper-income families.[2]

In-state tuition and fees (excluding room and board) for public, four-year schools average $6,185 for the 2007-2008 school year, up 6.6 percent from 2006-2007; out-of-state tuition averages $16,640. At private four-year schools, the average 2007-2008 tuition and fees is $23,712, up 6.3 percent from 2006-2007.[3] The cost of college has nearly doubled over the past 20 years, in inflation-adjusted dollars, and college tuition and fees have risen faster than inflation, personal income, consumer prices or even the cost of prescription drugs and health insurance.[4]

143

Student Debt Highest in New Mexico

College graduates in New Mexico in 2006 had an average of $28,770 in student debt, the highest of any state and almost 50 percent higher than the national average. Hawaii had the lowest average debt: $11,758 per student.

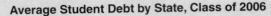

Average Student Debt by State, Class of 2006

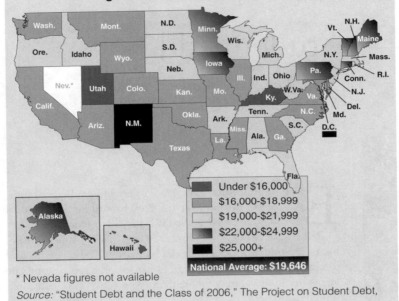

Under $16,000
$16,000-$18,999
$19,000-$21,999
$22,000-$24,999
$25,000+

National Average: $19,646

* Nevada figures not available

Source: "Student Debt and the Class of 2006," The Project on Student Debt, September 2007

The increasing difficulty of paying for college and the importance of lending are bringing new attention to financial-aid issues. Congress recently passed legislation boosting the value of grants for low-income students and trimming subsidies for private education lenders. New York State Attorney General Andrew M. Cuomo reached legal settlements last year with more than two-dozen colleges where financial-aid officials had accepted kickbacks in exchange for steering students toward particular private lenders; similar investigations are ongoing in other states.

In December, pricey Harvard University — which had already waived tuition payments for students whose families earn less than $60,000 a year, announced a boon for middle-earners, cutting costs by as much as 50 percent for families that earn between $120,000 to $180,000.[5]

If the main job of financial aid is to make college education more accessible to lower-income and minority students, current financial-aid and cost trends aren't helping, some analysts say.

Three decades ago, there was a 30-percentage-point gap in college attendance between low-income students and other students, says Donald E. Heller, director of the Center for the Study of Higher Education at Pennsylvania State University. "And since that time they haven't gained any ground," he adds.

"Only 7 percent of high-school sophomores from the lowest quartile of socioeconomic status eventually earn a bachelor's degree, compared with 60 percent of those from the highest quartile," according to Associate Professor of Public Policy Susan M. Dynarski and doctoral candidate Judith Scott-Clayton, both at Harvard. "Only 12 percent of Hispanics and 16 percent of African-Americans eventually earn a B.A., compared with 33 percent of non-Hispanic whites." Moreover, the gaps persist "even among well-prepared students," so difficulties paying for college are at least partly to blame, the researchers say.[6]

Federal Pell Grants for low-income students — a big part of the money problem — have lagged behind rising costs for decades. "The purchasing power of the Pell Grant is less than half what it was in the 1970s," Heller says.

Need-based aid offered by states has been dropping in many places over the past decade, says Ross Rubenstein, associate professor of public administration at Syracuse University. While need-based grants are still the largest share overall, a growing proportion of state grants are merit-based scholarships, a "huge middle-class entitlement" that shifts the focus of aid programs away from expanding access for the neediest, he says.

Merit aid has virtues, though. Reserving some scholarship aid for top students might improve student achievement in high school, Rubenstein explains, noting it has been shown to have modest positive effects on student achievement in Georgia, for example, which launched the first state merit program, HOPE, in 1993. "I wouldn't want to see a merit-based system replace a need-based one, but it has its place," he says.

A certain level of merit aid might not be a problem, Heller says; he cites Indiana's program, which has been successful in promising aid to all students who graduate from high school with a C average. "But I would caution against" including additional criteria "such as SAT scores or requiring higher grade averages," he adds, because of racial gaps in SAT scores and the general discouraging effect such criteria have on low-income students who are hesitant about their college chances anyway.

Merit aid may create heavier loan burdens for low-income students. The University of Maryland recently discovered that low-income students were graduating with more debt than middle- and high-income students and concluded that its grant program — which had 60 percent merit-based awards — should be revamped to include more need-based grants.[7]

The biggest trend in college financing is the heavy reliance on loans, which make up about 70 percent of higher-education financing today. The loans include both federally guaranteed and subsidized loan programs and, increasingly, completely private, non-federally subsidized loans with much higher interest rates. (*See graphs, p. 147.*)

"Loans are ubiquitous as more and more students are borrowing," says Laura W. Perna, an associate professor at the University of Pennsylvania Graduate School of Education.

The federal government makes some loans directly and subsidizes private lenders to handle others. The private lenders came under fire last year, as reports surfaced of outsized bonuses paid even at many nonprofit lenders and sweetheart deals that some lenders gave college financial-aid officials who steered students toward their services.

"Ninety percent of students who receive loans choose their lender based on their schools' recommendation," according to the Center for American Progress, a liberal think tank. With education debt high and

rising, "students should be able to count on their schools for impartial and helpful advice as they navigate a complicated and stressful process." Instead, "kickbacks" and "conflict of interest" put students at risk of doing business only with lenders who have offered the most blandishments to their colleges, the center said.[8]

But many college officials argue that students are getting excellent service from lenders and could suffer if they switched to direct government loans or if the number of private lenders was cut. "Private lenders tend to be more efficient, have better technology and are able to provide better services that aren't available from the government," said Seamus Harreys, dean of student financial services at Boston's Northeastern University.[9]

Disputes over private vs. public lending aside, however, mushrooming education debt is a trend that troubles many because of the financial burdens loans place on graduates.

Students don't understand the implications of the loans they apply for, according to the Federation of State PIRGs (public interest research groups). "The loan industry recommends that graduates . . . dedicate no more than 8 percent of their income to student-loan

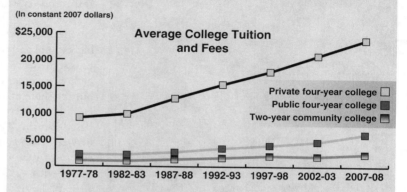

Costs Rose for State and Private Colleges

The average tuition and fees for private four-year colleges more than doubled in the past 30 years, to nearly $24,000 in the 2007-08 academic year.* During the same period, costs also doubled at public two-year schools and nearly tripled at public four-year institutions.

(In constant 2007 dollars)

Average College Tuition and Fees

Private four-year college ☐
Public four-year college ■
Two-year community college ▨

1977-78 1982-83 1987-88 1992-93 1997-98 2002-03 2007-08

* Tuition and fees constitute about two-thirds of the total budget for students at private four-year colleges but are just over one-third of the total budget for in-state students in public four-year colleges and less than 20 percent for public two-year college students.

Source: "Trends in College Pricing 2007," The College Board, 2007

AP Photo/Richard Drew

New York State Attorney General Andrew M. Cuomo, right, announces on May 31, 2007, an agreement between Columbia University and the National Association of Student Financial Aid Administrators to curtail improper student-aid practices. Investigations by Cuomo last year forced 10 private lenders to stop kickbacks to aid officials at two-dozen colleges in exchange for steering students to their firms. Similar investigations are ongoing in other states, including Missouri, Iowa and Pennsylvania.

repayment," according to PIRG analysts Tracey King and Ivan Frishberg, but students themselves "expected to contribute an average of 10.7 percent of future income." Furthermore, "students with larger debts more significantly overestimated the percentage of their income they could afford" for repayment.[10]

As students, college officials and political leaders wonder how future students will pay for higher education, here are some of the questions being asked:

Has the right balance been struck between merit-based and need-based aid?

Over the past decade and a half, public and private student-aid programs have increased the proportion of scholarships based on academic and athletic merit, many going to middle- and upper-class students. As a result, financially needy students have seen their share of the student-aid pie shrink.

Critics of the shift argue that achieving equity in education demands mostly need-based aid, since low-income and minority students still lag far behind in college attendance. But supporters of increased merit-based aid say merit scholarships spur students to work harder in high school and that, with college costs soaring, middle-class students deserve financial help, too.

Low-income students still get the most financial aid, but the share of aid claimed by students from wealthier families has increased in recent years. The Indianapolis-based Lumina Foundation for Education found that between 1995 and 2000 grants to students from families earning $40,000 or less increased by 22 percent, compared with a 45 percent boost for families making $100,000 or more.[11]

Many education analysts say merit-based grants, unlike need-based grants, don't increase low-income and minority access to higher education.

An adequate pool of need-based college aid actually improves high-school graduation rates, says Ed St. John, a professor of higher education at the University of Michigan. "Living in a state with more need-based aid increases the chances a low-income student will graduate," he says. "If a sophomore sees that there's aid, there's a bigger chance he'll finish."

"Poor kids don't think they can go to college, and their schools don't have the counselors" needed to explain how they can, St. John explains. A widely publicized need-based aid program can counter that perception. "Knowing you can afford to go allows people to do things they wouldn't do otherwise," he adds.

Merit-based aid does little to expand access for low-income students, says Rubenstein of Syracuse. "By targeting people with a B average, you're mostly targeting students who would go to college anyway," he says. State merit-based aid mainly helps the state keep its better students at in-state colleges, he says.

Often financed by lotteries, many state merit programs draw their funding from the mostly low-income people who play the games, while the aid flows mostly toward middle- and upper-income students, says Rubenstein. "There's no question that it's highly regressive."

Under a recent Massachusetts proposal, for example, more than 50 percent of students in the state's wealthiest school districts would qualify for scholarships compared to less than 10 percent in the poorest districts, according to Penn State's Heller.[12]

Similarly, private colleges' merit aid also boosts mainly the middle and upper classes. "If the private colleges don't refocus more dollars on students with high-level needs, they are going to become places that are totally closed to low-income students," said Sandy Baum, an economics professor at Skidmore College in Saratoga Springs, N.Y., and senior policy analyst for the New York-based College Board.[13]

Many low-income students are shut out from merit-based grants mainly because they attend very-low-performing schools, says Sara Goldrick-Rab, an assistant professor of education policy at the University of Wisconsin, Madison. "Essentially it's holding kids responsible for the poor K-12 program," she says.

Indeed, a recent analysis of a Michigan test that determines who gets state merit aid found that "schools with high numbers of minority students didn't even have the curriculum for the kids to take the whole test that would qualify them for the merit aid," says the University of Michigan's St. John.

Merit-based aid advocates contend that it exists for good reasons — to help deserving students get an education without plunging their families into debt. The high cost of college increasingly forces middle-class families to seek financial aid.

"The reality is even parents who work hard to save are coming up short" of college costs, said Mary Beth Moran, a financial adviser in Bloomingdale, Ill.[14]

"Middle-class families are being squeezed out," said Sharon Williams, college adviser at Elgin Academy in

Loans Are Now Biggest Source of Aid

In the past three decades federal education grants have been overtaken by federally guaranteed low-interest loans. Today, the ratio of loans to grants is seven to two, compared to just three to two in 1975-76 (top). Meanwhile, high-interest, wholly private loans have emerged in the past decade, largely in response to heavy marketing and the long, confusing federal loan applications. They represented 18 percent of all student loans in 2004-05, or double the percentage four years earlier (bottom).

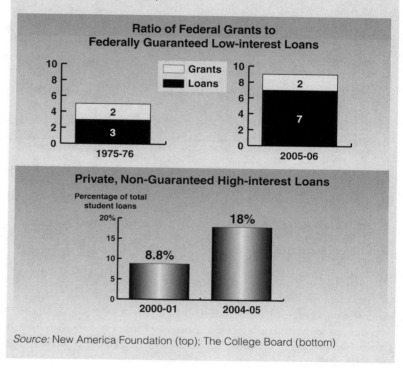

Ratio of Federal Grants to Federally Guaranteed Low-interest Loans

1975-76: Loans 3, Grants 2
2005-06: Loans 7, Grants 2

Private, Non-Guaranteed High-interest Loans
Percentage of total student loans

2000-01: 8.8%
2004-05: 18%

Source: New America Foundation (top); The College Board (bottom)

Elgin, Ill. "Need-based scholarships will not be there for them."[15]

Because of a growing perceived need among the middle class, but also because the public often doesn't like government programs that offer assistance without requiring something in return, merit-based aid is more politically popular, says William Doyle, an assistant professor of higher education at Vanderbilt University. That attitude shows up in public-opinion polls, he says. "People want reciprocity; they don't like entitlement-based grants."

"Mainly in response to middle-class demands, there was a huge surge in the 1990s and early 2000s of merit-based aid," says Rubenstein at Syracuse. The trend "has

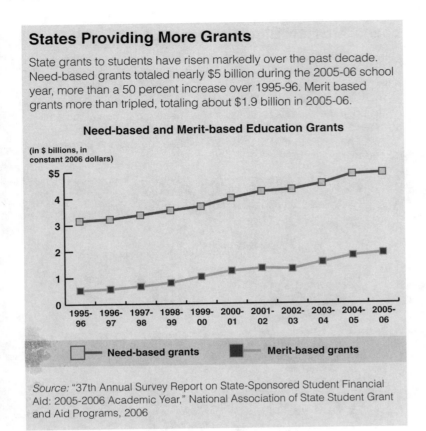

States Providing More Grants

State grants to students have risen markedly over the past decade. Need-based grants totaled nearly $5 billion during the 2005-06 school year, more than a 50 percent increase over 1995-96. Merit based grants more than tripled, totaling about $1.9 billion in 2005-06.

Need-based and Merit-based Education Grants

(in $ billions, in constant 2006 dollars)

Need-based grants   Merit-based grants

Source: "37th Annual Survey Report on State-Sponsored Student Financial Aid: 2005-2006 Academic Year," National Association of State Student Grant and Aid Programs, 2006

scholarships. Because their requirements are clear and simple, merit programs may encourage even low-income students to persist in preparing for college, explains the University of Pennsylvania's Perna. By comparison, the lengthy, complex applications of need-based programs "just leave a big question mark," as students remain in doubt for many months about whether they'll qualify for enough aid to allow them to afford their chosen school, she adds.

Georgia's HOPE Scholarships "provide an incentive for kids to work harder" in high school because they know for a fact they can claim the money if they attain the required grade average, says Rubenstein. High-school grades increased and not just because of grade inflation after the scholarship was launched, he says. In Georgia, many lower-income "kids really thought they couldn't go before the HOPE Scholarship established a simple rule: 'If I get that B, I can go,' " he says.

tapered off a bit, but I don't expect it to end. It's an extremely popular entitlement and hard to take away."

Some argue that merit-based aid ensures universities attract higher-quality students. "The incremental or marginal students that we have gained through [need-based] federal programs likely have extremely poor records with respect to college completion and probably shouldn't have been in college in the first place," said Richard Vedder, a former professor of economics at Ohio University and director of the Washington, D.C.-based Center on College Affordability and Productivity.[16]

While merit-based aid is skewed toward the middle class, it has helped students from poor families, Rubenstein points out. Georgia's merit-based HOPE Scholarship, for example, still increases aid to low-income students. With HOPE money in the system, all students receive substantially more aid than they did previously, he says.[17]

Merit-based programs have set criteria and specified award amounts, compared with most need-based

Merit-based scholarships can be designed to be more equitable, says St. John. For example, Texas has a plan in which state colleges must accept the top 10 percent of the graduating class from any state high school, he says. Besides giving students from low-income and high-minority schools an equal shot, the program indirectly encourages the schools to improve, he says.

Has the right balance been struck between grants and loans?

Since the late 1970s, federal student aid has been shifting from grants toward loans, most provided by private lenders that the government guarantees against loss should students default. Furthermore, over the last decade limits on the amount of guaranteed-loan funding each student can claim have lagged behind college costs, and more lenders have sought out student customers. As a result, more students also have turned to fully private loans — offered at regular consumer interest rates, unsubsidized by the government — as part of their college-funding package.

Forty years ago, the belief that expanding college access resulted in substantial public benefit led to the establishment of most federal student-aid grant programs, says James C. Hearn, a professor of higher education at the University of Georgia. Educators and officials reasoned that students shouldn't have to bear all their college costs because education serves many other goals, such as creating smarter voters and a more civilized and cultured citizenry, he says.

Since then, however, the reigning political philosophy has shifted toward economic conservatism. Today, the financial-aid landscape is shaped by the idea that education delivers primarily personal, not public, benefits, and that idea favors loans, Hearn says.

Nevertheless, many analysts say that expanding college access to those who couldn't afford it on their own should remain the key goal of financial aid and that grants, not loans, are the best tool.

"The fact is that loans of any kind don't improve college access for low-income students," says Heller of Penn State. For one thing, lower-income families tend to be more loan-averse and often lack the experience and information to successfully navigate a loan-based system, he explains.

"Limits on borrowing also are among the factors that make the loan-heavy aid system not work for low-income kids," whose total college-funding needs are greater and often outstrip the slow-rising limit on federally guaranteed loans, says the University of Michigan's St. John.

Today's graduates carry an average of about $20,000 in education debt, says Amaury Nora, a professor of education leadership and cultural studies at the University of Houston. (See map, p. 144.) Add to that credit-card debt for books and meals, and "a lot of times students say, 'I can't keep this up,'" he adds. Debt often becomes a compelling additional reason students leave college, initially intending to return. But in a 10-to-15-year study of such "stop outs," only a handful of students who left actually

returned, contrary to conventional wisdom in education-policy circles, Nora says. "Once they're out, they're pretty much gone."

Nonetheless, loans really expand the reach of federal aid in dramatic ways simply by adding more money to the student-aid system than taxpayers would ever pony up in the form of grants, Hearn says. Because there are finite resources for grants, other funding must be found, he says.

Loans also "provide a better opportunity for choice of college for upper- and middle-income students," says Heller.

The current loan-grant balance "is still a pretty good system for middle-class kids," says St. John.

Graduates who complain that education debt is an unfair burden ignore the benefits they've reaped, argues Radley Balko, a senior editor at the libertarian *Reason* magazine, citing two medical students profiled on CBS News as burdened with a "mountain" of debt. The two were shown drinking Starbucks coffee and Vitaminwater, pricey habits, in Balko's opinion, for a couple so worried about expenses. He estimated their lifetime earnings would be $8.2 million, or "a tidy $7.7 million profit on

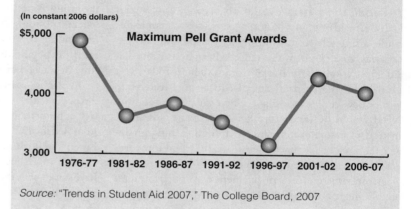

Size of Maximum Pell Grant Has Declined

The maximum Pell Grant award has declined nearly 20 percent in the past 30 years. The top grant today covers about 32 percent of average tuition, fees and room and board at public four-year colleges and 13 percent at private four-year schools. In 1986-87, maximum Pell Grants covered 52 percent of the costs at public four-year schools and 21 percent at private four-year institutions. Pell Grants are for low-income students and do not have to be repaid.

(In constant 2006 dollars)

Maximum Pell Grant Awards

$5,000 · 4,000 · 3,000

1976-77 · 1981-82 · 1986-87 · 1991-92 · 1996-97 · 2001-02 · 2006-07

Source: "Trends in Student Aid 2007," The College Board, 2007

the investment they made in their education. If only everyone had it so rough."[18]

Many analysts say a thoughtful balancing of loans and grants, plus greater flexibility on loan repayment, is the fairest way to distribute education funds. To make paybacks fairer, Congress passed the College Cost Reduction and Access Act in 2007, establishing new payback schedules for direct federal student loans that are "income-contingent," or based on graduates' earnings.

Basing the system mainly on income-contingent loans, instead of grants or non-income-contingent loans, could ultimately simplify applying for aid and perhaps do away with the complicated needs-analysis financial-aid application that scares off some students today, says Donald Hossler, a professor of educational leadership and policy studies at Indiana University, Bloomington. Need would be taken into account after graduation, instead of before enrollment, with high-earning graduates required to pay back their loans fully and low earners entitled to reduced payment.

An income-contingent payment scheme also might enable more students to select lower-paying, service-oriented careers like teaching, which some avoid because of worries about paying off their debts, Hossler says.

If money is tight, federal aid heavy on grants and income-contingent loans might be reserved for students in fields deemed to produce more public benefit — such as inner-city teaching — while students in fields that yield more individual benefit, like business, could rely more on traditional loans, says the University of Georgia's Hearn. "You can try to target your subsidies to public good" so taxpayers are on the hook only for educations that provide needed public services at a modest cost, he adds.

However, "some research shows we're not very good at playing the market" with targeted subsidies, Hearn says. In the 1960s and '70s, for example, the government used financial aid to lure people into nursing and engineering. But by the time the subsidies kicked in, market forces had erased the personnel shortages.

Do private lenders deliver good value for taxpayers and students?

Government-backed student loans come in two varieties: direct loans and loans offered by private lenders the government pays to participate in the program and guarantees against loss should students default. Debate has long raged over whether the government should remedy

perceived shortcomings in the federally backed loan program by beefing up its own direct-loan programs and making it tougher and less lucrative for private lenders to participate.

Supporters of retaining a large number of private lenders to students say the lenders aren't getting rich at the program's expense.

"No abnormal profits are being made in student loans," said Kevin Bruns, executive director of the industry group America's Student Loan Providers. Trimming federal subsidies to lenders could hurt students if banks respond by eliminating bonuses like discounted interest rates for payments that are made electronically or on time or waivers of some borrowers' fees, he said.[19]

Replacing loans made by private lenders with more direct government loans "would result in massive waste as well as burdensome red tape for students and parents," said John Berlau, a fellow in economic policy at the Competitive Enterprise Institute, a free-enterprise-oriented think tank. "The reason schools, with both programs to choose from, have stayed with the banks is that the banks offer the private sector's level of service," he said. "Some 600 schools have stopped participating in direct lending, and surveys of schools cite poor service as the primary reason."[20]

Some economic conservatives argue that all student aid should come in the form of private-sector loans or other private financial arrangements.

"The intellectual justification for expanded federal student-loan programs is extremely weak" because they've attracted only marginal students and encouraged colleges to be bigger spenders, said Vedder, at the Center on College Affordability and Productivity. "It is not clear that higher education has major positive spillover effects that justify government subsidies in the first place, and the private loan market that can handle anything from automobile loans to billion-dollar government bond sales can handle providing financial assistance to students."[21]

Late last year, lawmakers approved requiring lenders to bid for a limited number of spots in the loans-to-parents program, but private lenders generally oppose the move. Such an auction could backfire because unreliable lenders who give bad service might lowball the bidding and end up with all the business, said Bruns. "An auction creates that problem where everything is based on price," he said.[22]

But advocates of a strengthened government role argue that allowing large numbers of private lenders to offer federally backed loans costs a lot and is of questionable benefit.

The large number of lenders competing for student-loan business is evidence that private lenders have long received "excess subsidies," according to Michael Dannenberg, director of education policy at the liberal New America Foundation "There are reasons Sallie Mae's [Student Loan Marketing Association's] stock has increased by 2,000 percent in the last decade, and those reasons are a government guarantee against risk and very large government subsidies."[23]

Furthermore, the fact that the government guarantees private lenders 99 percent of a loan's value if a student defaults gives lenders "little reason to put resources into collecting payments from delinquent borrowers," he said.[24]

The idea that having more lenders in the program improves customer service through competition doesn't make sense, says Jason Delisle, research director for education policy at the New America Foundation. The loans are an identical commodity offered under government rules, so lenders have little room to customize loans or services, he says. In addition, lenders generally hold the loans on their own books for only a few months before selling them.

Because the subsidies are not market-based but negotiated by lenders and members of Congress, the private loan program has been plagued with "influence peddling" and "a dangerous amount of political influence," says Delisle, who sat in on discussions with lenders as a budget analyst for Sen. Judd Gregg, R-N.H.

"Without a market mechanism to set student-loan provider subsidy rates, banks will continue to inundate Congress and its staff with papers, meetings and phone calls pleading that a cut in that arbitrary subsidy rate would be 'catastrophic' to the lending business and that a lender-subsidy reduction means loans will no longer be made available to students," said Delisle.[25]

Education lenders complain to Congress that higher subsidies are necessary for their business to be profitable but often say the opposite to business associates, Delisle charges. "When Congress began considering this most recent round of subsidy cuts, Sallie Mae representatives told me and other congressional staffers that for the company to continue making federal student loans at the proposed lower subsidy rates, Sallie Mae would have to make the loans through its charity organization," he says.

"In other words, the proposed and now enacted subsidy cut" — in September 2007 Congress cut the subsidies by about half — "makes the federal student-loan business unprofitable, according to Sallie Mae," Delisle says. "But try to square that position" with recent Sallie Mae comments to a group that wanted to buy out the company. The potential buyers were worried that lower government subsidies would harm profitability, but "reports have those close to Sallie Mae" saying that the subsidy cuts' "effects on the company's earnings will be 'de minimis,' " Delisle says.

The buyout deal ultimately collapsed, however, and the company has recently reiterated to the U.S. Securities and Exchange Commission that Congress' 2007 subsidy cuts will substantially reduce profits from new loans and could potentially make the loans unprofitable.[26]

BACKGROUND

Gradual Beginnings

Financial aid to needy students dates back to 13th-century Europe.[27] But scholarship aid always has had mixed goals, which have complicated its development over the years. Some view student aid as mainly a boost for gifted but low-income individuals; others see it as a broad, public initiative for expanding access to education to the poor and to minorities. Many private and state universities have viewed it primarily as a means of boosting their own reputations by attracting higher-caliber students.

The first private colleges in America opened in the 17th century. Most then gave many grants to expand their enrollment and reputations. At Harvard in the 1700s endowed scholarships paid about half of school expenses for between a quarter and a third of students. New York University gave substantial grants to about half the students from the time it opened in 1831 through the mid-19th century; the grants enabled many to attend tuition-free.

But as private universities gathered students and solidified their reputations — and college education remained confined to the elite few — the days of generous grants waned.

Meanwhile, in the early days of state-funded colleges, financial aid as such was scarce, says Hearn of the University of Georgia. Instead, the states provided a virtually free education to all comers as a public good, he says.

C H R O N O L O G Y

1960s-1970s *As college prices rise, Congress creates loans and grants.*

1965 Federal Family Education Loan Program — then called the Guaranteed Student Loan Program — is launched.

1972 Congress creates Sallie Mae — the Student Loan Marketing Association — and establishes federal grants for low-income students, later named for Sen. Claiborne Pell, D-R.I.

1976 Congress allows states to issue tax-exempt bonds for education lending.

1977 College loans total $1.8 billion.

1978 Congress expands eligibility for Pell Grants to some middle-income students and makes federally guaranteed loans available to all income levels.

1979 Congress removes the cap on its subsidies to private education lenders, ensuring them good returns.

1980s *College costs rise, and pressure builds for aid to the middle class. With private lenders assured of high subsidies, loan volume explodes.*

1980 Parent Loans for Undergraduate Students — PLUS loans — allow parents at all income levels to borrow.

1986 Federal payments reimbursing lenders for education-loan defaults total more than $1 billion. . . . Private lenders barred from offering inducements to borrow, such as free appliances.

1989 College loans total $12 billion.

1990s *Private lenders begin marketing directly to students and parents. Merit-based scholarships proliferate in states.*

1990 A large guarantee organization, the Higher Education Assistance Foundation, collapses from loan defaults at for-profit colleges.

1992 Congress and the Clinton administration establish the federal Direct Lending Program to compete with private education lenders.

1993 Georgia uses lottery receipts to fund new merit-based HOPE scholarships for families with incomes below $66,000.

1994 College attendance peaks among welfare recipients.

1995 Georgia removes income restriction on HOPE Scholarships.

1996 Student loans total $30 billion. . . . Welfare-reform law directs low-income women toward jobs and away from college.

1997 Congress increases middle-class aid with tax credits, tax-free college savings accounts and tax deductions for student-loan interest.

2000s *More private lenders offer student loans at market rates without government guarantees, as rising tuitions see wealthier families seeking loans.*

2000 Private, non-government-backed loans total $4 billion. . . . Georgia allows HOPE Scholarship recipients to accept Pell Grants.

2002 Need-blind merit grants account for 24 percent of state education grants, up from 10 percent in 1992.

2005 Congress bars students from discharging most fully private education loans in bankruptcy.

2006 Non-federally guaranteed private loans for college education total over $16 billion. . . . Harvard waives tuition payments for families earning less than $60,000.

2007 New York Attorney General Andrew Cuomo investigates colleges that accept favors to steer students toward specific lenders. . . . Harvard expands tuition breaks to families earning up to $180,000. . . . Congress cuts subsidies to private lenders, ups Pell Grant funding and institutes income-dependent repayment for direct government loans. . . . Private investors back out of deal to buy student-lending giant Sallie Mae, which faces rising defaults after extending loans to risky borrowers.

2008 University Financial Services in Clearwater, Fla., settles charges it deceptively used Ohio University's logo and mascot to market loans.

While the picture varies from school to school, by the 1929-30 academic year, grant aid amounted to only 2.5 percent of U.S. college costs, including both tuition and living costs. Even after the Depression, when college charges fell and grant aid rose, grants amounted only to 3.6 percent of college costs by 1939-40.

Expanding Access

With the end of World War II, a new boom in financial aid began, along with new ideas about who would go to college. By the mid-20th century, large numbers of students became the first in their families to attend college.

In 1944, as the war was approaching its end, Congress enacted the Servicemen's Readjustment Act — the so-called G.I. Bill of Rights. As returning veterans signed up for classes, student aid quickly grew as a proportion of college costs, although the spurt was temporary. By 1949-50, student grants amounted to 55 percent of costs. That share fell to 14 percent by the 1959-60 school year, after G.I. benefits tailed off.

At the same time, colleges, especially state universities, were raising their fees — albeit gradually — after many decades of subsidizing all students by charging very low tuition.

"In the 1950s and 1960s, people began saying, 'We're subsidizing people who would go to college anyway. So why not charge something and aid students who can't afford it?'" says Hearn. Accordingly, at public colleges a new era of higher tuitions offset by financial aid for lower-income students replaced the old system under which states heavily subsidized lower tuition for all students in public universities.

With tuitions on the rise, the federal government as well as individual colleges launched new aid programs in the late 1950s. The new programs were influenced by the growing interest in seeing more students attend college, especially after the Soviet Union highlighted its national scientific prowess with the launch of the Sputnik satellite in 1957.

In 1954, a group of 95 private colleges and universities, mostly in the Northeast, formed the College Scholarship Service. Based on the philosophy that college aid should be largely based on need, the group set about developing standards and tools the schools could use to collect and assess a family's financial information and determine how much aid students required.[28]

In 1958, the National Defense Education Act financed low-interest college loans, with debt cancellation for students who became teachers after graduation. In 1965, Congress launched several aid programs, including a talent search to identify low-income students with academic ability and College Work-Study to subsidize schools' employment of needy students.[29]

By 1966-67, the poorest quarter of college students were getting 94 percent of their college costs paid for, 44 percent of it through grants, says Rupert Wilkinson, a former professor of history and American studies at England's University of Sussex. The aid dropped off steeply for the second-poorest quarter of students, however; only 38 percent of that group's needs were covered, 15 percent of it by grants, he says.

In 1972, Congress created Basic Grants, later renamed for Sen. Claiborne Pell, D-R.I. Initially proposed by the Nixon administration, the grants were designed to offer low-income students a basic subsidy large enough to ensure that they saw college as a possibility, but not so large that students wouldn't need to tap other sources.

Borrowers and Lenders

In addition to the various aid programs launched in 1965, Congress created the Federal Family Education Loan Program (FFELP) — then called the Guaranteed Student Loan Program — to offer federally backed loans through private lenders whom the government insured against default.

In the 1960s and early '70s student-lending programs didn't garner much attention, since college costs were still relatively low. Today, however, about 70 percent of federal student aid is in the form of government-guaranteed loans, while only about 20 percent is in the form of grants. Another 5 percent of aid is in the form of tax benefits, and the rest comes through various channels such as support for work-study programs.[30]

The growing importance of loans to students has become controversial over the last two decades. A "funding gap" between tuition and grant aid began in the early 1980s, "and that's where you begin to see students relying more on loans," says Karen Miksch, an assistant professor of higher education and law at the University of Minnesota. Since then, the funding gap has grown exponentially, she says.

Furthermore, with tuition rising and more students aspiring to college, middle-class families, which didn't

Low-Income Students Unaware of Aid

Clearer information is needed

College dreams for U.S. students — rich and poor — are at the highest levels ever. But when it comes to attendance and graduation, low-income students lag as far behind the middle class as they did 30 years ago.

More than 90 percent of students in all demographic groups now hope to attend college — the same expectation level for high-school graduation just a few years ago, says James C. Hearn, a professor of higher education at the University of Georgia. But while minority and low-income kids have quickly caught up to middle-class expectations, "big attainment gaps remain between what minority and low-income students aspire to do and what they actually do," he says.

For some of the students who hope to go to college but don't, the availability of aid is not the real problem, says Hearn. "Given the aid that's out there, there are still fewer low-income and minority students attending college than you'd expect," he says.

Some of the barriers between low-income students and college, such as bad schools and difficult family situations, are well-known and intractable. But many researchers are pointing to a hitherto unnoticed problem that's easier to fix: Low-income students and their families are less likely to know that aid is available for them, possibly causing many to give up on their college dreams.

"It really relates to how people grow up and whether they think of themselves as being able to go or not," says Ed St. John, a professor of higher education at the University of Michigan.

While middle- and upper-class students and their families believe college is in the future and prepare for it, many lower-income students doubt they can make it. "And if you can't imagine being able to pay for college, why would you prepare for it?" asks Karen Miksch, an assistant professor of higher education and law at the University of Minnesota.

Research shows that "middle- and upper-class kids get information from a whole variety of sources," says Sara Goldrick-Rab, an assistant professor of education policy at the University of Wisconsin, Madison, including the Internet, high-school counselors and college-educated family friends. But lower-income students, most of whom attend schools lacking guidance counselors, rely mainly on friends and family with little college experience, she says.

qualify for public need-based grants, were clamoring for aid.

In 1979-80, Pell Grants covered 99 percent of costs. Today a Pell Grant covers only 36 percent of tuition and on-campus room and board at the average four-year public institution, By the 1990s the gap between a Pell Grant and tuition at a public school was $4,000; today, it's more than $6,000, Miksch says.

"You're looking at two-thirds of the cost that now has to come from someplace else," Miksch adds.

For most, that "someplace else" is the private sector — specifically federally guaranteed private loans offered to both students and parents through several programs, such as federal Parent Loans for Undergraduate Students (PLUS), created in 1980.

Questions about how private lenders work in this arena stem mainly from the way the government pays them to participate, says the New America Foundation's Delisle.

When the program started, no banks were willing to lend to college students, he says, so Congress guaranteed lenders against default and also provided generous subsidies to entice lenders into the game. Because the government isn't set up to do the work of a bank, it uses private lenders' infrastructure to make and manage the loans.

Driven by concerns that not all students would get access to loans, Congress allowed as many lenders as possible to participate in the program and set subsidies high enough to attract many lenders.

Beginning in the 1980s, however, concerns grew in Congress about whether private lenders manipulated the system to push students into inadvisable loans. In 1986, Congress barred banks and other lenders from offering students "inducements" to borrow — such as toasters or other appliances.[31]

Despite tighter rules, however, by the early 1990s Congress and President Bill Clinton remained convinced

Informing students when they're in middle school that college aid will be available to them is crucial, because when students know they can go to college, they're more likely to stay in school and take courses that will prepare them for it, says Donald E. Heller, director of the Center for the Study of Higher Education at Pennsylvania State University.

He points to Indiana's successful Twenty-first Century Scholars Program targeting low-income eighth-graders. They are told that they can attend state colleges tuition-free or receive aid to attend a private college in Indiana if they graduate high school with a 2.0 average, use no illegal drugs or alcohol, commit no crimes and enroll in college within two years of graduation.[1]

Among 2,202 students enrolled in the program, 1,752 — nearly 80 percent — enrolled in a college in the state within a year of graduation, according to the Indianapolis-based Lumina Foundation for Education.[2]

Indiana also has a program to increase parental involvement, and it has worked, says Heller. It "creates a culture of college-going" and "gives the kids an incentive to prepare themselves," he says.

Congress could enact a similar early-commitment program, pledging federal Pell Grants to eligible middle-schoolers, says Heller. "By a conservative estimate, over 75 percent of students who get a free or reduced-price school lunch ultimately will be eligible for Pell," he says. "If they knew about that and kept working in school because of it, they'd get a grant that's more than the average tuition at a community college," he says.

The current system of applying for aid is complicated for a good reason: to target aid to the neediest students. But the complexity itself puts low-income families at a disadvantage because they are more likely to be daunted by the form and less likely to find good help to prepare it, according to a study by Susan M. Dynarski and Judith E. Scott-Clayton of Harvard University's Kennedy School of Government. Dropping the federal-aid application form from 72 questions to a more manageable 14 would result in virtually no change in eligibility for Pell Grants, they found in a recent analysis.[3]

Furthermore, programs such as Georgia's HOPE Scholarship and Social Security student benefits provide plenty of evidence that simplified aid programs can increase college enrollments significantly, they found.

---

[1] For background, see "Meeting the Access Challenge: Indiana's Twenty-first Century Scholars Program," Lumina Foundation for Education, August 2002, www.luminafoundation.org.

[2] *Ibid.*

[3] Susan M. Dynarski and Judith E. Scott-Clayton, "College Grants on a Postcard," The Hamilton Project, The Brookings Institution, February 2007. Dynarski is an associate professor of public policy at Harvard; Scott-Clayton is a doctoral student.

that lenders were unfairly getting rich off subsidies they received for offering federally guaranteed loans. In 1992, Congress created the Federal Direct Lending Program, which makes student loans without going through private middlemen, to compete with the private lenders in FFELP. In 1993, Congress called for direct government lending to make up 60 percent of federal student loans by 1998-99.

These moves didn't eliminate the private sector's role in the student-aid business. In fact, critics see education lenders' hot competition for borrowers as evidence that the federal subsidies remain too high.

Furthermore, nonprofit student lenders have been set up in most states, says Delisle, and "members of Congress — including Republicans — say we have to keep the subsidy high because we have to keep the nonprofits in business," although there's little evidence that those organizations serve the program any better than the hordes of private lenders.

Lenders Fight Back

When Republicans took control of Congress in 1995, they expressed dismay at the burgeoning government-run Federal Direct Lending Program. By then, 1,300 colleges — about a third of the nation's total — had switched to direct lending, taking that market share away from private lenders. Congressional leaders pressured Clinton to slow the program's growth.

They told him he would get no other legislation passed "if he continued to push direct lending," said Robert M. Shireman, a former top aide to Sen. Paul Simon, D-Ill., who sponsored the 1993 direct-lending expansion.[32]

Clinton agreed to back off, and private lenders stepped up their efforts to regain business. For example, private lenders such as Sallie Mae offered student borrowers discounted fees and interest rates that the federal government wasn't legally permitted to match.

Non-Traditional Students Face Pitfalls

Many feel disconnected from campus life

As more Americans aspire to higher education, the number of non-traditional students — active military, older workers, immigrants — has skyrocketed, causing problems for the system and the students.

The earliest large-scale federal student-aid program was the G.I. Bill of 1944, which helped World War II vets attend college by providing money and new education programs to suit these non-traditional, often older students.

Today, the Pentagon still touts education funding as a benefit of military service, but today's system may not be working as planned. Of the veterans who entered four-year colleges in 1995, only 3 percent had graduated by 2001, compared with about 30 percent of students overall. Several reasons account for this low rate. Universities don't have to refund tuition for soldiers who are pulled out in mid-semester for overseas deployment, and the schools are allowed to terminate veterans' student status if they don't immediately re-enroll when they return from deployment.[1]

Military assurances that soldiers can attend college online from their bases also may be unrealistic for many. "I don't know how they expect us to take classes in Iraq," said Alejandro Rocha, 23, a Marine from Los Angeles. "I manned Humvees and rolled around in Humvees. . . . When we were back in the U.S., we were just training

and training." Consistent study wasn't possible, Rocha said.[2]

Growing numbers of today's students are older and often work to support families, but financial-aid rules and practices sometimes make it difficult for them.

"A good 40 percent of Latino students are working off-campus," for example, says Amaury Nora, a professor of education leadership and cultural studies at the University of Houston. Working off-campus — which is often a side effect of inadequate aid — distances students from college life, with serious negative effects on their ability to persist in school, Nora says. "Students who work off-campus are 36 percent more likely to drop out."

Stress and disconnection are part of the reason, he adds. Emotional stress related to juggling off-campus jobs and classes and feeling strapped for money keeps students from fully participating in class and the other activities important to college life, he says. The less connected students feel, the more likely they are to drop out.

For most middle- and upper-income students, rising tuitions will only change which schools they attend, says William Doyle, an assistant professor of higher education at Vanderbilt University in Nashville, Tenn. "But for lower-income students, the price will cause them not to go to school, to work more or to go to school part time," all non-traditional

To attract universities, private lenders launched incentives such as the "school-as-lender" programs. In these deals, universities agree to stop offering federal direct loans to their professional- and graduate-school students and lend the money themselves, backed by a private lender. Then the university sells the loans back to the private lender for a profit.

In 2004, for example, the University of Nebraska made such a deal with the National Education Loan Network (Nelnet) and dropped direct federal loans. "The government could not match Nelnet's offer, which would provide the university with dollars it could apply to more-generous financial-aid packages for its students," said Stephen Burd, a fellow at the New America Foundation.[33]

Realizing that banks would pay for the right to be named a preferred lender, some universities actively sought out favorable deals from private lenders in return for abandoning the federal direct loan program. In 2003, for example, Michigan State University, the second-largest participant in the government-run program, asked private lenders to compete for its business.

The university openly asked lenders, "What will you do for us if we leave direct lending?" said Barmak Nassirian, associate executive director of the American Association of Collegiate Registrars and Admissions Officers. Other colleges soon followed suit, soliciting benefits from lenders such as staff support for financial-aid offices in return for dropping out of federal direct loans. "A giant sucking sound was inaugurated with this deal," Nassirian said.[34]

paths that lower students' chance of graduating, he says.

"Once you see yourself as somebody who's working and also going to school, your chances of graduating decrease," says Doyle. Part-time study "pushes down your credit hours, and that robs you of momentum," partly because it makes graduation seem very far off indeed. Taking six credits per semester means it takes between eight and 10 years to finish college, a daunting prospect for most, he adds.

On-campus work-study jobs are an exception to the rule, says Karen Miksch, an assistant professor of higher education and law at the University of Minnesota. Work-study students tend to say, "My job, it was great. The people were so helpful," she says. Bosses on work-study jobs tend to be another voice helping students navigate the shoals of college, she adds.

Financial-aid rules sometimes trip up working students in unexpected ways, says Sara Goldrick-Rab, an assistant professor of education policy at the University of Wisconsin, Madison. Today, students' earnings count in the income-qualification calculation for most aid, so that low-income students could literally work their way out of qualifying for aid, even though school would still be unaffordable, she

Military veteran Marc Edgerly, a sophomore at George Mason University, in Fairfax, Va., says he will have $50,000 in student loans when he graduates despite federal education assistance for vets.

says. "You can begin to work and find yourself without aid very quickly, and if you switch to part-time school, then your aid is cut."

The financial-aid system can work against non-traditional students in other ways, particularly if they are low-income students, Goldrick-Rab said. "We do not reward their choices. We penalize students who withdraw from school temporarily by stopping their financial aid and making it hard to restart. . . . We often fail to award full credit for courses our students take at other institutions, forcing them to repeat courses two or three times." College policies "are designed with traditional students, engaged in traditional attendance patterns, in mind."[3]

---

[1] Aaron Glantz, "Military Recruitment Lie: Pentagon's Education Pitch Is a Scam," *The Nation*, Nov. 29, 2007.

[2] Quoted in *ibid.*

[3] Sara Goldrick-Rab, "Connecting College Access With Success," *Wisconsin School Boards Magazine*, September 2005, p. 26.

Lenders also offered discounts, university staff support and other benefits as inducements for colleges to name them "preferred lenders." Students aren't required to pick a lender on a college's preferred list, but with students and parents often overwhelmed by the complexity of college applications and financing, the vast majority do.

"Lenders learned that it's much easier and more effective to market their goods and services to a couple of people at an institution than to thousands of customers," said Craig Munier, director of financial aid at the University of Nebraska, Lincoln, and chair of the National Direct Student Loan Coalition.[35]

With private lenders competing hard, direct government loan growth stalled, and private loans once again became the bulk of federal student lending. In 2005,

$287 billion in guaranteed private loans were outstanding, compared with $95 billion in direct loans.[36]

Meanwhile, as college costs continued to rise, private lenders also began offering non-government-guaranteed, wholly private education loans.

In 2005-06, lenders issued $17.3 billion in such loans, whose terms reflect those in regular consumer lending, including much higher interest rates than federally guaranteed loans.[37] In the past few years, the number of these loans has skyrocketed, making up 18 percent of all education loans by 2005, compared with 8.8 percent four years earlier.[38] Also in 2005, lenders won a provision in a new federal bankruptcy law that makes it extremely difficult to discharge the loans through bankruptcy, leaving many students with huge outstanding debt that will dog them for decades.

Only 12 percent of Hispanics and 16 percent of African-Americans eventually earn a B.A. degree, compared with 33 percent of non-Hispanic whites, according to researchers, who say difficulty in paying for college is partly to blame. Above, Hispanic students at Hayes High School in Birmingham, Ala. Federal Pell Grants for low-income students have lagged behind rising costs for decades.

The non-guaranteed private loans aren't subject to federal rules like the 1986 ban on offering "inducements" to borrow, and lenders have taken advantage of this to expand their markets. For example, Sallie Mae offered New York's Pace University a $4-million private-loan fund if it would agree to make Sallie Mae its exclusive lender.[39]

Analysts say two things are driving the growth of the private loan business: limits on the size of federally guaranteed loans that fall far short of covering annual college costs and the complexity of qualifying for federally backed loans.

"Borrowers have the private people coming to them" rather than having to seek them out, says Goldrick-Rab at the University of Wisconsin, Madison. Private lenders do direct mailings, and their application forms are faster and easier to complete than the Free Application for Federal Student Aid (FAFSA) forms, she says.

But she sees a big downside. "Financial-aid officers are seeing people getting private loans before they've maxed out" their eligibility for subsidized federal loans, and they could end up owing much more than they need to, she says.

Borrowers are "drawn to the '30 seconds and you'll be approved' type of approach we're seeing more and more

of now," said Robert Shireman, director of the nonprofit advocacy group Project on Student Debt.[40]

Also in the last decade-and-a-half, states have increasingly created mainly merit-based grant programs. The proportion of state aid that is merit-based rather than need-based had grown from 13 percent in 1994-95 to 27 percent by 2004-05.[41]

State merit-based aid wins political favor because it gives more support to the middle class.

The merit-based programs' main purpose may be state economic development, however. Such grants are pitched as economic development to attract good students to the state's schools and attract new residents, says Rubenstein of Syracuse. "We don't know if it's working," he says. "But now we have states competing against each other" to offer merit aid, which may reduce the grants' effectiveness in attracting upwardly mobile families to states.

Both state and private universities are currently in a merit arms race that may be cutting down what they're willing to spend on need-based aid, says Hossler of Indiana University. College rankings such as those published by *U.S. News & World Report* give big incentives for colleges to spend heavily on financial aid that could bring in students with higher SAT scores and a higher class rank, he says. Incoming freshmen with higher scores and grades mean "you automatically go up in the rankings," while with need-based scholarships or increased spending on faculty you don't, he says.

CURRENT SITUATION

In Congress

A full slate of financial-aid issues faces the new Democratic Congress and the Bush administration. Late in 2007, lawmakers modestly raised funding for Pell Grants, reduced interest rates on direct government loans, cut subsidies for private lenders and eased some students' loan burdens by making payments dependent on their incomes.[42]

The new law increases student financial aid by more than $20 billion over five years, paid for by cutting subsidies to lenders offering government-backed student loans.

The maximum annual Pell Grant for low-income students is slated to increase from $4,310 to $5,400 over

Should Congress do more to keep private lenders in the student-loan business?

YES   John Berlau
Director, Center for Entrepreneurship, Competitive Enterprise Institute

From the CEI Web site

Current federal student-loan programs are not perfect. They are a mishmash of subsidies and regulations that cause distortions, not the least of which is to raise the sticker price of tuition. That doesn't mean that it's not possible for the student-aid system to get worse. And that's what a bill from Massachusetts Sen. Ted Kennedy would likely do.

Under the plan, the government would basically bribe schools with extra federal aid to participate only in the Direct Lending program — rather than have subsidized dealings with private banks — a plan that would result in massive waste as well as burdensome red tape.

When signed into law by President Bill Clinton in 1993, "direct lending" was sold as a way to actually make money for the government by cutting out the "middleman." Since the 1960s, the government has subsidized banks and other firms that lend to students to make student loans more affordable. Advocates argued that the government would spend less money and could even profit if it made the loans and collected the interest itself.

As with many claims for government programs, direct lending hasn't yielded the benefits it promised. Not only is the program not making a profit, but over the decade it has been in existence the costs of direct lending have been coming in higher than the government's initial estimates, and these cost differences have been increasing.

At the same time, subsidized loans from banks have cost the government less than their estimated costs. The White House budget for fiscal year 2006 reported that direct lending has cost the government $7 billion more than initially predicted over the last decade, while subsidized loans have cost the government $5 billion less than they were estimated to cost. In the words of a report from the respected spending watchdog Citizens Against Government Waste, "the Direct Loan program flunks out."

Direct Lending has also failed in its promise to greatly reduce default rates. Department of Education statistics show a projected default rate above 15 percent for direct loans in 2005. The rate for subsidized loans was 13 percent.

In addition, some 600 schools have stopped participating in direct lending, and surveys of schools cite poor service as the primary reason.

If we were to make federal student loans totally government run, we would lose the innovation that private firms can show in servicing their customers, even in a situation that's not the free-market ideal.

NO   Sen. Edward M. Kennedy, D-Mass.
Chairman, Senate Health, Education, Labor and Pensions Committee

Written for *CQ Researcher*, January 2008

Millions of students face staggering tuition bills, and recent graduates juggle an average of about $20,000 in student debt. Congress should do more — not to keep private lenders in the loan business but to help students afford college and deal with debt. We should reduce unnecessary subsidies to private lenders and use the savings to increase aid to the neediest students.

Last year, Congress passed a bipartisan law raising federal aid to college students by $20 billion — the largest increase since the G.I. Bill. That's significant progress, but far from enough. Each year, 400,000 qualified students are still unable to attend a four-year college because of cost.

We should redouble efforts to ensure that the loan program is run as efficiently as possible. I'm eager to see the results of a pilot program that requires private lenders to bid for the right to offer federally subsidized parent loans. This mechanism will ensure that lenders compete for subsidies. The Congressional Budget Office estimates it will save $2 billion. If it's successful, we should expand it to all federal student loans made by private lenders.

We should examine the role of loan-guaranty agencies, which too often focus on aggressively pursuing borrowers who have defaulted, rather than preventing borrowers from defaulting. Colleges should be encouraged to switch from the privately funded loan program to the Direct Loan program, which time and again has been shown to be cheaper to taxpayers, and is untainted by the recent student-loan scandals.

Predictably, lenders claim the new law cut subsidies too deeply and is causing some to leave the program. Recently, the industry issued a flawed analysis suggesting that the privately funded loan program is less expensive than the Direct Loan program. Lenders have made these arguments before. The reality is that today more than 3,000 lenders participate in the federal program. Lenders make millions of dollars in profits from higher-interest private loans, which have grown tenfold in a decade and now account for almost a quarter of education loans.

The new law has begun to restore balance to the grossly unfair system by directing funds to students, not to banks. This year, we'll continue reform by finalizing new ethics rules. We'll ensure that banks treat students who take out private loans fairly and give them good terms and service. Most important, we'll keep the focus on students, so more can afford college and have a genuine chance at the American dream.

Joy Mueller displays her sense of humor during graduation at New York University. Mueller, who earned a master's degree in digital design, says the computer and the piggy bank signify the money she is going to have to earn with her computer skills to pay off her student loan. The average American owes $20,000 after graduation. Many students from low-income families — and graduate students in particular — also face large debt burdens.

five years. The reaction to the increase from college insiders was mixed.

"The increase in the Pell Grants is unprecedented in terms of its size," said Richard Doherty, president of the Association of Independent Colleges and Universities in Massachusetts.[43]

But Shelley Steinback, former general counsel of the American Council on Education, said that the increase would not have a really significant impact, given the likelihood of inflationary costs in everything that goes into providing higher education.[44]

Furthermore, whether the maximum authorized Pell Grant amounts are available each year will depend on whether funds are available in the federal budget. Already a budget squeeze has limited the funds for the first year of the expansion. For the 2008-09 school year, the new Pell maximum was set at $4,800, but in December Congress approved an omnibus spending bill that trims it to $4,731.[45]

For student borrowers, the new law temporarily cuts interest rates on both direct and federally guaranteed subsidized undergraduate loans — so-called Stafford loans — from 6.8 percent to 3.4 percent over four years. After the four years are up, it reverts to a higher rate, because Congress could not find a way to pay for a longer-lasting cut. And that draws a skeptical response from many congressional Republicans.

"Reducing student-loan interest rates is a good sound bite," but the per-student savings will be small; it "may be one latte, it may be two lattes. It's kind of hard to tell with today's market for coffee," said Wyoming Sen. Michael Enzi, top-ranking Republican on the Senate Health, Education, Labor and Pensions Committee.[46]

The new law also establishes the income-contingent loan repayment advocated by groups such as the Project on Student Debt for direct federal loans to students — not parents — and offers loan forgiveness for some graduates in public-service-oriented jobs in schools, charities and some government service.

Under income-based repayment, set to begin in July 2009, student-loan repayments in the direct lending program will be capped at a percentage of income, and most remaining balances will be canceled after 25 years. No payments will be required for people earning less than 150 percent of the federal poverty level — about $31,000 for a family of four — and payments will increase on a sliding scale above that level, with the highest earners' payments capped at 15 percent of family income. The plan will be available to all students, including those already holding the loans.

Public-service workers in both government and nongovernment organizations can have their loan debt from the government direct-lending program canceled after 10 years if they've made all the payments and been consistently employed in public service.

Meanwhile, loan experts say that even loan programs with lower student burdens probably won't improve college access much for lower-income students. And that still remains the goal of many higher-education analysts and advocates.

Willingness to borrow varies across groups, with low-income and minority families and students worrying more about the risks of loans, says the University of Pennsylvania's Perna.

In addition to concerns about filling out complicated loan applications and being able to repay in the future, many low-income and minority students know neighbors and family members who didn't complete college and are saddled with hefty loans they must repay on a high-school graduate's wages.

"The biggest problem is this uncertainty about the return on investment," and "part of this concern is definitely justified," Perna says. "Low-income and minority students are less likely to finish," while those who graduate often leave their old neighborhoods, skewing perceptions among remaining families, who only see dropouts around them.

Loan Scandals

Law enforcement authorities as well as lawmakers continue to take a keen interest in education aid following reports of illegal activities by lenders.[47]

Beginning in late 2006 and accelerating last year, reports surfaced of financial-aid administrators accepting payoffs and luxury gifts — like hefty consulting fees and Caribbean trips — from education lenders, presumably in exchange for listing the lenders as "preferred" in college financial-aid material. The flurry of stories stemmed from a massive investigation conducted by New York Attorney General Cuomo into what he called "an unholy alliance" between college officials and lenders involving kickbacks and other inducements for colleges to steer students toward certain lenders.

Indeed, financial-aid directors at Columbia University and the University of Southern California reportedly sat on the governing board of a New Jersey company, Student Loan Xpress, while reaping profits from selling company stock.[48]

Ironically, some of the revelations came from student journalists at the University of Texas, Austin. They found evidence that their school's financial-aid office kept tabs on how many gifts, "lunches, breakfasts and extracurricular functions" loan companies offered the university staff and factored the gifts into its decisions about which lenders to recommend. The office's director

was placed on leave in April 2007 when he was found to own at least 1,500 shares in the parent company of Student Loan Xpress, one of the lenders the university recommended.[49]

In April 2007 Cuomo announced the first settlements related to his investigation. More than two-dozen colleges and 10 lenders have signed pledges to abide by a new code of conduct that increases disclosure requirements on college-lending practices and bans gifts and compensation from lenders to college officials.[50] Colleges and lenders also have paid monetary settlements of more than $15 million as restitution to borrowers or as seed money for loan-education funds.

Investigations are ongoing in some other states.

Alarmed that Iowa students have a high debt burden — estimated at an average $22,926 for 2006 graduates, sixth-highest in the country — officials in 2007 investigated the nonprofit Iowa Student Loan Liquidity Corp., which the state created in 1979 to help make guaranteed loans available.[51] By 2007, the corporation was the state's dominant lender, with $3.3 billion in outstanding loans. Last fall, with the state attorney general investigating the agency's business practices, some Iowa lawmakers threatened to bar it from raising money for loans by issuing tax-free bonds.[52]

A similar nonprofit, the Pennsylvania Higher Education Assistance Agency, is also under scrutiny by state auditors after giving its senior employees $7 million in bonuses since 2004; the agency's chief executive resigned in October.

The Missouri Higher Education Agency also has paid millions of dollars in perks to senior executives. "They did not act like a state agency at all" but had "the mindset that they were a for-profit business," said state auditor Susan Montee.[53]

OUTLOOK

Generation Debt

As college costs soar, helping students pay for college will remain high on the agenda in the states and in Washington.

"We're at the intersection of two really important trends bumping against each other" that may change approaches to education finance, says Vanderbilt's Doyle. "One is the increasing importance of postsecondary education to attain a middle-class lifestyle." The other is a "bulge in the demographic pipeline," with the children of baby boomers — the baby-boom "echo" — set to create by 2009 the biggest generation ever with college aspirations.

Public pressure from all income groups for relief on college costs will only grow, and that likely means pressure on colleges to slow cost increases, says Doyle. "One thing we know is that states won't be doing a big tax increase" any time soon, he says. "So the question will be asked, 'Do the costs have to go up this fast?'"

"A couple of years of double-digit tuition increases and kids not getting the classes they need" will bring political pressure on the government and colleges to act, Doyle adds.

"Can institutions keep raising tuition without resistance? Probably not," says the University of Georgia's Hearn. "We may be hitting the point where public opposition" becomes a big factor in tuition decisions. "There may be some strong, strong resistance from students and families."

Already, the burden of education loans is heavy for many graduates in all income brackets, says Doyle. "I'm always surprised at the level of debt" even for "students from higher-income families," he says.

Most political trends point to a continuing evolution toward a system of loans and merit-based aid that helps the middle class, says Penn State's Heller. Political, not financial, imperatives are driving that shift, he emphasizes. "The merit-based aid that has cropped up in states in recent years could just as easily go to need-based aid, but political considerations work in the opposite direction."

The prospect of heavier reliance on loans and merit-based grants to help families with ever-rising college costs raises troubling questions about the future of college access for low-income, minority and immigrant students. The two fastest-growing parts of college aid — loans and merit-based grants — do little for those populations, as costs continue to rise.

Even colleges that currently make strong efforts to boost need-based grants and other assistance for low-income students probably won't do so indefinitely.

"Paying for 100 percent of need is expensive," said Yvonne Hubbard, director of student financial services at the University of Virginia. The school's need-based assistance program, AccessUVa, pays for everything for qualified students and costs the university about $20 million a year.[54]

The University of Illinois at Urbana-Champaign also has a strong program of need-based grants to improve access. Still, "What happens four years or 10 years down the road?" asked Daniel Mann, the school's director of financial aid, who doubts that future administrations will support the current ambitious aid program.[55]

The nation's burgeoning population of Hispanic immigrants also may mean that college attendance and graduation rates would decline, since Hispanics have "relatively low rates" of college attendance, says Penn's Perna.

"I'm a little skeptical of simple correspondences between college-graduation rates and money, but if nothing changes, we are going to see fewer college graduates" in years to come, says Indiana's Hossler.

NOTES

1. Quoted in Diana Jean Schemo, "Private Loans Deepen a Crisis in Student Debt," *The New York Times*, June 10, 2007.

2. For background, see Tom Price, "Rising College Costs," *CQ Researcher*, Dec. 5, 2003, pp. 1013-1044.

3. "Trends in College Pricing, 2007," The College Board, 2007, www.collegeboard.com/prod_downloads/about/news_info/trends/trends_pricing_07.pdf.

4. Jane V. Wellman, "Costs, Prices and Affordability," Commission on the Future of Higher Education, www.ed.gov/about/bdscomm/list/hiedfuture/reports/wellman.pdf.

5. Matthew Keenan and Brian Kladko, "Harvard Targets Middle Class With Student Cost Cuts," Bloomberg.com, Dec. 10, 2007.

6. Susan M. Dynarski and Judith E. Scott-Clayton, "College Grants on a Postcard: A Proposal for Simple and Predictable Federal Student Aid," The Hamilton Project, The Brookings Institution, February 2007.

7. Steven Pearlstein, "Cost-Conscious Colleges," *The Washington Post*, Nov. 16, 2007, p. D1.

8. Kate Sabatini and Pedro de la Torre III, "Federal Aid Fails Needy Students," Center for American Progress, May 16, 2007, www.americanprogress.org.

9. "What Financial Aid Officers Say," America's Student Loan Providers, www.aslp.us.

10. Tracey King and Ivan Frishberg, "Big Loans, Bigger Problems: A Report on the Sticker Shock of Student Loans," The State PIRGs, March 2001.

11. Quoted in Jay Mathews, "As Merit-Aid Race Escalates, Wealthy Often Win," *The Washington Post*, April 19, 2005, p. A8.

12. Donald Heller and Patricia Marin, "State Merit Scholarship Programs and Racial Inequality," The Civil Rights Project, 2004.

13. Quoted in Mathews, *op. cit.*

14. Quoted in Tara Malone, "Rising Tuition Hits Middle Class Hardest," [Chicago] *Daily Herald*, Nov. 21, 2004, www.collegeparents.org.

15. Quoted in *ibid.*

16. Richard Vedder, "The Real Costs of Federal Aid to Higher Education," *Heritage Lectures*, Jan. 12, 2007.

17. Ross Rubenstein, "Helping Outstanding Pupils Educationally," Education Finance and Accountability Project, 2003.

18. Radley Balko, "Government May Be Cause, Not Solution, to Gen Y Economic Woes," Cato Institute, July 12, 2006, www.cato.org.

19. Quoted in Larry Abramson, "Student Loan Industry Struggles Amid Controversy," "Morning Edition," National Public Radio, June 26, 2007.

20. John Berlau, "Ted Kennedy Says Eliminate Private Sector From Student Loans," Competitive Enterprise Institute Web site, April 4, 2007, www.cei.org/utils/printer.cfm?AID=5854.

21. Vedder, *op. cit.*

22. Quoted in Abramson, *op. cit.*

23. Michael Dannenberg, "A College Access Contract," New America Foundation Web site, www.newamerica.net.

24. Dannenberg, *op. cit.*

25. Jason Delisle, "The Business of Sallie Mae — Political Risk for Investors and Taxpayers," Higheredwatch.org, New America Foundation, Oct. 2, 2007, www.newamerica.net.

26. "Sallie Mae Decides To Be More Selective In Pursuing Loan Origination Activity," RTT News Global Financial Newswires, Jan. 8, 2008, www.rttnews.com/sp/breaking news.asp?date=01/04/2008&item=103&vid=0.

27. For background, see Rupert Wilkinson, *Buying Students: Financial Aid in America* (2005), and Tom Price, "Rising College Costs," *CQ Researcher*, Dec. 5, 2003, pp. 1013-1044.

28. "History of Financial Aid," Center for Higher Education Support Services (Chess Inc.), www.chessconsulting.org/financialaid/history.htm.

29. Lawrence E. Gladieux, "Federal Student Aid Policy: A History and an Assessment," in *Financing Postsecondary Education: The Federal Role* (1995), U.S. Department of Education, www.ed.gov/offices/OPE/PPI/FinPostSecEd/gladieux.html.

30. "Higher Education," New America Foundation, www.newamerica.net/programs/education_policy/federal_education_budget_project/higher_ed.

31. For background, see Kelly Field, "The Selling of Student Loans," *The Chronicle of Higher Education*, June 1, 2007.

32. Quoted in *ibid.*

33. Stephen Burd, "Direct Lending in Distress," *The Chronicle of Higher Education*, July 8, 2005.

34. Quoted in Field, *op. cit.*

35. Quoted in *ibid.*

36. Deborah Lucas and Damien Moore, "Guaranteed vs. Direct Lending: The Case of Student Loans," paper prepared for National Bureau of Economic Research conference, January 2007, www.newamerica.net/files/Guaranteed%20vs.%20Direct%20Lending.pdf.

37. "Private Loan Policy Agenda," Project on Student Debt, http://projectonstudentdebt.org/initiative_view.php?initiative_idx=7.

38. Aleksandra Todorova, "The Best Rates on Private Loans," *Smart Money*, June 20, 2006, www.smartmoney.com/college/finaid/index.cfm?story=privateloans.

39. Field, *op. cit.*

40. Quoted in Sandra Block, "Private Student Loans Pose Greater Risk," *USA Today*, Oct. 25, 2006.

41. "37th Annual Survey Report on State-Sponsored Student Financial Aid," National Association of State Student Aid and Grant Programs, July 2007, www.nassgap.org.

42. For background, see "Summary of The College Cost Reduction And Access Act (H.R. 2669)," News from NASFAA, National Association of Student Financial Aid Administrators, www.nasfaa.org/ Publications/2007/G2669summary091007.html.

43. Alex Wirzbicki, "$20.2 Billion Boost in Student Aid Approved," *The Boston Globe*, Sept. 8, 2007, www .boston.com.111.

44. Quoted in *ibid*.

45. Jason Delisle, "Pell Grants Cut," Higheredwatch.com, New America Foundation, Dec. 18, 2007.

46. Quoted in Libby George, "Broad Student Aid Overhaul Clears," *CQ Weekly*, Sept. 10, 2007, p. 2620.

47. For background, see "Special Report: Student Loan Scandal," Education Policy Program, New America Foundation, www.newamerica.net.

48. John Hechinger, "Probe Into College-Lender Ties Widens," *The Wall Street Journal*, April 5, 2007.

49. Josh Keller, "University of Texas Financial-Aid Office Took Gifts From Lenders, Student Journalists Report," *The Chronicle of Higher Education*, May 11, 2007.

50. Meyer Eisenberg and Ann H. Franke, "Financial Scandals and Student Loans," *The Chronicle of Higher Education*, June 29, 2007.

51. "Student Debt and the Class of 2006," The Project on Student Debt, September 2007.

52. Jonathan D. Glater, "College Loans by States Face Fresh Scrutiny," *The New York Times*, Dec. 9, 2007.

53. *Ibid*.

54. Quoted in Karin Fischer, "Student-Aid Officials Say Efforts to Expand Access Need Widespread Backing," *The Chronicle of Higher Education*, Sept. 22, 2006.

55. Quoted in *ibid*.

BIBLIOGRAPHY

Books

Getz, Malcolm, *Investing in College: A Guide for the Perplexed*, **Harvard University Press, 2007.**
An associate professor of economics at Vanderbilt University outlines the questions parents and students should ask about colleges and their financial-aid programs.

Vedder, Richard, *Going Broke by Degree: Why College Costs So Much*, **AEI Press, American Enterprise Institute, 2004.**
A former Ohio University economics professor argues that for-profit universities can provide badly needed price competition for traditional colleges, where tuitions are skyrocketing because the schools are inefficient and spend too much subsidizing non-instructional programs like sports.

Wilkinson, Rupert, *Buying Students: Financial Aid in America*, **Vanderbilt University Press, 2005.**
A former professor of American studies and history at Britain's University of Sussex details the social and economic history of student financial aid.

Articles

Field, Kelly, "The Selling of Student Loans," *The Chronicle of Higher Education*, **June 1, 2007.**
Beginning with the creation of the direct federal-loan program in the early 1990s, which set up a government competitor to private student-loan firms, lenders competed to be colleges' "preferred" loan sources, offering discounts, gifts and other favors to woo financial-aid officers.

Fischer, Karin, "Student-Aid Officials Say Efforts to Expand Access Need Widespread Backing," *The Chronicle of Higher Education*, **Sept. 22, 2006.**
Officials at selective universities say need-based grants and personal support are needed to expand enrollment of low-income students in top schools.

Schemo, Diana Jean, "Private Loans Deepen a Crisis in Student Debt," *The New York Times*, **June 10, 2007.**
Non-government-guaranteed loans are becoming a bigger part of the college financing picture as costs climb, and, unlike guaranteed loans, interest rates can be as high as 20 percent.

Reports and Studies

"Course Corrections: Experts Offer Solutions to the College Cost Crisis," Lumina Foundation for Education, October 2005, www.collegecosts.info/ pdfs/solution_papers/Collegecosts_Oct2005.pdf.
Analysts assembled by a nonprofit group suggest technological and organizational changes to control rising costs.

Recession, Retrenchment, and Recovery: State Higher Education Funding and Student Financial Aid, Center for the Study of Education Policy, Illinois State University, October 2006, www.coe.ilstu.edu/eafdept/centerforedpolicy/downloads/3R%20report10272006/3R_Final_Oct06_Updated%5B1%5D.pdf.

Analysts explore the consequences for higher-education funding and student aid in an era where state governments face recurring severe budget shortfalls.

"Student Debt and the Class of 2006," The Project on Student Debt, September 2007, http://projectonstudent-debt.org/files/pub/State_by_State_report_FINAL.pdf.

An advocacy group finds District of Columbia students have the highest debt — an average of $27,757 — and Oklahoma grads the lowest: $17,680 on average.

Cook, Bryan J., and Jacqueline E. King, "2007 Status Report on the Pell Grant Program," American Council on Education, June 2007, www.acenet.edu/AM/Template.cfm?Section=Home&TEMPLATE=/CM/ContentDisplay.cfm&CONTENTID=23271.

Analysts for a university membership alliance trace trends in the federal need-based grant program, finding that Pell grantees' median incomes are around $18,000, compared to $55,000 for other undergraduates.

Dynarski, Susan M., and Judith E. Scott-Clayton, "College Grants on a Postcard: A Proposal for Simple and Predictable Student Aid," The Hamilton Project, The Brookings Institution, February 2007, www.brookings.edu/~/media/Files/rc/papers/2007/02education_dynarski/200702dynarski%20scott%20clayton.pdf.

Public-policy analysts at Harvard argue that a drastically simplified federal aid-application process would significantly increase college enrollment among low-income and minority students.

Haycock, Kati, "Promise Abandoned: How Policy Choices and Institutional Practices Restrict College Opportunities," The Education Trust, August 2006, www2.edtrust.org/NR/rdonlyres/B6772F1A-116D-4827-A326-F8CFAD33975A/0/PromiseAbandonedHigherEd.pdf.

A nonprofit group says higher-education and financial-aid policies aren't helping low-income and minority students catch up to their higher-income white peers.

Wolfram, Gary, "Making College More Expensive: The Unintended Consequences of Federal Tuition Aid," Policy Analysis No. 531, Cato Institute, January 2005, www.cato.org/pubs/pas/pa531.pdf.

An analysis prepared for a libertarian think tank argues that college costs would decrease if private aid, rather than government aid, helped students fund their studies.

For More Information

American Association of Collegiate Registrars and Admissions Officers, One Dupont Circle, N.W., Suite 520, Washington, DC 20036; (202) 293-9161; www.aacrao.org. Provides information and professional education on college-admissions issues and policies.

America's Student Loan Providers, www.studentloanfacts.org. Advocacy group of banks and other organizations that make federally guaranteed college loans.

Education Sector, 1201 Connecticut Ave., N.W., Suite 850, Washington, DC 20036; (202) 552-2840; www.educationsector.org. A think tank providing research and analysis on education issues, including financial aid.

Lumina Foundation for Education, P.O. Box 1806, Indianapolis, IN 46206-1806; (317) 951-5300; www.luminafoundation.org. Supports higher-education research and projects to improve college access.

National Association for College Admission Counseling, 1050 N Highland St., Suite 400, Arlington, VA 22201; (703) 836-2222; www.nacacnet.org. Provides information and news updates related to college admissions.

National Association of State Student Grant and Aid Programs, www.nassgap.org. Provides information on state financial-aid programs.

National Association of Student Financial Aid Administrators, 1101 Connecticut Ave., N.W., Suite 1100, Washington, DC 20036; (202) 785-0453; www.nasfaa.org. Sets voluntary standards and provides professional education on student financial aid.

New America Foundation, 1899 L St., N.W., Suite 400, Washington, DC 20036; (202) 986-2700; www.newamerica.net. Liberal think tank provides research and analysis of higher-education issues, with a focus on financial aid.

Pell Institute for the Study of Opportunity in Higher Education, 1025 Vermont Ave., N.W., Suite 1020, Washington, DC 20005; (202) 638-2887; www.pellinstitute.org. Think tank that conducts research on college access and financial aid for low-income students.

Project on Student Debt, 2054 University Ave., Suite 500, Berkeley, CA 94704; (510) 559-9509; http://projectonstudentdebt.org. Provides advocacy and information on the growing debt load carried by U.S. students.

U.S. PIRG, 218 D St., S.E., Washington, DC 20003; (202) 546-9707; www.uspirg.org/higher-education. Consumer group advocating for more need-based financial aid and simplified aid-application processes.

Public-Works Projects

8

Do They Stimulate the Economy More Than Tax Cuts?

Marcia Clemmitt

Thirteen motorists died when Minneapolis' I-35 W Bridge over the Mississippi River collapsed in 2007, dramatizing the poor condition of the nation's highways, dams and other infrastructure. President Obama's stimulus plan is designed to pump additional demand for goods and services into the economy, largely by creating jobs and repairing and retooling American infrastructure for the future. Obama signed the measure into law on Feb. 17, calling it "the most sweeping economic recovery package in our history."

From *CQ Researcher*, February. 20, 2009.

AP Photo/St. Paul Pioneer Press/www.twincities.com/Scott Takushi

Matthew Sinkovec, a truck driver for an excavation company in northeastern Ohio, used to have plenty of work hauling materials to home-construction sites. But as the economic crisis battered the Cleveland area, work dried up.

"Instead of four jobs lined up, I'd get one and then wait for another," he said. Eventually, "the bottom just fell out," and he was laid off last November. Now, he mostly stays home, trying to conserve money, and hopes to find another construction job before his unemployment checks end. "I try not to think about this lasting too long," he said. "That is a real scary thought."[1]

The construction business was among the first to slow drastically as the United States entered a recession in December 2007, but most other industries now also have slowed — with accompanying layoffs. In January 2009 alone, 598,000 people nationwide lost their jobs, the biggest monthly total since 1974.[2] And in some places, net job loss has been a long-term phenomenon. Hard-hit Ohio, for example, has lost nearly a quarter of its manufacturing jobs since 2000 and 5 percent of its jobs overall.[3]

In response, President Barack Obama and congressional Democrats developed the American Recovery and Reinvestment Act and declared the so-called stimulus package their first major legislative priority of 2009. Enacted by Congress on Feb. 13 and signed into law by Obama on Feb. 17, the act is designed to pump additional demand for goods and services into the economy, partly by creating jobs and retooling infrastructure for the future. It is "the most sweeping economic recovery package in our history . . . putting Americans to work . . . in critical areas . . . that will bring real and lasting change for generations to come," Obama said.[4]

Spending Provides More Stimulus Than Tax Cuts

Stimulus items that increase spending provide more return — or "bang for the buck" — than tax cuts, according to Moody's economist Mark Zandi. Increasing spending on infrastructure, for example, is expected to yield $1.59 in revenue for every dollar spent by the federal government, compared with $1.22 for a lump-sum tax rebate.

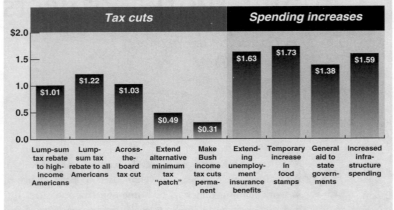

"Bang for the Buck" for Select Stimulus Programs

Source: Mark Zandi, testimony before House Committee on the Budget, January 2009

Parts of the bill's economic-stimulus provisions are familiar — tax cuts and boosts in aid programs such as unemployment benefits and Medicaid. But it's the first major economic-recovery plan to include federal spending for public works like highways and energy-efficiency upgrades for buildings since President Bill Clinton proposed, but ultimately dropped, such a proposal in 1993.[5]

Conservative analysts and most congressional Republicans were quick to denounce the public-works spending provisions as a waste of taxpayer dollars that won't help the economy.

"We have no evidence from recent or distant history" that public-works spending creates jobs or spurs the economy, says Ronald D. Utt, a senior research fellow at the conservative Heritage Foundation. The Obama plan is partly modeled on the Depression-era job-creation initiatives of President Franklin D. Roosevelt's New Deal. "I don't know that that many people were ultimately employed" by those 1930s programs, Utt says.

Obama says the programs will benefit the economy both in the current recession and long term. For example, grants to weatherize homes — modifying them to reduce energy consumption — will "immediately put people back to work," he told CBS News. "And we're going to train people who are out of work, including young people, to do the weatherization. "Not only are you immediately putting people back to work, but you're also [helping] families on energy bills and . . . laying the groundwork for long-term energy independence."[6] Other projects would repair highways, create high-speed rail systems and make public buildings energy-efficient.

A stimulus package similar to the one Congress approved should create between 3.3 million and 4.1 million jobs over the next two years — around 1.3 million of them from public-works programs — according to Christina Romer, chair of the White House Council of Economic Advisers, and Jared Bernstein, chief economist for Vice President Joseph Biden. More than 90 percent of the jobs created will be in the private sector, about a third of them in construction and manufacturing.[7]

Nevertheless, the plan can't possibly create enough jobs to offset the losses the economy is suffering, Romer and Bernstein caution.

About 11.6 million people were unemployed in January — 4.1 million more than a year earlier — and job losses are expected to continue into 2010, according to the U.S. Bureau of Labor Statistics.[8]

Our $15-trillion-per-year economy has taken at least a 5 percent hit — $750 billion — in the demand for goods and services that keeps business humming, says James K. Galbraith, an economist and a professor of government at the Lyndon B. Johnson School of Public Affairs at the University of Texas, Austin. "Infrastructure is not going to make up a $750 billion to $1 trillion hole in economic activity," he says.

Infrastructure experts also caution that the country's longtime neglect and underfunding of infrastructure maintenance and planning means that some public-works projects will suffer some delay before they're up and running.

"Up to a point, public works are a good way" to stimulate job creation, says Richard G. Little, director of the Keston Institute for Public Finance and Infrastructure Policy at the University of Southern California. However, in the present workforce we don't have the welders, the heavy equipment operators and other skilled workers we need, he says.

"A stimulus package with real sticking power should support training in the construction trades for the vast number of young and underemployed people for whom college is not the career solution," Little said.[9]

The key question for many is whether infrastructure spending that creates jobs quickly can also be visionary enough to strengthen the nation and economy long term.

Public-works spending can contribute toward stabilizing the economy in the short term, depending on how quickly the money can be spent, "but, much more important, the public works of today will redefine how we live in the future," says Galbraith, noting that the Interstate Highway System launched by President Dwight D. Eisenhower in the 1950s created America's suburbs.

When it comes to federal infrastructure spending, "forget about stimulating the economy over the next year," says Robert P. Inman, a professor of finance and economics at the University of Pennsylvania's Wharton School of Finance. "The rewards should be found in the project itself," and, ideally, the benefits should be national, he says.

For example, "if you give money to Pennsylvania to invest in education, their kids will be more productive, and they'll end up living everywhere in the country," benefiting the whole nation, he says. In the best-case scenario, the dollars would go to inner-city and other poor schools, he says, thus aiding a cause "that we value but that wouldn't have received help otherwise" — a good test for the worth of government spending, he adds.

There's tension between projects that will give local economies a quick boost and those that would best serve

Job Loss Likely to Exceed Job Creation

Economists predict the stimulus plan will create 1.3 million new public-works jobs and possibly two or three times as many jobs overall (right graph). U.S. employment has declined by 3.6 million since the start of the recession in December 2007. Nearly 600,000 jobs were lost in January — eight times the number lost in the same month a year earlier (left graph).

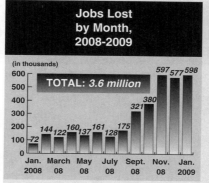

Jobs Lost by Month, 2008-2009

(in thousands)

TOTAL: 3.6 million

Jan. 2008: 72; March 08: 144; May 08: 122; 160; July 08: 137; 161; Sept. 08: 128; 175; Nov. 08: 321; 380; Jan. 2009: 597; 577; 598

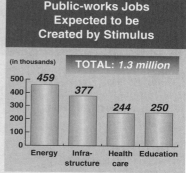

Public-works Jobs Expected to be Created by Stimulus

(in thousands)

TOTAL: 1.3 million

Energy: 459; Infra-structure: 377; Health care: 244; Education: 250

Sources: Bureau of Labor Statistics; Christina Romer and Jared Bernstein, "The Job Impact of the American Recovery and Reinvestment Plan," Offices of the President and Vice President, January 2009

future needs, says Anthony Shorris, a fellow at the liberal Century Foundation think tank and former executive director of the Port Authority of New York and New Jersey. "The fastest thing to build is a new road, but it's the opposite of everything we want" in the long run, producing more sprawling development and more carbon-emitting automobiles, he says.

"If the wrong things are done, they may do damage with this [stimulus] bill," says John Norquist, president of the Congress for the New Urbanism, a Chicago-based nonprofit that promotes walkable environments and sustainable development.

"It's crucial to think beyond the current crisis," says Guian A. McKee, an associate professor of history at the University of Virginia. "What do we want the structure of this economy to be 10 to 15 years from now? Do we want mass transit, alternative energy?" If so, then it's time to focus on such projects, he says. "While we need the shovel-ready stuff for the crisis, we shouldn't neglect the long-term things," he says.

At present, though, "we haven't really developed a vision of the 21st-century U.S. economy, so we don't

House and Senate Differ on Spending

Wide differences in funding separated many items in the House and Senate versions of the stimulus plan. Construction funds for schools and colleges took a big hit between the early and final versions of the measure. Federal-building construction and modernization was also cut, although defense-related spending, including construction, increased. Development of high-speed rail was boosted in the final bill, but greener investments in public transit received much less.

Public-Works Projects in Stimulus Plan
(in $ billions)

Item	House version	Senate version	Final version
Highway grants	$30 billion	$27.1 billion	$29 billion
Grants for public transit investment	$7.5	$8.4	$8.4
State and local government energy grants	$6.9	$4.7	$6.3
Defense environmental cleanup	$0.5	$5.5	$6
Advanced broadband program	$2.3	$6.6	$4.3
Modernization of defense facilities	$4.5	$2.5	$4.2
Energy efficiency for public housing	$2.5	$2.2	$2.2
Community health centers infrastructure	$1	$1.9	$2
K-12 school construction	$14	$0	$0
Public transit modernizations	$2	$0	$0
FEMA critical infrastructure construction	$0	$0.5	$0
State grants for high-speed and inter-city rail	0	$2	$8

Source: House Committee on Rules

"If we're going to have an infrastructure feeding frenzy, make sure government builds public works that will make us more productive as a nation," such as roads, bridges, mass transit, integrated information technology in public industries like health care, and military recapitalization, said New Hampshire Sen. Judd Gregg, top-ranking Republican on the Senate Budget Committee. "This is about bringing the nation out of this recession in a manner that makes us more competitive in the international market."[10]

As lawmakers, economists and infrastructure experts debate how to put Americans back to work, here are some of the questions being asked:

Will federal spending on public works create jobs?

The final stimulus plan banks heavily on projects such as expanding and repairing highway and rail systems and modernizing schools. But many conservative lawmakers and analysts argue that government spending cannot create jobs but only diverts money from the private sector, which they call the only true engine of job creation.

"Despite the rhetoric of 'government job creation,' economic logic denies the possibility that jobs can, on net, be created by government," Thomas J. DiLorenzo, a professor of economics at Baltimore-based Loyola College, wrote in an analysis for the libertarian Cato Institute think tank. That's because "government jobs programs . . . are usually financed by taxation or borrowing," diverting resources from private-sector business investment.[11]

Some infrastructure spending proposed by Democrats, especially so-called green projects, would not create jobs at all, Tyler Cowen, a professor of economics at George Mason University in Fairfax, Va., told National Public Radio. "Take solar energy," Cowen said. "The whole

know what infrastructure we need to support it," says Armando Carbonell, a senior fellow at the Lincoln Institute of Land Policy, a think tank in Cambridge, Mass. To create a vision for the transportation system, for example, "we need to know where the current system breaks down" and stymies important travels, he says. For example, highways and air travel are congested and frustrating in the Northeast Corridor, so we know that providing a rail alternative is a good possibility, he says.

"You need the target, and you need the vision, because tomorrow is going to be different from today," Carbonell says.

point of solar energy is that the sun does more of the work."[12]

New Deal government spending was not a great job creator, demonstrating that the Obama plan won't be either, argued Amity Shlaes, a columnist for Bloomberg News and senior fellow at the Council on Foreign Relations think tank. When FDR took office in 1933, unemployment was around 25 percent and remained above 14 percent between 1931 and 1940, not counting the jobs in federal public-works programs that Shlaes contends were temporary "make-work" jobs created only to provide paychecks, not for any real purpose.[13]

Most economists do count New Deal jobs as real employment, Shlaes acknowledged. Even then, she said, depression unemployment rates never get below 9 percent, but that's "hardly a jobless target to which the Obama administration would aspire."

For one thing, says the Heritage Foundation's Utt, federal jobs created won't necessarily match the skills of those who need work. "Will a roofer easily make the transition to laying concrete on highways?" he asks. "And finance people aren't going to get involved in working for the National Park Service."

Furthermore, a typical highway project spends only about 27 percent of its funding in the first year, according to the Federal Highway Administration, "so an infrastructure project will only make a little difference" in fighting the current recession, Utt says.

Even so-called shovel-ready projects take time to produce paychecks, acknowledges Alan Rubin, director of government relations for Buchanan Ingersoll, a national law and lobbying firm in Pittsburgh. For substantial projects — those creating 400 jobs or more — it takes about 90 days to hire a general contractor, says Rubin, then another month for the contractor to find a banking partner and meet bonding and insurance requirements and then two more weeks before workers get the first paycheck, he says.

"Infrastructure money won't have an impact until late 2009 or early 2010," so it likely won't shorten the recession, "although it may give some help" once the economy starts to rebound on its own, says Wharton's Inman.

But supporters of public-works initiatives reject the criticisms. For one thing, public spending doesn't divert resources from private business investment during a serious recession, said John D. Porcari, Maryland's secretary of transportation. "I hear from contractors all the time they don't have any private-sector work. . . . To be blunt about it, it's so bad out there that additional work right now would be the difference between holding onto employees and letting people go," Porcari said.[14]

And public-works jobs will offset some job losses even if the projects take time to get off the ground because in the last two recessions weaknesses in the labor market persisted well after the economy began its turnaround, John S. Irons, research and policy director at the liberal Economic Policy Institute, told the House Committee on Transportation and Infrastructure in October.[15]

Furthermore, economists now generally acknowledge that "this is going to be a long recession" that will continue to need stimulus months from now, says Roger Hickey, co-director of the progressive advocacy group Campaign for America's Future. "I think the public knows these layoffs they're experiencing now are just the beginning and that [earlier] tax cuts" — such as the tax rebates of 2008 and various other tax cuts before that — did not stop the downturn in the early 2000s, he says.

Economists estimate that $1 billion in construction spending creates between 14,000 and 47,000 jobs, so a $75 billion increase for infrastructure would support more than 1 million jobs in areas such as manufacturing, retail sales, scientific and technical jobs, administration and waste management, Irons said.[16]

In addition, the public needs to believe in something in order to spur businesses to take initiative again, and public-works spending that provides work and strengthens infrastructure can help build that belief, says McKee at the University of Virginia.

"This is the most perfect time . . . to lay out a plan to rebuild America, just like Roosevelt has done because it would stimulate the economy and it would create a tremendous amount of jobs," Gov. Arnold Schwarzenegger, R-Calif., told CNN. "It's not spending. It's an investment" in a stronger economy, he said.[17]

Some construction managers say they're counting on public-works funds to keep hiring. "I told my managers that based on the expectation of a stimulus plan, we should continue interviewing people at colleges, hoping to have work for them by the time they graduate in the spring," said Brian Burgett, president of a construction firm in Frederickton, Ohio.[18]

Contrary to what some conservative analysts argue, the New Deal did create jobs, says Jason Scott Smith, an assistant professor of history at the University of New Mexico and author of *Building New Deal Liberalism: The Political Economy of Public Works, 1933-1956*. Smith rejects Bloomberg columnist Shlaes' argument that public-works jobs are of limited value because they are mostly temporary. Temporary workers get and spend paychecks, helping boost business, he says.

"Except in the 1937-38 recession, unemployment fell every year of the New Deal," wrote Eric Rauchway, a professor of history at the University of California, Davis. Private, non-farm, non-governmental employment was "markedly lower under Roosevelt" than under Herbert Hoover, his predecessor. "The New Deal wasn't just offering make-work, it was stimulating the economy."[19]

Infrastructure must not only be built but maintained, and those jobs will be permanent and local, says the Century Foundation's Shorris. "You can outsource a new train, but not maintenance," he says.

With the right management strategies, it's even possible for infrastructure projects to defy the odds and become operational quickly, says William Eggers, global director for Deloitte Research — Public Services, a Washington branch of the international Deloitte consultancy. After the 1994 Northridge earthquake toppled a Los Angeles freeway, the road was rebuilt in two months — instead of a year or more as expected — "by putting in performance standards and changing procurement rules," Eggers says.

Does infrastructure construction strengthen the economy?

Proponents argue that well-targeted federal infrastructure spending can bolster the economy quickly as well as over the long term, but many critics say tax cuts are always a better economic booster.

"Americans might well rue the day when they trusted the federal government to spend the nation into prosperity," wrote Jacob G. Hornberger, president of the Future of Freedom Foundation, a conservative and libertarian advocacy group in Fairfax, Va. "It just isn't going to happen."[20]

The United States already spends substantial amounts annually on infrastructure — $102 billion in 2007 and $114 billion in 2008, for example, writes Lawrence Kudlow, a conservative economist and *National Review* columnist. "Didn't do much for growth, did it?"[21]

School modernization, touted by Democrats as a way to prepare future U.S. workers to meet economic challenges, won't accomplish its goal, said Lisa Snell, director of education research at the libertarian Reason Foundation. Stimulus funds could "more than double the current, total federal education budget" to more than $200 billion, but "unfortunately, this huge expansion is unlikely to spur improvements," she said. U.S. average per-pupil expenditure is already high, compared to other nations, yet student achievement lags behind many countries, and she notes that the plan would not require schools and teachers to adopt better practices.[22]

Japan has repeatedly tried infrastructure spending to end a nearly two-decade-long recession but failed, conservative commentators say.

An infrastructure-focused stimulus "may be intuitively attractive," but historical evidence suggests that its impact "is very modest and unlikely to enhance recovery or deter recession," said Utt at the Heritage Foundation. "The Japanese government implemented such a program during the 1990s, and the consequence was two decades of economic stagnation."[23]

Japan's infrastructure-building boom "yielded painfully little," said Bloomberg's Shlaes. "The Nikkei [stock average] stayed down. The country's standards of living failed to keep pace with the rest of the world's."[24]

But other analysts say massive government-funded infrastructure spending in Japan failed to restart the economy because the country didn't address its deeply flawed banking system, which crippled private lending and investment.

Japan's "banking sector was broken," says lobbyist Rubin, undercutting the ability of public-works spending to stimulate the economy. Similarly, he says, restoring the U.S. banking sector to good health will be vital to effective public-works spending.[25]

"Even when the government says it'll give you $50 million" for an infrastructure project, "they're not going to give you your payroll" in the form of upfront cash on day one, Rubin explains. Instead, "you get a government-backed guarantee" that the money will arrive in increments down the line. "So you need construction loans from banks to tide you over from one government check to the next," as well as to pay the architects and engineers who have to be on the payroll before building begins.

Thus, credit availability is crucial, even if the government is funding projects, he says.

In addition to a continuing banking crisis that crippled lending, Japan's small size — "smaller than Illinois" with a population nearly half that of the United States — led the government to invest in redundant projects, such as airports too close to each other, that didn't actually improve things, says the University of New Mexico's Smith. Unlike the United States, Japan also had not neglected its infrastructure in good economic times, so it had much less need for maintenance and updating projects, he says.

Increased infrastructure spending is "a particularly effective way to stimulate the economy," Mark Zandi, chief economist of Moody's Economy.com, told the House Committee on the Budget in January. "The boost to GDP from every dollar spent on public infrastructure is large — an estimated $1.59 — and there is little doubt that the nation has underinvested in infrastructure for some time, to the increasing detriment of the nation's long-term growth prospects."[26]

Nevertheless, virtually all analysts caution that a thoughtful vision of future needs must guide infrastructure spending if it's to benefit the economy long term.[27]

It's fine for the initial $50 billion to $100 billion disbursed for infrastructure to focus on projects that can be done quickly, but the next group "needs to be better thought out," says the University of Southern California's Little. Infrastructure built during Roosevelt's New Deal and the Eisenhower administration's federal highway initiatives helped bolster the U.S. economy for years, but all the projects "were guided by preexisting visions" of what future needs would be, he says.

"The middle of the 21st century won't look the same as the middle of the 20th, but so far I haven't seen a vision" articulated by the Obama administration, Little says. "We're not seeing somebody saying, 'Wow, we're going to retool this system,' but just, 'This is the quickest way to spend a couple of hundred billion dollars.'" Assessment is critical to build only what will be truly useful, he says. For example, Pittsburgh, population 350,000, has a water system that could serve a million people, based on erroneous estimates that the city would keep growing, Little says. "Those sunk costs last forever, and we don't want to repeat that too many times."

"There has to be some process for reviewing these projects," says Dean Baker, co-director of the liberal Center for Economic and Policy Research. Road-building projects of the past three decades "have promoted sprawl," for example, an outcome it's important to avoid this time around.

"We're not very good at making trade-offs across disciplines," says the Century Foundation's Shorris. For example, "a lot of the capacity of New York airports is taken up by people going to D.C. and Boston," he explains. When Germany's Frankfurt airport faced a similar situation, "they took the money and used it for trains." "But where in America do we get to have the

Transportation Spending to Focus on Roads

More than 75 percent of the stimulus funds that 19 states plan to spend on transportation target new highways while only 17 percent would go for "green" options such as public transit. Massachusetts, however, is targeting more than half its funding to transit and bike or pedestrian paths.

Anticipated Stimulus Spending on Transportation by 19 Select States

Bike/pedestrian **0.3%**

Aviation and other **5.2%**

Transit/intermodal (including all rail) **17.0%**

Roads **77.4%**

Examples of State Spending Plans

State	Roads	Transit/intermodal (including all rail)	Bike/pedestrian	Aviation and other
Ala.	100%	0%	0%	0%
Calif.	60.6%	37.1%	0%	2.3%
Florida	98.9%	1.0%	0%	0%
Mass.	29.7%	47.1%	2.2%	21.0%
N.C.	83.3%	10.2%	0%	6.1%
Texas	97.3%	2.3%	0%	0%

* Figures may not total 100 due to rounding.

Note: Only 19 states have released their transportation spending plans thus far.

Source: Phineas Baxandall, "Economic Stimulus or Simply More Misguided Spending?" U.S. PIRG Education Fund, January 2009

conversation" that leads to such a cross-sector solution? Shorris asks. The fragmented U.S. system, including 50 state governments that seldom think regionally, makes big-picture planning difficult, he says.

Funding infrastructure construction but ignoring future upkeep — which costs 4-6 percent of gross domestic product annually ($500-$700 billion) just for current infrastructure — also would diminish the success of public works, says Little.[28]

His suggestion: Trillions of dollars that sit in public-sector pension funds and in the federal Social Security Trust Fund awaiting workers' retirement could be lent to the states for revenue-producing projects such as highways, then repaid with interest. "Motorists using the facilities would also benefit because a portion of the tolls paid would support the long-term solvency" of the U.S. retirement system, says Little.[29]

Recession or no, federal "infrastructure spending makes a lot of sense on its own terms," provided it will provide benefits down the line to the nation in excess of what's paid for it, says Wharton's Inman. In the best-case scenario, when state governors submit lists of potentially fundable projects, "you would evaluate these things on their own terms, and if a project promises significant interstate value," then the federal government would consider funding it, he says. "Lacking that, the state government should do the project on its own."

In a recession, however, the question is a bit different, Inman continues. Economic theory holds that by stimulating "aggregate demand" for goods and services anywhere in the nation during a recession, the economy of the whole country can be revived, he says. If that's true, federal funding for a Pennsylvania infrastructure project that would benefit only Pennsylvanians could still be a good idea, since the spending itself would increase the nation's aggregate demand. However, "economists, in fact, know very little about this" or about the role of states in affecting the national economy and can't argue it conclusively, he says.

Is the Democratic public-works spending plan too big?

Many economists say the present recession is so dire that only massive government spending has a chance to jump-start the economy. However, many conservatives

argue that public-works spending at the levels proposed by Democrats will run up the national debt to disastrous proportions.[30]

The Democrats' stimulus plan will cause a dangerous expansion of the national debt, according to the House Republican Study Committee. The federal debt may grow by an unprecedented $2 trillion in 2009, an unsustainable burden given that it already grew by more than $2 trillion during the last two years, the panel argues.[31]

"In principle, every dollar spent by the government could cause national income to increase by more than a dollar it if leads to a more vibrant economy," noted Harvard University Professor of Economics N. Gregory Mankiw. But "the fly in the ointment . . . is the long-run fiscal picture. Increased government spending may be a good short-run fix, but it would add to the budget deficit," making it harder to fulfill other responsibilities like paying Social Security and Medicare benefits to the baby boomers now beginning to retire. So he thinks Democratic spending plans are too big.[32]

The federal government may not currently be equipped to manage spending of the scale contemplated, said Rudolph G. Penner, a senior fellow at the centrist Urban Institute and former director of the Congressional Budget Office. "It looks like the federal budget for highways would increase more than one-third," but "I do not hear of plans to increase the size of the federal bureaucracy temporarily by comparable amounts, and one wonders whether the existing civil service can provide adequate oversight."[33]

Big infrastructure requires both public and private players, and "it won't work unless everybody buys in and says, 'I'm going to do it straight,' " says lobbyist Rubin. To convince the public that "government needs to be part of the solution, government has to be responsible" and "be an honest broker so it doesn't turn people off," a necessity that has already been somewhat compromised by tax problems among some White House nominees, he says.[34]

But other analysts argue that, given the nearly unprecedented magnitude of the economic downturn, the proposed public-works spending may actually be too small.

The economy is on the way to losing between $1.1 trillion and $1.2 trillion in demand for goods and services this year, due to the decline in home and non-home construction and the loss of large chunks of many Americans' home equity and investments, said liberal economist Baker. Meanwhile, the two-year stimulus package as a whole — including all spending and tax cuts — would add back just under $400 billion per year, said Baker. "I don't think it's large enough."[35]

"Maybe private credit will recover faster than I think likely," so that some government spending will end up destructively competing with private spending, said University of Texas economist Galbraith. "But even allowing for this possibility, action now should be on a grand scale. It is far easier to trim back" provisions as the economy expands "than it will be to repeat the political effort of passing a large expansion package if the first one is too small."[36]

A sizable infrastructure-spending plan is warranted because of stingy U.S. spending on infrastructure for decades — 2.5 percent of gross domestic product annually — compared to the international average of 4 percent, says Carbonell at the Lincoln Institute of Land Policy. "Places that are growing tend to invest more" in public works, and, unlike some of our international competitors, "we haven't maintained our infrastructure or adapted it to changing conditions," he says.

Infrastructure spending in the House-proposed stimulus legislation is "almost minuscule," complained Rep. John L. Mica, R-Fla., top-ranking Republican on the House Transportation and Infrastructure Committee.

"They keep comparing this to Eisenhower, but he proposed a $500 billion highway system, and they're going to put $30 billion" into roads and bridges, said Mica. "How farcical can you be? Give me a break."

Instead of spreading funding thinly over many federal programs, Mica would have liked to see the administration focus efforts on some large goal, such as building high-speed rail in important travel corridors around the country.[37]

Only about $300 billion of the stimulus "goes to job creation — that's about 3.5 million jobs" or "only 2 percentage points on the unemployment rate," said Richard C. Levin, president of Yale University and a professor of economics. "That's not enough. There are lots of great public-works projects that would be well worth supporting. . . . What about . . . activities that put people to work right away, cleaning up public parks, weather-stripping homes, offices, schools, government buildings?"[38]

BACKGROUND
The Great Depression

Governments have long taken responsibility for building and maintaining society's infrastructure, from parks and libraries to highways and airports. Traditionally, however, governments generally did little or nothing to counteract economic downturns, relying on market forces to bring back prosperity, no matter how painful the consequences.

Confronting the Great Depression of the 1930s, President Franklin D. Roosevelt (FDR) made the first large-scale attempt to link infrastructure building with "economic stimulus" — putting money into people's pockets through tax cuts or government spending to pump up demand for goods and services.[39] But economists and politicians had already formulated a role for public-works spending in tough times.

In the first three decades of the 20th century, England, Germany and several other countries — along with Pennsylvania and California — enacted plans to use infrastructure building to stabilize faltering economies. So-called public-works "reserves" would list important infrastructure-building projects to be undertaken only in times of rising unemployment.

"The aggregate volume of public works . . . is so great that if . . . 20 percent . . . were deferred each year and the accumulation executed in a year of depression . . . the lifting power of public works would be at least one-third of the deadweight of . . . a depression" and would "almost iron out the fluctuations in unemployment," said the U.S. President's Conference on Unemployment in 1921.[40]

Columbia University economist Leo Wolman, who opposed the creation of an actual reserve, nevertheless said that if governments funded public-works projects of real economic importance during a recession, they could "constitute a stimulus to business and would lead, if done on a large scale, to the progressive diffusion of employment," although they would not provide jobs for the majority of the jobless.[41]

After the economy turned sour following the stock market crash in 1929, Republican President Hoover initially tried public-works spending. Between 1928 and 1932, private construction dropped off by 86.4 percent. In 1931, 1932, and 1933, Congress authorized boosts in federal highway-building grants from $75 million to $125

CHRONOLOGY

1920s–1930s *Economic boom, partly fueled by heavy borrowing by stock-market investors, is followed by the Crash of 1929. President Franklin D. Roosevelt's New Deal launches the first major government programs aimed at stimulating the economy.*

October 1929 Stock market crashes. . . . Banks begin to fail as debtors default on their loans, and panicking depositors try to withdraw their money.

1932 Industrial production falls 45 percent from 1929 levels as the Great Depression begins and spreads worldwide. . . . Roosevelt is elected to his first term.

1933 Twenty-five percent of all U.S. workers are unemployed. . . . Roosevelt and Congress establish Works Progress Administration, Civilian Conservation Corps and Tennessee Valley Authority to employ laid-off workers and create new public infrastructure.

1934 Heavy government spending enables Sweden to become the first nation to recover from the Depression.

1935 U.S. unemployment rate declines to 20 percent.

1936 Roosevelt is reelected in a landslide. . . . U.S. unemployment rate is just under 17 percent. . . . Military spending pulls Germany out of the Depression.

1937 Unemployment rate falls to 14 percent. . . . Rising federal deficits prompt Roosevelt to cut back on federal public-works spending.

1938 Economic growth slows again as unemployment rises to 19 percent.

1939 Economic growth revives, and unemployment falls to 17 percent as U.S. defense spending rises in anticipation of war.

1940s–1950s *Federal spending on World War II and Interstate Highway System helps keep unemployment low.*

1941 U.S. manufacturing output is up 50 percent over 1939.

1956 President Dwight D. Eisenhower commits $25 billion to building the new highway system, raising the federal funding share from 50 percent to 90 percent.

1960s–1980s *Amid rising prosperity, government job creation loses ground to tax cuts as the preferred way to stimulate the economy when business slows.*

1977 Congress approves $6 billion (around $21 billion in today's currency) in public-works spending proposed by President Jimmy Carter as an economic stimulus, but the proposals aren't enough to offset soaring unemployment.

1990s–2000s *Economists warn that rapid economic growth propelled by speculative "bubbles" like high housing prices isn't sustainable. Transportation engineers warn about infrastructure neglect.*

1993 President Bill Clinton proposes to spend $69 billion on infrastructure, housing and environmental retooling but withdraws the plan after criticism.

2002 Lawmakers mull infrastructure spending as economic stimulus but cut taxes instead.

2008 Recession that began in December 2007 deepens. . . . Tax rebates offered as economic stimulus don't reverse nation's economic slump, and Congress mulls public-works spending, asking governors to suggest "shovel-ready" projects. . . . China and European Union countries include infrastructure investment in economic-stimulus plans for the recession that's now spread worldwide. . . . American job losses hit 3.6 million.

2009 Newly inaugurated President Barack Obama and congressional Democrats propose public-works spending as economic stimulus; congressional Republicans fight to reduce such spending and add more tax cuts but mostly vote against it anyway. . . . Economists warn that unless the banking crisis is solved, neither spending nor tax cuts will reinvigorate the economy. . . . January job losses total 598,000. . . . Congress passes $787 billion stimulus bill on Feb. 13 and Obama signs it into law on Feb. 17.

million per year, and in 1930 Hoover requested still more public-works funding, in the form of an emergency appropriation of $150 million to keep the previously authorized infrastructure projects going longer.

Beginning in 1931, however, Hoover opposed additional public-works appropriations except to build revenue-producing projects. "The vice in . . . the proposals . . . is that they include public works of remote usefulness; they impose unbearable burdens upon the taxpayer; they unbalance the budget and demoralize government credit," he wrote in a 1932 letter to the American Society of Civil Engineers. "A larger and far more effective relief to unemployment at this stage can be secured by increased aid to 'income-producing works,' " such as "waterworks, toll-bridges, toll-tunnels, docks and any other such activities which charge for their service and whose earning capacity provides a return upon the investment."[42]

When Roosevelt took office in March 1933, the Depression had deepened and spread to Europe, partly because U.S. investment had played such a large part in sustaining European nations devastated by World War I. When U.S. prosperity and investment dipped, Europe's war-damaged economy fell still further. In response, FDR not only launched public-works projects in the United States but also joined with other international leaders in urging Europe to follow suit.

"In the period immediately before us, governments must employ such means as are at their disposal to relieve the unemployment by public works, and these efforts of individual governments will achieve their fullest effect if they can be made a part of a synchronized international program," Roosevelt and Italian Finance Minister Guido Jung declared at the 1933 World Monetary and Economic Conference.[43]

The pitch was less successful with Europe's cash-strapped governments than Roosevelt hoped, however, with Britain declining to participate at all. Britain had already tried to jump-start its economy with public-works expenditures and found it "unduly expensive" and therefore "an experiment we are not going to repeat," said British Board of Trade President Walter Runciman.[44]

Stimulus at Cooperstown

In the United States, Roosevelt launched several federal agencies to create jobs while producing public works for the nation. Winning public support for the initiatives was not always easy, however. In a way, baseball saved the day.

When Major League Baseball celebrated its 100th anniversary with a June 12, 1939, All-Star game between the American and National leagues and the opening of the National Baseball Hall of Fame and Museum, the event showcased not just the national pastime but brand-new Doubleday Field in Cooperstown, N.Y. Like thousands of other ball fields, stadiums and parks built during the 1930s, the field was a project of a New Deal agency, the Works Progress Administration (WPA), created in 1935 to generate public jobs for the unemployed.[45]

The popularity of baseball and other sports helped win public support for the WPA, which provided jobs for about 8.5 million people during its eight-year existence. The WPA created more than 125,000 public buildings, including hundreds of sports facilities, 8,000 parks, 75,000 bridges and some 650,000 miles of roads, at a total cost of about $11 billion.[46]

At the dedication of a Detroit stadium, Roosevelt noted that "some people in this country have called it 'boondoggling' for us to build stadiums and parks and forests to improve the recreation facilities of the nation. My friends, if this stadium can be called boondoggling, then I am for boondoggling, and so are you."[47]

Recreation projects aside, however, the WPA enjoyed less public support than some other infrastructure initiatives, such as the massive Tennessee Valley Authority (TVA), according to Jennifer Long, an associate professor of economics at the University of Science and Arts of Oklahoma. Many mistrusted "government involvement in the economy" and gave their wholehearted support mainly to programs whose results they saw as potentially significant for their future, she wrote.[48]

One of the earliest New Deal agencies, the TVA was created shortly after Roosevelt's 1933 inauguration. Rather than jobs, its primary purpose was to promote economic development in the poverty-plagued Tennessee Valley by damming flood-prone rivers to protect farms and generate electricity. Nevertheless, the unprecedented construction effort — building seven dams between 1933 and 1941 — employed some 50,000 people, mostly unskilled local workers doing manual labor.[49]

By rights, TVA's goal of electrifying the region "should have been subject to the same doubts about inappropriate involvement [in the economy] that haunted the WPA," since power production is often a private-sector enterprise, noted Long. However, the TVA was hailed by the public

'Transformational' Projects Focus on the Future

Goal is energy efficiency, not maintaining the status quo.

When public money is being spent on infrastructure improvements, it's vital to select "transformational" projects, or those that pave the way for a better future, says Michael A. Bernstein, a professor of history and senior vice president for academic affairs at Tulane University. In the 1930s and '40s, for example, transformational infrastructure spending supported the development of economy-transforming tools such as antibiotics, radio and television and air travel, he says.

Today, however, a transformational vision of infrastructure might mean "backing off from air travel and investing more in rail," Bernstein says.

One transformation crucial to developing infrastructure for the future is to seek energy efficiency, he says. But that vision runs headlong into the status quo when it comes to government funding for public-works projects, many analysts say. Last fall, for example, Congress asked state transportation departments to submit lists of "shovel-ready" infrastructure projects that could be funded in a stimulus package.

The 19 states that released their lists to the public mostly focused on projects that would ignore smart energy use, according to the consumer group U.S.PIRG (Public Interest Research Group). The states "prioritize new highways while paying relatively little attention to repairing crumbling bridges and roads and even less emphasis on forward-looking transportation options, such as public transit and intercity rail," says a January U.S.PIRG analysis.[1]

On average, the 19 states would spend more than 75 percent of the funds on highways and only 17 percent on public transit or intercity rail. "This would be a step backward from even the grossly inadequate 20 percent share received by transit in federal transportation laws since the 1970s," says U.S. PIRG.[2]

In search of shovel-ready projects, the federal government is opting to send funds to states with a long history of favoring limited-access highways such as freeways that bypass towns or neighborhoods, warns John Norquist, president of the Chicago-based Congress for the New Urbanism, which promotes walkable environments and sustainable development.

"Road builders could make money paving streets" rather than by building new highways, and, in fact, such projects are every bit as "shovel ready," Norquist says. Cities and villages should be getting a third of the money, at least, to improve their road networks, rather than funneling the money to new highways that encourage the building of ever-more-distant suburbs, he says. Rather than putting stimulus funds to work creating more energy-inefficient sprawl, planners should look to the European and Canadian examples, where highways link population centers but aren't used to bypass small towns or neighborhoods. "They haven't built a new freeway in Toronto since 1968," he says.

Beyond transportation, at least some states also are seeing the "jobs potential and the economic-development potential" in "green" energy projects, making them a good

in Tennessee as a success "before ground was broken for the first dam, partly because so many people stood to be helped," Long wrote. "People celebrated [it] as the beginning of a new era of economic prosperity" even as they viewed WPA's much smaller projects skeptically, as "necessary evils" to tide the region "over until normal times returned."[50] By 1939, the TVA brought electricity to some 288,000 homes[51] and today provides electricity to around 8.5 million customers in seven states.[52]

New Deal Economics

The New Deal convinced many doubters that the government could and should act to prevent the worst consequences of private-sector slumps. Opinions split, however, between supporters of British economist John Maynard Keynes, who embraced government spending on public-works programs — and those who favored tax cuts to encourage individuals and businesses to buy and invest.

In the 1930s, backers of government action to ease recession focused on employment as the key to stabilizing the economy.

"I believe in the inherent right of every citizen to employment at a living wage and pledge my support to . . . public works, such as . . . flood control and land reclamation, to provide employment for all surplus labor at all times," Roosevelt said in a 1932 campaign statement.[53]

target for economic-recovery funding, says Lew Milford, president of the Clean Energy States Alliance, which supports green projects.

A growing number of states have set up dedicated funding streams — from sources such as small charges on utility bills — to subsidize development of wind, biomass and other alternative energy sources, and a federal grants program in alternative energy could easily direct its grants to those funds, Milford explains. States realize that green energy "is no longer just an environmental thing" but a job maker, he says. "If you're putting up solar panels all the time, you generate skilled jobs for installers, and these jobs are local and non-exportable," he says.

Giving more Americans an equitable shot at economic advancement can and should also be accomplished by transformative public-works projects, says Judith Bell, president of PolicyLink, an Oakland, Calif., advocacy group that promotes social-equity projects, such as public-transit investments. It is especially helpful to low-income people trying to get and stay on their feet because low-income communities have the least access to cars, she says. Transit spending also produces 19 percent more jobs than highway spending, she says.

"The [Bush] administration turned its back on housing," and that's another area that the public-works spending should target, especially now, Bell says. "Giving cities and localities the ability to use resources to renovate the foreclosed homes helps cities," she says. School repair, especially in low-income communities, and broadband-Internet expansion in rural areas also would reap benefits across the socioeconomic spectrum, she says.

Many infrastructure analysts caution, however, that technology-related projects, like broadband expansion, require thoughtful planning, which may be tough to achieve through economic-stimulus legislation.

"The first rule of technology investment is you spend time understanding the end user," said Craig Settles, an Oakland technology consultant. "If you don't do this well, you end up throwing . . . potentially billions down a rat hole for things that people don't need or can't use."[3]

Updating the nation's electricity-transmission grid, for example, won't necessarily promote clean energy as many expect, said Patrick Mazza, research director for Climate Solutions, an anti-global warming advocacy group. "Better transmission can empower new coal plants as well as new wind farms, so we better make sure there is a green priority on any new lines," he said.[4]

For any public-works funding to succeed, America needs to change its attitude toward infrastructure, which has been "build it, then forget about maintaining it," says Michael A. Pagano, dean of the College of Urban Planning and Public Affairs at the University of Illinois, Chicago.

"If this new [stimulus] funding is going to add more capacity [without providing for maintenance], then we are in trouble because we have a long history of ignoring what we've got," Pagano says.

---

[1] Phineas Baxandall, "Economic Stimulus or Simply More Misguided Spending?" U.S. PIRG, Jan. 5, 2009, www.uspirg.org/uploads/75/pU/75pUUIAl1Eteahknhlx-OA/State-Stimulus-paper-FINAL2.pdf.

[2] *Ibid.*

[3] Quoted in David M. Herszenhorn, "Internet Money in Fiscal Plan: Wise or Waste?" *The New York Times*, Feb. 3, 2009, p. A1.

[4] Peter Mazza, comment posted in response to Mark Clayton, "For a Spiffier Electric Grid: $11 Billion," *The New Economy Blog, The Christian Science Monitor*, Jan. 28, 2009, http://features.csmonitor.com/economyrebuild/2009/01/27/for-a-spiffier-electric-grid-11-billion.

Critics point out that, despite Roosevelt's interventions, U.S. unemployment remained high until the 1940s, when the nation began manufacturing war materiel, and the remaining vast numbers of jobless were drafted into the military. Still, most analysts over the years have concluded that Roosevelt's infrastructure spending helped boost the Depression economy and provided lasting societal benefits as well.

Between 1933 and 1937, for example, the U.S. economy grew by around 9 percent each year, up from zero annual growth in the immediately preceding years, says the University of New Mexico's Smith.

It's unfair to criticize New Deal projects for failing to bring employment all the way back to pre-Depression levels since the employment drop-off from the boom years of the late 20s was so extreme, says the University of Virginia's McKee.

Moreover, the infrastructure programs were a boon to the private sector, not a bane, Smith says. The federal government worked with private contractors like the big construction firms Bechtel and Brown & Root (now part of KBR), which "stayed afloat through the Depression" thanks to public works, he says.

By the mid-1930s, the economy was growing again, with manufacturing output and productivity increasing

Solving the Infrastructure-Maintenance Dilemma

European governments turn to public-private partnerships.

With public coffers low in recent years, some states and localities have turned to public-private partnerships (PPP) to build and maintain infrastructure, such as offering 90-year leases to build and operate toll highways. The company usually pays an upfront, lump sum for the right to operate a road or other facility, then keeps the tolls collected.

But some PPP experts say the U.S. projects aren't really "partnerships" at all because they offer too few benefits to taxpayers, unlike PPPs in countries with more experience in public-private collaboration, especially the United Kingdom (U.K.).

The cash-strapped U.K. turned to PPPs in the 1980s to undertake infrastructure projects and has continued with the partnerships, says Irene Walsh, leader of the U.S. infrastructure and project-finance practice at Deloitte Financial Advisory Services.

Britain uses PPPs to build and operate schools, hospitals, rail lines, drinking-water systems and much more, says Stephen Harris, international development officer at the Tribal Group, a London firm that consults on private participation in delivering public services. Currently, about 15 percent of British capital projects are PPPs, which are limited to projects costing over 20 million pounds sterling (around $30 million U.S.), Harris says.

And, contrary to popular conception in the United States, PPP-run public services and facilities operate more efficiently and are "more consumer focused" than before the PPP conversion, Harris argues.

That success didn't come easily, however, Harris acknowledges. In the early years "we made every possible kind of mistake you could make" in structuring projects that could earn steady income for a company while serving the public good, he says.

Because PPP contracts are completely different from traditional government contracts, they pose the biggest problem for unwary governments, he says.

Mainly, a private company that wins a contract from a British town to build, operate and maintain a school for 25 years "has to raise all the money to build the thing," says Harris. "They don't get a single penny from the government until the children and teachers are ready to move in." Once the facility opens, the government pays the company set fees to operate it according to strict government standards, he says. If a light goes out in a classroom, the company has a limited amount of time to fix it, or their pay is docked.

A traditional government contract for a school, for example, might specify that it should be a building with

---

at pre-Depression rates, although unemployment remained high compared to the 1920s. In 1937, however, the economy took a new, shocking downturn, as manufacturing dropped off and unemployment shot up.

Economic historians don't agree about the cause of the slump.

"The conservative literature says that the 1936 recession came from government intervening too much in the market," says Michael A. Bernstein, a professor of history and senior vice president for academic affairs at Tulane University. "But I would say that it happened because FDR didn't do enough" spending to give the economy the full boost it needed.

Growth resumed in 1938, but boom times only returned beginning in late 1940, when the war already raging in Europe increased demand for American armaments and other materiel.

World War and Beyond

"World War II ultimately succeeded in ending the Depression because we spent so much," Bernstein says. "Nobody worried about the federal budget on Dec. 8, 1941," the day after the Japanese attacked Pearl Harbor, the big U.S. Navy base in Hawaii.

Construction of military bases and other infrastructure across the country fed the economic boom in the 1940s, thus "testing and proving" the proposition that infrastructure spending is stimulative, Smith says.

But the long-term economy-building effects of the New Deal are perhaps its most important — though overlooked — legacy, says McKee. Innumerable infrastructure projects provided a foundation for economic growth over decades, says Smith. For example, the New Deal built around 480 airports and constructed new

14 classrooms and two offices "and the doors should be this wide," he says. But Britain's PPP contracts use what he calls "output specifications" — a precise description of what the facility must accomplish for citizens, Harris explains. The contract might say something like, "Please provide educational facilities for a population of 5,500 for 25 years," says Harris. This kind of contract "brings in the innovation of the private sector," allowing a contractor to construct a smaller school building and accomplish some of the education via the Internet, for example, he says.

Furthermore, contracts "allow no deterioration whatsoever for 25 years" and penalize companies by cutting off government payments if standards aren't met, Harris says.

This solves an age-old infrastructure dilemma, says Harris. "If you just throw in money to build something and don't provide any money to maintain it, in 15 years you'll be right back where you started," with a decrepit building or water system, he says. "But once you've got a contract, it's the private sector's job to maintain it to a standard" that the government explicitly sets, he says.

The private sector benefits by getting "a guaranteed income for 25 years, which helps a company's share values," says Harris.

But unwary governments will face serious pitfalls. And there are "hardly any" governments currently PPP-savvy enough to write the contracts without getting the short end of the stick, Harris says. In America "there aren't any PPP experts," for example, he says. "In the U.S., people confuse privatization" — simply transferring facilities or services to private-sector operation — "with public-private partnerships," but "the issue with that is lack of control over quality," Harris says.

Privatized projects or PPPs with bad contracts can go horribly wrong, says Harris. For instance, a company running a PPP highway project in Toronto quickly jacked up tolls from $5 to $14, but when the government tried to sue it lost the case in court because the contract allowed the increase, he says. A water system in Australia became contaminated with cryptosporidium — a microorganism that causes severe gastrointestinal symptoms — but the locality couldn't hold the contractor responsible because cryptosporidium wasn't on a list of banned microorganisms in the contract, says Harris. "What the contract should have said was, 'At all times the water must be drinkable,' " Harris says.

With money tight and infrastructure deficiencies widespread, there's growing interest in pursuing PPPs, but Congress should create an expert PPP group to offer advice and assistance "before governors get screwed," says Frank M. Rapoport, chairman of the global infrastructure and PPP team at the national law and lobbying firm McKenna Long and Aldridge.

"Five years ago, if you asked [U.S. builders], 'Are you interested in sharing the risk?' " on a public-works project in the British fashion, most would have said, "No, I want to build it and walk away," as they've done under traditional government contract work, Rapoport says. But today a growing number of companies are open to new kinds of PPPs, he says.

schools in half of the nation's 3,000-plus counties, all of which "contributed greatly to the economic growth we saw after World War II," he says.

But many conservative analysts, then as now, viewed federal public-works spending as much inferior to tax cuts for building the economy. This view has gradually become the mainstream position in the past half-century, largely because it seems to many like a much more direct way to spur the increased demand for goods and services that prompt businesses to begin investing again.

A "defining moment" came when Congress failed to pass the Democrat-sponsored Full Employment Act of 1945, says McKee.[54] The bill was inspired by economist Keynes' idea that government action can boost demand for goods and services at times when fluctuating business investment creates unemployment and the belief — based on observation of historical business cycles — that a private-market economy will always have periods of high unemployment.

Based on the proposition that "all Americans able to work and desiring to work are entitled to an opportunity for useful, remunerative, regular, and full-time employment," the bill would have required the government to boost its own spending to increase employment during all business-cycle downturns. In 1946, Congress passed a watered-down version of the legislation, setting full employment as a national goal but not requiring any government action to produce it.[55]

"Had they passed the [original] bill — and it looked for a while as if they would — we would have had the government committed to a more activist employment policy," McKee says.

After World War II, the federal government continued substantial public-works spending, although, as Keynesian

Unemployment Hit 25 Percent in Depression

Unemployment shot up after the stock market crash in 1929. At the height of the Great Depression, nearly 25 percent of American workers were jobless — or about 13 million out of a labor force of 52 million. The jobless rate began dropping with President Franklin D. Roosevelt's New Deal public-works program, begun in 1933, then spiked again in the late 1930s when he scaled back the jobs program in an effort to balance the budget. Jobs created as the nation prepared for World War II finally brought unemployment under control.

Percentage of U.S. Labor Force Unemployed, 1929-1941

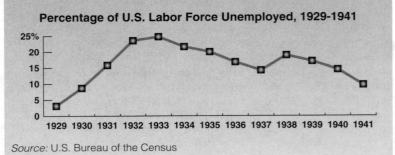

Source: U.S. Bureau of the Census

CURRENT SITUATION

Republicans Balk

A $787 billion stimulus bill — the American Reinvestment and Recovery Act — is now law. Congress passed the legislation on Feb. 13 and President Obama signed it into law four days later.

"What makes this recovery plan so important is not just that it will create or save three-and-a-half million jobs over the next two years, including nearly 60,000 in Colorado," Obama said from Denver after signing the measure. "It's that we are putting Americans to work doing the work that America needs done in critical areas that have been neglected for too long, work that will bring real and lasting change for generations to come."

Debates over infrastructure and other spending in the act have been intense and follow old patterns, for the most part.

The legislation includes $281 billion in tax cuts and $506 billion in spending — of which well under $300 billion is infrastructure-related — spread out over a two-year period, and passage did not come easily.[57]

Republicans in Washington balked when Obama and congressional Democrats first proposed a bill with substantial public-works components ranging from highways to mass transit to schools.

The bill "may create some government jobs, but actually . . . what it does is take money out of the economy," said Rep. Paul Broun, R-Ga., on Jan. 28. "It takes money away from those who are producing and gives it to government. . . . This is a huge leap towards socialism."[58]

"Let's get this notion out of our heads that the government creates jobs," newly elected Republican National Committee Chairman Michael Steele told CNN. "Not in the history of mankind has the government ever created a job. Small-business owners do, small enterprises do, not the government. When the government contract runs out, that job goes away."[59]

economics gradually fell out of favor, opponents increasingly blasted it as wasteful and worthless. In the 1950s, conservative Sen. Barry Goldwater, R-Ariz., branded President Eisenhower, a fellow Republican, a "dime-store New Dealer" when he used federal money to build the massive Interstate Highway System, says Smith. Still, most Americans backed the project.

Despite the success of the highway project, increasing worries about wasteful, ineffective "pork-barrel" spending by Congress created skepticism about whether federal infrastructure dollars could ever be spent appropriately.[56] Beginning in the late 1960s, the idea that the economy is best managed through tax cuts gradually became the standard approach, supported even by most liberals, says McKee.

Common wisdom doesn't necessarily survive hard times, however. Last year, as the economy faltered, then-candidate Obama and others began to propose infrastructure spending rather than tax cuts as a major component of an economic-stimulus and renewal plan. Their approach was bolstered by repeated warnings over the last several years that the United States had dangerously neglected upkeep on its roads and bridges.

Not all conservatives oppose government spending to spur renewed business activity in a recession, however.

"The only way to prevent a deepening recession will be a temporary program of increased government spending," said Martin Feldstein, a Harvard University professor of economics who was an adviser to Republican presidential candidate Sen. John McCain, R-Ariz. The lag time for getting infrastructure projects off the ground has meant they didn't provide much stimulus in previous recessions, but "this downturn is likely to last much longer" than the average 12 months, so public-works spending can help, Feldstein said.[60]

The spending "should include not only money for infrastructure such as bridges and roads but also for a wide range of equipment," such as "rebuilding some of the military capacity . . . depleted by the wars in Iraq and Afghanistan," said Feldstein. Shoring up federal research agencies such as the National Institutes of Health and the National Science Foundation also would help, Feldstein told PBS. That "wouldn't create a lot" of immediate jobs but "wouldn't cost a lot of money" either and would build the economy for the long term.[61]

Nevertheless, public-works provisions were among those that many congressional Republicans singled out as wasteful and ineffective, along with money for preventive-health programs such as smoking cessation and prevention of sexually transmitted disease, and extra funding for the 2010 census. As the bill began making its way through Congress, Republican lawmakers criticized a number of infrastructure-related items, including:

- $448 million to construct new headquarters for the Department of Homeland Security;
- $200 million to create public computer centers at community colleges;
- $500 million for flood-reduction projects on the Mississippi River;
- $6 billion to make federal buildings "green" and energy efficient;
- $500 million for state and local fire stations, and
- $850 million for Amtrak.[62]

As a result of the GOP opposition, the bill passed, 244-188, in the House on Jan. 28 without a single Republican vote; 11 Democrats also opposed it.[63]

Electricity-generating windmills rise 300 feet in Spanish Fork, Utah. The economic-stimulus package directs some funding to the alternative-energy sector — including wind power — which several state governments are banking on for long-term job creation.

Action on the legislation then moved to the Senate, where Republicans also battled infrastructure spending provisions and threatened to filibuster — stall a vote on the legislation with endless debate — until the spending items were reduced or removed. The 100-member Senate, where Democrats and two Democratic-leaning Independents hold 58 seats, requires 60 votes to pass legislation over a filibuster. The Senate's bill contained more tax cuts than the House version, including a $15,000 tax credit for buying a home.

In early February, in order to win the handful of Republican votes needed for a 60-vote majority, a group of centrist Republicans and Democrats, led by Sens. Susan Collins, R-Maine, and Ben Nelson, D-Neb., proposed partially or totally cutting many of the bill's spending provisions, including:

- watershed rehabilitation;
- distance-learning and telemedicine projects;
- broadband Internet;
- energy-efficiency modernization of federal buildings;
- new Coast Guard facilities;

AT ISSUE

Can government infrastructure spending boost the economy?

YES
Dean Baker
Co-Director, Center for
Economic and Policy Research

Written for *CQ Researcher,* February 2009

The Obama administration has wisely chosen to make rebuilding public infrastructure the centerpiece of its stimulus package. The proposal includes funding not only for traditional infrastructure, like repairing roads and bridges, but also for green infrastructure such as weatherizing and retrofitting buildings to reduce energy consumption and modernization of the nation's electric grid.

Infrastructure is extremely effective as a stimulus since shovel-ready projects can start to create jobs soon after they are authorized. These projects put money in workers' pockets that they will spend. As a result, infrastructure spending has a substantial multiplier effect.

Moodys.com estimated the multiplier for infrastructure projects at 1.59, meaning that we will get $1.59 of additional economic output for each dollar spent on infrastructure. By comparison, the multiplier for payroll tax cuts is less than 1.0 and for business tax cuts just 0.3. This gap is likely to be even larger now, since many families will use tax cuts to rebuild their savings after losing most of their wealth in the housing crash.

Ideally, the spending will go to projects that would have been undertaken in any case, even if not immediately. For example, thousands of schools across the country desperately need repairs, such as new roofs or plumbing. And a considerable backlog of repairs to roads and bridges can be drawn down through this stimulus package.

There are also a large number of energy-conserving improvements that can pay for themselves in three or four years. For example, standard home retrofits cost around $3,000, and typically produce annual savings in the range of $750 to $1,000. If the federal government can provide incentives to encourage individuals, businesses and governments to carry through retrofits, there will be enduring benefits in reduced energy costs and greenhouse gas emissions.

But two notable pitfalls endanger this path. First, there will be some waste and corruption. That happens when you spend hundreds of billions of dollars. President Obama is establishing mechanisms to minimize this problem by ensuring that the process of disbursing funds is as open as possible.

The other risk is that we will spend money on environmentally harmful projects, such as highways that encourage suburban sprawl. This can and must be prevented. We can tolerate some waste, since it is important that money be spent quickly. But it makes no sense to spend billions of dollars in ways that will worsen our environmental problems.

NO
Ronald D. Utt
Senior Research Fellow
The Heritage Foundation

Written for *CQ Researcher,* February 2009

There is good news and bad news in the American Recovery and Investment Act, proposed by the Democrats. The good news is the authors heed the skepticism of many fiscal conservatives and resist pressure for a massive commitment to transportation infrastructure. Conservatives argued that an infrastructure spending plan would have only a limited impact on the recovery because such projects take months to get off the ground and years to complete, offering little immediate relief in this most difficult time for the economy.

This bit of good news, unfortunately, is overwhelmed by the other 93 percent of the spending plans included in the proposal, making the plan little more than a massive bailout of ineffective federal programs and state governments. It also extends the federal government deep into new areas of responsibility like public-school construction and broadband investment, from which it may never extract itself. From public housing to the Economic Development Administration to the Rural Utility Service to Community Development Block Grants — and to a host of other petitioners — the authors of the bill appear to have been on a search for every questionable program they could find, and then threw billions at them.

The recovery act represents a massive and unprecedented peacetime transfer of wealth and income from the beleaguered taxpayers to government. And by its focus on the mediocre and the unnecessary, it also represents an unprecedented peacetime destruction of wealth our economy can't afford.

Moreover, at a time of economic peril, this plan will make things worse by absorbing massive volumes of scarce resources and credit that might better be deployed in the nation's already badly deprived, and federally mismanaged, credit markets.

A year from now we will be looking at a weaker economy, and Congress will be contemplating an even larger bailout of government. And soon we will find ourselves living in the kind of desultory economy more common to continental Europe than the vibrant economy that has made us the most prosperous and dynamic nation on earth.

With this at stake, Congress has offered a history-changing challenge to the new president, and in his first few weeks in office he must decide whether the voters put him there for a purpose no better than validating a massive congressional fraud on the nation's workers. Destiny beckons, Mr. President.

- construction of K-12 schools and, higher-education facilities, and
- modernization and energy-efficiency renovations to public housing. [64]

Passing the Stimulus

Republicans say even the final centrists' bill still contains too much spending. "We are going to amass the largest debt in the history of this country, and we are going to ask our kids and grandkids to pay for it," McCain told CBS News on Feb. 8.[65]

Many economists, however, say the centrists' infrastructure and other spending cuts have made the bill too small to either create enough jobs now or strengthen the economy long term.

The cuts in school construction spending, along with cuts in aid for states, "will ensure that we have at least 600,000 fewer Americans employed over the next two years," Princeton University Professor of Economics Paul Krugman, winner of the 2008 Nobel Prize for economics, wrote on Feb. 7.[66]

Analysts on the left argue that projects like energy-efficiency improvements and railroads would create a sounder foundation for economic growth than the "bubbles" that have driven economic growth in recent years — such as the unreasonably high share values awarded by Wall Street to so-called "dot.coms" during the 1990s and the now burst "housing bubble" of the 2000s.

History demonstrates a strong link between infrastructure investment and economic growth, according to an analysis prepared by University of Massachusetts Professor of Economics Robert Pollin and Associate Research Professor James Heintz for the Alliance for American Manufacturing. Between 1950 and 1979, U.S. public investment in infrastructure like transportation and electricity transmission grew by an annual average 4 percent, while annual growth in the gross domestic product (GDP) averaged around 4.1 percent, Pollin and Heintz said. By contrast, between 1980 and 2007 growth in infrastructure investment slowed to 2.3 percent annually while GDP growth slowed to an annual average of 2.9 percent. [67]

Federal spending on "green" projects, in particular, "will create about 17 jobs for every $1 million" spent, and the resulting jobs will be more sustainable in an energy-strapped future, Pollin said.[68]

Obama, infrastructure experts and many congressional Democrats argue that U.S. infrastructure badly needs upgrading if the country is to remain economically competitive as the world changes, and only public-works spending can make this happen.

"I never saw a tax cut fix a bridge," Rep. Barney Frank, D-Mass., chairman of the House Banking Committee, told ABC News.[69]

Meanwhile, as recession shakes economies worldwide, other countries also have turned to stimulus legislation that's heavy on public-works spending for a temporary boost. China recently proposed a $586 billion stimulus package, including railway, road and airport construction, and the European Union (EU) has adopted a $256 billion package with significant infrastructure investment by most EU countries.[70] A $27.7 billion Australian stimulus package includes $19 billion in infrastructure spending.[71]

OUTLOOK

Recovery Time

The fate of future attempts to link federal public-works spending with economic recovery may hinge on what happens over the next couple of years.

If the Congressional Budget Office (CBO) is correct in predicting recovery beginning late this year, "it is best to think of the stimulus not as something that will much limit the rise in unemployment but rather as something that will hasten the recovery," the Urban Institute's Penner told the Senate Budget Committee. Only time will tell whether CBO's forecast is correct or overly optimistic, however, he said.[72]

Many, but not all, economists expect the economy to experience an unusually long recession, with a turnaround not beginning until at least 2010.

"I'm on the optimistic side of the scale. I do think we have a pretty good shot at a reasonable form of recovery by the end of this calendar year," says Tara M. Sinclair, assistant professor of economics and international affairs at George Washington University in Washington, D.C. Americans will tighten their purse strings this year, but extra government spending on public works and other things will offset some of that current consumer caution, she says. Longer term, "we're Americans, and we love stuff," and aren't likely to fall into the long-term recessionary pattern of Japan, where the public has helped slow economic recovery for decades because of their unwillingness to spend money, Sinclair predicts.

A prolonged recession might make infrastructure spending more crucial to a stimulus, since it would be needed to boost employment during the darkest times, said Krugman. And that looks likely to happen, he said. Economic forecasters surveyed by *The Wall Street Journal* "predict, on average, that unemployment will reach 8.1 percent by [the] end [of] 2009 and peak at 8.4 percent" in 2010, Krugman said. "That's a forecast of what will happen with the stimulus plan, not of what would happen absent stimulus, which would presumably be considerably worse."

Ironically, however, that scenario might also put an end to any more attempts by Democrats to propose public investments in infrastructure, no matter how worthy, since it "could easily be spun by conservatives as a failure of Obama's policies," he said.[73]

NOTES

1. For background, see Janet H. Cho and Tom Breckenridge, "Ohio Has Lost 263,383 Jobs — 5% of its Work Force — Since 2000, Data Shows," *The* [Cleveland] *Plain Dealer*, Feb. 4, 2009, http://blog .cleveland.com.

2. Shobhana Chandra, "U.S. Economy: Jobless Rate Soars and Payrolls Plunge by 598,000." Bloomberg .com, Feb. 6, 2009, www.bloomberg.com.

3. Cho and Breckenridge, *op. cit.*

4. Quoted in David Young, "Economic Recovery Starts in Denver," *Loveland Connection*, Feb. 18, 2009, www.coloradoan.com/article/20090218/ LOVELAND01/90218004.

5. For background, see Richard Rubin, Benton Ives and Clea Benson, "Congress' Role: Help or Hinder," *CQ Weekly*, Jan. 21, 2008, p. 192.

6. Obama on Daschle: "I Messed Up," transcript, CBS News, Feb. 3, 2009, www.cbsnews.com.

7. Christina Romer and Jared Bernstein, "The Job Impact of the American Recovery and Reinvestment Plan," Transition Office of Barack Obama, Jan. 9,

2009, http://otrans.3cdn.net/45593e8ecbd 339d074_l3m6bt1te.pdf.

8. "The Employment Situation: January 2009," Bureau of Labor Statistics, Feb. 5, 2009, www.bls.gov/news .release/empsit.nr0.htm.

9. Richard G. Little, "Stimulus Has to Include More than Money for Building," *San Jose Mercury News*, Nov. 24, 2008, www.mercurynews.com/opinion/ ci_11064540.

10. Judd Gregg, "How to Make Sure the Stimulus Works," *The Wall Street Journal*, Jan. 4, 2009, http:// online.wsj.com.

11. Thomas J. DiLorenzo, "The Myth of Job Creation," Policy Analysis No. 48, Cato Institute, Feb. 19, 1984, www.cato.org.

12. Quoted in "The Problems with a Fiscal Stimulus," "Marketplace," National Public Radio, http:// marketplace.publicradio.org, Jan. 5, 2009.

13. Amity Shlaes, "The Krugman Recipe for Depression," *The Wall Street Journal*, Nov. 29, 2008, http://online .wsj.com.

14. Quoted in Lori Montgomery, "Critics Say Roads Projects Won't Jump-Start Economy," *The Washington Post*, Oct. 30, 2008, p. D1.

15. Testimony before House Committee on Transportation and Infrastructure, Oct. 29. 2008, http://transportation.house.gov/Media/File/ Full%20Committee/20081029/Irons.pdf.

16. *Ibid.*

17. Transcript, "Late Edition with Wolf Blitzer," CNN, Nov. 9, 2008, http://transcripts.cnn.com/ TRANSCRIPTS/0811/09/le.01.html.

18. Quoted in Jonathan Karp, "Construction Industry Counts on Obama," *The Wall Street Journal*, Jan. 14, 2009, p. C10.

19. Eric Rauchway, "FDR's Latest Critics: Was the New Deal UnAmerican?" *Slate*, July 5, 2007, www.slate.com.

20. Jacob G. Hornberger, "Obama's Public-Works Folly," "Hornberger's Blog," The Future of Freedom Foundation, Jan. 5, 2009, www.fff.org.

21. Lawrence Kudlow, "Infrastructure Spending Is No Cure-All," The Corner blog, *National Review* online, Dec. 9, 2008, http://cprmer.nationalreview.com.

22. Lisa Snell, "Huge Stimulus Plan Won't Change the Education System's Status Quo," Reason Foundation, Jan. 27, 2009, http://reason.org.

23. Ronald D. Utt, "Learning from Japan: Infrastructure Won't Boost the Economy," *Backgrounder No. 2222*, Heritage Foundation, Dec. 16, 2008, www.heritage.org.

24. Amity Shlaes, "The Perils of a Cement Tsunami," *The Washington Post*, Dec. 10, 2008, p. A25.

25. For background, see Thomas J. Billitteri, "Financial Bailout," *CQ Researcher*, Oct. 24, 2008, pp. 865-888, and Kenneth Jost, "Financial Crisis," *CQ Researcher*, May 9, 2008, pp. 409-432.

26. Testimony before House Committee on the Budget, Jan. 27, 2009.

27. For background, see Marcia Clemmitt, "Aging Infrastructure," *CQ Researcher*, Sept. 28, 2007, pp. 793-816.

28. Richard G. Little, "Not the Macquarie Model: Using U.S. Sovereign Wealth to Renew America's Civil Infrastructure," paper prepared for America 2050, Jan. 9. 2009, www.america2050.org/upload/2009/01/Paper_Richard_Little.pdf.

29. *Ibid.*

30. For background, see Marcia Clemmitt, "The National Debt," *CQ Researcher*, Nov. 14, 2008, pp. 937-960.

31. H.R. 1 — The American Recovery and Investment Act, Legislative Bulletin, Republican Study Committee, Jan. 27, 2009, http://rsc.price.house.gov/News/DocumentSingle.aspx?DocumentID=109553.

32. N. Gregory Mankiw, "What Would Keynes Have Done?" *The New York Times*, Nov. 30, 3008, p. BU4.

33. Rudolph G. Penner, "Addressing Short- and Long-Term Fiscal Challenges," testimony before Senate Budget Committee, Jan. 21, 2009, http://budget.senate.gov/democratic/testimony/2009/Penner-1-21-testimony.pdf.

34. See Anne E. Kornblut and Michael D. Shear, "Obama Says He erred in Nominations," *The Washington Post*, Feb. 4, 2009, p. A1.

35. Quoted in " Some of the Math Is Simple," "Talking Points Memo," Feb. 4, 2009, www.talkingpointsmemo.com.

36. Quoted in Mark Thoma, "Is the Deficit a Threat to a Future Recovery?" *Global Macro EconoMonitor, RGE Monitor*, Dec. 4, 2008, www.rgemonitor.com.

37. Quoted in Alec MacGillis, "Democrats Among Stimulus Skeptics," *The Washington Post*, Jan. 28, 2009, p. A1.

38. Quoted in Arianna Huffington, "Stimulus Package: If You Jump Halfway Across a Chasm You Fall Into the Abyss," *Huffington Post* blog, Feb. 2, 2009, www.huffingtonpost.com.

39. For background, see P. Webbink, "Federal Relief of Economic Distress," *Editorial Research Reports*, Dec. 8, 1930; B. W. Patch, "Public Works and National Recovery," *Editorial Research Reports*, July 18, 1933; M. Packman, "New Highways," *Editorial Research Reports*, Dec. 13, 1954; Kenneth Jost, "Stimulating the Economy," *CQ Researcher*, Jan. 10, 2003, pp. 1-24.

40. Quoted in Patch, *op. cit.*

41. *Ibid.*

42. Herbert Hoover, letter to Herbert S. Crocker of the American Society of Civil Engineers, May 21, 1932, "The Depression Papers of Herbert Hoover," www.geocities.com/mb_williams/hooverpapers/1932/paper19320521.html.

43. Quoted in Patch, *op. cit.*

44. *Ibid.*

45. Robert Kossuth, "Boondoggling, Baseball, and the WPA," *Nine: A Journal of Baseball History and Culture* (fall 2000), p. 56.

46. "Public Works Administration," *Encyclopedia Britannica Online*, 2008, www.britannica.com.

47. Quoted in Kossuth, *op. cit.*

48. Jennifer Long, "Government Job Creation Programs — Lessons from the 1930s and 1940s," *Journal of Economic Issues*, December 1999, p. 903.

49. *Ibid.*

50. *Ibid.*

51. "Early History of the Tennessee Valley Authority," Bruderheim Rural Electrification Association, www.bruderheim-rea.ca/ TVA.htm.

52. "TVA Board Approves Rate Adjustment," press release, Tennessee Valley Authority, July 22, 2005,

www.tva.gov/news/releases/julsep05/budget.htm.

53. Quoted in B. W. Patch, "Roosevelt Policies in Practice," *Editorial Research Reports*, Sept. 25, 1936.

54. For background, see B. W. Patch, "Full Employment," in *Editorial Research Reports*, July 30, 1945, available at *CQ Researcher Plus Archive*, http://library.cqpress.com.

55. Quoted in G. J. Santoni, "The Employment Act of 1946: Some History Notes," *Federal Reserve Bank of St. Louis Review*, November 1986, p. 5, http://research.stlouisfed.org/publications/review/86/11/Employment_Nov1986.pdf.

56. For background, see Marcia Clemmitt, "Pork Barrel Politics," *CQ Researcher*, June 6, 2006, pp. 529-552.

57. Andrew Taylor, "Congress Readies Final Vote on $787 Billion Stimulus Bill," The Associated Press, Feb. 13, 2009, www.google.com/hostednews/ap/article/ALeqM5gdDrWnoMueqVFI-Uo1ClxVZur22AD96ARQGO3.

58. *Congressional Record*, Jan. 28, 2009, p. H769.

59. Wolf Blitzer, "The Situation Room," transcript, CNN.com, Feb. 2, 2009, http://transcripts.cnn.com/TRANSCRIPTS/0902/02/sitroom.02.html.

60. Martin Feldstein, "The Stimulus Plan We Need Now," *The Washington Post*, Oct. 30, 2008, p. A23.

61. "A Conversation About Obama's Economic Stimulus Package," "Charlie Rose Show," PBS, Jan. 6, 2009, www.charlierose.com/view/interview/9899.

62. "What GOP Leaders Deem Wasteful in Senate Stimulus Bill," "CNN Politics," Feb. 3, 2009, www.cnn.com.

63. For background, see Jonathan Weisman, Greg Hitt and Naftali Bendavid, "House Passes Stimulus Package," *The Wall Street Journal*, Jan. 29, 2009, http://online.wsj.com/article/SB123315486943524321.html.

64. "FY 2009 Economic Recovery and Reinvestment Supplemental," Feb. 7, 2009, http://senateconservatives.files.wordpress.com/2009/02/nelson-collins-stimulus-final2.pdf.

65. Michelle Levi, "McCain: Stimulus Bill Is Generational Theft," "Political Hotsheet," CBS News, Feb. 8, 2009, www.cbsnews.com/blogs/2009/02/08/politics/politicalhotsheet/entry4783514.shtml.

66. Paul Krugman, "What the Centrists Have Wrought," The Conscience of a Liberal Blogs, *The New York Times* blogs, Feb. 7, 2009, http://krugman.blogs.nytimes.com/2009/02/07/what-the-centrists-have-wrought.

67. James Heintz, Robert Pollin and Heidi Garrett-Peltier, "How Infrastructure Investments Support the U.S. Economy: Employment, Productivity and Growth," University of Massachusetts Political Economy Research Institute, January 2009.

68. For background, see Marcia Clemmitt, "Mortgage Crisis," *CQ Researcher*, Nov. 2, 2007, pp. 913-936.

69. Quoted in James Gordon Meek, "President Obama to GOP: You Can Help Me Spend the Billions," *New York Daily News*, Feb. 1, 2009, www.nydailynews.com.

70. Eric Lotke, "Falling Apart, Falling Behind," "Blog for Our Future," Campaign for America's Future, Dec. 4, 2008, www.ourfuture.org.

71. "Factbox: Key Planks of Australia's Stimulus Packages," Reuters, Feb. 13, 2009, http://in.reuters.com/article/asiaCompanyAndMarkets/idINSYD36217320090213.

72. Penner, *op. cit.*

73. Paul Krugman, "Forecasts," The Conscience of a Liberal blog, *The New York Times blogs*, Jan. 6, 2009, http://krugman.blogs.nytimes.com.

BIBLIOGRAPHY

Books

Krugman, Paul, *The Return of Depression Economics and the Crisis of 2008*, W. W. Norton, 2008.
A Princeton economics professor and Nobel Prize winner argues that some of the same problems that led to the Great Depression have returned in our globalized world.

Leighninger, Robert D., Jr., *Long-Range Public Investment: The Forgotten Legacy of the New Deal*, University of South Carolina Press, 2007.
An Arizona State University sociologist catalogs the range of public works created in the New Deal and examines their impact on the U.S economy and infrastructure.

Smith, Jason Scott, *Building New Deal Liberalism: The Political Economy of Public Works, 1933-1956*, Cambridge University Press, 2005.
A University of New Mexico assistant professor of history argues that New Deal infrastructure investment paved the way for the economic growth that followed World War II.

Articles

Feldstein, Martin, "Defense Spending Would Be Great Stimulus," *The Wall Street Journal* online, Dec. 24, 2008, http://online.wsj.com.
The former chairman of President Ronald Reagan's Council of Economic Advisers argues that spending for military and research infrastructure would be an effective economic stimulus.

MacGillis, Alec, and Michael D. Shear, "Stimulus Package to First Pay for Routine Repairs," *The Washington Post*, Dec. 14, 2008, p. A1.
The need to create economy-stimulating jobs quickly may be at odds with hopes of creating public works that will benefit the economy long term.

Reports and Studies

Baxandall, Phineas, "Economic Stimulus or Simply More Misguided Spending?" *U.S. PIRG*, December 2008, www.calpirg.org/uploads/8B/eI/8BeIuP6wmZ1AR24UZUp6SQ/State-Stim-Report_CALPIRG_Final.pdf.
States' proposals for shovel-ready infrastructure funding put too much emphasis on highways, according to an analyst for a consumer group.

Edwards, Chris, "10 Reasons to Oppose a Stimulus Package for the States," *Tax & Budget Bulletin No. 51*, The Cato Institute, December 2008, www.cato.org/pubs/tbb/tbb_1208-51.pdf.
An analyst for a libertarian think tank argues that infrastructure investments would dangerously raise the federal debt and aren't needed because states have generously funded public works in recent years.

Heintz, James, Robert Pollin and Heidi Garrett-Peltier, "How Infrastructure Investments Support the U.S. Economy: Employment, Productivity and Growth," *Political Economy Research Institute, University of Massachusetts*, January 2009, www.americanmanufacturing.org/wordpress/wp-content/uploads/2009/01/peri_aam_finaljan16_new.pdf.
Economists argue that neglect of the nation's infrastructure, beginning in the 1970s, endangers future economic growth.

Pollin, Robert, and Jeannette Wicks-Lim, "Job Opportunities for a Green Economy," *Political Economy Research Institute, University of Massachusetts*, June 2008, www.peri.umass.edu/fileadmin/pdf/other_publication_types/Green_Jobs_PERI.pdf.
Economists argue that millions of manufacturing and construction workers would find permanent work in an economy focused on green energy.

Romer, Christina, and Jared Bernstein, "The Job Impact of the American Recovery and Reinvestment Plan," *White House*, Jan. 9, 2009, http://otrans.3cdn.net/45593e8ecbd339d074_l3m6bt1te.pdf.
Economic advisers to President Barack Obama and Vice President Joseph Biden estimate a stimulus package emphasizing infrastructure, energy projects and school repair would add or save three-and-a-half-million jobs through 2010.

Shorris, Anthony E., "Breaking Down Walls: Overcoming Institutional Barriers to Infrastructure Investment," *The Century Foundation*, 2008, www.tcf.org/publications/economicsinequality/shorris.pdf.
U.S. infrastructure has long been underfunded, from an economic-policy standpoint, but refocusing attention on public works would require overhauling antiquated government systems, writes a fellow at the liberal think tank.

Utt, Ronald D., "Learning from Japan: Infrastructure Spending Won't Boost the Economy," *Backgrounder No. 2222*, The Heritage Foundation, Dec. 16, 2008, www.heritage.org/research/economy/bg2222.cfm.
Japan's repeated attempts to boost its economy through public-works spending were a lengthy, expensive failure.

Utt, Ronald D., "More Transportation Spending: False Promises of Prosperity and Job Creation," *Backgrounder No. 2121*, The Heritage Foundation, April 2, 2008, www.heritage.org/Research/Budget/bg2121.cfm.
An economist at a conservative think tank argues that infrastructure spending occurs too slowly to stem a recession.

For More Information

America 2050, 4 Irving Place, Suite 711-S, New York, NY 10003; (212) 253-5795; www.America2050.org. A policy-analysis and advocacy group interested in overhauling U.S. infrastructure to meet future economic and environmental needs.

Blueprint America, www.pbs.org/wnet/blueprintamerica. The Web site of a series of PBS television documentaries on infrastructure, with interviews, links and reports on infrastructure and related economic issues, including stimulus proposals.

Center for American Progress, 1333 H St., N.W., 10th Floor, Washington, DC 20005; (202) 682-1611; www.americanprogress.org. Progressive think tank that analyzes economic issues and proposals on stimulus legislation and infrastructure.

The Century Foundation, 41 East 70th St., New York, NY 10021; (212) 535-4441; www.tcf.org. Liberal think tank that analyzes economic and infrastructure issues.

The Heritage Foundation, 214 Massachusetts Ave., N.E., Washington, DC 20002-4999; (202) 546-4400; www.heritage.org. Conservative think tank that analyzes economic issues, including infrastructure spending and the economic stimulus.

Lincoln Institute of Land Policy, 113 Brattle St., Cambridge, MA 02138-3400; (617) 661-3016; www.lincolninst.edu. Analyzes land-use issues, including infrastructure planning and economic development.

Manhattan Institute, 2 Vanderbilt Ave., New York, NY 10017; (212) 599-7000; www.manhattan-institute.org. Conservative think tank that analyzes economic issues, including proposals on infrastructure funding and stimulus.

Metropolitan Policy Program, The Brookings Institution, 1775 Massachusetts Ave., N.W., Washington, DC 20036; (202) 797-6000; www.brookings.edu. Research group at a centrist think tank that analyzes policy proposals affecting cities and metropolitan areas.

Political Economy Research Institute, University of Massachusetts, Gordon Hall, 418 N. Pleasant St., Suite A, Amherst, MA 01002; (413) 545-6355; www.peri.umass.edu. Analyzes policy ideas related to globalization, unemployment, economic development and the environment.

9

Universal Coverage

Will All Americans Finally Get Health Insurance?

Marcia Clemmitt

Many working Americans, like Daniel and Mindy Shea, of Cincinnati, are un- or under-insured. Young workers are hit especially hard: In 2004, a third of Americans ages 19-24 were uninsured. Only 61 percent of Americans under age 65 obtain health insurance through their employers. As health costs rise and incomes sag, more and more companies are dropping coverage, especially restaurants and small businesses.

AP Photo/David Kohl

From *CQ Researcher*,
March 30, 2007.

W hen Emily, a 24-year-old graduate student, discovered a lump on her thigh, her doctor told her to get an MRI to find out whether it was cancerous. But Emily's student-insurance policy didn't cover the $2,000 procedure, so she skipped it.[1]

Several weeks later, during outpatient surgery to remove the lump, Emily's surgeon found a rare, invasive cancer underneath the benign lump — with only a 20 to 40 percent survival rate. The skipped MRI could have detected the cancer much sooner, improving her chances for recovery.

Emily pieced together payment for her treatment from her school insurance, two state public-aid programs and a monthly payment plan that ate up more than 40 percent of her take-home income. But a year later she learned that annual health premiums for all students at her school would rise by 19 percent because a few, like her, had racked up high expenses. The price hike led many more students to skip purchasing the coverage altogether.

Advocates say such stories are a good reason why Congress should enact a universal health-insurance program. While Congress has been expanding public health insurance programs covering the very poor — especially children and their mothers — students and lower-income workers increasingly are losing coverage or are finding, like Emily, that they can't afford adequate coverage.

Today, 45 million Americans — about 15.3 percent of the population — lack health insurance, usually due to job loss, student status, early retirement or because they have entry-level jobs or work in a service industry or a small business. Only about

Cost of Premiums Rising Rapidly

The average annual cost of family health coverage has risen more than 50 percent since 2001, to $11,500, and is expected to exceed $18,000 in the next five years. Most of the cost is borne by the employer.

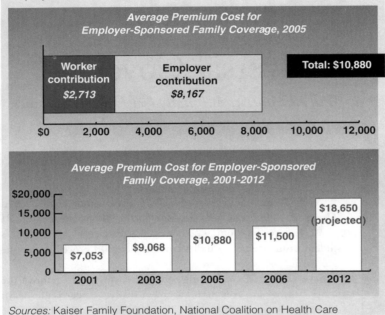

Average Premium Cost for Employer-Sponsored Family Coverage, 2005

Worker contribution $2,713 | Employer contribution $8,167 | Total: $10,880

$0   2,000   4,000   6,000   8,000   10,000   12,000

Average Premium Cost for Employer-Sponsored Family Coverage, 2001-2012

$20,000 — 15,000 — 10,000 — 5,000 — 0

$7,053 (2001) $9,068 (2003) $10,880 (2005) $11,500 (2006) $18,650 (projected) (2012)

Sources: Kaiser Family Foundation, National Coalition on Health Care

"Health insurance expenses are the fastest-growing cost component for employers," according to the National Coalition on Health Care. "Unless something changes dramatically, health insurance costs will overtake profits by 2008."[6]

"If there's one thing that can bankrupt America, it's health care," warns U.S. Comptroller General David Walker, chief of the Government Accountability Office, Congress' nonpartisan auditing arm. And in response to those who say the United States can "grow" its way out of uninsurance by creating more and better jobs with coverage benefits, he states flatly: "Anybody that tells you we are going to grow our way out of this . . . probably isn't very proficient at math."[7]

While the public often pictures the uninsured as being unemployed, the fact is that most un- and underinsured Americans have jobs. Only 61 percent of Americans under age 65 obtain health insurance through their employers — down from 69 percent in 2000. And as health costs rise and incomes sag, more and more companies are dropping coverage, especially very small businesses. Because insurers raise premium prices for high-cost groups — as happened with Emily's grad-school coverage — small companies whose employees get seriously ill or injured or pregnant often find themselves priced out of coverage altogether.

Now some states are trying to create new sources of affordable coverage. Massachusetts is launching a universal-coverage plan in 2007 that has bipartisan support, and Republican California Gov. Arnold Schwarzenegger hopes to enact a similar measure. At the federal level, no new initiatives are expected this year — except for a probable expansion of children's coverage — but many expect universal coverage to be a major theme in the 2008 presidential election.

America's creeping lack of health coverage constitutes a crisis for the uninsured, even as the skyrocketing cost of

40 percent of businesses employing low-wage or part-time workers offer health benefits, and at $11,480 a year, the average family's health-insurance premium now costs more than a minimum-wage worker makes in a year. Young workers are hit especially hard: In 2004, more than a third of Americans between the ages of 19 and 24 were uninsured.[2] And in the construction and service industries, only 80 percent of the managers have health coverage.[3]

And the situation is only expected to get worse. U.S. health spending is expected to double by 2015 — to more than $12,300 per person.[4] As health-care costs skyrocket, so does the cost of health insurance, whether purchased by individuals or by employers. Between 2000 and 2006, health premiums for employer-sponsored insurance jumped 87 percent, far outpacing inflation's 18 percent overall increase.[5]

health care makes it inevitable that even more people will be uninsured in the future. If health premiums continue rising at their current rate, about 56 million Americans are predicted to be uninsured by 2013 — 11 million more than today, according to a University of California at San Diego study.[8] The increase will cause 4,500 additional unnecessary deaths per year and $16 billion to $32 billion in lost economic productivity and other "human capital," the study says.[9]

The leading public myth about the uninsured is that "people without health insurance get the medical care they need," said Arthur Kellerman, chairman of Emory University's Department of Emergency Medicine and co-chairman of an Institute of Medicine (IOM) panel that has called for universal coverage by 2010.[10]

In fact, the uninsured seldom receive appropriate care at the appropriate time, said Kellerman. "The uninsured are less likely to see a doctor or be able to identify a regular source of medical care and are less likely to receive preventive services," he said.[11] And uninsured children admitted to a hospital due to an injury are twice as likely to die and 46 percent less likely to receive rehabilitation after hospitalization, according to a recent study by the consumer advocacy group Families USA.[12]

The growing number of uninsured Americans also pushes up the cost of publicly subsidized health insurance like Medicare, the panel said. Working-age uninsured patients with uncontrolled diabetes or high blood pressure eventually enter the health system sicker than they would have been had they been insured.[13] And about 20 percent of those with schizophrenia and bipolar disorder are uninsured and end up in jail or prison when their untreated conditions trigger illegal behavior, said the panel.[14]

Adding to the problem, manufacturing and unionized jobs were the mainstay of job-based coverage, but their numbers have been dropping for 20 years. "I suspect you're going to see wholesale withdrawal of employer-sponsored health care" for anyone earning less than twice federal poverty-level wages, said National Governors Association Executive Director Ray Scheppach.[15]

"Economic security, jobs, health care and retirement security — those are all now one and the same issue," says Henry Simmons, president of the National Coalition on Health Care, which includes employers, unions and academic and other groups advocating universal coverage.

Pension investment funds have recently realized that skyrocketing health-care costs could bankrupt Americans' future if they are not checked, says Simmons. Since Medicare covers only some of the health services needed by retirees, virtually all elderly people who can afford it also purchase private supplemental insurance to fill the gaps. But pension-fund investors are finding no investments that grow fast enough to allow retirees' savings income to keep up with the anticipated soaring cost of future Medicare and supplemental-coverage premiums.

While no one expects significant action from Congress until after the 2008 presidential election, federal

Americans Without Health Insurance

In 25 states, between 14 percent and 19 percent of the adults ages 18-64 did not have health insurance in 2005. States with high levels of uninsured residents typically have minimal state and employer insurance coverage.

Percentage of People Ages 18-64 Without Insurance, 2005

23% and over
19-22.9%
14-18.9%
Under 14%

Source: Kaiser Family Foundation, statehealthfacts.org, 2005

policymakers increasingly acknowledge a need for action. Consensus appears to be growing for some type of hybrid universal coverage that combines public and private insurance.

"In the past the debate got bogged down because different groups wanted their first-priority proposal only," says Ron Pollack, founding executive director of Families USA. "One group would say, 'Coverage must be financed through public programs,' while another would say, 'There should be no government action in the marketplace,' And since everyone's second-favorite program was the status quo, nothing happened."

But that decades-old logjam may be breaking up, as advocates on all sides of the issue creep closer to one another in their proposals. "The very grand visions on both sides" — a single-payer government system or relying on individuals saving money for their own care via Health Savings Accounts (HSAs) — "are both completely impossible in our political system," says Yale University Professor of Political Science Jacob Hacker.

Even President George W. Bush, a longtime proponent of individually purchased HSAs, softened that stance in his most recent proposal. Bush's fiscal 2008 budget plan would offer similar tax breaks to those buying all kinds of health insurance, not just HSAs, either as individuals or through employers.

Paul Ginsburg, president of the nonpartisan research group Center for Studying Health System Change, says critics rightly point out that Bush's tax break doesn't target lower-income people who are most in danger of losing coverage. Nevertheless, "eventually, the Democrats may see that the president has given them something — a revenue source" to help pay for expanding coverage, he says.

So far, every Democrat who has announced he or she will run for president has declared a commitment to universal coverage, tying the issue to the country's overall economic health, but only former Sen. John Edwards, D-N.C., has announced a specific coverage plan. "The U.S. auto industry is struggling, in part because of the rising cost of health care that this administration has done nothing to address," newly announced Democratic candidate Sen. Barack Obama, D-Ill., said last November. "I have long proposed that the government make a deal with the Big Three automakers that will pay for a portion of their retiree health costs if they agree to invest those savings in fuel-efficient technologies." Health-care costs

account for approximately $1,000 of the cost of each car produced by the America's largest automakers — more than they spend on steel.[16]

Among Republican presidential candidates, former Massachusetts Gov. Mitt Romney last year backed legislation intended to achieve universal coverage in his state, and former House Speaker Newt Gingrich, who has not yet thrown his hat into the presidential ring, also has called for systemic reforms in the health-care system.

While a consensus may be developing on the need for some kind of universal coverage, many contentious debates remain. For example, both the right and the left have criticized the California and Massachusetts plans for requiring individuals to buy health insurance, just as drivers are required to carry automobile insurance.

Nevertheless, many believe the country is on the verge of a focused, national debate on universal coverage. "In 2008, universal coverage will be up there with Iraq as top election issues," says Pollack.

As policymakers gear up for that debate, here are some of the questions being asked:

Can America afford universal health coverage?

Critics of universal-insurance proposals have long argued that while expanding coverage is desirable, covering everyone would simply cost too much.[17] Universal-coverage advocates, however, argue that current administrative expenses are high partly because the United States has a piecemeal system with many uninsured.

"It is impossible to get everybody covered," former Senate Majority Leader Bill Frist, R-Tenn., a transplant surgeon, said in 2004.[18] State efforts have shown that universal coverage is not financially feasible, he said.

For example, his home state of Tennessee managed to cover about 93 percent of residents — a national high — by having Medicaid cover both the uninsured and the uninsurable. But "in attempting to do this the state is going bankrupt," he said, "and there is a major effort to backtrack."[19]

Universal-coverage plans generally are "unrealistic," said former Health and Human Services Secretary Tommy G. Thompson. I just don't think it's in the cards. . . . "I don't think that administratively or legislatively it's feasible."[20]

While few politicians today will say America cannot afford universal coverage, both sides agree that the costs will be high. A study by the liberal Urban Institute says that in 2004 universal coverage would have added about $48 billion to the $125 billion the nation spent on health care for uninsured people — most of which was paid out of pocket by the uninsured or was delivered without compensation by doctors and hospitals.[21] A proposed universal-coverage plan for Maryland would raise that state's health spending by some $2.5 billion per year, while a Minnesota proposal to cover the state's 383,000 uninsured is projected to cost an estimated $663-$852 million in new annual funding.[22]

Most proposals for universal coverage call for increased government spending. And finding those dollars will be tough, given that the federal budget and many state budgets are facing substantial deficits.[23]

"The public sector has fewer resources" now compared to when the issue was debated previously, says Ginsburg, of the Center for Studying Health System Change. A lack of willingness or ability to commit new revenue has doomed at least one state plan, he adds. When Maine launched a universal-coverage initiative in 2003, the state "put almost no money in and got almost nothing out," he says.

The Democratic presidential hopefuls who have called for universal coverage "will be desperate for revenues when they put out their plans," says Robert Blendon, professor of health policy and management at the Harvard School of Public Health. "Money is going to be hard to come by. We've got defense costs that are very, very high."

A 2006 Massachusetts law requiring every resident to purchase subsidized coverage — unless that coverage is "unaffordable" — is already running into an affordability crisis, said Jonathan Gruber, a professor of economics at the Massachusetts Institute of Technology.[24] The state has said it would subsidize only those who earn up to three times the federal poverty level (about $30,000 for an individual and $60,000 for a family of four), said Gruber. But "at three times poverty, health insurance is still expensive. . . . It's not feasible to have someone spend 20, 30 or 40 percent of income on health insurance."[25]

While some say universal coverage is too expensive for America, many economists point out that industrialized nations with universal coverage spend less per capita on health care than the United States. In 2004, for instance, the United States spent $6,102 per capita on health care

while Australia spent $3,120; Denmark, $2,881; Germany, $3,043; Luxembourg, $5,089; Sweden, $2,825 and Switzerland, $4,077.[26]

Not only are Americans paying more for health care than those in any other industrialized country, but they are getting lower-quality care — by some measurements — than consumers in countries with universal coverage. While new drugs and technology have improved longevity and quality of life for many Americans, the United States is ranked 37th by the World Health Organization in overall quality of care, based on adult and infant mortality rates. The United States also ranks 24th among industrialized nations in life expectancy.[27]

Those stark realities, coupled with the fact that U.S. health-care costs are spiraling out of control, lead a growing number of analysts to argue that the United States can't afford not to have universal coverage.

"I've always believed universal coverage would carry significant costs and would bring daunting . . . economic challenges," said Harold Pollack, associate professor at the University of Chicago School of Social Service Administration.[28] But recently, he says, he has decided "there is no alternative" to pushing forward with universal insurance. "The current system is no longer able to accomplish important things we expect from our health care."[29]

Former Gov. John Kitzhaber, D-Ore. — an emergency-room physician who now heads the Oregon-based Archimedes Project, a health-reform initiative — agrees. Economic growth depends on good health for all, he said. Good health "is the first rung on the ladder of opportunity . . . the cornerstone of a democratic society, allowing people to . . . be productive and to take advantage of the opportunities of upward mobility."[30]

Others argue that having more than 15 percent of the population uninsured means that all Americans pay more for health care. "The uninsured are one of the inefficiencies" driving health costs into the unaffordable range, says Robert Greenstein, executive director of the liberal Center on Budget and Policy Priorities.

"Reining in health-care cost growth" — which soared by 7 percent last year alone — is a prerequisite for universal coverage, says Robert Laszewski, an independent health-care consultant and a former health-insurance executive. Health-insurance premiums grow even faster than costs, and neither government subsidies nor the

Universal Coverage Faces Financial Obstacles

Reducing health-care costs is the big challenge

Now that Americans appear to be reaching some consensus on the need for universal health coverage, major hidden obstacles — all involving money — must be overcome. Among those thorny financial issues are questions over who is going to pay for the coverage, how can affordable access be ensured for all and how can overall health-care costs be reduced.

Perhaps the most controversial issue is who will pay for the coverage. In 2005, employers paid 75 percent of workers' health-premium costs — about $500 billion compared to the approximately $170 billion that workers paid.[1] That's "most of the money outside the government that's spent on health care," says Stanley Dorn, a senior research associate at the liberal Urban Institute. To work, any universal-coverage plan will have to either continue to use those employer contributions or come up with a suitable replacement for them.

That's why many universal-coverage proposals ask employers for financial contributions. But making those contributions both fair and adequate is difficult, mainly because businesses vary so widely in what they pay today: Many contribute nothing, but others pay hundreds of millions of dollars each year.

Dorn says policymakers may want to consider asking all employers to pay a set amount into a general pool but vary the amount by companies' line of business and their geographical location. That way, companies that compete with one another would share the same burden.

Lawmakers also must figure out how to ensure affordable access to all. Many Republican proposals for expanding coverage rely on tax subsidies to help more people buy individual health policies. Because such coverage wouldn't be tied to a job, it would be "portable," so employees who switch or lose jobs would not be without insurance.

But buying individual health insurance can be far more expensive than purchasing through an employer because insurers don't "pool" risks the way they do for workers under employer-based policies. So individual purchasers pay based on their family's health status and age, which makes it the most expensive way to buy health insurance. Moreover, insurers won't even sell coverage to some people because the companies themselves consider it unaffordable.

"The words 'kinda crummy' come to mind when I think of the individual market," said former Maryland Insurance Commissioner Steve Larsen, now a private attorney. For example, a case of mononucleosis and a chronic condition like hay fever is enough for some insurers to deem a potential buyer unaffordable. "And if you have any type of serious mental illness, forget it," he said.[2] A study by the Georgetown University Institute for Health Care Research and Policy found that a 62-year-old overweight moderate smoker with controlled high blood pressure was deemed an unaffordable risk 55 percent of the times he sought individual health coverage.[3]

---

incomes of lower-wage working people can keep up with the current growth rate for long, he says.

If coverage expansion were accompanied by efforts to rein in spending and improve care, "there's absolutely no doubt that you can have universal coverage without substantially raised costs," says Simmons, president of the National Coalition on Health Care. "Every other country does it" already, he says.

Universal coverage is needed to "get a [health-care] market that works," he adds. "You can't fix the issue of cost" — which affects everyone, insured and uninsured — "without universal coverage," he says. Absent a universal-coverage requirement, "what markets do is avoid risk," such as when insurance companies develop marketing and risk-assessment procedures to avoid selling policies to

sick people. "It's an open-and-shut case that universal coverage is cheaper and better" than the status quo.

Getting everyone covered and specifying uniform benefit packages would create a huge, immediate, one-time financial saving, Simmons says. "Automatically, you're talking about hundreds of billions of dollars" in savings that "every other nation has already captured," partly accounting for lower costs abroad.

Should Americans depend on states to expand coverage?

In recent years, states have been far more active than the federal government in expanding health coverage. Massachusetts and Vermont passed universal-coverage laws in 2006, and Illinois created a program to cover all

Buying an individual policy is more affordable for the young and healthy, says health-care consultant Robert Laszewski. His 20-something son found an individual health policy for $150 a month several years ago, but "if he was 58 years old, his premium would have been $1,500," he says. "If you're going to do universal health care, you can't age-rate premiums or bar people based on pre-existing conditions."

No matter how widely risk is shared, however, behind the high cost of insurance lurks the ever-rising cost of health care. "The 10,000-pound elephant in the room is cost," says Laszewski.

Health-care costs have been growing faster than the entire economy or any other sector in it for the past 45 years, says Gail Wilensky, a senior fellow at the nonprofit health-education foundation Project HOPE and former head of Medicaid and Medicare. "They can't go on doing it for the next 30 years" without crippling other parts of the economy, she says.

U.S. health-care costs are the world's highest because of insurers' high administrative and marketing costs and because American doctors and medical suppliers enjoy higher profits and salaries than their counterparts in other industrialized countries.[4]

In today's fragmented insurance system, insurers' efforts to attract the healthiest, cheapest customers add extra overall costs, point out Paul Menzel, a philosophy professor at Pacific Lutheran University in Tacoma, Wash., and Donald W. Light, professor of comparative health care at the University of Pennsylvania.[5] For example, they wrote, a Seattle survey found that 2,277 people were covered by 755 different policies linked to 189 different health-care plans.

"The $420 billion (31 percent!) paid [annually] for managing, marketing and profiting from the current fragmented system could be drastically cut" if insurers had to take all comers rather than carefully jiggering their policies, premiums and marketing strategies to attract only the healthiest, least expensive buyers, they said.

A key is to cut spending on care by "learning more about what works for whom," says Wilensky. But getting that information requires investment, she says.

In addition to cutting excess services, says Laszewski, making coverage affordable will ultimately mean sacrificing some of the health-care industry's high profits and salaries. International comparisons show that other countries spend less on health care while delivering the same amount or even more services to patients.

Americans don't understand that controlling cost is crucial to sustaining the health system, let alone expanding coverage, says Laszewski. "I'll bet you if you told consumers that if they lost their jobs, replacing their insurance would cost $15,000 or $16,000 a year, they'd understand that," he says.

---

[1] Aaron Catlin, Cathy Cowan, Stephen Heffler and Benjamin Washington, "National Health Spending in 2005: The Slowdown Continues," *Health Affairs*, January/February 2007, p. 148.

[2] Quoted in "Reinsurance for Individual Market Pricks Up Many Ears," *Medicine & Health*, "Perspectives," Oct. 28, 2002.

[3] "Hay Fever? Bum Knee? Buying Individual Coverage May Be Dicey," *Medicine & Health*, June 25, 2001.

[4] For background, see Marcia Clemmitt, "Rising Health Costs," *CQ Researcher*, April 7, 2006, pp. 289-312.

[5] Paul Menzel and Donald W. Light, "A Conservative Case for Universal Access to Health Care," *The Hastings Center Report*, July 1, 2006, p. 36.

children. Other proposals are being discussed in state legislatures this spring.[31]

States' uninsured populations vary widely around the country, so they are the natural venue for expanding coverage, say some analysts. "All states face different challenges in reducing the number of uninsured residents," so "imposing a one-size-fits-all program" at the federal level "will not work," said Arthur Garson, dean of the University of Virginia School of Medicine.[32]

With no national consensus emerging on how to cover the uninsured, encouraging state action is the only way forward, said Stuart M. Butler, vice president for domestic and economic policy at the conservative Heritage Foundation. "Successful welfare reform started in the states," and coverage could be expanded by removing federal roadblocks and offering federal incentives to states "to try proposals currently bottled up in Congress."[33]

In California over the past year, the Republican governor, Democratic legislators and top executives from the state's largest private insurer all proposed universal-coverage plans, though none has yet been enacted. The California-based Kaiser Foundation Health Plan offered a plan that would provide "near-universal coverage" within two years to California's 5 million uninsured — who represent a whopping 10 percent of all uninsured Americans. "Despite the greater dimensions of the problem in California, we believe that a state-based solution is possible," wrote Kaiser executives.[34]

Private Insurance Coverage Dropped

The percentage of people with private health insurance dropped by 8 percentage points from 1987 to 2005 (right). At the same time, the percentage of people insured by either government or private insurance dropped 3 percentage points (left). As people lost private coverage, government picked up the slack to keep as many people insured as possible.

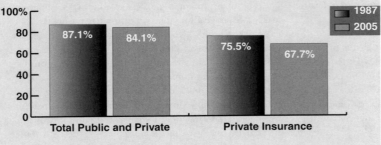

Percentage of Uninsured People by Type of Coverage

Source: U.S. Census Bureau, "Historical Health Insurance Tables"

While federal laws restrict states' ability to expand Medicaid and set rules for employer-sponsored coverage, that hasn't stopped some states from expanding coverage, says Stanley Dorn, a senior research associate at the liberal Urban Institute. In the early 1990s, for example, Minnesota and Washington state both "implemented coverage systems that succeeded brilliantly," he says.

Regardless of whether they succeed completely, state initiatives provide models and impetus for future national efforts, say many analysts. "The state action provides great momentum," says Dorn.

"States are hugely important," says Hacker at Yale University. "When two Republican governors" — Romney and Schwarzenegger — "break with the national party to propose universal coverage, that's a huge boost," he says.

Ginsburg, of the Center for Studying Health System Change, says the 2006 Massachusetts law has been a catalyst — "the answer to political gridlock." It is supported by both Republicans, who've traditionally been skeptical of universal coverage, and liberal Democrats who favor a single-payer system, he says.

The flurry of major state proposals shows the nation is ready for change, he says, even though "the federal government has been dysfunctional on domestic issues

for many years." He predicts "a few more states" will expand coverage soon, but many states are limited in what they can accomplish.

All states can't emulate the Massachusetts model, said James J. Mongan, chief executive of the New England-based hospital and physician network Partners HealthCare, because Massachusetts is very different from most other states.[35]

"We started with half the problem solved," said Mongan, a former congressional health aide who also worked in the Carter administration. Only about 10 percent of Massachusetts residents were uninsured, compared to uninsurance rates in other states of 25 percent and higher. And the state was already spending more than $500 million annually to compensate hospitals for treating the uninsured.[36]

"Federal action ultimately [will] be essential," said Shoshanna Sofaer, professor of health policy at Baruch College in New York City and a member of the IOM insurance panel. States don't have the steady financing or the legal flexibility to expand coverage to all of their residents. One roadblock, she said, is the federal Employee Retirement Income Security Act (ERISA), which limits states' power to control insurance.[37]

"The best thing states can do is set up role models," says Brandeis University Professor of Health Policy Stuart Altman. "You can't design true national health insurance state by state, because you'd get past a few states, then stop."

While states' efforts are important, says the National Coalition on Health Care's Simmons, "we don't think that any single state, no matter how large," can accomplish universal coverage of its residents "without major federal policy changes." Many governors agree and acted on their own only because they're frustrated with a lack of federal action, he says.

Furthermore, even if all states achieved universal coverage, the result would be a cost-increasing nightmare — the last thing the health system needs, says Simmons. "If you think we have administrative complexity now, imagine 50 individual state programs."

Should individuals be required to buy health insurance?

At the turn of the new century, few people were advocating that all Americans be required to buy health insurance, but in recent years such voices have grown louder. With interest growing in a system that subsidizes the cost of private coverage, advocates say unless everyone participates no functioning insurance market can develop. Insurance is designed to even out annual health costs for everyone by having everyone pay similar amounts into an overall pool each year, whether they are healthy in that particular year or facing an unexpected sickness or injury.

But opponents on the left say mandating insurance is unfair to lower-income families who can't afford even heavily subsidized private insurance. And conservative critics say a requirement to purchase is undue government intrusion into private life.

"You can talk until you're blue in the face about risk pools and actuarial tables and all the green-eyeshade reasons that the health insurers need everyone to participate in order to write affordable policies. I understand all that, and I basically don't care," wrote lawyer and policy blogger David Kravitz about Massachusetts' new buying requirement. "It is fundamentally wrong to force people to buy an expensive product in the private market, simply as a condition of existing in this state," he wrote on the Blue Mass Group policy blog.[38]

Monitoring who is obeying the requirement and determining subsidy sizes creates "one more aspect of citizens' lives" that government would monitor, complains Michael D. Tanner, director of health and welfare studies at the libertarian Cato Institute. A mandate would also be extremely difficult to enforce, he says.

"An individual mandate crosses an important line: accepting the principle that it is the government's responsibility to ensure that every American has health insurance," said Tanner. "In doing so, it opens the door to widespread regulation of the health-care industry and political interference in personal health-care decisions. The result will be a slow but steady spiral downward toward a government-run national health-care system."[39]

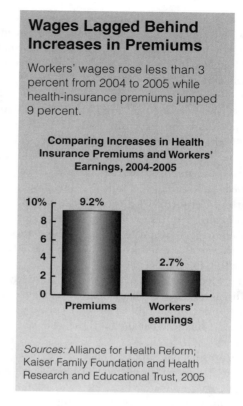

Wages Lagged Behind Increases in Premiums

Workers' wages rose less than 3 percent from 2004 to 2005 while health-insurance premiums jumped 9 percent.

Comparing Increases in Health Insurance Premiums and Workers' Earnings, 2004-2005

Sources: Alliance for Health Reform; Kaiser Family Foundation and Health Research and Educational Trust, 2005

Advocates of a mandate argue that if government can require automobile insurance to ensure that costs are paid when drivers cause accidents, then health insurance shouldn't be any different, notes Tanner. But economists say there are some key differences between the two kinds of coverage. For example, few people will drive more recklessly just because they have auto insurance. But the prevalence of generous health insurance has been shown to encourage patients to seek — and doctors to prescribe — more and sometimes unnecessary or unduly expensive treatments.[40]

Nevertheless, policymakers from both parties increasingly consider mandating health insurance "an essential accommodation to limited public resources," explains Ginsburg of the Center for Studying Health System Change. In 2004, for instance, then-Senate Majority Leader Frist said "higher-income Americans have a societal and a personal responsibility to cover in some way themselves and their children."[41] If those who can afford coverage don't enroll, the government should enroll them automatically in a high-deductible insurance plan that covers catastrophic expenses and obtain the payment for the premiums at tax time, Frist said. And the mandate should apply to the "very, very rich" initially, then expand over time, he said.[42]

Requiring everyone to buy coverage ensures that those with lower medical needs will pay premiums alongside those with expensive illnesses, analysts point out. "If the government says an insurance company must take whoever comes their way, they couldn't predict risk and might go broke" if only sicker people enrolled, says Marian R. Mulkey, senior program officer at the California Healthcare Foundation, which funds health-care research. A mandate like the one Schwarzenegger proposes "relieves this concern of insurers, who are businesses and must be on solid financial footing to offer benefits."

The Medicare drug benefit works similarly, says health-care consultant Laszewski. Medicare doesn't require seniors to enroll but strongly encourages it by imposing a financial penalty on those who wait to sign up, he says.

The drug benefit "created a huge pool of people from age 65 to 95," he explains. "And you allowed people in at the same rates no matter what their age or pre-existing condition, so long as they signed up as soon as they became eligible."

The proof of that approach is in the pudding, he says. Private insurers have "flooded the market with plans," and people are not faced with steeply escalating premiums as they age or their health worsens, he says.

BACKGROUND

America vs. Europe

From the beginning, America differed sharply from other industrialized nations in its approach to health insurance. While Europe turned to social insurance, in which all residents pay into a common fund that provides population-wide benefits, American physicians resisted, fearing such an approach would encourage government influence over the practice of medicine.[43]

The development of the American workplace-based insurance system echoed "themes that distinguish the more general history of the United States," wrote Rosemary A. Stevens, University of Pennsylvania professor emeritus of the history of science. Social insurance was trumped by "the commitment to private solutions to public needs" and "the belief in local initiatives wherever possible."[44]

As the 19th century ended, Europeans leaned more toward "social democracy" — the belief the free market cannot supply certain human necessities, such as a minimum income to purchase food, clothing, housing and access to health services. Governments were seen as necessary to guarantee those needs, explains Thomas Bodenheimer, adjunct professor of community medicine at the University of California, San Francisco.[45]

In the late 1800s a conservative German government enacted the first social-insurance programs in hopes of heading off a wholesale movement toward more radical socialism with government ownership of industries. Supported by mandatory contributions from all citizens, the first programs paid out when people lost their livelihood through unemployment, disability or retirement. In 1883, Germany added health care to its social-insurance offerings, though with a twist. Unlike other programs, health insurance was run by privately operated "sickness funds." Social insurance, including for health care, soon became the European norm.

In the United States, lawmakers debated social insurance for decades and ultimately used it for a few programs. But health insurance remained a voluntary purchase, managed by private companies.

American liberals argued that social insurance for health would unite the entire population into a single risk pool and serve everyone's long-term interest, according to Bodenheimer. Though younger people would pay for older people, and healthy people for sick people, this would even out in the end, progressives argued, since the young will one day be old and the healthy injured or sick.[46]

But conservatives answered that it's unfair to force young people to subsidize health care for older, sicker neighbors and that people will spend more prudently on medical care if they buy their own.

Sickness Insurance

The private market for what we call health insurance today — policies that pay medical bills — grew slowly, mainly because health costs were low, even in the early 20th century. Before 1920, there were virtually no

CHRONOLOGY

1880s-1920s *Most European countries adopt compulsory health insurance.*

1895 German physicist William Roentgen discovers the X-ray, ushering in the age of modern medicine and rising health-care costs.

1920 Public commissions in California, New Jersey, Ohio and New York recommend universal state health coverage.

1926 The private Committee on the Cost of Medical Care (CCMC) endorses developing private health insurance; American Medical Association (AMA) opposes the idea.

1929 The first hospital prepayment insurance plan is launched for school teachers in Dallas, Texas.

1930s-1940s *Private hospital prepayment insurance spreads around the country, as hospitals worry they'll go under when poor patients don't pay. Congress and legislatures in at least eight states debate but don't enact compulsory health insurance.*

1935 Attempts to include health coverage in the new Social Security Act are unsuccessful.

1943 The first measure calling for compulsory national health insurance is introduced in Congress. . . . National War Labor Board declares employer contributions to insurance income-tax free, enabling companies to offer health insurance to attract workers.

1950s-1960s *Health spending and consumption rise rapidly, and workplace-based health insurance spreads. Medicare and Medicaid are enacted for the poor and elderly.*

1970s-1980s *Worries grow about health care becoming unaffordable. Presidents Nixon and Carter propose universal health coverage and health-care price controls. Cost controls reduce federal spending on Medicare, but doctors and hospitals shift their costs to employers.*

1990s *Federal government expands Medicaid and enacts a new children's health program, but employers begin dropping health benefits. Washington lawmakers shy away from large-scale coverage expansion after President Clinton's ambitious attempt to enact universal coverage fails.*

Sept. 22, 1993 Clinton unveils sweeping plan to reform U.S. health-care system.

Sept. 26, 1994 After a year of fierce debate, Senate leaders declare Clinton's bill dead.

1996 Congress enacts Health Insurance Portability and Accountability Act to make employer-provided coverage transferable between jobs and more accessible to the self-employed.

1997 Congress enacts State Children's Health Insurance Program (SCHIP) to help states cover children from low-income families.

2000s *As health costs and the ranks of the uninsured rise, Congress mulls new tax deductions and credits to help consumers buy coverage; interest grows in compulsory insurance.*

2002 Congress enacts Health Care Tax Credit, available to those who lose their jobs due to foreign competition.

2006 Massachusetts enacts universal-coverage plan requiring all residents to buy health insurance. . . . Vermont enacts voluntary coverage plan with subsidized insurance and medical cost trimming. . . . Maryland plan to force large employers to supply coverage or pay into a state insurance pool is struck down in federal court.

2007 President Bush proposes replacing the tax break received by those with employer-based coverage with a tax deduction available to everyone. . . . Gov. Arnold Schwarzenegger, R-Calif., proposes universal, state-subsidized health insurance. . . . Advocates press Congress to expand SCHIP to more children and parents.

Looking Into the Future of Health Coverage

New proposals offer new approaches

With the number of uninsured Americans creeping inexorably upward, universal coverage is likely to become a hot political issue in the 2008 presidential campaign. While former Sen. John Edwards, D-N.C., is the only candidate to have offered a specific plan so far, the state and federal plans being considered contain some new wrinkles that might help policymakers reach a compromise on how to expand coverage.

Massachusetts — A 2006 state law requires all residents to buy insurance, beginning this year, or pay a penalty. Massachusetts will subsidize premiums for those earning under 300 percent of poverty level (about $60,000 for a family of four) and waive the coverage requirement if no "affordable" policies are available.

Coverage will be sold through a state-operated market, the Massachusetts Health Insurance Connector, and the state is negotiating with insurers to get affordable premiums for comprehensive policies, something that's proven to be more of an uphill struggle than lawmakers imagined.

"Massachusetts decided consciously not to grapple with rising health-care costs and decided to do it later," says Paul Ginsburg, president of the Center for Studying Health System Change, a nonpartisan research group. "Now they're having a problem with the bids coming in higher than expected."

That decision may have doomed the plan, says Robert Laszewski, a consultant and former insurance executive. Annual health-insurance premiums for the average Massachusetts family had already reached $15,000 a few years ago — higher than the current national family average of $12,000 — in part because of the state's high-cost

academic medical centers and plethora of physicians, he points out. "Yet the Massachusetts legislature came up with $200 a month" — $2,400 a year — "as a reasonable premium for their plan," he laments. "The chances that the law will ever be implemented are slim."

California — In January, Gov. Arnold Schwarzenegger unveiled a universal-coverage plan that also would require all residents to buy a minimal level of coverage. Public programs would be expanded to cover the lowest-income Californians, and subsidies would help others buy private insurance.

Insurers would offer policies to all comers, at state-approved rates. Employers with 10 or more workers would pay at least 4 percent of payroll for health insurance or pay that amount into a state pool. To trim costs, insurers would be required to spend at least 85 percent of every premium dollar on patient care.

To entice more hospitals and doctors to participate in California's subsidized Medi-Cal program, the state would increase payments to participating providers. This would also eliminate what Schwarzenegger calls the "hidden tax" — low public-program payments and uncompensated care for uninsured people that providers now pass along as higher prices to paying patients. The Medi-Cal pay boost would be funded by a tax on non-participant doctors and hospitals.

Some employers are skeptical of the plan, which must be approved by the California legislature. The plan would help companies that already provide health benefits because it would force their competitors to ante up for health care also, said Scott Hauge, president of the advocacy group Small Business California. But that could be perceived as

---

antibiotics and few effective drugs, and X-rays had been discovered only in 1895. Most of the financial burden from illness was due to lost wages, so insurers sold income-protection "sickness" or "accident" insurance.

The first such policy was sold in 1850 by the Massachusetts-based Franklin Health Assurance Company. For a 15-cent premium, the policy paid $200 if its holder was injured in a railway or steamboat accident.[47]

Some employers offered sickness insurance as a worker benefit. In 1910, the catalog store Montgomery

Ward and Co. established a group insurance plan to pay half of an ill or injured employee's salary.[48] In 1918, the Dallas, Texas, school system established sickness insurance to protect teachers against impoverishment during the great influenza epidemic.[49]

The mining, railway, and lumber industries led the way in establishing insurance plans more similar to modern HMOs (health maintenance organizations), paying medical costs. Their workers faced serious health risks and labored in remote locations where traditional care wasn't

unfair by some companies with young workers, whose "invincibility-of-youth syndrome" means they'd prefer cash to health benefits they believe they don't need, he said.[1]

Insurers are expected to balk at being forced to spend 85 percent of premiums on patient care, says Laszewski. "Wellpoint, California's biggest [for-profit] insurer, puts 80 cents on the dollar toward care, holding on to a full 20 cents for profits and administration," he says. Nevertheless, "everybody knows that it can be done for less. In Medicare, 95 cents on every dollar goes to patient care."

President George W. Bush — The president wants to replace the current unlimited government subsidy for employer-sponsored health coverage with a flat standard deduction available to everyone who buys at least catastrophic health coverage on their own or through an employer. Federal funds would be available for states to improve their markets for individual health policies, where people would shop for non-workplace coverage.

Economists praise Bush for proposing to replace the government's current subsidy for health insurance — the exclusion from taxable wages of employer-sponsored coverage — with more widely available assistance. But some critics on both the left and right agree the proposal doesn't target the people most in need of subsidies and doesn't help create enough affordable coverage for them to buy.

"Replacing the current tax treatment with a new standard deduction is a big step in the right direction," said Heritage Foundation Vice President Stuart M. Butler and Senior Policy Analyst Nina Owcharenko.

Former Sen. John Edwards, D-N.C., is the first presidential candidate to propose a detailed universal health-care plan.

Nevertheless, "an even better step would be to replace it with a tax credit," which would help lower-income families who are least likely to have insurance, they said. Unlike tax credits, which benefit everyone equally, Bush's proposed deduction has a much higher dollar value for higher-income people.[2]

The Edwards plan — The presidential candidate also proposes an individual mandate, but the requirement would only kick in once new, affordable coverage options are available and employers are either contributing to a general pool or helping their own workers buy coverage through new, regional nonprofit purchasing pools known as "health markets." The federal government would help states or groups of states set up such health markets, which would offer a choice of competing health policies. Unlike most current proposals, the health markets would offer all buyers — in addition to private coverage — a public-insurance plan modeled on Medicare.

"Let's have real competition between public and private systems," says Yale University Professor of Political Science Jacob Hacker, who consulted with Edwards on the proposal. "If you put a level playing field between the public and private sectors" — as in the health markets — the public programs "might turn out to be cheaper. If that happens over time, people would vote with their feet," he says.

[1] Renuka Rayasam, "Schwarzenegger Health Plan Raises Doubts." *U.S. News & World Report*, Jan. 10, 2007, www.usnews.com/usnews/biztech/smallbizscene/070110/schwarzenegger_health_plan_rai.htm.

[2] Stuart M. Butler and Nina Owcharenko, "Making Health Care Affordable: Bush's Bold Health Tax Reform Plan," *WebMemo No. 1316*, Heritage Foundation, Jan. 22, 2007.

---

available. So companies established clinics that prepaid doctors fixed monthly fees to provide care.

Nevertheless, between 1910 and 1920, near the end of the so-called Progressive Era in American politics, "government-sponsored health insurance seemed a practical possibility in the United States," according to the University of Pennsylvania's Stevens. In 1920, expert panels in four large states — California, New Jersey, Ohio and New York — recommended universal state-sponsored health insurance.[50]

However, doctors, hospitals and insurance companies feared if universal coverage was adopted they would lose control and cash. The chairman of Ohio's commission complained about "the confusion into which the public mind had been thrown by the misleading, malicious and false statements emanating from an interested and active commercial insurance opposition."[51]

Soon, popular support for government-sponsored insurance dropped to a low level again, as financial worry receded in the 1920s economic boom. In the overall

Stakeholder Groups May Balk at Changes

They fear paying more, losing coverage

As costs and the ranks of the uninsured soar, there's plenty not to like about the current health-care system. Nevertheless, many longtime stakeholders fear change. As has often happened in the past, insurance companies, health providers, employers and those with expansive work-based health coverage all may balk at the changes universal coverage may bring.

"You often see interest groups wearing a cloak of ideology," saying they oppose a reform plan on economic or philosophical grounds when they're really protecting their money, says Stanley Dorn, a senior research associate at the liberal Urban Institute.

For example, he says, during the bitter debate over President Bill Clinton's universal-coverage plan in the early 1990s, "you had companies that didn't provide insurance to their workers and knew they could lose money" if the proposal succeeded. "But they didn't talk about that. They talked about how evil it would be for the government to take over the health system."

Various employer groups are likely to weigh in on both sides of the debate. Those who offer health coverage as a benefit today are more likely to embrace the change, although

they may still be hesitant to endorse all universal-coverage proposals, says Paul Ginsburg, president of the Center for Studying Health System Change, a nonpartisan research group. "They'd like to get out of the business of coverage long term, but the issue has always been whether they'd end up paying more in taxes" for a new universal coverage system than they spend now to provide benefits, he says.

Large, unionized employers like U.S. automakers initially supported the Clinton plan in 1993, said Walter Mahan, former vice president for public policy of DaimlerChrysler Corp.[1] Employers who didn't offer health benefits strongly opposed the Clinton plan, which, like many universal-coverage plans today, asked all businesses to chip in, including those that didn't offer health benefits before.

Caught somewhat off guard by ferocious opposition from businesses that didn't offer coverage — like restaurants and soda manufacturers — architects of the Clinton plan reduced the payments required from companies that had not previously offered coverage and hiked the amount asked from employers who offered coverage. Complaining of unfairness, unionized employers then pulled their support, said Mahan. The bad news for the new crop of reformers:

prosperity of that decade, the medical system flourished, and hospitals built new wings in the mood of general optimism.

By 1929, however, more than a third of hospital beds were empty, and many hospitals struggled to pay off the loans that had funded expansion. Baylor University Hospital in Dallas, for example, had $1.5 million in overdue loan payments for construction and was behind in other bills. "Baylor was just 30 days ahead of the sheriff," said one observer.[52] Baylor's crisis led to health insurance as we know it today.

In search of cash, Baylor made common cause with local employers. In late 1929, the Dallas school system set up a hospital-service prepayment plan that operated alongside its sick-benefit fund. For a monthly premium of 50 cents, teachers would get free hospitalization for 21 days and a one-third discount on additional days. Benefits became effective on Dec. 20, 1929, less than two months after the stock market crash.[53]

A few days later, elementary-school teacher Alma Dickson slipped on an icy sidewalk and broke her ankle.[54] Hospitalized with a cast, Dickson became the first patient in the first prepaid hospitalization plan, the forerunner of today's Blue Cross system.[55]

By 1935, 19 such plans had been created in 13 states, as hospitals struggled to stay afloat during the Great Depression.[56]

But many influential physicians argued that "prepayment" threatened professional independence. Recommendations that the nation adopt insurance to protect people against the rising cost of care amounted to "socialism and communism — inciting to revolution," wrote Morris Fishbein, editor of the *Journal of the American Medical Association.*[57]

Workplace Plans

During the 1930s, as businesses folded and millions sank into poverty, the United States made its largest-ever foray in social insurance.

More and more companies have been dropping coverage since then, so the constituency of businesses not offering coverage "is stronger now," he said.[2]

Insurers may have the biggest stake in the current system. Most analysts say proposals that would abolish private insurance in favor of a government-run universal plan modeled on Medicare are politically impossible today. However, most reform plans would force insurers to cover potentially sicker beneficiaries than most do today and would tighten rules for selling and marketing insurance policies.

Many insurers mistrust changes because the current employer-based system works well for them by weeding out the sickest populations, said former Rep. William Thomas, R-Calif., who chaired the House Ways and Means Committee. Employed people "have to get up every morning, go to work and carry out difficult and complex tasks." They're essentially prescreened to be, on average, healthier than the general population and thus easier to cover and still earn profits, he said.[3]

Insurers also distrust government-run "insurance exchanges" in many universal-coverage plans that would establish standard benefit packages, ensure affordability and replace insurers' marketing with government-scrutinized plan descriptions. Insurers have "traditionally hated" government limits on their marketing, says Dorn.

Finally, virtually all economists say any health system reform must include cost cutting, including reining in salaries and profits of doctors, hospitals and drug manufacturers. Some proposals ask providers to put money in up front to support coverage proposals. Providers always push back against such steps.

For example, when Democratic Maine Gov. John Baldacci unveiled a universal-coverage program in 2003, he included a tax on insurance premiums along with both voluntary and mandatory price caps on many health services, without which the governor said the program could not survive. Maine's hospitals said they couldn't survive having prices capped.

"That cannot happen . . . without irreparably harming Maine's hospitals," said Warren Kessler, a consultant and former head of the Maine General Medical Center in Augusta.[4]

Finally, those who currently have good coverage are sensitive to any proposal that might make their own insurance worse or cost more. Interest groups like insurers and doctors who oppose any new plan "just have to play on the public's fear of losing what they now have," says Dorn.

---

[1] "Universal Coverage: It Can't Happen Here . . . Or Can It?" *Medicine & Health Perspectives*, March 31, 2003.

[2] Quoted in *ibid.*

[3] Quoted in "Thomas Takes Aim Again at Tax-Favored Employer Coverage," *Medicine & Health*, Feb. 16, 2004.

[4] Quoted in "Baldacci Says Everyone Must Give a Little to Fund Care," *Medicine & Health*, May 12, 2003.

Developed by Democratic President Franklin D. Roosevelt and enacted in 1935, Social Security is a mandatory, universal system that provides income support for retirees, severely disabled people, widows and under-age bereaved children. During the debate over passage, activists argued for including health insurance, but the administration declined, in part because it feared the contentious health issue might doom the whole plan.[58]

Later, members of Congress made unsuccessful attempts to extend social-insurance to health in 1943, 1945, 1947, 1949 and 1957. Nevertheless, by 1966, 81 percent of Americans had hospitalization insurance — mostly offered through their workplaces and often as a result of labor union demands — compared to only 9 percent in 1940.[59]

Unlike today, from 1940 to 1966 large unionized companies dominated the economy. Offered as a worker benefit, employer-sponsored health plans successfully pooled the risk and contributions of many employees in order to keep individuals' costs low and uniform, even in years when they had accidents or illness. And, since the sickest people are unlikely to be employed, relying on workplace-based plans allowed private insurers to more easily predict and control costs.

As the primary source of Americans' health insurance, the still union-dominated U.S. auto industry has evolved over the years into "a social-insurance system that sells cars to finance itself," said Princeton University economics Professor Uwe Reinhardt.[60]

But even in the early days, employer-provided insurance had limits. Many retired people, very low-income families and the disabled never had workplace-based insurance and were too poor to buy individual policies, for which they would be charged premiums based on health status.

After several years of debate, Congress in 1965 enacted a new compulsory, universal insurance plan — the Medicare program — to provide health coverage for elderly and some disabled people and Medicaid to provide health care for the poorest mothers with children, elderly and the disabled.

Coverage Declines

With Medicare and Medicaid in place, most Americans had access to health care.

Nevertheless, health spending was rising sharply, and Presidents Richard M. Nixon and Jimmy Carter both proposed reforms to keep care affordable, including universal coverage. Neither plan gained traction, however.

Gradually, the higher costs and the changed nature of American business began to erode the work-based insurance system.

"Forty years ago, the largest private employer was AT&T, a regulated monopoly with guaranteed profits," wrote Stanford University Professor Emeritus Victor Fuchs and Ezekiel Emanuel, chairman of clinical bioethics at the National Institutes of Health. "If health-insurance premiums rose, they could easily be passed on to telephone subscribers."[61]

That changed, however, as union membership began declining in the 1980s, and manufacturing jobs began migrating overseas and U.S. companies had to compete with foreign competitors that don't offer health benefits. More and more Americans ended up working in the largely non-unionized service industry, which offered few benefits.

"Today, the largest private employer is Wal-Mart, which despite its size faces intense competition daily from a host of other retail outlets," Fuchs and Emanuel wrote. "When they offer health insurance, it must come out of their workers' wages; for minimum-wage employees, this is not possible."[62]

Over the past two decades, employer-sponsored coverage has gradually waned, along with the number of insured Americans. Government programs have grown and picked up some of the slack, however.

In 1987, fully 87.1 percent of Americans were insured, with 75.7 percent insured through private, mostly employer-sponsored, coverage. By 1999, the percentage of insured Americans had dropped to 85.5.percent, 71.8 percent through private coverage. In 2005, the overall percentage had dropped to 84.1 percent — 67.7 percent with private insurance.[63] (*See graph, p. 198.*)

In the face of declining coverage, proposals to expand coverage have been advanced repeatedly by the White House, members of Congress, state and local governments and others. Only some small-scale efforts have gone anywhere, however.

In 1994, Tennessee used federal Medicaid dollars and state funding to create TennCare. State officials hoped money-saving HMOs could provide coverage to many lower-income people and sicker Tennesseans, who were ineligible for Medicaid and couldn't afford insurance on their own.

For a few years, the program saved money and enrolled 500,000 residents who would otherwise have been uninsured. But the federal government had agreed to contribute funding for only 1.5 million people, and when enrollment exceeded that cap, TennCare refused to accept new applicants and struggled financially. For the past several years, TennCare has fought to survive, plagued by charges of poor care at its HMOs and disputes with the federal government over funds.

Clinton Plan

The highest-profile recent effort to enact a universal health care plan was President Bill Clinton's ambitious proposal to restructure the nation's health care system, unveiled on Sept. 22, 1993. His Health Security Act was proposed at a time when the uninsured ranks had swelled to 40 million, and polls showed that up to two-thirds of Americans favored tax-financed national health insurance.[64] Yet, within a year Senate Democrats had pronounced the plan dead, the victim of bruising attacks by business, insurers and medical providers.[65]

Five days after his inauguration, Clinton announced that first lady Hillary Rodham Clinton would chair a health-care task force made up of Cabinet members and White House staffers. It held hearings for a year and produced a plan to attain universal coverage mainly through expanded private coverage. It aimed to offer people a choice of affordable coverage while maintaining the existing private insurance industry and holding down health-cost growth.

To do that, the Clinton panel proposed creating regional government-managed insurance markets to negotiate health-care and premium prices and insurance-benefit packages and to oversee insurance marketing. It also called for annual caps on health-coverage cost increases, and a requirement that all employers contribute to the cost of coverage.

But opposition soon grew from businesses that believed they had more to lose than to gain from change. Employers who didn't offer coverage balked at proposed fees to help finance the plan. Insurers objected to regulations aimed at keeping them from skimming off healthy customers. After 10 months of strenuous campaigns by opponents, public approval had dropped to a lukewarm 40 percent.[66] (*See sidebar, p. 204.*)

Former first lady Clinton — now the Democratic senator from New York who is running for president — has assured voters she still believes in universal health coverage, but she has not yet announced a specific plan. "I think she's learned her lesson" and likely will propose something "not quite as big and ambitious" this time, says Brandeis' Altman.

For the next decade the dramatic failure of the Clinton plan frightened lawmakers away from the issue, while conservative lawmakers said the booming 1990s economy would enable the United States to "grow its way" out of uninsurance by creating more and better jobs with coverage benefits.

But that did not turn out to be the case. From 1997 to 2001, the economy boomed and jobs were created, but rates of employer-sponsored health insurance did not rise. The late '90s experience "tells us that relying on economic growth alone to reduce the number of uninsured won't work," said Ginsburg at the Center for Studying Health System Change.[67]

Since Clinton's efforts, Congress enacted two coverage expansions. The State Children's Health Insurance Program (SCHIP) was enacted in 1997. The Clinton administration and a bipartisan group of lawmakers led by Sens. Edward M. Kennedy, D-Mass., and Orrin G. Hatch, R-Utah, gave states federal matching funds to expand coverage for children in low-income families. Today, SCHIP operates in all states, making nearly all otherwise uninsured children with family incomes up to twice the poverty level eligible for public coverage.

With Republicans dominating the White House and Congress, most recent debate over coverage has focused on tax incentives to help Americans buy insurance. Criticized by lawmakers of both parties for offering too-small tax breaks in its early proposals, the Bush administration has gradually expanded its plan each year but has seen none enacted.

The only federal health-coverage expansion enacted in this decade was a tax credit to assist workers unemployed due to competition from international trade, enacted in 2002 after a long contentious delay. But the credit has reached only 10 to 20 percent of those eligible for it, says the Urban Institute's Dorn, which he calls a "tragic" outcome for states like North Carolina, where it was intended to help people facing "the largest layoff in the state's history — the closing of the textile mills."

The program failed to catch on because its premiums are too high, he says. "It's not realistic to ask people to pay 35 percent of premiums when they're not working" when working people pay only 15 to 25 percent of theirs, he says. In addition, the tax credit in most states could only be used for individual policies, whose premiums generally are based on age and health status. "Even with a 65 percent subsidy, people were facing an unaffordable $1,000 a month premium."

Over the past decade, some congressional Republicans also have proposed allowing business and professional groups to offer association health plans (AHPs), which would enable small businesses and the self-employed to band together to buy health insurance free from the state regulations that apply to individual and small-group insurance plans, which AHP advocates say unduly drive up coverage costs for small business.

In the 1970s Congress waived state insurance regulation for large employers to encourage them to provide coverage for workers. But today both Democratic and some Republican lawmakers staunchly oppose allowing AHPs the same freedom. AHP opponents argue that it is too easy for such loosely formed groups to skim off workers most likely to be healthy and low-cost, which would raise premium costs even higher for those left behind.

Meanwhile, outside of legislative chambers advocates increasingly have been calling for universal coverage. In 2004, a three-year-long Institute of Medicine study declared that eroding coverage poses such a threat that the federal government must launch a "firm and explicit" plan to achieve universal coverage by 2010.[68]

Many analysts agree that universal coverage has been stalled not because of a lack of knowledge of how to accomplish it but because lawmakers lack the will to demand sacrifices. (*See sidebar, p. 196.*)

There are "at least four ways" to get universal coverage, says Simmons of the National Coalition on Health Care. "This problem is solvable. It does not require atomic science."

CURRENT SITUATION

Interest Grows

A few states are moving forward with universal-coverage plans, but little action is expected on Capitol Hill this year. Meanwhile, all Democratic contenders for the 2008 presidential nomination have advocated universal coverage, although only former Sen. Edwards has offered a specific plan so far. (*See sidebar, p. 202.*)

While few expect federal action until after the next president takes office in 2009, many Washington hands think the tide finally may be turning.

"One big difference between now and several years ago is that there is a loss of faith in employer-provided coverage as capable of covering everyone, including from unions and key business groups" who have been its strongest supporters, says Yale's Hacker.

Coalitions of interest groups have come together in 2007 to announce support for universal coverage. In January, the Health Coverage Coalition for the Uninsured (HCCU) advocated a phased-in approach to universal coverage, beginning with an expansion of SCHIP and creation of tax credits for families with incomes up to about $60,000, and then creating similar programs for childless adults.[69]

The coalition includes groups that have traditionally sparred over health care, including Families USA, the retiree organization AARP, the American Medical Association, the American Hospital Association and the health-insurance lobby America's Health Insurance Plans. "Organizations that have never spoken to one another in a friendly manner are now talking about this, and that has transformed the debate," says Pollack of Families USA.

Others aren't convinced. The HCCU's "rhetoric was wonderful," says Altman of Brandeis. "But the result shows how little they actually agree on."

But Pollack says the coalition's proposal is a "sequential" plan that will follow expansion of the children's program — expected to be enacted this year — with a move to universal coverage after that.

In February the Service Employees International Union joined with Wal-Mart, an employer whose limited health benefits have been sharply criticized by the union, to form the Better Health Care Together group, calling for "quality, affordable" universal health care by 2012. The group plans a national summit this spring to rally support but has not announced a proposal, saying only that it supports joint public and private-sector efforts.[70]

But Dana Rezaie, a Wal-Mart shelf stocker in Fridley, Minn., says, "anybody can say they support something. They need to show they really do." After six years at the store, the widowed mother of three says she can't afford Wal-Mart's health plan.[71]

Congress is reviewing President Bush's fiscal 2008 budget, which proposes a new version of his tax-based coverage-expansion proposals. Bush would ditch the current tax break Americans receive for employer-sponsored coverage and replace it with a more general tax break that would apply both to employer coverage and to insurance purchased individually.

The plan gets points from Greenstein at the Center on Budget and Policy Priorities for tackling the unfairness of the current tax treatment of insurance, which penalizes those who purchase insurance on their own. But the plan "has an Achilles' heel," Greenstein says, since it doesn't encourage pooling sicker and healthier people to spread costs and skews its tax benefits toward higher-income people.

Nevertheless, Pollack does not expect "a serious productive debate on universal coverage" in 2007, with a presidential campaign heating up. However, he does expect Congress to reauthorize — and possibly expand — the 10-year-old SCHIP program "before the end of the calendar year."

The Bush budget recommends funding SCHIP leanly, by not offering federal assistance, for example, to the 14 states anticipating shortfalls in their 2007 SCHIP budgets. But both political parties strongly support SCHIP and are likely to ride to the rescue. Sen. Gordon Smith, R-Ore., has called for doubling the federal cigarette tax to pay for the aid.[72]

State Steps

As health insurance gains momentum as a public issue, many states are flirting with expanding coverage, and three are struggling to get universal coverage off the ground.

Massachusetts was first out of the gate, enacting a plan in 2006 that will require residents to buy health insurance, often with government assistance. A state-operated clearinghouse — the Health Insurance Connector — will help consumers comparison-shop for affordable coverage. So

Should Congress enact President Bush's tax proposal for expanding health coverage?

YES
Stuart M. Butler, Vice President
Nina Owcharenko, Senior Policy Analyst
The Heritage Foundation

From the foundation's Web site, January 2007

President Bush's proposal to reform the tax treatment of health care takes a bold step toward fixing America's health system by widening the availability of affordable and "portable" health plans and by defusing some of the pressure that currently leads to higher health costs.

Although some Americans would have more of their compensation subject to taxes, this proposal is no more a tax increase than limiting or ending tax deductions to move toward a flatter tax system. It would remove distortions and inequities and make tax relief for health insurance more widely available.

While the proposal can be improved in ways that would further reduce uninsurance, it is a big step toward sound tax and health policy. It would treat all Americans equally by ending the tax discrimination against families who buy their own health insurance, either because they do not have insurance offered by employers or because they prefer other coverage.

Ending that discrimination would have the added advantage of stimulating wider choice and greater competition in health coverage, which will help moderate the growth in costs. It would also make it easier for families to keep their chosen plan from job to job, reducing the loss of coverage that often accompanies job changes.

The president's proposal could be improved. While replacing the tax treatment with a new standard deduction is a big step in the right direction, an even better step would be to replace it with a tax credit more like the current child tax credit — at least for those buying health coverage outside their place of work. A tax credit would especially help lower-income families. With a deduction, many families would still be unable to afford basic coverage, but a credit set at a flat dollar amount or a high percentage of premium costs would make coverage more affordable.

A tax credit could be grafted onto the president's current proposal and would strengthen it considerably.

By taking this step, Congress can help make the tax treatment of health care more equitable and efficient, help more Americans choose the coverage they want and retain it from job to job and begin to reduce the tax-break-induced pressure that is a factor in rising health costs.

NO
Karen Davis
President, The Commonwealth Fund

From the fund's Web site, January 2007

While it is encouraging that President Bush made health care a theme of the State of the Union address, his proposal to offer tax deductions to those who buy health insurance would do little to cover the nation's 45 million uninsured.

Under the president's proposal, Americans with employer-provided health insurance would have the employer contribution counted as taxable income. But anyone with health coverage — whether provided by an employer or purchased individually — would have the first $7,500 of income excluded from income and payroll taxes or, in the case of families, the first $15,000 of income.

Those purchasing coverage in the individual market would get a new tax break, as would those whose employer contribution currently is less than the new standard deduction for health insurance.

The proposal would increase taxes on workers whose employers contribute more to health insurance than the premium "cap" allows, such as those that serve a large number of older workers. The administration estimates this change would translate into a tax increase for about 20 percent of employees. However, this could rise to more than half of employees by 2013, if increases in health-insurance premiums continue to outpace general inflation. In addition, the president proposes diverting federal funds from public hospitals to state programs for the uninsured.

Although the plan would offer subsidies to people looking to buy insurance on the private market, it would fail to assist most of the uninsured. Insurance premiums would still be unaffordable for Americans with modest or low incomes. And the tax increase for employees would likely lead to the erosion of employer-sponsored health insurance over time.

The proposal wouldn't do anything to make individual coverage available or affordable for those with modest incomes or health problems. The Commonwealth Fund found that one-fifth of people who had sought coverage in the individual health-insurance market in the last three years were denied coverage because of health problems or were charged a higher premium. The proposal, unlike plans in California and Massachusetts, does not require insurers to cover everyone.

Nor would the proposal likely help the currently uninsured. More than 55 percent of the uninsured have such low incomes that they pay no taxes, while another 40 percent are in the 10-to-15-percent tax bracket and would not benefit substantially from the tax deductions.

far, however, the state is struggling to define benefit packages that insurers can sell at "affordable" prices.

Meanwhile, the California legislature is considering Gov. Schwarzenegger's proposal to require individuals to buy coverage. The plan would be funded with contributions from multiple sources, including government, individuals, employers, insurers and health-care providers.

Vermont's new Catamount Health program will focus first on promoting information technology and other reforms to shave administrative costs and an evidence-based standard of care "community by community," says Emory University's Thorpe, who consulted on the program. Then the state will turn to expanding coverage.

In Texas, where more than 25 percent of the population is uninsured, Republican Gov. Rick Perry is looking for revenue sources to subsidize more coverage. In February he proposed selling off the state lottery and putting part of the proceeds in an endowment fund to expand insurance coverage.[73]

Last year, Rhode Island began requiring insurers to develop "wellness benefit" policies to help individuals and small businesses afford at least basic coverage.[74]

OUTLOOK

Health Politics

Consensus has been building around proposals that link public subsidies to private coverage. But it remains an open question whether Congress will finally enact a universal-coverage plan, which undoubtedly would shift some resources and benefits away from currently insured people and health providers.

Some interest groups that helped bring down the Clinton plan have softened their stance, says Ginsburg at the Center for Studying Health System Change. In Massachusetts, he points out, insurers have accepted the new state-run insurance marketplace, even though in the past they would have preferred to send out their own people to market policies and avoid head-to-head consumer comparisons of plans. But insurers realize that their long-time bread and butter — employer-sponsored coverage — "has topped out," Ginsburg says, so they anticipate no growth unless they embrace government-sponsored expansions.

"There's [been] a dramatic change in national political attitudes," says Simmons of the National Coalition on Health Care. One "truly remarkable thing is that every Democratic and some Republican candidates now say we have to achieve universal coverage."

But others say the country may still not be ready to make the concessions needed.

"It's not clear to me that life has changed very much," says Altman of Brandeis. Forces that have resisted change in the past "are stronger today," and "you have very weak leadership" from the White House and Congress.

Endorsement by Democratic presidential hopefuls doesn't necessarily mean much, says Harvard's Blendon. "Democratic primary voters disproportionately care about this," he says. But different priorities will prevail in the general election. "The biggest thing on everyone's mind is casualties in Iraq."

Furthermore, Americans generally "do not want an alternative health system," he continues. "They want to fix the one they have."

Unfortunately for politicians, the simplest, catchiest sound bite on health reform involves covering the uninsured, but that doesn't "play politically," says Blendon. While people do want everyone to have access to health care, "what they want most is cheaper premiums for themselves."

NOTES

1. Jay Himmelstein, "Bleeding-Edge Benefits," *Health Affairs*, November/December 2006, p. 1656.

2. Jeffrey A. Rhoades, "The Uninsured in America, 2004: Estimates for the U.S. Civilian Non-institutionalized Population Under Age 65," *Medical Expenditure Panel Survey Statistical Brief #83*, June 2005, Agency for Healthcare Research and Quality.

3. Diane Rowland, executive vice president, Henry J. Kaiser Foundation, "Health Care: Squeezing the Middle Class With More Costs and Less Coverage," testimony before House Ways and Means Committee, Jan. 31, 2007; for background, see Keith Epstein, "Covering the Uninsured," *CQ Researcher*, June 14, 2002, pp. 521-544.

4. Christine Borger, *et al.*, "Health Spending Projections Through 2015: Changes on the Horizon," *Health Affairs* Web site, Feb. 22, 2006.

5. Rowland testimony, *op. cit.*

6. "Facts on Health Care Costs," National Coalition on Health Care, www.nchc.org.

7. Quoted in Steven Taub and David Cook, "Health Care Can Bankrupt America," CFO.com, March 6, 2007, For background, see Michael E. Chernew, Richard A. Hirth and David M. Cutler, "Increased Spending on Health Care: How Much Can the United States Afford?" *Health Affairs*, July/August 2003.

8. Todd Gilmer and Richard Kronick, "It's the Premiums, Stupid: Projections of the Uninsured Through 2013," *Health Affairs*, April 5, 2005, www.health affairs.org.

9. *Ibid.*

10. "Coverage Matters: Insurance and Health Care," statement of Arthur L. Kellerman, co-chairman, Consequences of Uninsurance Committee, Institute of Medicine, www7.nationalacademies.org/ocga/ testimony/Uninsured_and_Affordable_Health_ Care_Coverage.asp.

11. Quoted in "IOM Uninsured Report Cites Rising Costs, Attacks Myths." *Medicine & Health*, Oct. 15, 2001.

12. "The Great Divide: When Kids Get Sick, Insurance Matters," Families USA, March 1, 2007, www .familiesusa.org/assets/pdfs/the-great-divide.pdf.

13. "Expanding Coverage Is Worth It for All, IOM Panel Insists," *Medicine & Health*, June 30, 2003.

14. *Ibid.* For background, see Marcia Clemmitt, "Prison Health Care," *CQ Researcher*, Jan. 5, 2007, pp. 1-24.

15. Quoted in "States Scramble for Ways to Cover Working Uninsured," *Medicine & Health*, "Perspectives," Feb. 8, 2005.

16. Barack Obama, "Obama Statement on President's Meeting with Big Three Automakers," press release, Nov. 14, 2006, http://obama.senate.gov.

17. For background, see Marcia Clemmitt, "Rising Health Costs," *CQ Researcher*, April 7, 2006, pp. 289-312.

18. Quoted in "Frist: 100 Percent Coverage Impossible, 93 Percent Not Working So Well Either," *Medicine & Health*, Feb. 9, 2004.

19. Quoted in *ibid.*

20. Quoted in "Who Should Pay for Health Care?" PBS Newshour Extra online, Jan. 19, 2004, www.pbs.org/ newshour/extra/features/jan-june04/uninsured_1-19 .html.

21. Jack Hadley and John Holahan, "The Cost of Care for the Uninsured: What Do We Spend, Who Pays, and What Would Full Coverage Add to Medical Spending?" The Kaiser Commission on Medicaid and the Uninsured, May 10, 2004, p. 5.

22. "Maryland Universal Coverage Plan Estimated to Cost $2.5 Billion," *Healthcare News*, News-Medical. Net, Feb. 21, 2007, www.news-medical.net; also see "How Much Would It Cost to Cover the Uninsured In Minnesota? Preliminary Estimates," Minnesota Department of Health, Health Economics program, July 2006.

23. For background, see Marcia Clemmitt, "Budget Deficit," *CQ Researcher*, Dec. 9, 2005, p. 1029-1052,

24. "Universal Coverage Rx: Tax-Code Changes, Money, Insurance Pools and a Mandates," interview with Jonathan Gruber, "On My Mind: Conversations with Economists," University of Michigan Economic Research Initiative on the Uninsured, www.umich.edu.

25. *Ibid.*

26. "Health Expenditure," Organization for Economic Cooperation and Development, www.oecd.org/doc ument/16/0,2340,en_2649_37407_2085200_1_1_ 1_37407,00.html; also see Rhoades, *op. cit.*

27. See Clemmitt, "Rising Health Costs," *op. cit.*

28. "Pushed to the Edge: The Added Burdens Vulnerable Populations Face When Uninsured," interview with Harold Pollack, "On My Mind: Conversations With Economists," University of Michigan Economic Research Initiative on the Uninsured, www.umich.edu.

29. *Ibid.*

30. John Kitzhaber, "Why Start With the Health Care Crisis?" The Archimedes Movement, www.JoinAM .org.

31. "Access to Healthcare and the Uninsured," National Conference of States Legislatures, www.ncsl.org/ programs/health/h-prmary.htm.

32. Arthur Garson, "Help States Cover the Uninsured," *Roanoke Times*, May 26, 2006.

33. Stuart M. Butler, "The Voinovich-Bingaman Bill: Letting the States Take the Lead in Extending Health Insurance," *Web Memo No. 1128*, The Heritage Foundation, June 15, 2006.

34. George C. Halvorson, Francis J. Crosson and Steve Zatkin, "A Proposal to Cover the Uninsured in California," *Health Affairs*, Dec. 12, 2006, www .healthaffairs.org.

35. Quoted in Christopher Rowland, "Mass. Health Plan Seems Unlikely to Be U.S. Model," *The Boston Globe*, April 14, 2006.

36. *Ibid.*

37. Quoted in "IOM Panel Demands Universal Coverage by 2010," *Medicine & Health*, "Perspectives," Jan. 19, 2004.

38. David Kravitz, "The Individual Mandate Still Sucks," Blue Mass Group, Jan. 30, 2007, www .bluemassgroup.com.

39. Michael D. Tanner, "Individual Mandates for Health Insurance: Slippery Slope to National Health Care," *Policy Analysis No. 565*, Cato Institute, April 5, 2006, www.cato.org.

40. "Problems of Risk and Uncertainty," The Economics of Health Care, Office of Health Economics, p. 26, www.oheschools.org/ohech3pg3.html.

41. Quoted in "Frist: Limit Tax Exclusion for Employer-Based Coverage," *Medicine & Health*, July 19, 2004.

42. *Ibid.*

43. For background, see Anne-Emmanuel Birn, Theodore M. Brown, Elizabeth Fee and Walter J. Lear, "Struggles for National Health Reform in the United States," *American Journal of Public Health*, January 2003, p. 86; Laura A. Scofea, "The Development and Growth of Employer-Provided Health Insurance," *Monthly Labor Review*, March 1994, p. 3; Thomas Bodenheimer, "The Political Divide in Health Care: A Liberal Perspective," *Health Affairs*, November/December 2005, p. 1426.

44. Rosemary Stevens, foreword to Robert Cunningham III and Robert M. Cunningham, Jr., *The Blues: A History of the Blue Cross and Blue Shield System* (1997), p. vii.

45. Bodenheimer, *op. cit.*, p. 1426.

46. *Ibid.*, p. 1432.

47. Scofea, *op. cit.*, p. 3.

48. *Ibid.*

49. Cunningham and Cunningham, *op. cit.*, p. 5.

50. Stevens, *op. cit.*, p. vii.

51. Quoted in Scofea, *op. cit.*

52. Quoted in Cunningham and Cunningham, *op. cit.*, p. 4.

53. *Ibid.*, p. 6.

54. "Dallas School Teachers, 1928," Rootsweb.com; http://freepages.history.rootsweb.com/~jwheat/teachersdal28.html.

55. Cunningham and Cunningham, *op. cit.*, p. 6. For background, see also "Sickness insurance and group hospitalization," *Editorial Research Reports*, July 9, 1934, from *CQ Researcher Plus Archive*, http://library.cqpress.com.

56. Scofea, *op. cit.*

57. Quoted in Cunningham and Cunningham, *op. cit.*, p. 18.

58. For background, see "Federal Assistance to the Aged," Nov. 12, 1934, in *Editorial Research Reports*, available from *CQ Researcher Plus Archive*, http://library.cqpress.com.

59. Cunningham and Cunningham, *op. cit.*

60. Quoted in Danny Hakim, 'Health Costs Soaring, Automakers Are to Begin Labor Talks," *The New York Times*, July 14, 2003, p. C1.

61. Victor R. Fuchs and Ezekiel J. Emanuel, "Health Care Reform: Why? What? When?" *Health Affairs*, November/December 2005, p. 1400.

62. *Ibid.* For background, see Brian Hansen, "Big-Box Stores," *CQ Researcher*, Sept. 10, 2004, pp. 733-756.

63. "Historical Health Insurance Tables," U.S. Census Bureau, www.census.gov.

64. Bridget Harrison, "A Historical Survey of National Health Movements and Public Opinion in the United States," *Journal of the American Medical Association*, March 5, 2003, p. 1163.

65. For background, see "Health-Care Debate Takes Off," *1993 CQ Almanac*, pp. 335-347, and "Clinton's Health Care Plan Laid to Rest," *1994 CQ Almanac*, pp. 319-353.

66. Harrison, *op. cit.*

67. "Rising Tide of Late '90s Lifted Few Uninsured Boats," *Medicine & Health*, Aug. 6, 2002.

68. "IOM Panel Demands Universal Coverage by 2010," *op. cit.*

69. "Unprecedented Alliance of Health Care Leaders Announces Historic Agreement," Health Coverage Coalition for the Uninsured, press release, Jan. 18, 2007, www.coalitionfortheuninsured.org.

70. Dan Caterinicchia, "Rivals Want Health Care for All," *Columbus Dispatch* [Ohio], Feb. 8, 2007.

71. Quoted in *ibid.*

72. Alex Wayne, "War Supplemental To Include Money for Children's Health Insurance Program" *Congressional Quarterly Healthbeat*, Feb. 27, 2007.

73. Quoted in The Associated Press, "Texas Governor Has Funding Idea: Sell the Lottery," *The Washington Post*, Feb. 7, 2007, p. A7.

74. "Rhode Island: Making Affordable, Quality-Focused Health Coverage Available to Small Businesses," *States in Action: A Bimonthly Look at Innovations in Health Policy*, The Commonwealth Fund, January/ February 2007.

BIBLIOGRAPHY

Books

Derickson, Alan, *Health Security for All: Dreams of Universal Health Care in America*, Johns Hopkins University Press, 2005.
A professor of history at Pennsylvania State University examines the ideas and advocates behind the numerous 20th-century proposals for universal health care in the United States.

Funigello, Philip J., *Chronic Politics: Health Care Security from FDR to George W. Bush*, University Press of Kansas, 2005.
A professor emeritus of history at the College of William and Mary describes the politics behind a half-century of failed attempts at major health reform.

Gordon, Colin, *Dead on Arrival: The Politics of Health Care in Twentieth-Century America*, Princeton University Press, 2003.

A professor of history at the University of Iowa explains how numerous private interests — from physicians desiring autonomy to employers seeking to cement employer-employee relationships — have helped halt development of universal health coverage in America.

Mayes, Rick, *Universal Coverage: The Elusive Quest for National Health Insurance*, University of Michigan Press, 2005.
An assistant professor of public policy at Virginia's University of Richmond explains how politics and earlier policy choices regarding the U.S. health system shape the range of possibilities available for future reforms.

Richmond, Julius B., and Rashi Fein, *The Health Care Mess: How We Got Into It and What It Will Take to Get Out*, Harvard University Press, 2005.
Two Harvard Medical School professors recount the history of American medicine and trends in financing health care and conclude that the United States could afford universal health coverage.

Swartz, Katherine, *Reinsuring Health: Why More Middle-Class People Are Uninsured and What Government Can Do*, Russell Sage Foundation, 2006.
A professor of health policy and economics at the Harvard School of Public Health argues that more people could buy insurance and coverage would be cheaper if the federal government offered insurance companies financial protection for the highest-cost illnesses.

Articles

Appleby, Julie, "Health Coverage Reform Follows State-by-State Path," *USA Today*, April 5, 2006.
States take different approaches to expanding health coverage as worry over lack of insurance grows.

Gladwell, Malcolm, "The Moral Hazard Myth," *The New Yorker*, Aug. 29, 2005.
Some fear that large-scale expansion of health coverage would encourage patients to rack up higher amounts of useless health-care spending.

Holt, Matthew, "Policy: Why Is Fixing American Health Care So Difficult?" *The Health Care Blog*, Oct. 16, 2006; www.thehealthcareblog.com/the_health_care_blog/2006/10/abc_news_why_is.html#comment-2418315.

An independent health-care consultant — along with blog comments by analysts, businesspeople and members of the public — describes and discusses the interest-group politics that shape the universal-coverage debate.

Holt, Matthew, "Risky Business: Bush's Health Care Plan," *Spot-On Blog,* **Jan. 25, 2007; www.spot-on.com/ archives/holt/2007/01/bush_tax_deductions_and_the_ lo.html.**
An independent health-care consultant explains the concept of risk-pooling for insurance and the current tax break already enjoyed by workers with employer-sponsored coverage.

Reports and Studies

"Covering America: Real Remedies for the Uninsured," Vols. 1 and 2, *Economic and Social Research Institute,* **June 2001 and November 2002.**
Economists assembled by a non-partisan think tank analyze multiple proposals for achieving universal coverage.

"Insuring America's Health: Principles and Recommendations," *Institute of Medicine Committee on the Consequences of Uninsurance, National Academies Press,* **2004.**
In its sixth and final report, an expert panel urges federal lawmakers to create a plan for insuring the entire population by 2010.

Burton, Alice, Isabel Friedenzoh and Enrique Martinez-Vidal, "State Strategies to Expand Health Insurance Coverage: Trends and Lessons for Policymakers," *The Commonwealth Fund,* **January 2007.**
Analysts summarize recent state initiatives to extend health coverage to more adults and children.

Haase, Leif Wellington, *A New Deal for Health: How to Cover Everyone and Get Medical Costs Under Control, The Century Foundation,* **April 2005.**
A health analyst for the nonprofit group outlines cost, quality and coverage issues that the group says make it necessary for the United States to switch to universal coverage.

For More Information

Alliance for Health Reform, 1444 I St., N.W., Suite 910, Washington, DC 20005; (202) 789-2300; www .allhealth.org. Nonpartisan, nonprofit group that disseminates information about policy options for expanding coverage.

Economic Research Institute on the Uninsured, www .umich.edu/~eriu. Researchers at the University of Michigan who conduct economic analyses of the hows and whys of uninsurance and coverage-expansion proposals.

Families USA, 1201 New York Ave., N.W., Suite 1100, Washington, DC 20005; (202) 628-3030; www.familie susa.org/contact-us.html. A nonprofit group that advocates for large-scale expansion of affordable health coverage.

The Health Care Blog, www.thehealthcareblog.com. Blog published by health-care consultant Matthew Holt; analyzes coverage proposals and other insurance issues.

Heritage Foundation, 214 Massachusetts Ave., N.E., Washington, DC 20002-4999; (202) 546-4400; www .heritage.org. Conservative think tank that supports state-organized purchasing groups for health care.

Kaiser Family Foundation, 1330 G St., N.W., Washington, DC 20005; (202) 347-5270; www.kff.org. Nonprofit private foundation that collects data and conducts research on the uninsured.

National Coalition on Health Care, 1200 G St., N.W., Suite 750, Washington, DC 20005; (202) 638-7151; www .nchc.org. Nonprofit, nonpartisan group that supports universal coverage; made up of labor, business and consumer groups, insurers and health providers' associations.

Physicians for a National Health Program, 29 E. Madison, Suite 602, Chicago, IL 60602; (312) 782-6006; www.pnhp.org. Nonprofit group that advocates for single-payer national health insurance.

10

Television's Future

Will TV Remain the Dominant Mass Medium?

Alan Greenblatt

Increasing numbers of viewers are watching TV programs via Internet streams and iPod downloads, and millions more are hooked on user-generated videos on sites such as YouTube. Despite all the changes, people are watching more television than ever, including today's most popular show, "American Idol," shown here as host Ryan Seacrest, left, prepares to tap singer Taylor Hicks (right) as the winner of the show's 2006 competition.

From *CQ Researcher*,
February 16, 2007.

A dvertisers always pay dearly to run TV commercials during the Super Bowl, and this year was no exception — $2.6 million for a 30-second spot. The commercials themselves are often expensive, over-the top extravaganzas like the famous Michael Jackson ads for Pepsi. But among this year's crop were several that had been shot on cheap digital-video cameras by consumers encouraged by Alka-Seltzer, Chevrolet and Doritos to submit homemade spots.

"These are people who are [accustomed to] personalizing what's important to them, whether it's through their MP3 playlist or their social-site profile," said Doritos spokesman Jared Dougherty. "We wanted to bring that to Doritos, to let them express their love for Doritos in 30 seconds."[1]

In a sense, this was advertising imitating art. Doritos and the other companies were hoping to imitate the sense of free-form buzz that had been generated a few months before by "The Diet Coke & Mentos Experiment," a celebrated Internet film created by a juggler and a lawyer, which showed them sticking mints into plastic bottles of soda and creating fountain effects worthy of the Bellagio in Las Vegas.

Originally distributed on Revver, a video file-sharing Web site, the three-minute film was soon being viewed by millions via such popular video sites as YouTube and MySpace. It also drew enormous "old media" attention and spawned countless imitators. Perhaps most significantly, the ad increased Mentos sales by 15 percent. Partly as a result, the trade publication *Advertising Age* named consumers themselves its "Agency of the Year."[2]

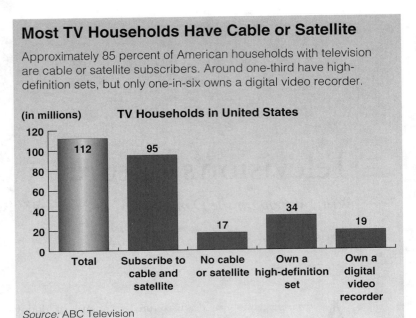

Most TV Households Have Cable or Satellite

Approximately 85 percent of American households with television are cable or satellite subscribers. Around one-third have high-definition sets, but only one-in-six owns a digital video recorder.

TV Households in United States

(in millions)

Total	Subscribe to cable and satellite	No cable or satellite	Own a high-definition set	Own a digital video recorder
112	95	17	34	19

Source: ABC Television

Self-generated video content has become one of the biggest fads of the moment, with hundreds of millions of people around the world spending hours viewing and sharing short films on the Web. *Time* magazine went so far as to name "You," as in YouTube, its "Person of the Year" — an honor traditionally bestowed on the person who has done the most to shape the news.[3]

Today's proliferation of media platforms is triggering an explosion of user-generated content and challenging the television industry's historic business model. Instead of the age-old concept of "one to many" — in which a network broadcasts a show watched by millions simultaneously — "one" person now can watch a program in "many" different ways, whether downloaded onto a video iPod, clipped into smaller bites for "snacking" on cell phones or YouTube or "streamed" over the Internet on a network's own Web site.* Using digital video recorders (DVRs), such as TiVo, viewers also can decide when to watch a program. Traditional "appointment television" — with

---

* The flow of content across multiple media platforms and the willingness of media audiences to use any platform to enjoy the kind of entertainment they like is known as convergence.

viewers settling down at 8 or 9 p.m. to watch a particular program — increasingly appears to be a hopelessly dated concept.

"The old idea of a station having to pick up a signal from the network by landline and delivering it to people by antennas on a big hill in your community is about as antiquated as the blacksmith shop," says Robert J. Thompson, founding director of the Center for the Study of Popular Television at Syracuse University.

Clearly, the major networks' near-total hegemony over video distribution is ending. But it's unclear whether they will surrender their dominant position in producing content for most Americans.

The broadcast networks' audience share may be shrinking, and people may be uploading 70,000 videos a day onto YouTube, but even the most popular "viral videos," which spread like a virus — such as the Mentos experiment — are watched by fewer people than a moderately successful network television show. In fact, despite all the new television-viewing devices and platforms, people are watching more traditional TV — game shows, reality TV shows, sitcoms and dramas — than ever before. During the 2005-2006 season, for instance, the average household had the television running for eight hours and 14 minutes a day — up one hour from 1996, according to Nielsen Media Research.[4]

Some industry experts predict that the new platforms — rather than "cannibalizing" traditional TV — will increase its popularity by allowing people to catch up on shows they missed during their regular broadcast time. Television executives' new mantra is that viewers can watch "what they want, when they want it and how they want it."

By failing to embrace such a strategy, the music business squandered a big percentage of its business to illegal file-sharing, says David Poltrack, chief research officer for CBS Corp.[5] "The lesson that the television industry has learned from the music industry is that you can't keep the technology down, so let's use it to make our product ubiquitous," he says. CBS and its competitors are aggressively negotiating with Internet and cell phone companies and anyone else who can keep them a step or two ahead of consumers wanting to watch popular shows like "House" or "Lost" on a new platform.

"We are not a television company. We are a sports media company," said John Skipper, ESPN's executive vice president for content. "We're gonna surround consumers with media. We're not gonna let them cut us off and move away from our brand."[6]

Such bold talk has become commonplace, but television types nonetheless remain nervous. After all, there's no reliable way yet to measure how many people have watched an episode of a program recorded by TiVo — or played on an iPod, both of which allow the viewer to easily fast-forward past the commercials. That makes it difficult for advertisers to feel confident that the networks have delivered the promised number of viewers. And that's a serious issue, because advertising is by far the industry's leading source of revenue.

"The marketing community is somewhat dazzled and confused right now," Bob Liodice, president of the Association of National Advertisers, said during a discussion on the future of television in January 2007 in Brooklyn, N.Y. "All the media are shifting, and advertisers are not certain about which ones are working."

TV and ad executives may sound optimistic when they gamely predict that the new platforms will only create additional opportunities for advertisers, but even more neutral observers say predictions about the demise of traditional television will turn out to be misguided.

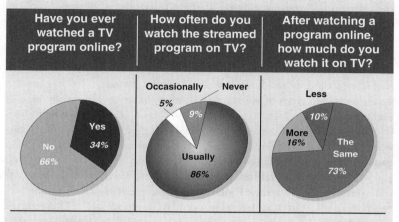

A Third of Watchers View TV Shows Online

About one-third of regular TV viewers have watched a program on the Web (left), with most viewing a "streaming" program that they regularly watch on TV (middle). Only 10 percent of online viewers watch a program less on TV after viewing it on the Internet (right).

Have you ever watched a TV program online?	How often do you watch the streamed program on TV?	After watching a program online, how much do you watch it on TV?
No 66% / Yes 34%	Occasionally 5% / Never 9% / Usually 86%	Less 10% / More 16% / The Same 73%

Not all percentages total 100 due to rounding.

Source: CBS Television City New Technology Focus Groups, December 2006

"We're in a moment in time when media power operates top down from corporate boardrooms and bottom up from teenagers' bedrooms," says Henry Jenkins, director of the comparative media-studies program at the Massachusetts Institute of Technology.

And while there clearly will be no decrease in media — or media choices — in the foreseeable future, the ultimate effect the proliferation of new forms of content and distribution models will have on TV viewing is as yet unknown. "If anyone tells you what the television business is going to look like a decade out," said Dick Wolf, creator of the "Law & Order" franchise, "they are on drugs."[7]

As television continues to evolve and change, here are some of the questions people in the industry are asking:

Will the Internet kill television?

No one doubts that technology is rapidly changing the television landscape. But many question whether the new ways to watch video content really threaten traditional television.

"There is so much competition for the time people spend in front of a screen," said Daniel Franklin, executive editor of *The Economist*. "The Internet has behaved like a serial killer. First, the print media suffered, and then the music industry suffered. Perhaps television is next."[8]

Besides making a surfeit of amateur videos universally available, the Internet now provides increased access to professional content. And thanks to its virtually unlimited storage capacity and flexibility, the Web offers better access to certain types of video programming than normal television.

Take sports, for instance. Subscribers to MLB.tv can watch 2,500 baseball games a year and have up to six games showing on their screens at any given time. The site will even send subscribers personalized alerts when pitchers or batters from their Fantasy or Rotisserie league teams are playing in other games not displayed on their screens.*

"People tend to watch on TV," says Michelle Wu, chief executive officer of MediaZone, which offered Internet-based coverage of the Wimbledon tennis tournament last year. But because her service covered more than 300 matches, "People can watch a lot more and at a more convenient time."

The Internet is supplanting traditional TV in numerous other ways as well. For instance, on July 2, 2005, more people worldwide watched the "Live 8" concerts (to fight poverty in Africa) online than via TV. Netflix, the Internet-based DVD-rental company, in January began offering a limited selection of movies and TV shows for direct viewing over the Internet. And the wide availability of video on the Internet in general threatens to end the traditional content monopoly enjoyed by broadcast stations within their particular markets.

For instance, relatively few people sat through the entire Golden Globes telecast this year, writes columnist

Andrew Sullivan. "The next day, however, many downloaded clips of the more embarrassing acceptance speeches or the more touching moments. It's more efficient."[9]

What's more, those who depend on the Internet for video are no longer limited to watching on their computers. Apple announced in January that it would soon release a device, called Apple TV, which will wirelessly move video from the Internet onto regular TV sets. Other companies, such as Sony, Microsoft and Sling Media, are working on comparable tools.

"The consumer demand that has currently and historically been served by broadcast television is increasingly being served in other ways by technologies that have more capacity and permit a much more diverse menu of choices," says Bruce M. Owen, a Stanford University economist and author of *The Internet Challenge to Television*. "As a result, the former medium is shrinking."

Those in the television industry are growing more confident about their ability to meet the Internet challenge, because they own the content that people most want to watch, despite the current fad of watching amateur videos on YouTube and other sites. And television presents content in the most watchable format: high definition, which offers a picture quality unequaled on computer screens.

In fact, much of the most popular content on Internet video sites comes from traditional television sources. "So much of YouTube content is still based on repurposing traditional network programming or access to old programming," says Syracuse University's Thompson.

So, rather than television being swallowed by the Internet, things may be going the other way around, says Chris Pizzurro, vice president of digital and new media advertising sales and marketing for Turner Entertainment. Computers and iPods that play video are becoming, in effect, TV sets. "The TV is actually growing to other devices," he says. "It's because of the programming."

In addition, programming is beginning to migrate from the Web to traditional TV outlets. NBC broadcasts a program based on the popular iVillage site for women, while Warner Brothers just made a deal with Fox to

---

* Fantasy and Rotisserie leagues involve virtual teams in which players are drafted from multiple teams but their scores depend on real-life statistics.

distribute a television version of its even more popular entertainment Web site, TMZ.com.

The networks and production companies are negotiating with YouTube, Yahoo and other Internet companies to show content or at least excerpts of their programs. "Everything is geared toward more individualized consumption," said comedian Jon Stewart, host of Comedy Central's "The Daily Show." "Getting it off the Internet is no different than getting it off TV."[10] Stewart's program has been among the most-watched shows, in clip form, on the Internet.

But the networks are jealously guarding control of their copyrighted material. On Feb. 2, the media conglomerate Viacom — which owns the MTV and Nickelodeon networks — demanded that YouTube remove 100,000 clips of Viacom content from its site, including "The Daily Show." In addition, Paramount and 20th Century Fox each recently subpoenaed YouTube, forcing the site to disclose the identities of users who had uploaded copyrighted material. Other video-sharing sites have been targets of litigation as well.

Still, given the flow of content between television and computers — and the emergence of devices to expedite that flow — it appears likely that these media will continue to merge, with people watching TV programming via computers and Web sites showing ever more video.

But that doesn't mean TV will disappear. "There will always be a market for consumers who want to watch television in the traditional way, in their living room at a particular time," says Brian Dietz, vice president for communications at the National Cable & Telecommunications Association. "The vast majority of the content that people want is available through the traditional television-viewing service."

Will television remain a viable medium for advertisers?

Many people in the television industry are wondering whether their longstanding approach — aggregating many viewers to watch a particular show and selling their attention to advertisers — is sustainable.

"That sort of grand bargain — we'll show you the program if you watch the advertising — is breaking down," said *The Economist's* Franklin.

While television's venerable paradigm may not have broken down completely, it clearly is under pressure due to the changing nature and fragmentation of the audience. Viewers, especially younger ones, are increasingly willing to migrate to new platforms and sources — often allowing the user to bypass commercials — in pursuit of entertainment and information.

New technologies and sources have been chipping away at the broadcast networks' monopoly since the advent of cable. With the explosion of cable channels over the last 30 years, networks today are down to about 42 percent of the viewing audience compared to 80 percent in the 1970s.[11] Today's most popular network shows are lucky to draw half the audience enjoyed by '70s-era TV hits.

During the 1960s, an advertiser could reach 80 percent of adult American women by simply buying a prime-time spot on all three networks; today reaching the same group would require advertising on 100 different channels.[12] The audience is fracturing even more with the advent of new video devices.

"Advertising is suffering because of the sheer amount of it, the lack of innovation within traditional advertising formats and the power that media fragmentation and technology give to consumers to tune out the noise," writes Tom Himpe, author of *Advertising Is Dead: Long Live Advertising.*[13]

For a couple of years now, advertisers have worried that more viewers will use TiVos and other DVRs to skip past their messages. A Ball State University study last year found that viewers do not watch commercials all the way through 59 percent of the time, either because of impatient channel surfing, bathroom breaks or because they have TiVos or other DVRs.[14]

"Only a small percentage of the audience is there during the commercial," says Jeffrey Cole, director of the Center for the Digital Future at the University of Southern California's Annenberg School. "It's not the Internet, it's the bathroom and the remote control."

However, the phenomenon naturally has networks nervous. They also worry about the difficulty of measuring

Home-Grown Videos Open New Frontiers

Anybody can be a producer — or a journalist

About a year ago, Will Albino and Brian Giarrocco brought a video camera to Padua Academy, a college-preparatory school for girls in Wilmington, Del. Albino asked students to sign a petition to end women's suffrage, correctly figuring they would mistake the right to vote for "suffering."

They edited the video footage of their prank down to three minutes, posted it on the Internet, and instantly had a hit.

"You hardly need a modicum of talent to create a really entertaining video," says Todd Herman, online video strategist for Microsoft. "That clip in one day sucked 8 million minutes out of people's lives."

Video is becoming about as big a presence on the Web as search engines, turning amateurs into producers, and in some cases, journalists. Last October, Google bought YouTube, the leading video-sharing site, for $1.65 billion. Soon after, *Time* magazine named "You," as in YouTube, its "person of the year." A recent survey found that 38 percent of Americans wanted to create or share content online.[1]

Clearly, the days when Hollywood produced most video entertainment are ending. But what does it mean for everyone in the world to be a potential producer?

A few clips, such as "Evolution of Dance," Judson Laipply's six-minute compression of the last 50 years of steps, have become hugely popular. Laipply's clip has been streamed more than 40 million times. Like most blogs, though, most posted videos get little or no attention. Few people have any real interest in dull footage of cats playing piano.[2]

"The barrier to entry" — the once-formidable cost of shooting and editing footage — "has almost disappeared," says Jeffrey Cole, director of the University of Southern California's Center for the Digital Future. "But what's still in short supply, and always will rise to the top, is good ideas."

Some of those possessed of both camcorders and talent are getting serious offers. Studios and talent agencies are actively scouting and hiring online talent — and sometimes getting turned down. "Hollywood tried to court us," said Kent Nichols, one of the producers of the popular "Ask a Ninja" online series, "and we're like, 'What are you talking about? We already have an established fan base.'"[3]

In addition to serving as a possible farm team for Hollywood, the world of Internet films is opening up production possibilities for professionals. All the TV networks and other content creators are thinking hard about producing shorter films for distribution over Web sites and cell phones.

Merely cutting longer programs or 30-second ads doesn't always work, but the approach pioneered by successful Internet films — great concepts that are instantly understandable — provides a perfect template.

the audience that is watching specific episodes of shows on DVRs, iPods, Internet streams and other new platforms. "It's something everybody in the business is watching, both at the network and affiliate level," says Dennis Wharton, executive vice president for media relations at the National Association of Broadcasters. "We recognize that the model has to evolve."

But he remains optimistic — and with reason. Despite the competition from new media, viewership of traditional television programming continues to rise, and with it advertising revenues. Total TV advertising amounted to $47.2 billion during the first nine months of 2006 alone, according to TNS Media Intelligence — an increase of 5.2 percent over the same period the previous year.[15]

Still, networks and advertisers are experimenting with myriad ways to get viewers to think about brands and products. Some, for instance, are turning to product placement — paying to have their products appear as an embedded part of a show. Long common in movies, product placements pop up in many television shows today, often without any subtlety. (*See list, p. 227.*)

Television executives are also trying to "create a good environment for advertising" on new platforms, says Rick Mandler, vice president of digital media advertising at Disney/ABC Media Networks. Several ABC programs are available in their entirety on the company's Web site hours after they have aired on television. They are presented with few commercial interruptions by a single sponsor, but the sponsor's logo appears above the program-viewing window throughout the entire show.

"It's an opportunity for bigger companies to present material to an audience where the stakes are not so high," says Sue Rynn, vice president for emerging technologies at Turner Broadcasting. "You can migrate successful ideas from some of these niche opportunities into the more traditional space" of broadcast television.

If video entertainment continues to be an area of experiment and uncertainty, it's clear that home-grown video information is having a big impact. Sens. Hillary Rodham Clinton, D-N.Y., and Barack Obama, D-Ill., made their initial presidential campaign announcements on videos they posted on Web sites. They had clearly learned lessons from the paths taken by "viral videos," watching their messages rapidly spread via e-mail as well as replays on traditional media outlets.

In addition, having millions of eyewitnesses capable of capturing what they see on video is changing the nature of news coverage and events. Ordinary people have shot some of the most compelling news footage in recent months, from the famed "Macaca" video of then-Sen. George Allen, R-Va., insulting a dark-skinned volunteer for his opponent's campaign, to the cell phone footage of comedian Michael Richards slinging racial

The producers of the popular Ask a Ninja online series were courted by Hollywood.

slurs in a nightclub. The careers of Allen and Richards were brought to a halt in ways that would never have happened before YouTube.

Citizen journalists are now regularly sending footage to Web sites such as Scoopt and NowPublic. From the 2005 London subway bombings to last year's crash of Yankee pitcher Cory Lidle's plane into a Manhattan apartment building, images shot by amateurs dominate news coverage of some events.

"In 1991, when a bystander videotaped the beating of Rodney King in Los Angeles, the incident was almost unbelievable — not the violence but the recording of it," writes media critic James Poniewozik in *Time*. Today, incidents such as the Richards meltdown "are wearing away the distinction between amateur and professional photojournalists."[4]

---

[1] Bob Garfield, "YouTube vs. Boob Tube," *Wired*, December 2006.

[2] For background, see Kenneth Jost and Melissa J. Hipolit, "Blog Explosion," *CQ Researcher*, June 9, 2006, pp. 505-528.

[3] Matthew Klam, "The Online Auteurs," *The New York Times Magazine*, Nov. 12, 2006, p. 83.

[4] James Poniewozik, "The Beast With a Billion Eyes," *Time*, Dec. 25, 2006-Jan. 1, 2007, p. 63.

Viewers can fast forward through the program itself, but they can't skip past the ads.

The ads run for the traditional 30 seconds but present more information than a traditional commercial. Viewers can click on images showing the products being advertised, which will take them to the sponsor's own Web site. These examples illustrate a paradoxical point that Mandler makes. In an age of proliferating video, advertisers need to think of ways of presenting their messages that are not simply 30-second videos.

Ad agencies have said, " 'OK, we'll give you our 30,' " Mandler says. "And we've said, 'We don't want your [traditional] 30.' We want you to create something that leverages the interactivity of the platform."

Interactive advertising, in which viewers navigate their way through a virtual marketplace, has long been

talked about but is just now being tried on cable and satellite systems. (It's more common in the United Kingdom and other countries.)

But some television executives believe the real future of TV advertising lies in its past. In the early days of the medium, shows were presented in their entirety — and sometimes were even produced — by the sponsors. That's why, for instance, comedian Milton Berle's show in the '40s and '50s was not named for him but was called "Texaco Star Theater." Something similar may happen again, with a single sponsor presenting a show, with fewer commercial interruptions, and presenting it not only on network TV but also on every device and Internet platform.

Something like that has already happened with Ford, which has presented two season premieres of "24" without

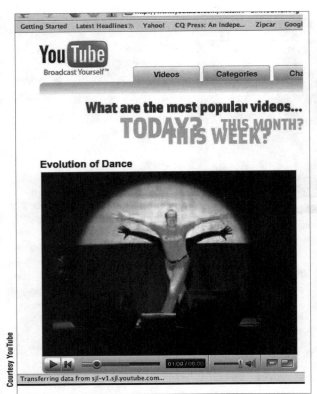

Courtesy YouTube

Most homemade videos posted on the Web get little attention, but "Evolution of Dance," created by 30-year-old Cleveland comedian Judson Laipply, became a global sensation. The hilarious six-minute compression of the last 50 years of dance steps has been streamed on YouTube more than 40 million times.

commercial interruption, running two- or three-minute ads at the start and finish of the broadcasts. The carmaker found that consumers had unusually high recall of those ads, despite the fact that they only appeared once. In addition, star Kiefer Sutherland drove a Ford Expedition as part of the show, while other Ford vehicles were woven into the story. "Basically, we own the show," said an advertising executive working with the company.[16]

Sponsors are also heavily integrated into the most popular current TV show, "American Idol," which was pre-sold to sponsors before it won a spot in the Fox network lineup. Contestants appear in little music videos extolling Ford, the judges are never seen without Coke cups nearby, and much viewer voting is done using AT&T's text-messaging service. AT&T Wireless reported

that about a third of those voters had never sent a text message before. "Our venture with Fox has done more to educate the public and get people texting than any marketing activity in this country to date," said a company spokesman.[17]

Are scripted TV shows obsolete?

New media have breathed new life into some network series. Fox began producing fresh episodes of its canceled "Family Guy" cartoon series after it sold well on DVD, and NBC has kept the relatively low-rated comedy "The Office" on the air in large part because it has sold well, at $1.99 an episode, on Apple's iTunes online store.

"I'm not sure that we'd still have the show on the air" without that boost, said Angela Bromstead, president of NBC Universal Television. "When it went on iTunes and really started taking off, that gave us another way to see the true potential other than just Nielsen Media Research."[18]

Typically, though, the shows that perform best in the new platforms are the same ones that are most popular in their regular broadcast time slots. The question facing television executives today is how many stragglers they can afford to keep producing.

With the TV audience continuing to fracture, advertising revenues — even though they are rising — are not keeping up with growing production costs. That leads some in the industry to conclude that the heyday of scripted programming — hour-long dramas and 30-minute situation comedies — may be over.

Scripted shows, after all, are much more expensive to produce than game shows or reality programs. The average scripted drama costs $2.6 million per episode, while the popular "Deal or No Deal" game show, which NBC airs three times a week, only costs about $1 million each, including the prize money.[19]

In October 2006, NBC announced that it would no longer offer scripted shows during the first hour of prime time most nights of the week. Many took it as a sign of things to come. "It's absolutely more bad news for me," TV writer Tim O'Donnell said. "It just eliminates the shelf space available for networks to put on what I pitch."[20]

Although NBC soon backed away from that policy, the network — like its sisters — is producing less scripted programming than a few years ago. The networks have 22 hours of prime time to fill (three hours a night, four on Sunday), and more and more of that time is filled either with game or reality shows or reruns of scripted shows that aired earlier in the week.

Because of dwindling weekend audiences, the networks have essentially ceased offering original programming on Saturday nights and are easing away from broadcasting new material on Fridays as well. "Writers have a lot of reason to be anxious," said Dean Valentine, former head of Disney's television unit. "The world they've been living in no longer exists."[21]

To make matters more uncertain, the contracts covering all the major creative guilds — writing, acting and directing — will expire over the next 18 months. The writers' guild contract expires this summer, and there is a lot of talk in Hollywood that a strike is imminent. Writers and other creative talent — who generally feel they did badly during earlier rounds of negotiation over DVD revenues — now are eyeing a bigger share of the income generated by new media.

A Canadian guild went out on strike in January over similar issues. "We want to be compensated fairly for use of our work on the Internet," the union announced. "We will not have our work put on the Internet for free."[22]

As part of their contingency planning, the networks are developing more reality and game shows. All of the major networks now have multiple prime-time game shows either in production or development. "It's really cheap, and it's a quick way to build an audience," says media consultant Phillip Swann, author of *TV.com*.

But even though the major networks are producing fewer scripted shows, more networks are producing scripted television than ever before. Cable networks such as HBO have stepped up production, as have second-tier broadcast networks such as TNT and The CW (created by CBS and Warner Brothers).

"Yes, CBS and ABC are showing fewer hours a day in scripted entertainment than a decade ago, but there are more networks producing scripted shows today," says MIT's Jenkins. "They're just dispersed across more networks."

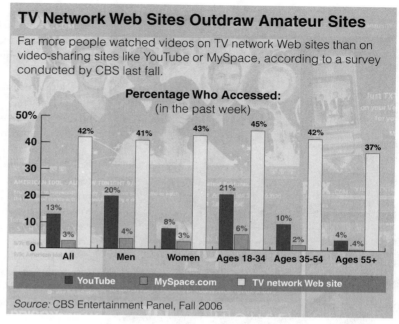

TV Network Web Sites Outdraw Amateur Sites

Far more people watched videos on TV network Web sites than on video-sharing sites like YouTube or MySpace, according to a survey conducted by CBS last fall.

Percentage Who Accessed:
(in the past week)

	YouTube	MySpace.com	TV network Web site
All	13%	3%	42%
Men	20%	4%	41%
Women	8%	3%	43%
Ages 18-34	21%	6%	45%
Ages 35-54	10%	2%	42%
Ages 55+	4%	.4%	37%

Source: CBS Entertainment Panel, Fall 2006

What's more, although sitcoms seem to have lost their luster, many hour-long scripted dramas are huge hits. "The success of 'Lost' and 'Desperate Housewives' and 'Heroes' and 'Ugly Betty' suggests there is still an enormous economic value and public interest in scripted programs," Jenkins says.

In fact, many TV people think today's scripted shows are as good or better than ever before. Under challenge from new media, networks have stepped up their game to produce or showcase programs that, in some cases, rival feature films in quality of production and entertainment value.

"Scripted shows to a certain extent are making a comeback," says Wharton, of the National Association of Broadcasters. "Just from the point of view of the craft of television drama, today's shows hold up as well as any shows in the history of television."

Moreover, while reality shows and game shows have proven to be surprisingly durable since their emergence in prime time, even the most popular examples have tended to enjoy fairly limited shelf lives. (A notable exception is the hugely popular "American Idol," averaging about 36 million viewers per episode; the show is in its sixth season.) The limited runs make network executives nervous about abandoning scripted shows, which — when they become hits — can run for years.

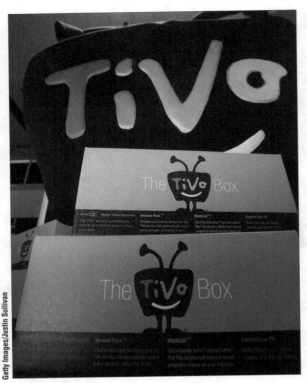

The TiVo recording device helped revolutionize TV watching by allowing viewers to record their favorite shows and skip over commercials.

Thus, networks realize they must maintain a strong lineup of scripted shows, which have longer half-lives than even the most popular reality shows — whether in syndication, on DVD or across new platforms such as Internet streaming or iTunes downloads.

The shows most often set for automatic, season-long recording on digital video recorders are all scripted programs, including "Law & Order" and "CSI."

"It's a case of the rich getting richer," says Mark Loughney, vice president of sales and strategy research for ABCTV. "If people are passionate about your programming, they will watch it and find ways of watching it."

BACKGROUND

Early TV

The history of television is replete with examples of new devices — color sets, high-definition sets, videotape recorders,

flat screens — that burned through years and many iterations before becoming popular. Television itself has been the most important mass medium for so long that it's easy to forget that it, too, was once a newcomer threatening the financial health of older forms of communication.

Television distribution and production were built on the back of radio, with the radio networks — particularly NBC and CBS — seamlessly dominating the newer technology, using many of the same programs and stars.[23]

The concept of sending images over long distances had been discussed for decades prior to the advent of regular television broadcasting. In the 1870s, American inventor Thomas Edison coined the term "telephonescope" to describe the process of converting light and shade into electrical signals that could be transmitted wirelessly.[24] A crude color system was demonstrated as early as 1928.[25]

Television grew rapidly after World War II, which had interrupted its commercial development. In 1946, there were just two TV networks (NBC and CBS) providing 11 hours of programming to a handful of stations that could be watched on one of the 10,000 TV sets then in use. In 1950, four networks (including ABC and DuMont) were sending out 90 hours of weekly programming to about 100 stations serving 10.5 million sets. By the end of the 1950s, 87 percent of U.S. households had a television.[26] National advertisers migrated to the new medium from both radio and from general-interest magazines (*The Saturday Evening Post* and *Collier's* started to fold) and began cutting into the audience for comic books and movies.

Between 1949 until 1952, the FCC protected the oligopoly NBC and CBS had enjoyed in radio by freezing new licenses for broadcasters, essentially making wider competition within the TV industry difficult. The two networks enjoyed complete dominion over 80 percent of the 60-odd TV markets until 1962, when Congress mandated that all new sets be able to receive both UHF and VHF signals, allowing more competition.

ABC, which had struggled (it had been created in 1943 after a government antitrust suit forced NBC to divest itself of its Blue Network, as ABC was called at the time), came into its own during the 1960s, building an audience through heavy coverage of weekend sports. Fox pursued a similar niche strategy in the 1980s by appealing to viewers ages 18-34 and by broadcasting sports.

For most of their history, however, broadcasters tried to appeal to as wide an audience as possible, pursuing the doctrine of "least objectionable programming" — the

CHRONOLOGY

1940s-1960s *Television explodes in popularity.*

1941 Federal Communications Commission (FCC) authorizes commercial TV operations to start the following year on 10 stations.

1948 FCC freezes new TV licensing because of flood of applications, effectively preserving oligopoly of NBC, CBS and their affiliates.

1949 The first cable systems are created in Oregon and Pennsylvania, allowing rural communities too far from broadcast stations to receive programming.

1952 FCC lifts its licensing freeze, approving plan for 82 new channels (12 on VHF and 70 UHF), opening up the possibility of hundreds of new broadcast stations.

1965 FCC begins regulating cable TV.

1970s-1990s *New competitors challenge broadcast network dominance over home video content.*

1975 Home Box Office becomes the first cable network to use satellites to send its signal to cable operators, pioneering the concept of a pay-TV channel.

1984 Congress approves Cable Communications Policy Act, deregulating cable rates and codifying the FCC's ban on telephone companies operating cable systems within their service areas.

1988 Cable penetration reaches half of U.S. households, with 42.8 million people subscribing to 8,500 cable systems. . . . Congress allows cities to regulate cable systems and home viewers to receive programming via satellite.

1992 FCC permits local telephone companies to operate video-delivery systems. . . . With cable rates rising sharply, Congress reregulates cable industry — overriding a presidential veto — in order to provide "increased consumer protection."

1996 For the first time in 62 years, Congress overhauls telecommunications law in effort to increase competition for new broadcasting, cable and other video services. Critics say the new law is a failure.

1997 DVD players are introduced in the U.S. market.

1999 TiVo machines (digital video recorders) are released to the public, giving television viewers an easy way to record, store and play back their favorite shows.

2000s *The video audience continues to fragment as the Internet opens up new delivery systems and promotes user-generated content.*

2002 For the first time, cable networks combined draw more viewers than broadcast networks. . . . Major broadcasters lose a record 30 percent of their viewers in the summer.

2005 More people watch the July 2 "Live 8" charity concerts online than on TV. . . . On Oct. 12 Disney and Apple announce pact to sell TV episodes for downloading and replaying on video iPods. . . . On Nov. 23, the video-swapping Web site BitTorrent agrees to remove links to unlicensed copies of Hollywood films.

2006 Sen. George Allen, R-Va., is captured on tape on Aug. 11 calling a young man of Indian descent "Macaca"; footage is quickly uploaded onto YouTube and helps to doom Allen's re-election. . . . On Oct. 9, Google buys YouTube for $1.65 billion. . . . NBC Universal announces it will end most scripted programming during the first hour of prime time and cut 700 jobs, applying savings to digital-media investment. It later reconsiders the decision. . . . On Dec. 20, the FCC limits local governments' ability to regulate video services.

2007 Apple announces on Jan. 9 it will soon release Apple TV, a device to store digital video and transfer it wirelessly from computers to TV sets. . . . On Feb. 2, Viacom demands that YouTube remove more than 100,000 clips of its content from the video-sharing site.

Feb. 17, 2009 All local TV stations must switch from analog to digital broadcasting, freeing up the analog spectrum for emergency use.

Teens' Media Multitasking Raises Questions

Do they learn? Retain what they learn?

Several years ago, an ad showed a young man, TV remote control in hand, saying, "You've got 3 seconds. Impress me."[1]

It was meant to underscore the impatience of habitual channel surfers. But what's striking about the ad today is how focused the youth looks. He's only holding a remote control. Where is his iPod, his cell phone and his laptop?

"This is the M Generation," says Ian Rowe, vice president of strategic partnerships and public affairs for MTV Networks. "They want all media all the time."

Cell phones and computers seem to encourage media multitasking. Today's young people watch TV with a computer on their laps and a cell phone by their side, checking their Facebook pages, monitoring eBay auctions, playing fantasy football or video games, sending text messages and phoning their friends. And the young — who grew up using computers from an early age — are more likely than adults to embrace newer media devices and technology.

According to a Kaiser Family Foundation study released in December 2006, anywhere from a quarter to a third of 7th-to-12th-graders say they multitask "most of the time." A majority of kids multitask some of the time while fewer than 20 percent say they never do.[2]

While today's teens and "tweens" absorb more media than ever and often are interacting with more than one medium at a time, how that affects their attention spans and their ability to learn and retain what they've learned — or whether they will keep up such distractible habits as adults — are all debatable. While multitasking among the young has become a field of serious academic study, most of the basic questions are unanswered.

"One of the great myths is that the 16- or 18-year-old doing six things at once will still be consuming their media that way when they're 30," says Phillip Swann, president of TV Predictions Inc., a consulting firm in North Beach, Md., outside Washington.

"Are we going to reach the point in our culture where people have a hard time listening to a two-and-a-half-minute pop song without channel surfing?" asked New York-based composer R. Luke DuBois. "I see people do that on their iPods all the time. They'll listen to songs only through the first chorus, and then they'll switch to another song."[3]

That kind of behavior is bad news for brain development, says Jordan Grafman, chief of the cognitive neuroscience section at the National Institutes of Health. He argues that the inability to focus is a modern version of a primitive response of the brain's frontal lobe to be attracted to novel stimuli.

Some studies suggest multitaskers don't retain as much of what they learn as they would when they are more focused. Divided attention "doesn't allow you to do any deep deliberation," Grafman says. "If you have to invent something, if you have to design something new, if you have to have a different take on an issue, there is no way you are going to be able to do that effectively if you are multitasking."

But research also suggests that young people who have grown up in a media-saturated age "can toggle back and forth between things better than those who are older," says Lee Rainie, director of the Pew Internet & American Life Project. "If you're motivated, you can learn a lot more than you used to, because it's so much easier."

Another school of thought holds that the skills promoted by the convergence of new media — including creativity, peer-to-peer learning and, yes, multitasking — are becoming necessary for success in the modern world. The MacArthur Foundation announced in October 2006 it will devote $50 million over five years to the study of digital media and learning.

Henry Jenkins, a media studies professor at MIT who is working with the foundation, scoffs at those who think multitasking is a bad habit that will be shed with age. "We all live in a world where multitasking is an essential skill," he says. Given the current bombardment of media, "If we can't shift our attention from one piece of information to another, we really will cease to function."

One thing appears to be certain. Despite the way today's young people flock to the latest gadgets and features, including Facebook and MySpace, their primary loyalty is to television. The average young person watches nearly four hours of TV a day — compared to 49 minutes playing video games — and is more likely to give television his or her undivided attention rather than any other media device, according to the Kaiser study.[4]

"Television still completely dominates kids' time with media and is eight times more likely to be a primary activity than a secondary activity," meaning their main focus will be on the TV set rather than on the cell phone or computer, says Ulla G. Foehr, author of the Kaiser study. "So anyone who thinks TV is becoming irrelevant should think again."

---

[1] Henry Jenkins, *Convergence Culture* (2006), p. 64.

[2] Ulla G. Foehr, "Media Multitasking Among American Youth," Kaiser Family Foundation, December 2006, p. 7.

[3] Quoted on "Studio 360," Public Radio International, Jan. 26, 2007.

[4] Foehr, *op. cit.*, p. 8.

belief that out of three choices, viewers will select the one they are least likely to hate. That style of programming fell out of favor by the 1980s, however, when cable began providing viewers with many more options.

Cable and Video

Cable television had been around, in embryonic form, since the late 1940s. Initially, it evolved as a way to bring television to remote areas that were too far away from the stations to receive broadcast signals. In 1949, E. L. Parson, a radio-station owner in Astoria, Ore., erected an antenna system to receive the signal of a TV station in Seattle, 125 miles away, which he then shared with 25 "subscribing neighbors" via a network of wires.[27] Most early cable systems were similar mom-and-pop operations; by 1957, there were 500 such systems, averaging about 700 subscribers each.[28] A few years later, a San Diego cable operator decided to bring in TV stations from Los Angeles. For the first time, cable service was being offered in cities already served by three or more broadcast stations.

In 1975, Home Box Office (HBO) became the first cable network to use satellites to transmit programs to cable operators, demonstrating the viability of offering a premium channel that subscribers would pay extra to receive. It soon attracted competitors, such as Showtime. Cable revenues grew from $900 million in 1976 to $12.8 billion in 1988 — $3 billion of that from so-called pay-TV channels alone.[29]

Congress had deregulated the cable business in 1984 to offer it protections from certain local government mandates (it would re-regulate the industry in 1992). By 1988, half of U.S. households with television were wired for cable. That year, Congress passed legislation to permit continued transmission of programming to owners of home satellite dishes.[30]

The year 1988 was also a "tipping point" for sales of videocassette recorders (VCRs) — nearly 12 million in the United States alone. Ampex had introduced a videotape

TV Advertising Changes With the Times

Changes in television technology — notably the development of devices that allow viewers to bypass commercials — have led some companies to abandon traditional commercials (left) and instead to pay producers to write their products into the show itself (right).

Top 10 Traditional TV Advertisers (in $ billions spent)		Top 10 Product Placement Brands (by most appearances)
1. Procter & Gamble	$2.9	1. Coca-Cola
2. General Motors	2.0	2. Chef Revival Apparel
3. AT&T	1.4	3. Nike Apparel
4. Ford	1.4	4. 24 Hour Fitness
5. DaimlerChrysler	1.3	5. Chicago Bears
6. Time Warner	1.2	6. Cingular Wireless
7. Verizon Communications	1.1	7. Starter
8. Toyota	1.1	8. Dell
9. Altria Group	1.0	9. SLS Electronic Equipment Speakers
10. Walt Disney	1.0	10. Nike Footwear

Source: Nielsen Monitor-Plus, 2006

Source: Nielsen Product Placement, 2006

machine as early as 1957 and had not been able to keep up with consumer demand — despite a hefty $50,000 price tag. In 1976, Sony introduced its Betamax machines in the United States. Prices had come down a bit — to $1,295 for the machine and $15 for one-hour cassettes. Despite its headstart in the marketplace and superior picture quality, however, Sony lost the "format war" to VHS.

Believing that VCRs would lead to piracy and cut into profits, production companies such as Universal and Disney lobbied Congress and filed lawsuits, claiming their copyrights were being violated.

Eventually, VHS would lose its preeminence to DVD players, which first became available in the U.S. market in 1997. DVD sales and rentals quickly became major sources of profit for television and movie production companies.

Regulating Competition

By the 1990s, cable systems had attracted a new competitor: the telephone industry, which began clamoring to provide video services to its customers. The FCC had banned telephone companies from owning cable systems

Most Adults Have Digital TV, Broadband

About two-thirds of U.S. adults have digital televisions and/or high-speed (broadband) Internet connections. Only 35 percent have analog TVs and no broadband or cable connections.

Percentage of Adults with Digital TV, Broadband
(spring 2006)

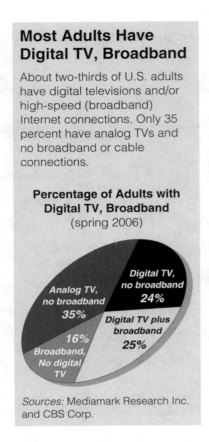

Sources: Mediamark Research Inc. and CBS Corp.

in 1970 out of fear that "telcos" would refuse to let a competing cable system use their telephone poles to hang its wires. Congress effectively codified its ban as part of the 1984 cable deregulation act, aiming to prevent discrimination against cable systems and to promote diversity in media ownership.

By the late 1980s, however, Washington policymakers were beginning to advocate allowing telephone companies to provide cable services. The FCC permitted phone companies to engage in video programming and recommended that Congress repeal the cross-ownership ban.

Impatient with the lack of action in Congress, Bell Atlantic asked a federal judge to rule that the ban violated the First Amendment. A judge so ruled in 1993, prompting several other telcos to seek legal redress. By

late 1994, the legal barriers to competition between telcos and cable had fallen.

In 1996, Congress systematically revamped telecommunications law for the first time in 62 years. The Telecommunications Act of 1996 was intended to open up new delivery systems for both television and telephone to increased competition, lifting most barriers to media ownership. While limiting companies and individuals from owning TV stations that reached more than 35 percent of the national audience, the act abolished many of the cross-market barriers that had prohibited dominant players from one communications industry, such as telephone, from providing services in other areas, such as cable.[31]

The law's general approach was to replace government regulation with competition as the chief way of assuring that telecommunications services are delivered to customers cheaply and efficiently. It imposed new regulations to help open markets and equalize the burden on competitors, but it also lifted many price controls and other regulations — in some cases before local monopolies are broken. The law eliminated the provision of the 1984 cable act that barred local phone companies from entering the cable-TV market in their service areas, while easing or eliminating the price controls on cable companies.

The law allowed competition, but it did not effectively foster it. Telcos still lag well behind cable companies in providing multichannel video services to consumers.

"In truth, the bill promised the worst of both worlds: More concentrated ownership over communications with less possibility for regulation in the public interest," wrote Robert W. McChesney, a University of Illinois communications professor, in his 2004 book *The Problem of the Media*. "Accordingly, both the cable and telecommunication industries have become significantly more concentrated since 1996, and customer complaints about lousy service have hit all-time highs. Cable industry rates for consumers have also shot up, increasing some 50 percent between 1996 and 2003."[32]

CURRENT SITUATION

Policy Debates

Regulation typically lags behind technological changes, and that is likely to remain the case for television, since it appears unlikely that the 110th Congress will address the changing landscape of television and telecommunications policy in a comprehensive way.

The House passed a major telecom bill in 2006 that would have made it easier for phone companies to enter the video business, but the Senate never completed its version.[33] Since then the political landscape on Capitol Hill has changed dramatically, with Democrats taking control of both chambers in the 2006 congressional elections. While Congress may take up some narrow issues, "the consensus is that there's won't be a comprehensive bill," says James Brad Ramsay, general counsel for the National Association of Regulatory Utility Commissioners.

Meanwhile, the issues that dominated the telecommunications debate last year have migrated to the Federal Communications Commission (FCC) and the states. For instance, Congress sought to streamline existing rules that require video operators — anyone offering video through cable, broadband or fiber-optic phone lines — to negotiate franchising deals with individual cities and towns. The big phone companies, such as AT&T and Verizon, which are spending billions on new fiber-optic networks to provide broadband video services, say current franchise processes slow the rollout of such services and help preserve a competitive advantage for cable companies.

The FCC on Dec. 20 issued new guidelines that require municipalities to respond to TV franchise requests from phone companies within 90 days. The agency also restricted the limits local governments can place on new video service "build-outs," as well as the franchise fees they can charge. Congress is expected to review the guidelines, which local governments have roundly criticized.

Municipalities say they need discretion to determine which companies would best use public rights of way while digging up roads to install their cables and building their infrastructure, since many vendors are competing for the same valuable space. Many cities also want to require new entrants to guarantee access to low-income or underserved areas.

"We want to be able to negotiate these things, not to have them taken off the table by regulatory framework," says Carolyn Coleman, director of federal relations for the National League of Cities.

Telephone companies have also brought their case to numerous states. Since 2005, nine states have streamlined local franchise rules. The laws vary, but in essence the states

Apple CEO Steve Jobs introduces the new iPhone at Macworld on Jan. 9, 2007 in San Francisco. The revolutionary device combines a mobile phone and widescreen iPod with touch controls and an Internet communications device with the ability to use e-mail and Web browsing. It starts shipping in the U.S. in June 2007.

have made it easier for the telephone companies to get their franchises and override the protests of local governments.

Not surprisingly, cable companies aren't happy about these new arrangements. "It's not what you think of as a new entrant needing regulatory protection," says Rick Cimerman, vice president for state government affairs at the National Cable & Telecommunications Association. "AT&T alone dwarfs the entire cable industry."

The telephone companies' hope of getting a national franchise bill that would allow them to compete on favorable terms with cable foundered on their own

The Top 10 TV Programs

There's something for almost everyone among the most popular regularly scheduled television shows, from amateur performers to medical and crime dramas to sports and suburban seductresses.

1.	American Idol — Tuesdays (Fox)
2.	American Idol — Wednesdays (Fox)
3.	Dancing with the Stars (ABC)
4.	CSI (CBS)
5.	Dancing with the Stars Results Show (ABC)
6.	NBC Sunday Night Football (NBC)
7.	CSI: Miami (CBS)
8.	Desperate Housewives (ABC)
9.	House (Fox)
10.	Deal or No Deal — Mondays (NBC)
10.	Without a Trace (CBS)

Source: Nielsen Media Research, 2006

objections — shared by the cable industry — to a provision known as "net neutrality." They even object to the term, saying that it really represents governmental intrusion into the telecommunications business.

Net neutrality bills, which continue to be introduced both in Congress and in a handful of states, would block Internet service providers (ISPs) from charging content providers for priority access.[34] Some companies — including some television companies — want to be able to pay extra to make sure their data can get through quickly and not be slowed by heavy Internet traffic (which can cause video streams to appear jerky). Proponents of the measure say net-neutrality policies would ensure equal access to Internet distribution, preventing a large company from buying up bandwidth and thus blocking or slowing access for competitors or small sites that can't afford to pay for premium treatment.

Another debate likely to rage in Washington this year involves media-ownership rules. In the 1996 Telecommunications Act, Congress barred media companies or individuals from owning television stations serving more than 35 percent of the U.S. population and required the FCC to review media-ownership rules every four years. In 2003, the commission voted to ease the restrictions and allow

station owners to reach a combined 45 percent of the national audience. A federal appeals court then struck down the deregulation as "arbitrary and capricious."[35]

As the FCC undertakes its quadrennial review in 2007, broadcasters hope the agency will ease ownership rules. Because the media landscape is shifting so rapidly, traditional TV providers cannot be handcuffed if they are to remain competitive, argues Wharton of the National Association of Broadcasters. "We have asked the government to explore some modest changes to some of the rules that would preserve and enhance free, over-the-air broadcasting," he says, "such as rules that bar a newspaper owner from owning stations in the same market, or partnering with another station in same market."

FCC Chairman Kevin J. Martin is sympathetic to calls for deregulation and has said he hopes to revisit the newspaper-broadcast cross-ownership ban. But a strong backlash against the 2003 deregulation has put pressure on Martin and other commissioners to be more sensitive to public opinion. Moreover, Democrats in Congress will be skeptical about attempts to ease media-ownership rules, with House Telecommunications Subcommittee Chairman Edward J. Markey, D-Mass., particularly vocal about the issue. (*See "At Issue," p. 231.*)

"The paradox is that you'll have some people telling you there are fewer and fewer companies owning more and more of the media, so there is really very little access to diversity, yet other people are telling you that we have a world without gatekeepers, that anyone can post anything they want on the Internet," says Jenkins of MIT.

"Both of those statements are true."

All-You-Can-Eat TV

Outside of the regulatory and policy arena, technology appears to be putting everything about the nature of television up for grabs. A medium that has long been primarily underwritten by advertising is watching its audience crack and break into a million pieces. Viewers accustomed to flipping through channels now are finding

Should the government preserve limits on media ownership?

YES
Rep. Edward J. Markey, D-Mass.
Chairman, House Subcommittee on Telecommunications and the Internet

From keynote address, National Conference for Media Reform, Jan. 13, 2007

It is often said that our system of democratic self-government relies on an informed citizenry. Informed citizens need to know enough to make decisions in a democracy. And they need to know [not only] raw information but also context as well as the history of issues.

Media ownership is a key tool utilized in this policy context. And that's because diversity of ownership has historically been used as a proxy for diversity of viewpoints and diversity of content. Simply put, therefore, elimination of ownership limits eradicates an important tool we have to help ensure that the public has access to a wide array of viewpoints in local news and information.

In 2003, we were challenged. We were challenged by the drastic and indiscriminate elimination of mass-media ownership rules proposed by the previous Federal Communications Commission [FCC] under its former chairman. In response to pressure from special political and corporate interests, the FCC, on a 3-to-2 vote, rammed through changes to media ownership that would have eviscerated the public-interest principles of diversity and localism.

The FCC's plan did not create more entertainment and information sources for consumers. Nor did it enhance the ability of the broadcasting medium to meet the informational and civic needs of the communities it serves. Instead, it threatened to intensify control of information and opinion in entire cities and regions of the country. The aggregate effect would have encouraged the rapid consolidation of mass-media ownership in this country and the elimination of diverse sources of opinion and expression.

The good news is that the challenge was answered. People took notice, took action and went to court. Congress also responded and enacted some limits, and the court shut down the rest of the sweeping changes and sent the plan back to the FCC. But it was only a temporary reprieve. Today, the FCC is embarked upon another round of analysis and is re-examining whether to change the media-ownership rules.

The communications revolution has the potential to change our society. Unless we continue to revere localism and diversity, we risk encouraging a new round of "communications cannibalism" in mass-media properties on both the national and local levels that would put real progress in bolstering minority ownership of media even further away.

NO
Bruce M. Owen
Director, Public Policy Program Stanford University

Written for *CQ Researcher*, Feb. 15, 2007

Despite the activist hype about "media power," the number of alternatives for viewers is increasing, not decreasing. Economic competition among media for viewers and advertisers is greater now than at any time in history. So is economic competition among the wired and wireless pipelines that carry content to the home.

Competition in the marketplace of ideas, one of the keys to democracy, is more robust than ever, thanks to the ubiquitous Internet. FCC ownership rules, which only apply to the old, regulated media, make it harder for the older media to compete with the new. This will ultimately disadvantage those consumers who still prefer the traditional technologies.

Antitrust laws protect consumers against mergers that reduce competition. The way to tell if competition is threatened is to ask whether consumers will end up with too few choices. Advertisers already have many alternatives to regulated broadcast media, even without considering online advertising. TV viewers use cable or satellite or DSL and online services, each with essentially unlimited channel capacity. These multi-channel media compete among themselves and with newer, wireless high-speed services such as EV-DO and Verizon's BroadbandAccess.

Anyone who thinks media moguls are monolithic gatekeepers standing between freedom of speech and the public should consider how Sen. Hillary Rodham Clinton recently chose to announce her candidacy for the presidency: on her Web site, with an online video. Other candidates did the same. One reason: To bypass any "spin" imparted by traditional media reporting. More generally, as the Federal Election Commission proclaimed last spring: "[T]he Internet's near infinite capacity, diversity and low cost of publication and access has democratized the mass distribution of information, especially in the political context. The result is the most accessible marketplace of ideas in history."

Neither FCC rules nor new laws restricting media ownership make sense. We have perfectly good antitrust laws and enforcers, including the courts, to deal with threats to economic competition in the advertising and video entertainment markets.

Catering to misinformed populist ideas is not a costless indulgence. Traditional media are trying to remain relevant in the new media world. Some consumers would be inconvenienced by their premature demise. This is not the time to increase their costs of operation by maintaining or even increasing regulatory constraints that make them less competitive.

CQ Press/iTunes Screenshot

Shows like ABC's wildly popular "Ugly Betty," starring America Ferrera (right), face new competition for viewers. With higher production costs and a declining viewership, sitcoms and other scripted TV programs are being pushed aside by reality shows such as "American Idol."

that their primary loyalty may be to specific programs, which in turn may be delivered to them through any number of different media. And watching television, which has always been primarily a passive, relaxing activity, is becoming more interactive, with more viewers potentially engaging with content — and trying to find it.

One of the major questions facing the television industry today is whether its traditional means of delivery — networks of affiliated stations — can survive the changes.

"When we see that big box in the living room, we think of channels," says Andrew Kantor, a technology reporter for the *Roanoke Times* and columnist for USAToday.com. "There's no reason for television to be divided by that, other than convention."

It's possible now for viewers to access their favorite shows without giving any thought to what channel it's on. It might help them find a show like "Lost" on the Internet if they know what network presents it, but they certainly need no longer know that it's scheduled to be on their local Channel 5 on Wednesdays at 10 p.m. If they have a TiVo or DVR, they can program it to record "CSI" automatically by series name, not by channel or broadcast time.

"We don't watch Fox, we watch '24,' " says Kantor. "The only reason to have channels today is to help us navigate the content. Who cares if the show is on Fox or CBS or Jimmy's FunStuff channel?"

Some niche networks are considering becoming broadband channels, allowing viewers to watch their programming directly via the Internet without their even having a presence on cable or satellite systems. Companies such as Wal-Mart and Amazon are also lining up to send content directly to viewers. Access will expand even further as new devices become available that can send Internet content to television sets. Apple is expected to release its Apple TV device this month, with major competitors putting out their versions soon. Video game players have already grown accustomed to switching seamlessly from games to live network video feeds on their televisions.

But all the longstanding television services — networks, local affiliates and cable and satellite — have advantages that offer them hope of thriving well into the future, assuming they can adapt to changing circumstances. For instance, while viewers may soon be able to use the Internet or other means to bypass local affiliate stations, it will take time for traditional viewing habits to change. Local stations are a familiar and easy way to find television shows, and they provide local news, emergency and weather information that Internet providers do not.

As part of a 2005 budget bill, Congress set 2009 as the deadline for broadcasters to switch from analog to digital broadcasting. Lawmakers wanted both to promote digital TV, which provides a crisper picture, and to free up broadcast spectrum for wireless and emergency communications systems.

Although broadcasters opposed the imposition of a specific deadline, they now say it provides them with a competitive advantage. The move to all-digital television broadcasting will enable individual stations to send out multiple signals, essentially providing multiple channels. So, in addition to airing network programs, the station can simultaneously broadcast news, weather and sports coverage produced in-house.

"TV networks have the franchise content people want to watch," says CBS' Poltrack. "The local station is the best way to get it."

That remains true despite the present industry upheaval, Poltrack says, largely because of the advent of high-definition television, which provides a much clearer picture than standard television. Watching a television show via Internet streaming cannot match the picture quality offered by TV now that most prime time shows are being produced using HDTV technology.

"I can't tell you how many people have told me they'd rather watch grass grow in HD than their favorite show in standard television," says ABC's Mandler.

And networks themselves maintain advantages in the new age, he argues. While content producers may try to sell their products directly to viewers — bypassing the networks and other traditional middlemen — the costs of marketing individual shows can be prohibitive, he points out. While producer-to-consumer programs could become instant word-of-mouth hits just as low-budget videos get e-mailed to millions today, he concedes, they would be exceptional.

Networks also own the rights to decades' worth of old programming. While some people enjoy owning DVDs of their favorite shows, as video-on-demand services become more common more people will prefer to dial up old episodes of "I Love Lucy" or "My Mother the Car" owned by the networks and their partners. Cable services such as Comcast already make hundreds of movies available to subscribers at any time "on demand."

And as the viewing audience continues to fragment, Mandler says, "the folks who can stitch together a national audience and sell it to advertisers are the networks." Even cable and satellite companies — and relatively new broadband services offered by telephone companies — are too fragmented to promise a truly national audience.

Nielsen Media Research, the leading TV audience-measurement company, is working to find new ways to measure the fracturing audience. It will soon offer minute-by-minute ratings that also measure ad and DVR playback viewership, although some in the TV industry are skeptical they can deliver this as quickly as promised. The company is also experimenting with a 400-member video iPod viewership panel and is "moving rapidly" on measuring video viewership on the Web, according to Scott Brown, a Nielsen senior vice president.

"If it's measurable, advertisers will want it," says Jin Kang, an executive producer with NDS, a digital video company. "As long as you can measure viewership, advertisers don't really care how they looked at it."

When industry officials first began to notice the fracturing of the TV audience, the changes in video delivery technology and the growth of user-generated content, "There was a certain amount of panic in the room," concedes Peter Olsen, senior vice president of national ad sales for A&E Television Networks. "Now, there's more enthusiasm. We

Time selected "You" — as in YouTube — as its 2006 Person of the Year, reflecting the growing popularity of user-generated video content and the increasing participation of everyday citizens in the next Internet generation.

feel like what we do well — which is produce great content — consumers will continue to enjoy."

OUTLOOK

Shifting Landscape

The television industry today feels a bit like the dot-com boom a decade ago. There is endless experimentation, rapid technological change, a host of start-ups and a scramble among established companies to form new partnerships and adapt to a rapidly shifting landscape. There will be plenty of false starts — and big wins, like the sale of YouTube to Google for $1.65 billion last October.

"The changes of the next five years will dwarf the changes of the last 50," said Jeff Zucker, chief executive of NBC Universal's television group, as he announced a restructuring of the company last fall.

But no one really knows what those changes will look like. Viewers may use new devices to access nearly any video content they want, and networks and cable companies may make more and more content available "on demand." Almost everyone agrees that television and the Internet will offer increasingly overlapping experiences.

But all the new technology involves, well, new technology. For now, watching videos through the Internet remains more cumbersome than watching on traditional TV — which also delivers superior picture quality. And millions of people like watching television by simply turning it on and watching.

Some people predict that the big money in future television will be in search engines. "It's great to be given the keys to the Library of Congress, but if there's no card catalog, it's not much use," says Todd Herman, a new-media strategist for Microsoft.

For now, despite the advent of consumer-created video and the wide array of choices, most viewers are watching more TV than ever and are using new devices and platforms to keep up with favorite shows by watching at a more convenient time or when "nothing's on."

And networks and advertisers appear increasingly confident that they will be able to measure viewership, and advertisers hope they will be able to more accurately target the right viewers for messages about the appropriate products.

Meanwhile, interactive TV — talked about for years — is just taking root in this country. Interactive TV allows a viewer to make selections to "go" to different places on their set, similar to selecting the movie, special features or language features while watching a DVD. Not too many people are choosing to navigate through an advertiser's showplace on their television set, but enough are starting to do so to make it worthwhile.

But how will the splintering of viewership affect the nation and its culture? "Ultimately, the biggest story of the 21st century will be the fracturing of the 20th-century audience," says Thompson, of the Center for the Study of Popular Television. "We spent the first eight decades of the 20th century putting together the biggest mass audience of all time. You had virtually everybody — old and young, rich and poor — feeding from the same cultural trough at least a few hours a week, if not a few hours a day."

Thompson doesn't think it's a coincidence that the citizenry has become more divided politically at the same time that its main source of popular culture and information has become more splintered. NBC News anchor Brian Williams seems to agree. "It is now possible — even common — to go about your day in America and consume only what you wish to see and hear," he writes.[36]

But others point out that the Internet has also made it easier for people of like-minded interests to find each other and form communities.

And many viewers prefer to make their own entertainment choices, rather than relying on the programming judgments of network executives. "The common culture of my youth is gone for good . . . splintered beyond repair by the emergence of the Web-based technologies that so maximized and facilitated culture choice as to make the broad-based offerings of the old mass media look bland and unchallenging by comparison," writes critic Terry Teachout. "For all the nostalgia with which I look back on the days of the Top 40, the Book-of-the-Month Club and 'The Ed Sullivan Show,' I prefer to make my own cultural decisions, and I welcome the ease with which the new media permit me to do so."[37]

It's an age of choice, and no one is certain which formats viewers will ultimately favor. But whether people are watching programs on demand, via the Internet or on their local station, it's clear that Americans will continue watching lots of television into the foreseeable future.

"If I had to describe the future of TV in one word," says Mike Bloxham, director of research at Ball State University's Center for Media Design, "it would be 'more.'"

NOTES

1. Jennifer Mann, "Ads Mimic 'Viral Videos,' " *The Kansas City Star*, Feb. 3, 2007, p. C1.

2. Matthew Creamer, "John Doe Edges Out Jeff Goodby," *Advertising Age*, Jan. 7, 2007, p. S-4.

3. Lev Grossman, "Person of the Year: You," *Time*, Dec. 25, 2006-Jan. 1, 2007, p. 40.

4. Gary Holmes, "Nielsen Media Research Reports Television's Popularity Is Still Growing," Nielsen Media Research, news release, Sept. 21, 2006.

5. For background, see Alan Greenblatt, "Future of the Music Industry," *CQ Researcher*, Nov. 21, 2003, pp. 988-1012.

6. Frank Rose, "ESPN Thinks Outside the Box," *Wired*, September 2005, p. 113.

7. Brian Steinberg, " 'Law & Order' Boss Dick Wolf Ponders Future of TV Ads," *The Wall Street Journal*, Oct. 18, 2006.

8. "Forum With Michael Krasny," KQED, Dec. 16, 2006; available for streaming at www.kqed.org/epArchive/R612131000.

9. Andrew Sullivan, "Video Power: The Potent New Political Force," *Sunday Times of London*, Feb. 4, 2007, p. 4.

10. Thomas Goetz, "Reinventing Television," *Wired*, September 2005, p. 104.

11. Kevin Downey, "Milestone: Cable Widens Lead in 18-49s," *Medialife Magazine*, April 21, 2006.

12. Henry Jenkins, *Convergence Culture* (2006), p. 66.

13. Quoted in Lewis Lazare, "Changing Face of Ads Examined," *Chicago Sun-Times*, Dec. 4, 2006, p. 63.

14. Marc Ransford, "New Study Has Good and Bad News for Television Advertising Industry," press release, Ball State University, Sept. 26, 2006, www.bsu.edu.

15. Lisa Rockwell, "Web Sparks Advertising Revolution," *Austin American-Statesman*, Dec. 17, 2006, p. H1.

16. Frank Rose, "The Fast-Forward, On-Demand, Network-Smashing Future of Television," *Wired*, October 2003.

17. Jenkins, *op. cit.*, p. 59.

18. Verne Gay, "How iTunes Saved 'The Office,' " *Newsday*, Nov. 1, 2006, p. B21.

19. Gary Levin, "Networks Have Eyes on the Prize," *USA Today*, Dec. 18, 2006, p. 1D.

20. Richard Verrier, "No Time for Making New 'Friends' at NBC," *Los Angeles Times*, Nov. 7, 2006, p. C1.

21. *Ibid.*

22. "Key Issues in ACTRA's Strike," www.actra.ca/actra/control/feature14.

23. For background, see "Radio Development and Monopoly," *Editorial Research Reports*, March 31, 1924; available at *CQ Researcher Plus Archive*, http://library.cqpress.com.

24. Andrew Crisell, *A Study of Modern Television* (2006), p. 17.

25. For in-depth background on the development of television, see "Television," *Editorial Research Reports*, July 12, 1944, available at *CQ Researcher Plus Archive*, http://library.cqpress.com; and Andrew F. Inglis, *Behind the Tube* (1990), p. 237.

26. George Comstock and Erica Scharrar, *Television: What's On, Who's Watching, and What It Means* (1999), p. 6.

27. Inglis, *op. cit.*, p. 360.

28. *Ibid.*, p. 365.

29. *Ibid.*, p. 385.

30. For background, see the following *CQ Researchers*, available at *CQ Researcher Plus Archive:* "Cable Television: The Coming Medium," Sept. 9, 1970, "Television in the Eighties," May 9, 1980; "Cable TV's Future," Sept. 24, 1982; "Cable Television Coming of Age," Dec. 27, 1985; "Broadcasting Deregulation," Dec. 4, 1987; Kenneth Jost, "The Future of Television," Dec. 31, 1994, pp. 1129-1152.

31. For background, see David Masci, "The Future of Telecommunications," *CQ Researcher*, April 23, 1999, pp. 329-352.

32. Robert W. McChesney, *The Problems of the Media* (2004), p. 53.

33. Joelle Tessler, "2006 Legislative Summary: Telecommunications Overhaul," *CQ Weekly*, Dec. 16, 2006, p. 3370.

34. For background, see Marcia Clemmitt, "Controlling the Internet," *CQ Researcher*, May 12, 2006, pp. 409-432.

35. Joelle Tessler, "Diversity Debate Shapes Media Ownership Rules," *CQ Weekly*, Jan. 27, 2007, p. 302.

36. Brian Williams, "Enough About You," *Time*, Dec. 25, 2006-Jan. 1, 2007, p. 78.

37. Terry Teachout, "Culture in the Age of Blogging," *Commentary*, June 2005, www.terryteachout.com/archives20070204.shtml#108419.

BIBLIOGRAPHY

Books

Carter, Bill, *Desperate Networks*, Doubleday, 2006.
The *New York Times* TV reporter recounts behind-the-scenes stories of how many of today's hottest shows made it on the air (and, in many cases, almost didn't).

Jenkins, Henry, *Convergence Culture: Where Old and New Media Collide, NYU Press*, **2006.**
The director of MIT's comparative media-studies program looks at how content is flowing across multiple media platforms and how audiences are migrating to watch and interact with it.

Marc, David, and Robert J. Thompson, *Television in the Antenna Age: A Concise History, Blackwell Publishing,* **2005.**
The authors, both affiliated with Syracuse University, summarize both technological evolution and the content presented during TV's first 50 years.

Articles

Creamer, Matthew, "John Doe Edges Out Jeff Goodby," *Advertising Age*, **Jan. 8, 2007, p. S-4.**
User-generated videos that feature popular consumer products are, in cases such as "The Diet Coke & Mentos Experiment," doing a better job of selling those products than paid advertising, making the consumer the "agency of the year."

Garfield, Bob, "YouTube vs. Boob Tube," *Wired*, **December 2006.**
Advertising Age's editor-at-large argues the fractured media universe means TV will lose its ad-dollar dominance but makes it clear that no one is sure how to make money sponsoring user-generated content.

Grossman, Lev, "Person of the Year: You," *Time*, **Dec. 25, 2006-Jan. 1, 2007.**
The next generation of the Web involves more participation and creativity from users, greater consumer control of video and text content and more news being made and covered by average people.

Levin, Gary, "Networks Have Eyes on the Prize," *USA Today*, **Dec. 18, 2006, p. 1D.**
With scripted-programming costs rising, all the major networks have multiple game shows in production or development.

McHugh, Josh, "The Super Network," *Wired*, **September 2005, p. 107.**
The writer argues that Yahoo! has taken the lead in formulating intuitive ways to help people search through millions of hours of programming in a comprehensible way.

Rockwell, Lisa, "Web Sparks Advertising Revolution," *Austin American-Statesman*, **Dec. 17, 2006, p. H1.**
With so many viewers watching video via the Internet and new devices, advertisers wonder how long a 50-year-old business model built on expensive TV advertising will be sustained.

Sullivan, Kevin, "Regular Folks, Shooting History," *The Washington Post*, **Dec. 18, 2006, p. A1.**
Devices such as cell-phone cameras are making it easier for non-journalists to capture images of the news as it happens, and there is an increasing market for their pictures and videos online and through established news outlets.

Verrier, Richard, "No Time for Making 'Friends' at NBC," *Los Angeles Times*, **Nov. 7, 2006.**
NBC's decision to cut back on scripted programming in prime time is a signal that such programming, which is expensive to produce, is fading fast — particularly sitcoms.

Reports and Studies

Berman, Saul J., Niall Duffy and Louisa A. Shipnuck, "The End of Television As We Know It," *IBM Institute for Business Value*, **2006.**
The generations-old model of a TV audience happily embracing scheduled programming is coming to an end. Over the next several years, programmers, networks and their competitors will have to stay ahead of "gadgeteers," who will lead passive viewers into a more interactive future.

Foehr, Ulla G., "Media Multitasking Among American Youth: Prevalence, Predictors and Pairings," *Kaiser Family Foundation*, **December 2006.**
More than 80 percent of 7th-to-12th-graders use more than one media device at a time on a regular basis, but their primary loyalty is to television.

Roberts, Donald F., Ulla G. Foehr and Victoria Rideout, "Generation M: Media in the Lives of 8-18 Year-Olds," *Kaiser Family Foundation*, **March 2005.**
A national survey of children and teens finds that most live in homes with access to multiple media outlets. Since there are only so many hours in a day, kids increase their already heavy "media diets" through multitasking — using a computer or listening to music while watching TV, rather than devoting their attention to one device at a time.

For More Information

Association of National Advertisers, 708 Third Ave., New York, NY 10017; (212) 697-5950; www.ana.net. A trade association for the marketing community that follows industry trends, offers networking and training opportunities and lobbies on behalf of advertisers.

Center for the Digital Future, University of Southern California Annenberg School for Communication, 300 S. Grand Ave., Suite 3950, Los Angeles, CA 90071; (213) 437-4433; www.digitalcenter.org. A research and policy institute devoted to the study of mass media and evolving communication technologies.

Center for the Study of Popular Television, S. I. Newhouse School of Public Communications, Syracuse University, Syracuse, NY 13244; (315) 443-4077; http://newhouse.syr. edu. An academic center that supports research into all aspects of television and popular culture.

Comparative Media Studies Program, Building 14N-207, Massachusetts Institute of Technology, 77 Massachusetts Ave., Cambridge, MA 02139; (617) 253-3599; http://cms. mit.edu. Sponsors conferences and encourages students to understand media changes that cut across delivery techniques and national borders.

Federal Communications Commission, 445 12th St., S.W., Washington, DC 20554; (888) 225-5322; www.fcc.gov. The federal agency charged with regulating interstate and international communications by radio, television, wire, satellite and cable.

National Association of Broadcasters, 1771 N St., N.W., Washington, DC 20036; (202) 429-5300; www.nab.org. A trade association that lobbies Congress, the FCC and the judiciary on behalf of more than 8,300 local radio and television stations and the broadcast networks.

National Association of Television Program Executives, 5757 Wilshire Blvd., Penthouse 10, Los Angeles, CA 90036; (310) 453-4440; www.natpe.org. Serves as a clearinghouse for information and convenes meetings for professionals involved in the creation, development and distribution of TV programming.

National Cable Television Association, 25 Massachusetts Ave., N.W., Suite 100, Washington, DC 20001; (202) 222-2300; www.ncta.com. The principal trade association of the cable television industry, representing 200 cable networks and cable operators who serve more than 90 percent of the nation's cable TV households.

11

Drinking on Campus

Have Efforts to Reduce Alcohol Abuse Failed?

Barbara Mantel

Phanta "Jack" Phoummarath, right, was 18 when he died from alcohol poisoning at a University of Texas fraternity party last December. His parents, left, sued the fraternity, Lambda Phi Epsilon, alleging Jack had been forced to drink heavily during a hazing ritual. "We don't want any family to go through what we have had to go through," said his sister. More than 1,700 college students die annually in alcohol-related incidents.

High-school softball player Laura Smith hoped to play on a varsity team in college, and like thousands of other athletes, she visited various schools during her senior year. But during a visit last March to California State University at Chico, the 17-year-old's college career almost ended before it started.

While attending a party hosted by team members, Smith (not her real name) became unresponsive after heavy drinking. Luckily, someone called 911. At 1 a.m., paramedics rushed her to a hospital to be treated for alcohol poisoning.* Smith was released five hours later.

Following Smith's close call, Chico State Athletic Director Anita Barker promptly canceled the remainder of the softball season and kicked the players who attended the party off the team, several of whom were underage. "This is not the kind of behavior we expect," Barker said. "The magnitude and gravity of the situation warrants swift and decisive action that sends a message this behavior is unacceptable."[1]

In fact, several alcohol-related deaths have occurred at Chico State. Last year, a 21-year-old fraternity pledge died during a hazing ritual. In 2001, a 19-year-old was hit by a train after passing out drunk on the tracks. And the year before, an 18-year-old student died after drinking a bottle of brandy at a fraternity party.[2]

While the deaths at Chico and other campuses capture media and public attention, they only hint at the extent of heavy drinking

From *CQ Researcher*,
Augest 18, 2006.

---

* Consuming large amounts of alcohol can lead to unconsciousness and then death because vital organs, such as the heart and lungs, can be slowed to the point of stopping.

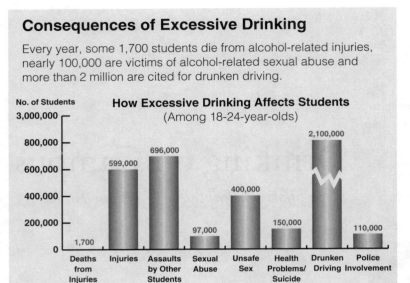

Consequences of Excessive Drinking

Every year, some 1,700 students die from alcohol-related injuries, nearly 100,000 are victims of alcohol-related sexual abuse and more than 2 million are cited for drunken driving.

No. of Students

How Excessive Drinking Affects Students
(Among 18-24-year-olds)

3,000,000		
800,000		2,100,000
600,000	696,000	
	599,000	
400,000		400,000
200,000		150,000
0	1,700	97,000	110,000

Deaths from Injuries — Injuries — Assaults by Other Students — Sexual Abuse — Unsafe Sex — Health Problems/ Suicide Attempts — Drunken Driving — Police Involvement

Source: National Institute on Alcohol Abuse and Alcoholism

among college students and the harm that results. In the early 1990s, Harvard University's School of Public Health showed that 44 percent of students questioned in a nationwide survey reported binge drinking — consuming five or more drinks in a row for men and four or more for women — during the two weeks prior to the survey. Other studies have shown similar results.[3]

Since then, both houses of Congress have passed resolutions asking college presidents to address the problem, and colleges have instituted various initiatives to reduce college drinking. The National Institute on Alcohol Abuse and Alcoholism formed a special task force in 1998, and the surgeon general in 2000 set a goal of reducing college binge drinking by 50 percent within 10 years.

Surveys show that goal will be difficult, if not impossible, to meet: Heavy, episodic drinking continues to be a stubborn problem on campuses across the United States.

"Drinking is a very big part of the American college experience," says Emma Ross, a rising senior at the University of Chicago. "Rarely does someone participate in an evening activity (aside from going to the movies) that doesn't involve drinking, and when they do drink it is not uncommon for someone to get incredibly drunk. Also, large college parties are generally not very fun if

one isn't drunk. They are loud and impersonal, so no one would want to be at a party and not be drinking."

Depending on the study, the number of college students who binge drink is holding steady or falling only slightly.

For instance, Harvard's "College Alcohol Study" showed that "remarkably similar proportions of students were classified as binge drinkers in 2001, as in previous survey years (44.4%)."[4] The University of Michigan's "Monitoring the Future" survey finds a similar percentage: 41.7 percent in 2004, roughly the same as in the early 1990s but down slightly from 1980, when the survey began. (*See graph, p. 251.*)

"I would definitely say that there is a binge-drinking culture at the University of Southern California," says Sean Fish, a senior at the Los Angeles campus. "Many students, many of my friends and myself, are guilty of binge drinking." But Fish considers it a problem only when it happens two or more times a week.

While studies show that frequent binge drinking is becoming more common, paradoxically the percentage of students abstaining from alcohol is also increasing, indicating that student drinking habits are becoming more polarized between those who don't partake and those who do — heavily.

How alarming are these trends? That depends on whether one focuses on the sizable minority who binge drink or the majority who don't.

"Alcohol use on college campuses is certainly a problem, but hardly the epidemic it is made out to be," writes Aaron White, an assistant professor of psychiatry and an alcohol researcher at Duke University Medical Center. "Not all college drinkers get out of hand, drink to get drunk or require treatment for alcohol poisoning."[5]

It may not be an epidemic, but it is a chronic problem, says Ralph Hingson, a professor of public health at Boston University and an investigator at its Youth Alcohol Prevention Center. "If over 40 percent of college students are consuming five or more drinks on any one occasion, I consider that a very substantial minority," says Hingson. "And remember, drinkers don't just harm themselves."

Excessive drinking causes a variety of negative consequences, including blackouts, injury, unintended and unprotected sex, impaired driving, poor grades, property damage, fights, arrests and sexual violence. (*See graph, p. 240.*) In the 2001 "College Alcohol Study," about 30 percent of students who drank in the past 30 days reported missing a class because of alcohol, 21 percent had unplanned sex because of alcohol and about 13 percent reported getting injured — all somewhat higher than in 1993.[6]

There are also secondhand effects, as Hingson points out. About 48 percent of students had to care for a drunken student, 60 percent had their sleep or study interrupted and about 20 percent experienced an unwanted sexual advance.[7]

The fatal consequences of heavy drinking are also on the rise. The number of alcohol-related deaths among college students has increased faster than the college population. In 2001, an estimated 1,717 college students ages 18 to 24 died from alcohol-related accidental injuries, including motor vehicle crashes; that's an increase of 6 percent (adjusted for college population growth) over the 1,575 deaths in 1998.[8]

"How can this be if the percentage of students that binge drink has remained relatively stable?" Duke University's White asks rhetorically. He explains that categorizing students as either binge or non-binge drinkers does not reveal *how much* they actually drink. For instance, a female student would be considered a binge drinker if she had four drinks or 40 drinks, but the danger from consuming 40 drinks would be quite different.

"It is entirely possible that peak levels of consumption beyond the binge threshold have been skyrocketing for years, and we have simply missed this fact," says White.[9]

He recently reported that roughly half of all males categorized as binge drinkers actually consumed 10 or more drinks, twice the binge threshold, at least once in the two weeks before being questioned.[10] However, the study — using data from a single, online college survey — did not measure whether the amount consumed by bingers has changed over time.

Whether they drink five or 20 drinks in a single sitting, bingers are not a representative slice of the college population. "I see problem drinking more among

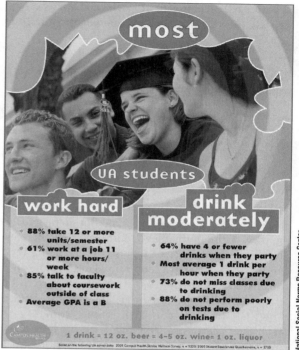

Posters displayed at the University of Arizona use social norms marketing to change students' misperception about their peers' drinking habits. An increasing number of schools are using social norms marketing to reduce social pressure to binge drink.

freshmen," says Caroline Stevens, who graduated last spring from Beloit College in Wisconsin. "By the time you graduate, drinking is not such a big deal, and it's much less common to get drunk."

Stevens' impressions are statistically correct. According to several studies, the students who drink the most are males, whites, fraternity members, athletes and first-year students. Students who drink the least are those who attend two-year institutions, religious schools, historically black colleges and universities and commuter schools.[11]

Even though many colleges and universities have tailored and directed their alcohol messages to those most at risk, the rate of heavy, episodic college drinking shows little or no improvement.

"For the most part, schools have not based their prevention efforts on strategies identified and tested for effectiveness by research," according to a report from the National Institute on Alcohol Abuse and Alcoholism.[12]

In today's environment of heightened awareness but unimpressive results, here are some of the questions being debated:

Do social norming and online tools convince students to drink more responsibly?

"I honestly am not sure what a college can do about drinking. I think a lot of students just view it as part of the college experience."

— *Julie Price, senior, The College of William and Mary, Williamsburg, Va.*

For decades schools have tried to educate students about the risks of excessive drinking. But simply providing them with information doesn't seem to work. Now an increasing number of schools are changing their approach by turning to a controversial method known as social norms marketing.

The technique is based on a simple premise. Studies show that students consistently overestimate how much their peers drink and underestimate how many abstain. By giving students accurate information, such as that less than a majority — 44 percent — of students nationwide binge drink, social norms marketing campaigns hope to reduce the social pressure to drink to excess.

Like many students, William and Mary's Price overestimates her peers' drinking habits. "I would say that binge drinking probably occurs among a majority of students on a regular basis," she says.

Social norms marketing campaigns usually involve a series of ads in various campus venues. At Florida State University (FSU), for example, the campaign features ads on campus shuttle buses, billboards and radio and TV spots, says Mary B. Coburn, vice president for student affairs. "A typical ad has been a picture of actual students debunking the myth that everybody is doing it," she says.

FSU's campaign, begun in 2000, is funded by the Anheuser-Busch Foundation. The alcoholic-beverage industry often funds social norms campaigns, as do federal and state agencies. Proponents say social norming celebrates what students are doing right rather than denouncing what they are doing wrong, and that it works.

"We have case studies at different universities showing that when a norm campaign corrects misperceptions of drinking . . . we get parallel reductions in actual heavy drinking behavior and in negative consequences," says Michael Haines, director of the National Social Norms Resource Center, at Northern Illinois University in Dekalb.

But Henry Wechsler, director of Harvard's "College Alcohol Study," calls the case studies "very rudimentary." For instance, they used no control group for comparison, says Wechsler, and they did not take into account other alcohol programs on campus that may influence drinking behavior. FSU, for instance, also has limited students' access to alcohol and has implemented programs to educate students about the risks of drinking.

Wechsler's own study — comparing the results of the "College Alcohol Study" at schools that say they use social norms marketing with schools that don't — concluded that there is no difference in student drinking behavior between the two. "If you really believe that the industry is going to back a method that is going to cut down on sales of alcohol, then you're an idealist," says Wechsler. "Their job is to sell alcohol."

But proponents of social norms marketing attacked Wechsler's study. "Given the prestigious academic platform of a Harvard project, the public would logically assume that the study has merit," wrote researchers at the Higher Education Center for Alcohol and Other Drug Abuse and Violence Prevention, a nonprofit funded by the U.S. Department of Education. "But it does not."[13]

Wechsler's study, the researchers said, failed to distinguish between schools doing legitimate social norms marketing and those that said they were doing the marketing but in fact were not making a serious effort. Putting up a few posters, these researchers said, does not constitute a social norms marketing campaign.

Experimental studies are needed, but so far only a limited one has been conducted. Researchers at San Diego State University introduced a social norms marketing campaign into one residence hall at a large university and used another residence hall as a control for comparison. "The campaign successfully corrected students' misperceptions of drinking norms but had no effects, or counterintuitive effects, on drinking behaviors," the study concluded.[14] However, the campaign was short, just six weeks, and the study took place at only one school.

The results of the first national experimental study of social norms marketing will be published in the *Journal of Studies on Alcohol* in a few months. Project Director Laura Gomberg Towvim, of the Higher Education Center, says a social norms marketing campaign implemented for three years at nine schools produced "a protective effect against increases in alcohol consumption" compared to nine control schools. In other words, drinking did not decline at the schools where the campaign was implemented, but neither did it increase, as it did at schools without a campaign.

Given that only a protective effect was observed, experts expect the debate about the efficacy of social norms marketing to continue.

Meanwhile, colleges increasingly are using online tools to encourage students to reduce excessive drinking. Nearly 2,400 college campuses are using an interactive course called "Alcohol 101 Plus," created by the Century Council, a nonprofit funded by American distillers that promotes responsible drinking. And more than 500 colleges and universities, up from 23 three years ago, now require students to take "AlcoholEdu," developed by Outside the Classroom, a private company that receives no alcohol-industry money.

"We have two simple goals: to promote abstention and reduce risk for drinkers," says Brandon Busteed, Outside the Classroom's founder and CEO. "But nowhere in the course do we tell a student not to drink." Instead, it explains the risks of behaviors like drinking and driving, chugging and playing drinking games.

Students begin by answering confidential, pre-survey questions about their drinking behaviors. They then take the course, which uses audio-visual presentations and is customized depending on the survey answers.

Students enter their height, weight, sex and typical drinking pattern and are told how soon they will reach levels of impairment. They also are shown videos of drinking situations and asked to make decisions about when to drink and how much. There are short quizzes, a final exam and a short post-course survey. The entire process takes about three hours. Students are given a follow-up survey two months later.

According to a recent study, a group of fraternity and sorority members who took the online course had

West Virginia University students gamble at one of the school's Up All Night events. The popular program offers alcohol-free weekend entertainment alternatives to the heavy-drinking college party scene.

modest reductions in heavy drinking and its negative consequences compared to a control group that did not take the course. But, overall, research about such online programs is scant, and even the author of the study says he's surprised by AlcoholEdu's rapid adoption, given the lack of hard evidence that it works.[15]

"There is still plenty to prove, and we are aware of that," says Busteed.

Ralph Blackman, president and CEO of the Century Council, agrees. "But that doesn't mean you don't do it," he says. "Education is the key to helping students make responsible decisions about alcohol."

Focusing on the individual decisionmaker, however, appears not to be enough.

"By just focusing on the individual, we ignore the broader environment that encourages heavy alcohol use by students," says Traci Toomey, director of the alcohol epidemiology program at the University of Minnesota.

Signs of Alcohol Poisoning

Alcohol is a depressant and slows down many of the body's vital functions. Consuming large amounts of alcohol can lead to unconsciousness and then death because the heart and lungs can be slowed to the point of stopping.

Alcohol poisoning may have occurred if there are signs of:

- Mental confusion, stupor, coma or inability to be roused
- Vomiting
- Seizures
- Slow breathing (fewer than eight breaths per minute)
- Irregular breathing (10 seconds or more between breaths)
- Hypothermia (low body temperature), bluish skin color, paleness

If alcohol poisoning is suspected:

- Call 911. Do not try to gauge drunkenness. Do not wait for all symptoms to be present.
- Try to wake the person.
- Turn the person on her/his side so that if vomiting occurs, the airway will not be blocked. Stay with the person so they don't roll over onto their back.
- Check skin color and temperature and monitor the breathing. If the skin is pale, bluish and clammy then the person is not getting enough oxygen.
- Be aware that a person who has passed out may die.

If alcohol poisoning goes untreated:

- Hypothermia (low body temperature) can occur.
- Hypoglycemia (too little blood sugar) can lead to seizures.
- Severe dehydration from vomiting can cause seizures, permanent brain damage or death.

Source: National Institute on Alcohol Abuse and Alcoholism

Should colleges ban or restrict alcohol on campus?

Studies indicate that the college environment promotes excessive drinking. College-bound high-school students drink less than their non-college-bound peers, but the trend reverses itself once they reach college, where they end up drinking more.[16]

In an effort to change the physical, social, economic and legal environment in which college students make decisions about drinking, many schools are adopting a strategy called environmental management.

A majority of schools, for instance, now offer alcohol-free or substance-free housing (*see Sidebar, p. 254*). Others sponsor alcohol-free social events. For instance, the West Virginia University Up All Night program offers free food and activities every Thursday, Friday and Saturday night.

"They have cool things like arts and crafts, laser tag, pool and bowling," says Jared Russell, a sophomore. "When I'm hungry late at night, I go for the free food. I've gone a lot, actually, and I think a lot of people use it as an alternative to going to bars and drinking."

But when schools move beyond offering alcohol-free options to limiting access to alcohol, the response is often quite different, as Robert Carothers discovered. When Carothers first arrived at the University of Rhode Island 15 years ago as the new president, he was given the task of improving academics.

"I found I wasn't going to achieve that with a significant portion of the community hung over from Thursday to Monday," he says. "So that had to be taken on."

Ten years ago, Carothers convinced the university to adopt a policy of no alcohol at campus events, including faculty dinners, university fundraisers, parties at fraternity and sorority houses, parties in dormitories, sporting events and even during homecoming weekend. This was a dramatic change for a school once known as URHigh.

"It was very, very difficult to stay the course in the face of criticisms," recalls Carothers. Some alumni refused to attend homecoming, and "we had a terrible time when we went through this with the fraternities. I closed 10 fraternities because they just didn't get it." Many donors, several of whom had been fraternity members, weren't pleased.

But binge drinking has dropped significantly at the university since then, according to Carothers, and donations are up, as are SAT scores of applicants. The school, he says, is no longer attracting the hard-partying crowd.

According to a recent study, one in three schools now bans alcohol on campus for all students, regardless of age. And more than 40 percent restrict alcohol use at athletic contests, homecoming, tailgate parties, dances, concerts and other events.[17]

But critics say these policies just drive student drinking off campus or underground. When the University of Southern California banned alcohol sales at sporting events a few years ago, it didn't stop the drinking, says senior Sean Fish. "A lot of people have made the decision to get more intoxicated before the game since they can't drink at the game, and that leads to rowdiness at the gates."

"Data suggest that the crackdowns as they're practiced now change the location of drinking without changing the behavior, making it more dangerous," says Haines, of the National Social Norms Resource Center. "The heavy drinking is taking place off campus in a non-monitored environment."

But that argument assumes that drinking on campus — at sporting events, tailgate parties, fraternity parties or in campus bars — is properly monitored and inherently safer. Many researchers say that is not the case and that underage and intoxicated students often have no trouble being served alcohol anywhere.

Still, no one wants to see drinking just change locations. Toomey of the University of Minnesota says colleges need to work with their communities to crack down in town as well. "We need to put controls in as many places as possible. There is not one thing that is going to fix it all," says Toomey.

Proponents of environmental management recommend that colleges:

- Work with restaurants, bars and taverns to train servers to check age identification, to recognize the signs of intoxication and to have the confidence to refuse an alcohol sale;
- Ask police to conduct regular compliance checks of alcohol-serving establishments;
- Work with town leaders to reduce the number of bars near campus;
- Help the town develop ordinances and a program to crack down on rowdy off-campus parties.

Duke University lacrosse player David Evans proclaims his innocence after being indicted with two other team members for allegedly sexually assaulting an exotic dancer at a team party in May. The team has a reputation for wild parties, and in the past three years 15 team members have faced alcohol-related charges, including underage alcohol possession and public urination.

"When you have a private house that's used for loud and noisy parties, the police put a big, red circle right on it," says Carothers of the University of Rhode Island, which pays for some of the extra town police needed for the job. Two or three red circles can lead to fines for the landlord, who then usually voids the students' leases.

"Hey, this is New England," says Carothers. "We use the big, red scarlet letter here."

The number of colleges trying comprehensive environmental management has not been tallied. But there are many barriers to its adoption, including the cost of extra policing, the challenge of community organizing, the need for strong leadership and the common belief that such an approach punishes the majority for the transgressions of a few.

Bob Saltz, a senior research scientist at the Prevention Research Center of the Pacific Institute for Research and Evaluation, in Berkeley, Calif., refutes the notion that banning alcohol unfairly punishes moderate drinkers. Using data from a student survey, he says, "We added up all the problems by the type of drinker, and there were so many more light and moderate drinkers than extreme drinkers, that — even though their risk of problems is lower for individuals — as a group, they were responsible for the bulk of the problems."

But the research on environmental management is far from extensive. Saltz is conducting an experiment at 14 universities in California, and Harvard's Wechsler studied 10 schools that implemented similar strategies in a program called "A Matter of Degree."

Overall, Wechsler found no statistically significant change in drinking or in negative consequences due to drinking. But at the five colleges with the most comprehensive environmental-management programs, he found statistically significant declines in alcohol consumption, alcohol-related harms and secondhand effects compared to control schools.

Should colleges restrict alcohol advertising and marketing to students?

Advocates of environmental management also recommend reducing the advertising and marketing of alcohol to college students. Distillers voluntarily no longer advertise in college newspapers, and brewers do so only with school permission. But the biggest controversy is over alcohol advertising at or during college sporting events.

"The time has come to sever the tie between college sports and drinking — completely, absolutely and forever," said University of Miami President Donna Shalala.[18] Three years ago, the school phased out stadium signs from Bacardi, Budweiser, Coors and Miller and replaced them with the logos of banks and soft drinks.

The University of Florida went further and banned alcohol commercials during local broadcasts of its games. In fact, 246 colleges — about a quarter of the members of the National Collegiate Athletic Association (NCAA) — have banned such ads.[19] Between 2001 and 2003, alcohol ads on college sports broadcasts dropped from 5,737 to 4,747.[20]

These colleges have signed "The College Commitment," created by the Center for Science in the Public Interest (CSPI), a consumer-advocacy group. They agree not only to ban alcohol advertisements on local sports broadcasts but also pledge to pressure their athletic conferences to ban alcohol ads from conference broadcasts. So far the Big South and Ivy League conferences have climbed on board.

And in June, the Big Ten Conference signed an agreement with Fox Cable Networks to create the Big Ten Channel, a national network that will not accept alcohol advertising. "This is the first time that a major conference has told the telecaster there will be no alcohol ads principally because the ads are incompatible with conference values," says George Hacker, director of the CSPI's alcohol policies project.

But beer companies, which lead the alcohol industry in college advertising, are not thrilled. "They're certainly welcome to do what they want, although we'd rather they didn't do that," says Jeff Becker, president of The Beer Institute, a trade group for brewers. "A lot of our consumers are sports-minded people."

Banning alcohol ads on college sports broadcasts is not the way to attack underage drinking, says Becker. Most viewers of college sports, he points out, are adults. According to Nielsen Media Research, 89 percent of college football viewers and 86 percent of college basketball fans are 21 or older.[21] "The advertising targets adults, period," says Becker. "We don't think youth drinking has anything to do with advertising."

But research seems to show a link. The latest study, a survey of individuals 15 to 26 years of age, concluded that youths who saw more alcohol advertisements drank more, on average.[22] "These studies are all finding statistically significant correlations" between advertising and youth drinking, says David Jernigan, executive director of the Center on Alcohol Marketing and Youth at Georgetown University. They also contradict the industry argument that alcohol advertising only causes brand switching.

Others say that local marketing by bars and liquor stores may have more of an impact on college drinking than televised beer ads. "There are a lot of bar specials here," says Russell, the West Virginia University sophomore. "I've memorized them all." For instance, if a bar is offering free Jack Daniels and Coke until 11 p.m., Russell says, his friends will usually go there and get drunk. "Then if they have money, they'll spend it buying more drinks after 11 p.m."

Recent research shows that the greater the number of drink specials and price discounts at local bars and liquor stores, the more binge drinking there is at nearby colleges. However, such studies show only a correlation, not causation. Heavy drinking by students could be inducing bars and restaurants to compete for their patronage by lowering prices, instead of the other way around. But researchers point out that under the laws of economics, high demand usually causes prices to rise, not fall.[23]

CHRONOLOGY

1970s *As 18-year-olds are being drafted for the Vietnam War, states lower the minimum drinking age.*

1971 Federal government creates National Institute of Alcohol Abuse and Alcoholism to combat alcohol abuse. . . . On July 5, Congress gives 18-year-olds the right to vote, prompting many states to lower the drinking age.

1980s *Anti-alcohol-abuse movement gains momentum; minimum drinking age is raised.*

1983 College organizations create the Inter-Association Task Force on Alcohol and Other Substance Issues, which sponsors National Collegiate Alcohol Awareness Week.

July 17, 1984 Minimum Uniform Drinking Age Act requires states to set their minimum drinking age to 21 or risk losing federal highway funds.

Oct. 27, 1986 Drug-Free Schools and Communities Act establishes and expands drug abuse and prevention programs in schools.

Nov. 18, 1988 President Ronald Reagan signs omnibus anti-drug bill requiring drug-free workplace policies and health labels on alcoholic beverages.

December 1988 By now, all 50 states have raised the minimum drinking age to 21. Legal challenges to the laws fail.

1990s *Federal government steps up its campaign against underage drinking and drunken driving.*

Sept. 15, 1990 National Commission on Drug-Free Schools calls alcohol and tobacco the most misused drugs and criticizes the alcohol and tobacco industries for targeting youth.

Jan. 27, 1992 White House releases anti-drug strategy that for the first time addresses underage drinking.

Dec. 7, 1994 Harvard University's "College Alcohol Study" finds that 44 percent of surveyed students binge drink.

1995 Federal Zero Tolerance Law mandates that by Oct. 1, 1998, states must pass "zero-tolerance" laws — prohibiting anyone under 21 from driving with any

measurable blood alcohol content (BAC) — or risk losing highway safety funds.

March 1997 Two national fraternities — Phi Delta Theta and Sigma Nu — announce they will ban alcohol from chapter houses beginning July 1, 2000. Within five years, both grade-point averages and membership levels at Phi Delta rise, while insurance premiums drop.

August-November 1997 After alcohol-related deaths of students at several colleges, including MIT, the nation becomes more aware of the scope of the campus binge-drinking problem.

1998 By now, zero-tolerance legislation has passed in all 50 states.

2000-Present *States toughen drunken-driving laws as more colleges try to reduce student alcohol use.*

Oct. 23, 2000 Congress passes and President Bill Clinton signs law redefining drunken driving by lowering the legal BAC level from 0.10 percent to 0.08. By year's end, 21 states lower their limits to the new threshold.

2001 An estimated 1,717 college students between ages 18-24 die from alcohol-related unintentional injuries, including motor vehicle crashes; by comparison there were 1,575 such deaths in 1998.

March 2002 Study shows that 44 percent of surveyed students say they are binge drinkers — unchanged from eight years earlier — while the percentage of students abstaining from alcohol has risen.

April 2002 National Institute on Alcohol Abuse and Alcoholism releases "A Call to Action," advocating adoption of research-based programs to reduce alcohol use on campus and for increased research into the problem.

July 2004 By now, all 50 states and the District of Columbia have set 0.08 BAC as the threshold for drunken driving.

June 21, 2006 The Big Ten Conference and Fox Cable Networks announce that their new Big Ten Channel, devoted to Big Ten athletic and academic programs, will not broadcast alcohol ads.

Do European Students Really Drink Less?

Julie Price, a senior sociology major at the College of William and Mary in Williamsburg, Va., has often heard the argument: American kids binge drink because alcohol is a forbidden fruit until age 21, and kids rebel against the restriction by drinking to excess. European youths, on the other hand, learn to drink alcohol in moderation because they have wine at the dinner table with their parents from an early age. The minimum drinking age across most of Europe is 18 — 16 in Spain — so alcohol is not a forbidden substance there.

As a result, the theory goes, European youths do not abuse alcohol as a way to challenge authority. Thus, if the minimum drinking age in the United States were lowered from 21 to 18, there would be less sneaking around and less binging among American college students, say those who follow this reasoning.

"But given what I saw in Scotland," says Price, who spent her spring semester studying there, "I'm not so sure." Her Scottish classmates — who can drink legally at age 18 — were drunk more often and "much earlier in the day" than her fellow students are at William and Mary, she says.

Although there is sparse data comparing the drinking patterns of U.S. and European college students, detailed surveys comparing the drinking patterns of 10th-graders tend to disprove the theory. The surveys show that in comparison to American 16-year-olds, more European 16-year-olds drink — and drink heavily.[1]

For instance, 35 percent of the U.S. 10th-graders surveyed reported drinking in the past 30 days. In many European countries — including Austria, Germany, Denmark, Ireland and the United Kingdom — more than 70 percent of the surveyed students reported drinking in the past 30 days. Only Turkey, a Muslim country, had a lower rate of 20 percent.

As for binge drinking, 22 percent of those surveyed in the United States reported recently drinking five or more drinks in one sitting, compared to 40 percent reporting heavy drinking among respondents in many European countries.[2]

Moreover, while 18 percent of the American 16-year-olds reported being drunk in the past 30 days, only six European countries — Turkey, France, Romania, Portugal, Greece and Cyprus — had lower percentages. Italy's percentage was nearly identical to America's. Most countries had more than a quarter of their 16-year-olds reporting intoxication, with the highest numbers in Northern Europe: 61 percent in Denmark; 53 percent in Ireland; 48 percent in Austria and 46 percent in the United Kingdom.[3]

"This idea that all we have to do is lower the drinking age to solve our problem is not supported by this data,"

In any case, a growing number of colleges are cutting back on the amount of alcohol marketing allowed on campus. FSU, for instance, outlawed flyers from bars and liquor stores after such handbills "littered the campus uncontrollably," said Michael Smith, director of the school's Florida Center for Prevention Research.[24]

Other schools are working with town leaders, local bars and liquor stores to eliminate alcohol specials and promotions. In Lincoln, Neb., for instance, business owners, City Council members, prevention specialists, law enforcement officials and alcohol distributors formed a coalition in the 1980s to reduce the irresponsible sale and serving of alcohol in town. Recently, the University of Nebraska joined in. The school works with the town to enforce state laws restricting happy hours, free drinks and one-price drink specials. Most bars comply, say college officials, but there are always some that try to skirt the law.

"A bar can have a policy that says the first drink is $3 and the second is 1 cent," says Linda Major, director of student involvement at Nebraska. "They aren't technically violating the law because the second drink isn't free." Other times, bar managers allow servers to make up their own specials on the spot. "They'll flip a coin with a customer and the drink is half price," says Major.

The university — working with the City Council, other bar owners, law-enforcement officials and the business-improvement association — will then pressure the bar owner to eliminate alcohol specials that violate the spirit of the law. "The chief of police is the co-chair of our coalition," says Major's colleague Tom Workman.

says Joel Grube, director of the Prevention Research Center at the Pacific Institute for Research and Evaluation. "It's a much broader and complex cultural issue."

The European researchers who studied the survey data did not offer explanations for the differences in drinking patterns between Northern and Southern European countries. But Grube says it is interesting that the countries with the lowest intoxication rates are Mediterranean countries.

"To some extent, I think people who argue that there is a different culture in those countries — that alcohol is more integrated into the meals — are absolutely correct."

But that culture is changing, Grube says. In Spain, for instance, there is anecdotal evidence that teenagers are beginning to drink like Northern Europeans. They are switching from wine to beer, and their intoxication levels are rising. "It's hard to know why," he says.

European Teens Drink More

Youths in Europe drink nearly twice as heavily as teens in the United States.

Teen Drinking in Europe and the U.S.
(among 10th-graders)

	U.S.	Many European nations
Drank in last 30 days	35%	70%
Recently binged	22%	40%

Scale: 0% 10 20 30 40 50 60 70 80

Source: U.S. Department of Justice, May 2005

[1] "Youth Drinking Rates and Problems: A Comparison of European Countries and the United States," Office of Juvenile Justice and Delinquency Prevention, U.S. Department of Justice, May 2005, p. 2.

[2] *Ibid.*, p. 3.

[3] *Ibid.*, p. 4.

Northern European drinking patterns also are changing, Grube says. "Ireland, contrary to people's perception, was a low-alcohol-consuming country for youth about 20 years ago, and now it's among the highest," he says, adding that the reason for the change is unclear. One theory posits that a vastly improving economy has put more discretionary income into the hands of youths. In addition, it has become more socially acceptable for young women in Ireland to drink.

"And they are definitely drinking more," says Grube.

"In order to do any of this, you have to have a whole community ready."

But when there are no laws restricting happy hours and drink specials, the situation can get complicated. For instance, in 2004, when Madison, Wis., was considering requiring bars to close earlier and to pay for extra law enforcement in order to cut down on rowdy drinking by University of Wisconsin students, bars owners volunteered instead to eliminate all happy hours and drink specials on weekends.

But eight months later, three university students sued 24 bars and the local Tavern League — and eventually the university and the town council — for price fixing. The students won, and the case is now on appeal in federal district court. So far, legal fees for the bar owners exceed half a million dollars.

BACKGROUND

Temperance and Prohibition

Alcohol has long been a part of American life, but Americans can't seem to make up their minds about its proper role. From Colonial times to the temperance era, and from Prohibition to the present, attitudes toward alcohol have undergone major shifts.

For early settlers, a stiff drink "kept off chills and fevers," and a few glasses "aided digestion," according to the social history *Drinking in America*. Beer and cider were the preferred beverages, and even children were served alcohol at meals. "Simply stated, most settlers drank often and abundantly."[25]

But drunkenness was a crime throughout the Colonies, and the penalties could be severe, including

Newsmakers/Joe Raedle

A college student in Texas chugs beer through a bong during spring break. The rate for female binge drinkers has risen slightly in recent years.

fines, imprisonment, whipping and serving time in the stocks, which a historian has called "the Colonial era's equivalent of the alcoholism-treatment facility."[26]

By the early 19th century, Americans were drinking nearly three times as much as they do today, and most continued to believe that alcohol imparted health, eased fevers, cured colds and even relieved snakebite. But as consumption rose to unprecedented levels, "an awareness of the dangers of drink began to emerge, and the first American temperance movement took hold," wrote historian David Musto.[27]

In 1826, one of the most dynamic speakers of the time, the Rev. Lyman Beecher, called for a crusade against alcohol. By 1855, about a third of the nation's 40 states and territories had banished the sale of alcohol. "Alcohol consumption fell to less than a third of its pretemperance level," wrote Musto, "and never again reached the heights of the early republic."[28]

But abstinence fell out of favor during the Civil War, and some states repealed their alcohol prohibitions in the 1860s. In other states, the laws fell into disuse or the courts found them unconstitutional.

By the turn of the century, the abstinence movement was again gaining momentum, culminating in ratification of the 18th Amendment to the U.S. Constitution in 1919 barring the "manufacture, sale or transportation of intoxicating liquors" within the United States as well as all imports and exports of alcoholic beverages. The

Prohibition amendment became effective in January 1920 and remained in effect nearly 14 years.

On the positive side, medical evidence suggests that Prohibition reduced mortality from alcoholism and cirrhosis of the liver. But that doesn't mean that Americans stopped drinking. Many stockpiled alcohol before Prohibition took effect. There were rural and urban stills, "medicinal liquor," industrial alcohol, underground breweries and imports smuggled from Canada, Britain and the Caribbean. Prohibition fostered the growth of entrenched criminal organizations and "secret" drinking establishments known as speakeasies.

Nevertheless, Prohibition did significantly reduce alcohol consumption among adults and youth. "College students remained fairly temperate, at least through the mid-1920s — the wild sprees depicted in the novels of F. Scott Fitzgerald notwithstanding."[29] It was visiting alumni who did most of the excessive drinking. Contemporary surveys showed considerable support for Prohibition among college students at the time.

But the success of Prohibition in cutting alcohol consumption contained the seeds of its eventual defeat. As problem drinking waned, so did the public's concern. Failures of enforcement, growing violence and the Great Depression all caused the nation to overwhelmingly reject Prohibition in 1933 and return to the states the broad power to regulate alcohol.

Minimum Drinking Age

After Prohibition was repealed, most states established 21 as the minimum legal drinking age, but as adult consumption of alcohol rose, so did concerns about college drinking. In response, researchers Robert Straus and Selden D. Bacon undertook one of the first comprehensive surveys of college drinking in the early 1950s.

"Stereotypes of college drinking include the belief that most students drink, that they do so heavily and frequently, and that dangerous and disgraceful behavior often ensues," they wrote. However, based on their research, the authors observed that adult perceptions of college drinking were "very far from the reality."[30]

In fact, they reported, only 21 percent of college men and 10 percent of the women reported drinking more than once a week. When they did drink, the amounts were usually modest. Only 9 percent of men and 1 percent of women reported drinking large amounts of beer, and only 4 percent of men and less than half a percent of women reported drinking large amounts of wine. Liquor, however, was more popular: 29 percent of men and 7 percent of women reported drinking large amounts of spirits.[31]

But as the decades passed, drinking among college students became more common. Then in 1971, as young men were being drafted during the Vietnam War, enactment of the 26th Amendment to the U.S. Constitution lowered the voting age from 21 to 18. Many states then lowered their minimum legal drinking ages to 18 or 19, based on the argument that if young people were responsible enough to vote and to go to war, they were responsible enough to drink.

In the 1970s and '80s, however, studies began to show that alcohol-related traffic crashes were increasing significantly among young people ages 18 to 20 in states with lower drinking ages. The public — led by the powerful grass-roots group Mothers Against Drunk Driving (MADD) — began to demand that the drinking age be raised again to 21.

"It now appears that a third era of temperance is under way in the U.S.," Musto wrote in 1996.[32]

MADD was established in 1980 by a mother whose 13-year-old daughter had been killed by a drunken driver. By the end of 1982 it boasted 100 chapters. Pushing for tougher laws at the state and federal level, MADD marshaled support for the more than 100 new state anti-drunken driving laws passed by the end of 1983.

Largely as a result of MADD's lobbying, Congress in 1984 enacted the Minimum Uniform Drinking Age Act, which required that a portion of federal highway funds be withheld from states that do not raise the minimum drinking age to 21. Four years later, every state had complied.

"Raising the age to 21 led to an immediate decline of roughly 15 percent or so in teen drinking," says Alexander Wagenaar, professor of epidemiology and health

Frequent Bingeing Increased

Binge drinking among college students remained virtually unchanged between 1993 and 2001, but frequent bingeing (more than once a week) increased among both men and women. Meanwhile, the gap increased between those who don't drink and those who binge frequently (top). While occasional bingeing rose among blacks and Asians, it decreased among Hispanics and Native Americans (bottom).

	1993	2001
Prevalence of Drinking Among College Students (by percentage)		
Binge drinkers, total	43.9%	44.4%
Abstainers, total	16.4	19.3
Frequent bingers, total	19.7	22.8
Female bingers	17.1	20.9
Male bingers	22.4	25.2
Binge Drinkers Among College Students (by percentage)		
Whites	49.5%	50.2%
African Americans	16.7	21.7
Hispanics	39.7	34.4
Native Americans	39.3	33.6
Asians	23.1	26.2

Source: Henry Wechsler, et al., "Trends in College Binge Drinking During a Period of Increased Prevention Efforts," Journal of American College Health, March 2002

policy research at the University of Florida. "And this is despite weak enforcement." The National Highway Traffic Safety Administration (NHTSA) estimates that the laws save nearly 1,000 lives a year.

But assessing the impact on college students alone is difficult. There are few high-quality studies focusing so narrowly, but of those that do, none have been able to link higher-minimum drinking age laws to lower alcohol consumption among college students. "If one assumes that the minimum legal drinking age is less effective on college campuses," researchers wrote, "perhaps it is due to lax enforcement and particularly easy access to alcohol by underage youth in such settings."[33]

Others argue that the higher drinking age has increased the allure of alcohol on college campuses by making it a "forbidden fruit" and has caused students to "pre-game:" load up on alcohol before going out.

Delta Sigma Phi, University of Virginia (both)

Party Time

University of Virginia students play popular drinking games, including flip cup and beer pong, at the Delta Sigma Phi fraternity house (top). UVa students get ready to party (bottom). Students at many schools "pre-game," or drink in their rooms before going out.

"Before the age was increased, we had a very different environment," said Ronald Liebowitz, president of Vermont's Middlebury College. "You had kids drinking beer and getting sick on beer, but you didn't have gross alcohol poisoning and binge drinking."[34]

Liebowitz is among many college administrators who advocate a lower minimum drinking age. In a 2000 survey of 330 directors of student affairs, 39 percent preferred a drinking age below 21.[35] No state, however, is seriously considering such a proposal.

Instead, states have adopted a range of additional policies to attack underage and excessive drinking. Many states have toughened penalties on the use of false identification or withdrawn driving privileges of minors guilty of any alcohol violation. All states have adopted laws that set blood-alcohol concentration (BAC) levels at almost zero for drivers under 21.

In another MADD-promoted campaign, states also lowered the BAC limit for all drivers, after President Bill Clinton in 2000 signed a law requiring states to reduce the legal BAC from 0.10 to .08 or risk losing federal highway funds. Despite heavy lobbying against the measure by the alcohol industry, all states had complied by July 2004.[36]

'Responsible Drinking'

As the "third era of temperance" took hold in the 1980s, the alcohol industry developed its own programs to encourage "responsible drinking." They generally involved distributing brochures for parents and training material for alcohol servers, providing funds for campus programs and launching national advertising campaigns.

Anheuser-Busch, for instance, says it has spent more than $500 million since 1982 on such programs. Its latest advertising campaign, "Responsibility Matters," extols "the good practices of adults who exercise personal responsibility by designating a driver, calling a cab when they or a friend is too drunk to drive and talking to their children about underage drinking.[37]

"No company benefits when its products are misused, something acknowledged by the company's thousands of employees who are raising children, living and working in this society and driving on the same highways as everyone else," says Anheuser-Busch's Web site.[38]

But some critics — such as the Marin Institute, an industry watchdog in San Raphael, Calif. — call industry-sponsored responsibility advertising a "public relations ploy." The industry is trying to "avoid responsibility for the consequences of its products and shift blame to individual consumers," the institute said.[39]

Research shows that youths ages 16-22 are not impressed by responsibility advertising from beer companies. In one study, participants rated brewer-sponsored ads as "less informative, believable, on-target, and effective" than conventional public-service announcements sponsored by the government or advocacy groups.

And when asked why brewers sponsor responsibility advertisements, respondents ranked the "improvement of the company's image and selling its beer" higher on the companies' agendas than preventing drunken driving.[40]

CURRENT SITUATION

Greek Scene

When it comes to drinking, fraternity and sorority members are way ahead of the rest of the college pack. Three-quarters of fraternity or sorority house residents reported binge drinking in a 2001 survey of 119 colleges — far higher than other student groups.[41]

"Alcohol consumption is the fraternity's social lubricant," writes Harvard alcohol researcher Wechsler, lead author of the study. "Most students who join fraternities expect alcohol to be central to their experience, even though they are likely to be legally underage."

As a fraternity member told Wechsler:

"I am not an advocate for responsible college drinking; in fact, I am quite the opposite. I party hard and I party a lot, but what the 1980s labeled as a 'party animal' has now taken on the label of 'binge drinker.' So what if I bong three beers at a time and often play drinking games that drain a case of beer between four people in less than 75 minutes? I drink often and a lot, but I know my limits and I don't wake up each morning needing a drink. If you look hard and ask students, you'll see that alcohol has become an institution at parties. A party is not a party without it. . . . By the way, I am only 20 years old and my parents know full well how much I drink. In fact, more than 90 percent of the parents of the men in our fraternity know how much their kids are drinking, and they aren't worried about it."[42]

Fraternity members are paying a price for heavy drinking. Nearly twice as many fraternity residents as non-fraternity residents fall behind in schoolwork, argue with friends, damage property, have unprotected sex or suffer injuries. They also are more likely to drink and drive or ride with a drunken driver.[43]

While some fraternity members may bring their heavy drinking habits from high school, many others acquire them in the fraternity. "For students who do not already drink heavily when they begin college, joining a Greek organization is a sure way to start," writes Wechsler. In his study, three-quarters of fraternity and sorority residents who had not binged in high school became binge drinkers in college.[44]

In the past decade, fraternities and college administrators have taken steps to curb excessive drinking within the Greek system. Most sororities have been dry for decades, but not fraternities.

But that is beginning to change. Eleven of the 68 national fraternities active in the U.S. have gone dry, including Phi Delta Theta, Phi Kappa Sigma, Delta Sigma Phi and Theta Chi. They allow no alcohol in chapter houses, even during parties and even among upperclassmen of legal drinking age. Local chapters that want to serve alcohol at parties have to hold those events elsewhere. Often they will rent out the back room of a bar.

In part, rising insurance premiums have triggered the change. A study by an insurer of fraternities found that alcohol was involved in:

- 95 percent of roof/window falls;
- 94 percent of fights;
- 93 percent of sex-abuse incidents;
- 88 percent of fatalities;
- 87 percent of auto incidents.[45]

Another reason to go dry is to return the fraternity to its roots. Ron Binder, director of Greek Affairs at Bowling Green State University in Ohio and president of the Association of Fraternity Advisors, says fraternities and sororities were founded on four core values: scholarship, service, leadership and brotherhood. "What I've seen in the past 10 years is a renewed focus on those values," he says.

Initially, many fraternities that went dry saw their membership drop, but membership has since rebounded, and the type of member they attract "has changed pretty dramatically," says Binder. "They're attracting people whose sole focus isn't social."

College administrators sometimes take the matter into their own hands. On July 1, Rensselaer Polytechnic Institute in Troy, N.Y., began severely curtailing drinking at Greek houses. Alcohol is allowed only in the rooms of students 21 years of age or older and forbidden in common areas. All house parties must be alcohol-free.

"They just kind of popped it on us," said Rory Arredondo, a Sigma Alpha Epsilon brother. "It's going to kill Greek life."[46]

But imposing an alcohol-free policy on fraternities is not an option for many colleges. Only about a third of Greek societies have on-campus residences owned by the college. The rest are privately owned and are often located off campus, raising questions about how far the university can extend its reach.

Substance-free Dorms Offer Peace and Quiet

During her freshman year at Vassar College in Poughkeepsie, N.Y., Kathryn Thomas chose to live in a "wellness" corridor in her dorm. "I wanted a quiet place for studying," she said. In fact, her floor eventually became a refuge for other students "who come and stay until they know the party on their floor has died down."[1]

Her corridor is part of a growing trend. Colleges and universities across the country have been offering students the option of living in so-called substance-free housing, where no alcohol, tobacco and other drugs are used. Institutions ranging from small, liberal arts colleges like Franklin & Marshall to large public universities like the State University of New York have reserved halls and suites for students — most often freshmen — who sign up. Upperclassmen, who usually choose their own roommates, have less of a need for the school to set the rules.

Parents like having the option. "Sometimes we get students who complain that their parents filled out the housing form," says Katherine Steele, dean of students for residential life at the University of the South in Sewanee, Tenn., a liberal arts college where 6 percent of the 1,400 students choose substance-free housing.

But Steele believes that most students choose substance-free housing on their own. Otherwise, she thinks she would see more violations of the policy. This year the school had only one or two violations, she says.

"Students understand that it is a commitment, and it's a commitment to each other," says Steele. Like most schools, the University of the South relies on head residents and student proctors to enforce residence hall policies and report violations.

In a survey at the University of Michigan, students said they chose substance-free housing in order to:

- Avoid roommate problems associated with alcohol or drug use (78 percent),
- Live in an atmosphere conducive to studying (59 percent),
- Satisfy parents (26 percent),
- Abide by religious preferences (22 percent).

Another 6 percent said they made the choice because of a family member with a substance-abuse problem.[2]

Some parents may wonder why colleges need substance-free housing — especially for younger students — when, by law, all residences with younger students should be alcohol free.

"We have an alcohol policy that says if you're under 21, you have to adhere to state and federal laws," says Rob

---

The University of Nebraska's privately owned Greek houses must be alcohol-free because they admit freshmen and therefore are considered part of campus housing. The university began serious enforcement in 1998, and by 2000 half of all fraternities were on some form of judicial sanction. One fraternity lost its housing status and ended up closing.

At that point, "our Greek leaders recognized that things were going to have to change; it was getting ugly," says Workman of the university's Office of Student Involvement. Funded by a grant from the U.S. Department of Education, the school and its fraternities have been working to change the drinking culture.

Other schools are not going quite so far as to ban alcohol in fraternity houses but are tightening up enforcement of rules that govern party size, underage drinking, noise and security.

There are some signs that these changes are having an impact on drinking, if only a small one. The percentage of residents of fraternities and sororities who binge drink was slightly lower in 2001 than in 1993, according to Harvard's "College Alcohol Study." And the study found that fewer students are attending fraternity and sorority parties, and when they do, the percentage who binge drink is slightly lower.

However, students apparently are taking their heavy drinking elsewhere. The percentage of students attending off-campus parties has risen along with the percentage who binge drink when they get there.[47]

State Actions

States play an important role in the fight against underage and excessive drinking, passing laws, enforcing compliance and providing guidance to local communities. However, the number of laws, the severity of penalties and the level of enforcement vary from state to state, and — not surprisingly — so does the level of drinking.

Wild, an assistant director of residential life at Washington University in St. Louis, Mo., where 20 percent of first-year students request the special housing. "But we know that students make choices about alcohol; we try to limit the really dangerous and disruptive behaviors."

For instance, Wild says, a head resident is not going to cite a 19-year-old for drinking a beer in her dorm room on a Friday night, unless she lives in a substance-free hall.

Students who choose to live in substance-free housing do not promise to abstain completely, only in the residence. Nevertheless, a Harvard study shows that students living in substance-free dorms were three-fifths less likely than students living in unrestricted residences to engage in heavy, episodic drinking and were also less likely to fall behind in schoolwork, ride with a drunken driver, get in trouble with police or damage property.[3]

"You're less likely to have the occasional broken window, or a student shooting off a fire extinguisher or vomiting," says Steele. "It's really frustrating how living spaces can get treated sometimes."

Does substance-free housing have an impact on overall campus culture? "That's a good question," says Wild, and he doesn't have an answer. Even the Harvard researchers found it difficult to distinguish cause and effect.

"While the difference in heavy, episodic drinking rates may be due to self-selection of determined non-drinkers into these residences, the lower rates [of heavy drinking] may also

> "You're less likely to have the occasional broken window, or a student shooting off a fire extinguisher or vomiting"
>
> — *Katherine Steele,*
> *Dean of Students for Residential Life,*
> *Univ. of the South, Sewanee, Tenn.*

be due to the influence that such an environment has on students," said the researchers.[4]

Nevertheless, Wild and a growing number of college administrators believe that offering substance-free housing is an important option for students who want to avoid the noise, mess and disruptions associated with heavy drinking.

---

[1] Tamar Lewin, "Clean Living on Campus," *The New York Times*, Nov. 6, 2005.

[2] "Preventing Alcohol-Related Problems on Campus: Substance-Free Residence Halls," Higher Education Center for Alcohol and Other Drug Prevention, 1997, p. 4.

[3] Henry Wechsler, et al., "Drinking Levels, Alcohol Problems, and Secondhand Effects in Substance-Free College Residences," *Journal of Studies on Alcohol*, January 2001, press release.

[4] *Ibid.*

Slightly more than 33 percent of college students binge drink in states with four or more laws restricting promotion and sales of high volumes of alcohol, according to one study. But in states with fewer laws, the binge-drinking rate was just over 48 percent.[48]

According to the Alcohol Policy Information System, a government Web site that monitors state alcohol policies, more and more states are passing laws designed to restrict underage drinking and service to intoxicated individuals.[49] For instance:

- 27 states have keg-registration laws, up from 20 in 2003. An identification number is attached to kegs exceeding a specified capacity (two-to-eight-gallon minimum depending on the state). The retailer records the purchaser's identifying information and may collect a refundable deposit. The intent is to make it difficult for underage youth to obtain kegs for parties.

- 28 states have happy-hour restrictions — such as prohibitions against free drinks, price discounts, unlimited drinks for a fixed price and giving alcoholic beverages as prizes — up from 27 states in 2003.

- 22 states have laws requiring beverage-server training, up from 15 in 2003; liquor-license holders, managers or servers must attend training to prevent alcohol sales to minors and intoxicated customers.

- 20 states have social-host liability laws, up from 16 in 2003; individuals (social hosts) can be held criminally responsible for underage drinking events on their property.

But just having a law on the books does not mean it is enforced or that its penalties are serious enough to actually matter. For instance, 47 states prohibit serving alcohol to someone obviously intoxicated. "But those laws are basically ignored in all states," says Jim Mosher,

Should federal and state excise taxes on alcohol be raised to curtail underage drinking?

YES
David L. Rosenbloom
Director, Youth Alcohol Prevention Center, Boston University School of Public Health

Written for *CQ Researcher*, July 2006

Higher taxes on alcohol will yield important public health and safety benefits. Since taxes on beer range from 2 cents a gallon in Wyoming to $1.07 a gallon in Alaska, it is possible to know the real-world consequences of these taxes. For example, the five states with the lowest beer taxes have teen binge-drinking rates that are twice as high as the five states with the highest beer taxes. Higher alcohol tax states have lower alcohol-related teen driving deaths. Research has also shown that increasing the total price of alcohol decreases drinking and driving among all age groups. Today, there are about 600,000 alcohol-related violent incidents a year on American college campuses. Researchers estimate that a 10 percent increase in the price of alcohol would reduce problem drinking enough to avoid about 200,000 of these incidents every year.

Increasing alcohol taxes will not have much effect on most people. About a third do not drink any alcohol. Many adults drink and enjoy alcoholic beverages in moderation without harm to their health and safety. However, drinking teenagers and alcoholics consume so much that even a modest increase in the price of each drink will reduce their total consumption — and that is a good thing. Recent research has shown that alcohol has damaging long-term effects on adolescent brain development. Increasing the price through taxes is the quickest and most effective way to avoid some of this damage. Similarly, making it more expensive for alcoholics to continue their self-destructive, compulsive drinking may encourage some of them to seek the effective treatment that is now available — and increased alcohol taxes might pay for it.

In most of the country, alcohol taxes no longer provide any of these benefits because their impact has been eaten away by inflation. Many states have not raised their taxes in decades. In Massachusetts, the beer tax was set at 11 cents a gallon in 1975 — about 3 cents a gallon in current terms. State and federal governments need to raise alcohol taxes to a meaningful level and then index them to inflation to make sure they continue to save lives in the future.

A significant majority of all voters — including those who drink — consistently favor increasing alcohol taxes. The only opposition is from the alcohol industry and politicians who have taken the pledge against taxes — but not rum.

NO
Roger Brinner
Chief Economist, The Parthenon Group, Boston

Written for *CQ Researcher*, July 2006

Would higher alcohol taxes work as well as other alternatives to reduce underage or abusive drinking? The clear answer from the most careful economic research is "No."

The best study — a meticulous econometric analysis by Thomas Dee of the U.S. government's "Monitoring the Future" student data from 1977-1992 — found beer taxes were not statistically related to either "moderate" or "binge" drinking rates.

Research appearing to find substantial deterrence effects of alcohol taxes is badly flawed. It erroneously attributes declining alcohol consumption trends (across states or across time) to price increases without accounting for the correct explanations of legal penalties and enforcement, server training, parental involvement, counseling and peer norms programs.

Quite simply, higher taxes are relatively impotent policy tools. Abusive drinkers either ignore taxes or seek cheaper alcohol options. Potential teen drinkers know they risk parental punishment, significant fines, community service, lost privileges in school activities and driver's license suspension. A tax mainly encourages responsible, moderate adult drinkers to reduce their purchases, trade down to lower-priced brands or swap lower alcohol-concentration products for those with higher alcohol content — like beer for distilled spirits. Moderate-drinking adults — clearly not the policy target — pay the vast majority of alcohol taxes, rather than problem drinkers.

Moreover, the collateral damage from excise taxes is great. As moderate-drinking adults reduce their purchases, good-paying jobs are lost at the producer, distributor and retail levels, as well as by farmers, truckers and suppliers. Rather than contributing to the state's economy, these individuals then become a burden to the state through unemployment benefits paid by all state taxpayers — drinkers and non-drinkers alike. Finally, beer is already among the most highly taxed consumer categories, with more than 40 cents of each beer dollar attributable to taxes, 68 percent higher than the average consumer product.

Why use the blunt weapon of taxes, which is unfair and economically damaging to all citizens regardless of whether or not they drink, when effective, focused policy alternatives are working well? According to the Partnership for a Drug-Free America, all types of teen alcohol use in 2005 were significantly down from 1998, and other well-known government data show the lowest levels of underage drinking since surveys began tracking it decades ago. It's not a difficult decision to reject excise taxes.

director of the Center for the Study of Law and Enforcement Policy at the Pacific Institute for Research and Evaluation. "Research shows that if you send in actors who feign intoxication, they will be served 70 percent of the time."

States need to send more undercover agents to bars and publicize the crackdowns, Mosher says. "This is particularly important in college communities where we see bars conducting these very questionable sales practices." But alcoholic-beverage control agencies in most states are underfunded and understaffed.

In addition, the fines for selling alcohol to intoxicated customers or to minors are often paltry. "Many states haven't adjusted their fines in 30 years," says the University of Florida's Wagenaar. "The fines can be just $100, or maybe $200 for a second offense."

But when states try to toughen penalties, the alcohol industry "often sends their lobbyists to state legislatures to oppose the actions," says Wagenaar. "So we have a slow process of change to modernize these regulations."

OUTLOOK

Student Support?

Colleges' efforts to curb alcohol abuse have produced both hopeful and worrisome results. On the plus side, the percentage of students abstaining from drink has increased, and binge drinking at fraternities has decreased. There are more "frequent" binge drinkers, however, and more binge drinking at off-campus parties. Moreover, the rate for female binge drinkers has risen slightly over the years, particularly at all-women colleges. And the percentage of African-American students who report binge drinking has risen significantly, although it is still half that of white students.[50] (*See graph, p. 251.*)

Colleges clearly need to enlist the support of students to reverse these trends. The latest Harvard "College Alcohol Survey" showed that the vast majority of students support some alcohol-control policies, such as clarifying the alcohol rules, providing more alcohol-free recreational and cultural opportunities and offering more alcohol-free residences — all measures supported by at least 89 percent of students.[51]

But, not surprisingly, support declines when restricting access to alcohol is proposed. For instance, only 56 percent

To Drink or Not to Drink

Students sip wine in a dining hall at Colby College in Waterville, Maine, which serves beer and wine to students 21 and older to teach them about responsible drinking (top). Members of Phi Delta Theta fraternity hit the books at the University of Cincinnati. It's one of 11 national fraternities that bar alcohol in chapter houses (bottom).

of the students surveyed support cracking down on drinking in Greek houses, 60 percent support prohibiting kegs on campus and 63 percent support stricter enforcement of campus alcohol rules.[52] Yet, supporters point out, those are still majorities of the students surveyed.

Sometimes the minority, however, will "fight for their right to party." Riots or public disturbances occurred at several schools when alcohol restrictions were tightened, including the University of Colorado, Syracuse and Michigan State.

College administrators are struggling to convert the heavy-drinking minority to the majority view. The most effective alcohol policies may be those crafted in collaboration with students and alumni, according to Harvard's Wechsler. "Students are key contributors to the success of any prevention efforts," he says.[53]

At Florida State, for example, the social norms marketing campaign is not a top-down program. Students submitted proposed ads, says Coburn, which she feels made them more effective.

Similarly, the University of Nebraska's ambitious NU Directions program to reduce high-risk drinking tries to involve students at every step. The university held a series of meetings between students, neighbors and landlords in order to improve relations between students and town residents, and the university regularly holds discussion groups for students on alcohol policies.

Wechsler advises colleges to include students in all efforts, such as helping to develop campus codes of conduct, joining alcohol task forces and campus-community coalitions and serving on judicial review boards that adjudicate alcohol infractions. Most important, Wechsler advises administrators not to give up.

"The problem of college binge drinking took decades to develop," he said. "You won't get rid of it overnight."[54]

NOTES

1. Quoted in Christine Vovakes, "Chico's softball gets ax," *Sacramento Bee*, March 31, 2006, p. B1.

2. *Ibid.*

3. For previous coverage in *CQ Researcher*, see Karen Lee Scrivo, "Drinking on Campus," March 20, 1998, pp. 241-264; David Masci, "Preventing Teen Drug Use," March 15, 2002, pp. 217-241; Sarah Glazer, "Preventing Teen Drug Use," July 28, 1995, pp. 657-680; Charles S. Clark, "Underage Drinking," March 13, 1992, pp. 217-240; J. Rosenblatt, "Teen-Age Drinking," in *Editorial Research Reports*, May 15, 1981, available at *CQ Researcher Plus Archive*, CQ Electronic Library, http://library.cqpress.com.

4. Henry Wechsler, *et al.*, "Trends in College Binge Drinking During a Period of Increased Prevention Efforts," *Journal of American College Health*, March 2002, p. 207.

5. www.duke.edu/~amwhite/College.

6. Wechsler, *et al.*, *op. cit.*, p. 210.

7. *Ibid.*, p. 211.

8. Ralph Hingson, *et al.*, "Magnitude of Alcohol-Related Mortality and Morbidity Among U.S. College Students Ages 18-24," *Annual Review of Public Health*, 2005, p. 265.

9. www.duke.edu/~amwhite/College.

10. Aaron White, *et al.*, "Many College Freshmen Drink at Levels Far Beyond the Binge Threshold," *Alcoholism: Clinical and Experimental Research*, June 2006, p. 1008.

11. National Institute of Alcohol Abuse and Alcoholism, "A Call to Action," April 2002, p. 8.

12. *Ibid.*, p. 2.

13. "Harvard Study of Social Norms Deserves 'F' Grade for Flawed Research Design," Higher Education Center for Alcohol and Other Drug Abuse and Violence Prevention, p. 2.

14. John D. Clapp, *et al.*, "A Failed Norms Social Marketing Campaign," *Journal of Studies on Alcohol*, May 2003, p. 409.

15. David Kesmodel, "Schools Use Web to Teach About Booze," *The Wall Street Journal*, Nov. 1, 2005.

16. www.duke.edu/~amwhite/College.

17. Wechsler, *et al.*, "Colleges Respond to Student Binge Drinking: Reducing Student Demand or Limiting Access," *Journal of American College Health*, January/February 2004, p. 161.

18. "Guest Editorial: College sports must end ties with alcohol," *NCAA News Comment*, Oct. 12, 1998, www.ncaa.org/news/1998/19981012/comment.html.

19. Campaign for Alcohol-Free Sports TV, http://cspinet.org/booze/CAFST/index.htm.

20. Center on Alcohol Marketing and Youth, "Alcohol Advertising on Sports Television, 2001 to 2003," p. 12.

21. alcoholstats.com, Anheuser-Busch Companies, www.alcoholstats.com/mm/docs/2754.pdf.

22. Leslie B. Snyder, *et al.*, "Effects of Alcohol Advertising Exposure on Drinking Among Youth," *Archives of Pediatrics and Adolescent Medicine*, January 2006, p. 18.

23. Meichun Kuo, *et al.*, "The Marketing of Alcohol to College Students," *American Journal of Preventive Medicine*, 2003, p. 210.

24. Christina Hoag, "More colleges are turning off tap for booze advertising," *The Miami Herald*, Dec. 12, 2005, p. A1.

25. Mark Edward Lender and James Kirby Martin, *Drinking in America* (1987), pp. 2, 9.

26. "The History of the NIAAA," National Institute of Alcohol Abuse and Alcoholism, June 2002, p. 1.

27. David Musto, "Alcohol in American History," *Scientific American*, April 1996, p. 78.

28. *Ibid.*, p. 81.

29. Lender and Martin, *op. cit.*, p. 144.

30. Robert Straus and Selden D. Bacon, *Drinking in College* (1953), p. 100.

31. *Ibid.*, p. 103.

32. Musto, *op. cit.*, p. 78.

33. Alexander Wagenaar and Traci Toomey, "Effects of Minimum Drinking Age Laws: Review and Analyses of the Literature from 1960 to 2000," *Journal of Studies on Alcohol*, Supplement No. 14, 2002, p. 219.

34. Pam Belluck, "Vermont Considers Lowering Drinking Age to 18," *The New York Times*, April 13, 2005, p. A13.

35. George Mason University, "The 2000 College Alcohol Survey," March 2001, p. 13.

36. For background, see Kathy Koch, "Drunken Driving," *CQ Researcher*, Oct. 6, 2000, pp. 793-808.

37. www.beeresponsible.com/home.html.

38. *Ibid.*

39. The Marin Institute, "Alcohol Industry 'Responsibility' Advertising," p. 1.

40. Gina Agostinelli and Joel W. Grube, "Alcohol counter-advertising and the media: a review of recent research," *Alcohol Research and Health*, winter 2002, p. 18.

41. Wechsler, *et al.*, 2002, *op. cit.*, p. 208.

42. Henry Wechsler and Bernice Wuethrich, *Dying to Drink* (2002), p. 35.

43. *Ibid.*, p. 38.

44. *Ibid.*, p. 37.

45. *Ibid.*, p. 43.

46. Kenneth Aaron, "College restricts Greek lifestyle," *The [Albany, N.Y.] Times Union*, April 15, 2006, p. A1.

47. Wechsler, *et al.*, 2002, *op. cit.*, p. 212.

48. Toben F. Nelson, *et al.*, "The State Sets the Rate," *American Journal of Public Health*, March 2005, p. 443.

49. Alcohol Policy Information System, www.alcoholpolicy.niaaa.nih.gov.

50. Wechsler, *et al.*, *op. cit.*, p. 208.

51. Wechsler, *et al.*, *op. cit.*, p. 213.

52. *Ibid.*

53. Wechsler and Wuethrich, *op. cit.*, p. 227.

54. *Ibid.*, p. 237.

BIBLIOGRAPHY

Books

Lender, Mark Edward, and James Kirby Martin, *Drinking in America: A History, The Free Press*, 1987.
Professors at Kean University in New Jersey (Lender) and the University of Houston (Martin) describe drinking in America, from Colonial times through the 1980s.

Seaman, Barrett, *Binge: What Your College Student Won't Tell You, John Wiley & Sons*, 2005.
A retired reporter and editor at *Time*, who spent two years investigating campus life, describes students who are overextended, isolated by technology, drink too much and study too little.

Wechsler, Henry, and Bernice Wuethrich, *Dying to Drink: Confronting Binge Drinking on College Campuses, Rodale*, 2002.
The director of the "College Alcohol Study" at Harvard's School of Public Health (Wechsler), and a science writer (Wuethrich) warn that campus drinking is taking a bigger toll than the public realizes and offer some possible solutions.

Articles

Butler, Katy, "The Grim Neurology of Teenage Drinking," *The New York Times*, July 4, 2006, p. F1.
Mounting evidence suggests that drinking alcohol causes more damage to the developing brains of teenagers than was previously thought.

Horovitz, Bruce, Theresa Howard and Laura Petrecca, "Alcohol Makers on Tricky Path in Marketing to College Crowd," *USA Today*, **Nov. 17, 2005, p. B1.**
Industry watchdogs, lawmakers, parents and college administrators are scrutinizing how alcohol companies market to college students.

Kesmodel, David, "Schools Use Web to Teach About Booze," *The Wall Street Journal*, **Nov. 1, 2005.**
Colleges trying to curb drinking are increasingly requiring first-year students to take an online class about alcohol.

Roan, Shari, "Threat Behind the Party-Girl Image," *Los Angeles Times*, **May 8, 2006, p. F4.**
Recent surveys suggest young women today are drinking more and earlier than previous generations.

Thornburgh, Nathan, "Taming the Toga," *Time*, **Feb. 20, 2006, p. 52.**
As campuses fight boorish behavior, the nation's largest fraternity seeks a manners makeover.

Reports and Studies

"Blueprint for the States: Policies to Improve the Ways States Organize and Deliver Alcohol and Drug Prevention and Treatment," *Join Together*, **June 2006.**
A Boston University program dedicated to developing community-based alcohol- and drug-abuse prevention and treatment programs describes how states can raise funds and improve delivery of substance-abuse prevention and treatment.

"A Call to Action: Changing the Culture of Drinking at U. S. Colleges," *National Institute on Alcohol Abuse and Alcoholism*, **April 2002.**
The institute's Task Force on College Drinking reports on the consequences of college drinking and recommends research on the best way to change the drinking culture on college campuses.

Hingson, Ralph, *et al.*, **"Magnitude of Alcohol-Related Mortality and Morbidity Among U.S. College Students Ages 18-24: Changes from 1998 to 2001,"** *Annual Review of Public Health*, **April 2005, pp. 259-279.**
Alcohol-related deaths and injuries among college students increased between 1998 and 2001.

Kuo, Meichun, *et al.*, **"The Marketing of Alcohol to College Students,"** *American Journal of Preventive Medicine*, **2003; 25(3): 204-211.**
Regulating sale prices, local promotions and advertisements may help reduce binge drinking among college students.

Wagenaar, Alexander C., and Traci L. Toomey, "Effects of Minimum Drinking Age Laws: Review and Analysis of the Literature from 1960 to 2000," *Journal of Studies on Alcohol*, **Supplement No. 14: 206-225, 2002.**
Raising the legal age for purchase and consumption of alcohol to 21 has been the most successful effort to date in reducing drinking among teenagers.

Wechsler, Henry, *et al.*, **"Trends in College Binge Drinking During a Period of Increased Prevention Efforts,"** *Journal of American College Health*, **March 2002.**
Researchers summarize the results of four national surveys of drinking patterns on college campuses conducted by the Harvard School of Public Health's "College Alcohol Study."

Ziegler, D. W., *et al.*, **"The Neurocognitive Effects of Alcohol on Adolescents and College Students,"** *Preventive Medicine*, **40 (2005), pp. 23-32.**
Underage alcohol use is associated with brain damage and neurocognitive deficits, with implications for learning and intellectual development.

For More Information

Alcohol Policy Information System (APIS), www.alcoholpolicy.niaaa.nih.gov. An online government resource that provides detailed information on state and federal alcohol-related policies.

American Beverage Licensees, 5101 River Rd., Suite 108, Bethesda, MD 20816; (301) 656-1494; www.ablusa.org. A trade association for nearly 20,000 U.S. retail liquor license holders, including bars, taverns, restaurants, casinos and package stores.

Association of Fraternity Advisors, 9640 North Augusta Dr., Suite 433, Carmel, IN 46032; (317) 876-1632; www .fraternityadvisors.org. An international organization providing resources, recognition and support for campus fraternity and sorority advisers.

Beer Institute, 122 C St., N.W., Suite 350, Washington, DC 20001; (202) 737-2337; www.beerinstitute.org. A trade association that represents brewers before Congress, state legislatures and public forums across the country.

Center on Alcohol Marketing and Youth, Health Policy Institute, Georgetown University, 3300 Whitehaven St., N.W., Suite 5000, Washington, DC 20057; (202) 687-1019; www.camy.org. Monitors the alcohol industry's marketing practices.

The Century Council, 1310 G St., N.W., Suite 600, Washington, DC 20005; (202) 637-0077; www.century-council.org. Funded by American distillers, a nonprofit organization dedicated to fighting drunken driving and underage drinking.

College Drinking: Changing the Culture, www.college-drinkingprevention.gov. A government Web site providing comprehensive research-based information on issues related to alcohol abuse and binge drinking among college students.

Distilled Spirits Council of the United States, 1250 I St., N.W., Suite 400, Washington, DC 20005; (202) 628-3544; www.discus.org. A trade association representing producers and marketers of distilled spirits and importers of wine sold in the United States.

Higher Education Center for Alcohol and Other Drug Abuse and Violence Prevention, 55 Chapel St., Newton, MA 02458; (800) 676-1730; www.higheredcenter.org. Funded by the U.S. Department of Education; provides support to institutions of higher education in their efforts to address alcohol and other drug problems.

National Center on Addiction and Substance Abuse at Columbia University, 633 Third Ave., 19th Floor, New York, NY 10017; (212) 841-5200; www.casacolumbia .org. Studies and combats substance abuse in all sectors of society.

National Social Norms Resource Center, Social Science Research Institute, Northern Illinois University, Dekalb, IL 60115; (815) 753-9745; www.socialnorms.org. Supports, promotes and provides technical assistance in the application of the social norms approach to a broad range of health, safety and social-justice issues.

Pacific Institute for Research and Evaluation, 11710 Beltsville Dr., Suite 125, Calverton, MD 20705; (301) 755-2700; www.pire.org. A nonprofit organization focusing on individual and social problems associated with the use of alcohol and other drugs. The organization's Prevention Research Institute is in Berkeley, Calif.

12

Juvenile Justice

Are Sentencing Policies Too Harsh?

Peter Katel

Alice Smith takes her son Erik home after his release from a juvenile prison in Corsicana, Texas, last year. She said Texas Youth Commission prison guards stood by while he was physically abused by other inmates. Last year the Dallas Morning News revealed brutality, sexual abuse of inmates and cover-ups at several commission facilities. Abuses have also been revealed at juvenile correctional facilities in California, Maryland and other states in recent years.

From *CQ Researcher*,
November 7, 2008.

W ashington, D.C., lawyer Matthew Caspari has developed some strong feelings about punishing teenage criminals since last August. That's when he wrestled with a knife-wielding 17-year-old who'd been harassing one of his neighbors on Capitol Hill.

Caspari had been taking a walk with his wife and their 6-month-old daughter when he saw a neighbor in trouble. As he was calling 911, the young man threatened him, and they began to fight. When Caspari's dropped cell phone picked up his wife's screams, police raced to the scene and arrested the man.

But what happened afterwards was equally disturbing, Caspari told a City Council hearing in October. After a Family Court judge released the youth while he awaited sentencing, he was back on the street hanging out with a tough crowd, Caspari said. That's why he said he opposed legislation to rescind the U.S. attorney's sole power to try teenagers 15 and older in adult court for violent crimes.

"Family Court is no deterrent," said Caspari. "Punishment and consequences are simply not taken seriously by the offenders. If you want to instill a sense of accountability in these teens and provide therapy and services — there's no reason why you can't provide that in the adult system — while protecting the community."

Democratic Councilman Phil Mendelson, who is co-sponsoring the proposal to reign in the U.S. attorney, says statistical evidence shows adult-court prosecution tends to reinforce — rather than diminish — young offenders' criminal tendencies.

"The inclination is, if somebody commits a crime, particularly a violent crime, then lock 'em up," Mendelson told the

Cut-Off Age for Juvenile Courts Is Typically 17

Children through age 17 must be tried in juvenile court in 39 states and the District of Columbia. The cut-off age is 16 in nine states and 15 in two — New York and North Carolina.

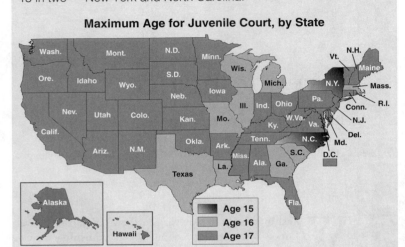

Maximum Age for Juvenile Court, by State

Age 15
Age 16
Age 17

Source: Sarah Hammond, "Adults or Kids?" *State Legislatures*, April 2008

Experts say they haven't determined how many convicts are serving time for crimes committed before they were 18. But the Campaign for Youth Justice, a Washington-based advocacy group, estimates that on any given day 7,500 youths under 18 are in jail or awaiting trial or transport to prison or juvenile detention.

Adult court sentences often are tougher than those in juvenile courts. Until 2005, they could include the death penalty, which the U.S. Supreme Court then banned for anyone who committed a capital crime before turning 18.

The backdrop to that decision was a decline in youth crime, and the drop continues. According to the most recent statistics, the 2007 arrest rate for youths ages 10-18 was down to fewer than 300 per 100,000 — the same level as in 1982.[1]

hearing. "And the research shows that is statistically counterproductive."

Mendelson's comment echoed the views of a growing number of juvenile justice experts and activists. With violent juvenile crime trending downward for the past 13 years, they say it's time to replace the tough sentences that state lawmakers enacted in the 1980s and '90s and handle more youth cases in juvenile court. The hard-line policies reflected skyrocketing juvenile crime and the prediction — later proved baseless — that violent, young "superpredators" would take over the nation's inner cities. (*See "Background," p. 271.*)

The get-tough measures eased the transferring of juveniles to adult courts where they faced tougher sentences. Some states allowed prosecutors to "direct file" juvenile cases in adult court; others left the decision to a judge, or made transfers automatic for certain charges.

But standards differ on when courts legally recognize that adulthood begins. In most states — especially those striving for more rehabilitation — 18 is the threshold age. In 10 states — Georgia, Illinois, Louisiana, Massachusetts, Michigan, Missouri, New Hampshire, South Carolina, Texas and Wisconsin — teens become adults at 17; in New York and North Carolina, it's 16.

To counter assertions by prosecutors that tougher laws brought crime rates down, opponents of harsh penalties point to studies showing that juveniles tried as adults come out of prison more dangerous than when they went in, and hence more prone to become adult criminals. A nationwide Task Force on Community Preventive Services, appointed by the U.S. Centers for Disease Control and Prevention, concluded in late 2006: "Overall, available evidence indicates that use of transfer laws and strengthened transfer policies is counterproductive for the purpose of reducing juvenile violence and enhancing public safety."[2]

Indeed, at a recent conference on juvenile rehabilitation at the Brookings Institution, Bart Lubow, director of programs for high-risk youth at the Annie E. Casey Foundation, said the punitive laws of the 1980s and '90s had "resulted in the criminalization of delinquency." The Baltimore-based nonprofit is advising 100 cities and counties on how to reorganize their juvenile systems so that they rely less on incarceration.

Many prosecutors say they also want to channel more juveniles into detention alternatives — but not all of them.

In Oregon, says Clatsop County District Attorney Joshua Marquis, "We went from an extreme — 'everyone needs a hug and cup of Ovaltine' — to a more nuanced system. Delinquents who need a minimum of incarceration and a maximum amount of structure get treated one way. And then there are the young criminals who for all intents and purposes are young adults — they don't act like children, don't respond like children and you can't treat them like children."

Oregon voters approved the present system in 1994, when the tough-on-crime approach was sweeping the nation. Measure 11 stiffened sentences for certain violent offenses and applied them to defendants as young as 15.

By 2003, 31 states had passed laws requiring juveniles charged with certain crimes to be tried as adults. Also during the '90s, 13 states lowered the top age for juvenile court jurisdiction to 15 or 16. As a result, the number of inmates serving life without parole for crimes committed when they were under 18 began climbing; today 2,484 youthful offenders are serving such sentences.[3]

But rollback advocates have scored a few successes. Connecticut last year raised its age threshold for adult court from 16 to 18. In 2006, Colorado abolished juvenile life without parole. In addition, several states have restricted adult-court transfers, and advocates are readying legislation for introduction in other states next year (*see p. 266*).[4]

Hard-liners can claim some victories as well. This year, a California proposal to abolish life without parole for juveniles failed to get the required two-thirds majority needed for passage. And in Colorado, Democratic Gov. Bill Ritter Jr., a former district attorney, vetoed a bill that would have stripped prosecutors of their sole authority to charge juveniles in adult court.[5]

"They wanted to take away our discretion — there's still a movement in our state to do that," says Denver

Youths Get Life in Prison in 31 States

Judges in 31 states must sentence juveniles to life in prison without parole (LWOP) if they are convicted of first-degree murder or certain other offenses; judges in 14 states have sentencing discretion. Five states and the District of Columbia do not permit juvenile LWOP. Pennsylvania has 444 youths serving life without parole — more than any other state.

* Ban on sentencing to LWOP is under court challenge

Source: "The Rest of Their Lives," Human Rights Watch, May 2008

District Attorney Mitch Morrissey. "They wanted to have more hearings and more experts and cost a lot more money."

Morrissey and other supporters of tough laws argue that prosecutors use them sparingly. In the suburbs of Minneapolis-St. Paul, Dakota County Prosecutor James C. Backstrom tells of resisting heavy pressure in 2006 to press for life without parole for two 17-year-olds who gunned down one of the boys' parents in cold blood. Instead, the prosecutor accepted pleas to a charge that didn't carry the no-parole proviso, giving them a chance to apply for release after 30 years.

"They knew right from wrong; there was no question they should be convicted of first-degree murder," Backstrom says, "but they had no criminal history whatsoever. I just did not feel that locking them up for the rest of their natural lives was the right thing to do. They'll have a chance to salvage some part of their lives. There were some strong disagreements, even from the victims' family."

AP Photo/Waco Tribune-Herald/Duane A. Laverty

Texas state Rep. Paula Pierson talks with an inmate at the Texas Youth Commission facility in Marlin in March 2007 in the wake of a scandal involving the sexual abuse of incarcerated youths.

Prosecutors everywhere can recall horrendous cases that warranted tough sentences. But rollback advocates argue such cases tend to obscure the fact that more than half of juvenile cases that end up in adult court don't involve crimes against people.

"You could certainly say that when you expand the use of adult court transfer you are likely to capture more serious offenders," says Jeffrey A. Butts, a research fellow at the University of Chicago's Chapin Hall Center for Children. "But it's a blunt instrument, so you pull a lot of youth into that pathway in the attempt to grab all serious offenders."

According to the Justice Department's Office of Juvenile Justice and Delinquency Prevention (OJJDP), about 51 percent of all 6,885 juvenile cases transferred ("waived") to adult court in 2005 (the most recent figures available) involved "person" offenses — that is, crimes against individuals. The rest were property crimes (27 percent), drug offenses (12 percent) and public order violations (10 percent), such as weapons, sex or liquor violations.[6] (*See graph, p. 268.*)

No national statistics exist on the total number of juveniles tried in adult court. The closest estimate, based on calculations by Butts, is 200,000 a year.

To be sure, statistics don't capture the nitty-gritty of crime in the streets. Lawyer Caspari says the teen who pulled a knife on him wasn't eligible for transfer to adult court because Caspari was never cut or stabbed. But he could have been.

That's why Caspari opposes allowing judges — instead of prosecutors — to send cases to adult court. The relative speed of the present system, he says, tells young offenders that they'll be held accountable quickly. "The practical reality is the defendant's lawyer can gum up the system by requesting it go back down to juvenile court, and that's another nine months," he says. "Is that the message you want to send to these kids?"

As prosecutors and experts debate the nation's juvenile justice policies, here are some of the key questions:

Should states roll back their tough juvenile crime laws?

When youth crime skyrocketed in the late 1980s and early '90s, legislatures across the country took a new approach toward handling young people charged with crimes. Lawmakers carved out major exceptions to practices designed, broadly speaking, to rehabilitate rather than to punish.

"Today we are living with a juvenile justice system that was created around the time of the silent film," Sen. John Ashcroft, R-Mo. (later U.S. attorney general in the first George W. Bush administration), complained to the Senate in 1997, reflecting a widely held sentiment. It's a system "that reprimands the crime victim for being at the wrong place at the wrong time, and then turns around and hugs the juvenile terrorist, whispering ever so softly into his ear, 'Don't worry, the State will cure you.' . . . Such a system can handle runaways, truants and other status offenders, but it is ill-equipped to deal with those who commit serious and violent juvenile crimes repeatedly."[7]

The new get-tough approach, adopted with variations in all states and Washington, D.C., focused on easing the process by which juveniles accused of homicide and other violent offenses could be tried in adult court. In some states, those convicted would do their time in adult institutions.

At least two states turned the corner ahead of the others. In New York, following two random murders by a 15-year-old in the New York City subway in 1978, the legislature gave automatic jurisdiction to the adult court system in violent crimes involving defendants as young as 13. Three years later, Idaho enacted a law that automatically sent youths 14 to 18 to adult court for murder and four other violent crimes.

A drop in violent crime by both adults and juveniles that began in the early 1990s and continued into the new century seemed to validate the hard-line laws. Yet, criminologists argued that the drop would have happened anyway for a variety of reasons, including the waning of the crack boom.

"Most systematic analyses show that the crime rate is much less sensitive to crime policy than most people think," says Laurence Steinberg, a psychology professor at Temple University in Philadelphia and a specialist in adolescent development. In any event, he and others have said, juveniles handled in adult courts were more likely to return to crime upon release than those handled in juvenile court.

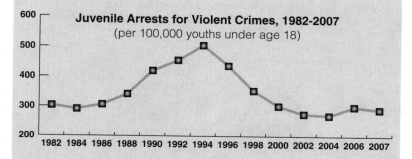

Arrest Rate Declining for Juveniles

The arrest rate for violent crimes committed by juveniles has steadily declined since peaking in 1994. The rate in 2007 roughly equaled the level 25 years earlier. Prosecutors say crime is down because laws are tough. Youth advocates say crime in general has been dropping and that harsh laws only cause more recidivism.

Juvenile Arrests for Violent Crimes, 1982-2007
(per 100,000 youths under age 18)

Source: Jeffrey A. Butts, "Juvenile Arrest Rates 1982-2007," presentation at the University of Chicago, September 2008

The effects of old-school confinement for young people is also being questioned in states that run juvenile institutions patterned on adult prisons. In California, a state judge in February ordered the Corrections Standards Authority to improve its reporting on conditions at the institutions, which failed to cite beatings and other mistreatment that federal investigators had uncovered. And in Texas, a major scandal over sexual and other abuses led to enactment of a new law that imposes new standards on youth prisons, including removing juveniles charged with misdemeanors from the institutions.[8]

Studies of juvenile recidivism often focus on adult court transfers. In a Justice Department-funded study in Florida, researchers reported in 2005 that 49 percent of juveniles transferred into the adult court system committed new crimes after release, compared with only 35 percent of the offenders who were kept in the juvenile system. Among violent offenders, recidivism ran to 24 percent and 16 percent, respectively.

"Juveniles exiting the adult criminal justice system are more likely — not less likely — to re-offend than juveniles who committed the same crimes and had comparable criminal histories," Steinberg says. "And those coming out of the adult system re-offend sooner and more seriously."

Young convicts who return from prison have serious effects on communities, Steinberg says. "Juvenile offenders have a lower success rate in the transition to adulthood than any other group of disadvantaged individuals," he says. "Our current policy, which presumably is supposed to reduce crime, actually makes our neighborhoods more dangerous."

But Oregon District Attorney Marquis says that juvenile advocates who focus on recidivism overlook a key fact — imprisoned criminals don't hurt anyone while locked up. "Incapacitation" is the law-enforcement term for that outcome, and, "That's not a small thing," says Marquis, a member of the National District Attorneys Association's Executive Committee.

Oregon's Measure 11 requires long prison sentences for 16 violent and sex-related crimes for all perpetrators age 15 and older. "The most effective thing that is done, realistically, is incapacitation," Marquis says. "In Oregon they actually counted up the number of people not raped, beaten or robbed as result of Measure 11." According to Crime Victims United, a citizens' group, the measure prevented 67,822 robberies, aggravated assaults, forcible rapes, manslaughters and murders through 2006.[9]

However, a 2004 Justice Department-funded study by the nonprofit RAND Corp. concluded the

Many Youths Land in Adult Courts for Committing 'Non-Person' Crimes

Half of the juveniles referred to adult criminal courts in 2005 were charged with property, drug and so-called public order offenses, or roughly the same percentage referred for crimes against persons.

Referral Offenses for Juveniles, 2005
(by percentage referred to adult courts and number of offenses)

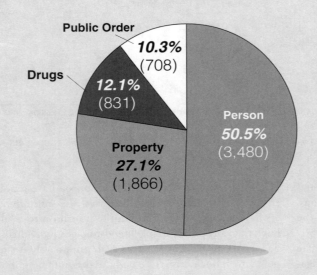

Public Order
10.3%
(708)

Drugs
12.1%
(831)

Property
27.1%
(1,866)

Person
50.5%
(3,480)

Source: "Easy Access to Juvenile Court Statistics: 1985-2005," National Center for Juvenile Justice, 2008; Office of Juvenile Justice and Delinquency Prevention

the juvenile codes in the early-to-mid-'90s were long overdue," says Minnesota prosecutor Backstrom. "In most Minnesota jurisdictions, 1-2 percent end up in adult court."

Did tough laws lower crime rates?

Juvenile crime began falling nationwide just as the last states to enact measures treating some juveniles as adult criminals were falling into line with the national trend.

The juvenile violent crime boom hit its peak in 1994. From that year to 1996, juvenile arrests for violent crimes declined by 12 percent, according to the Justice Department. Overall, juvenile arrests increased by 3 percent from 1995-1996, to 2.8 million, but drug crimes along with curfew violations and other "status" offenses largely accounted for the increase.[11]

To be sure, the pattern didn't hold true throughout the country, as is typical of all crime trends. And juvenile crime in the late 1990s remained far above its early-1980s level. Violent crime arrests began climbing steadily in 1988 — just as a virulent crack cocaine epidemic began hitting the nation's inner cities — from about 350 per 100,000 10-to-17-year-olds in the population to a peak of about 525 per 100,000 in 1994.[12]

But the decline in juvenile crime continued well into the new century. In 1995-2004, arrests of suspects age 18 and under fell 22 percent. (Adult crime remained essentially flat, registering a 1 percent drop, during the same period.)[13]

As early as 1996, Georgetown's Bilchik, then head of the Justice Department's Office of Juvenile Justice and Delinquency Prevention, noted that forecasts of an ever-rising wave of juvenile violence had been wrong. Instead, the crime numbers had started heading down. "The predictions of an onslaught of violent crime have been proven wrong two years in a row," he wrote in the

incapacitation effect was lessened because more small-scale offenders were being imprisoned along with violent criminals. In 1994, 24 percent of the defendants sent to prison for the crimes that would later be covered by Measure 11 had clean records. In 1999, when the law had kicked in, 36 percent had no prior offenses.[10]

Some leaders of the rollback movement favor certain exceptions. "I'd have no hesitancy even today to send some kids to adult court," says Shay Bilchik, a former Miami prosecutor who now directs Georgetown University's Center for Juvenile Justice Reform. "But that's a small minority of cases, probably less than 5 percent of kids who get transferred."

Prosecutors typically argue that a greater share of young offenders deserve transfer. "The changes incorporated in

department's annual statistical report, in a tacit swipe at the "superpredator" thesis.[14]

As violent juvenile crime continues to decline, however, hard-liners cite the downward trend as evidence that the tough laws of the 1980s and '90s delivered on their promise.

But even as the punitive approach took hold, some cities and counties used the flexibility in some laws to channel delinquents into rehabilitation-oriented programs. The outcomes have been positive, says the Annie E. Casey Foundation's Lubow. "Nobody's suffered, there's been no great public safety risk."

But backers of the tougher approach argue that juvenile crime responded to tougher laws just as adult crime trended downward in states that adopted laws requiring prison time after a third felony conviction.[15]

"You can compare the result to adult crime after we passed the three-strikes law in California," says Nina Salarno-Ashford, a former prosecutor who headed California's Office of Victims' Services. "We're taking the worst off the streets, and it does lower re-offending. Some do recidivate, but the heavier sentences for top-end offenders help in the decline." Salarno-Ashford's family founded Crime Victims United after her older sister was murdered in 1979.

Rollback supporters note that, despite the tougher laws, an uptick of violent crime from 2004-2006 briefly interrupted the downward slide. "I would venture that few of these get-tough reformers are willing to take credit for the increase in crime that has taken place in the last several years," Temple University's Steinberg told the Brookings youth rehabilitation conference.

Some on the law-enforcement side of the debate agree that simple explanations for crime upsurges and declines should be treated with some skepticism. But supporters of the tougher laws say they're willing to accept some uncertainty about what brought crime down — as long as it went down.

"Something's working," says Denver District Attorney Morrissey. "If it is because these laws got passed, and we treated violent offenders differently, I think that's good to see. Fewer people are getting victimized."

Morrissey says it would take a thorough statistical analysis to identify a direct connection between declining crime and a 1987 Colorado law that expanded prosecutors' power to transfer juveniles to adult court.

Just as important, he suggests, are Colorado's rehabilitative programs for juveniles in detention institutions. "They tend not to go to prison" as adults, he says.

Crime-trend analysts on the youth advocate side of the debate have been arguing for years that the causes of crime surges and declines have little to do with law and policy changes. "If we go back to the 1970s and '80s, when New York was expanding the use of adult courts and prisons for juveniles, do you see a corresponding decline for youth crime in New York? No," says the University of Chicago's Butts, summarizing research by criminologist Simon Singer of Northeastern University.

Conclusive cause-and-effect evidence is virtually impossible to find, Butts says. "You'd need a study that is impossible to do — take a big sample of youth who don't know anything about criminal justice and expose some of them to information about adult transfer, and keep the others in a bubble," he says. Tracking the number from each group who got into trouble with the law would provide definitive statistics, he says.

Does the prospect of facing the adult court system deter juveniles from crime?

A key argument for tougher laws holds that many young, potential criminals are scared "straight" at the thought of going to adult court — and possibly adult prison.

"Proponents of the latest reform proposals espouse a philosophy of retribution and punishment — insisting that the juvenile court and its sanctions do not deter juvenile crime," the Office of Juvenile Justice and Delinquency Prevention said in summarizing a 1996 conference in Washington.[16]

In Idaho, the main author of a 1995 state law proclaims that the deterrent effect of his state's tougher approach is palpable. "Before, it was no big deal to go to juvenile court," says Republican state Sen. Denton Darrington. "Now, kids don't like to go before a judge who has control over their lives. He has a lot of options at his disposal: He can bind them over to adult court. He can put them in a local juvenile detention center. He can put them on probation and dictate the terms."

While the Idaho law stepped up penalties and eased the transfer of juveniles to adult court, it also expanded or created treatment programs for juveniles who weren't sentenced to detention.

Darrington, who logged 33 years as a junior high school history teacher, says he's certain young peoples'

After a Youth's Death, Who Pays?

Would jail help rehabilitate a young shooter?

Like his father before him, Airrion "Ali" Johnson was in the wrong place at the wrong time. The 16-year-old Washington, D.C., youth was hanging out with some friends late one night last year during the Labor Day weekend. He had his mother's permission to be out, but he wasn't where he said he'd be.

Instead of his best friend's apartment, Johnson and his pals were at another, unsupervised apartment, and one of the teenagers there, 18-year-old David Williams, had a pistol. Playing around, a 15-year-old girl grabbed the gun and — thinking it was unloaded — pulled the trigger.

"My mom is going to be so mad," Johnson said just after the bullet hit him in the chest. He was pronounced dead a little while later.

Johnson's mother was angry indeed. But Theresa Norville reserves her strongest outrage for the city's Department of Youth Rehabilitation Services (DYRS), which she says has failed to provide long-term supervision and counseling for the girl who fired the fatal shot.

"I'm in therapy once a week, and I know that there's no way this child could be rehabilitated in six months," says Norville, 35, an account analyst at Children's Hospital. She thinks the girl should be incarcerated — for her own good — not out walking around. "Rehabilitation can't be roaming the streets," Norville says, noting that Williams, the gun owner, pleaded guilty to manslaughter and is serving a 36-month sentence.

In fact, DYRS statistics show that young people who don't serve time at detention facilities but are supervised in "community-based placements" — which can include remaining at home — had a recidivism rate of 28 percent in fiscal 2007. By comparison, the recidivism rate for young people who had been confined was 16 percent.

Johnson's death provides a window into the real-life circumstances that lie behind arguments about punishment versus rehabilitation. Though Norville's criticism of the youth agency might have been expected from a victim's mother, she insists that she's not out for vengeance. What she says she wants is intensive counseling and monitoring for the girl, pointing out that only confinement would ensure that she gets help.

Indeed, Norville has visited Williams in prison and says he's gotten better rehabilitative attention behind bars than the girl has received from the DYRS.

Norville tells of hugging the girl after she broke down in the courthouse shortly after the shooting. "I told her, 'I don't hate you, I don't want you locked up for the rest of your life, but you did what you did, and everybody has to pay for their actions in life.' "

Shootings in Washington are often deliberate, so Johnson's accidental death wasn't a typical crime. But it did stem from possession of a handgun — which was illegal in Washington at the time. Williams had been carrying the pistol, he said, for protection from enemies in the often rough Shaw neighborhood, Norville says.

That same neighborhood, in fact, is where Johnson's 26-year-old father, also named Airrion, died in 1997 after being hit by a stray bullet fired during a shootout between rival gangs. A *Washington Post* report on the trial of some of the accused shooters confirms Norville's account of her husband. "He had a son, Ali, whom he adored, and a job working construction. He had nothing to do" with the gangs, the newspaper reported.[1]

Covering the trial three years later, reporter Neely Tucker wrote that immediately after being shot, Airrion "sat on the pavement saying, 'I'm okay. No, really, hey, I'm all right.' Then he died."

"Ali was six or seven at the time," Norville says. "We talked about it often." She used his father's death, she says, as a lesson about the importance of trying to be in the right place at the right time.

---

[1] See Neely Tucker, "Revenge on Trial," *The Washington Post*, April 6, 2000, p. A1.

determination to avoid the expanded juvenile system has played a major part in the juvenile crime decline. From 1994 through 2004, Idaho's juvenile arrests fell 27 percent — from 23,170 to 16,747 — even as the under-17 population grew 8 percent — from 158,005 to 170,936.

The rollback advocates don't quarrel with some aspects of the Idaho program and others that resemble it. But Idaho also allows imprisoning youths in adult prisons if they're convicted in adult court, though that step isn't mandatory.

But youth advocates draw the line at confining youths with adults, arguing that no deterrence or other purpose is served. "The more punitive the response, the more juvenile offenders re-offend," says Temple University's Steinberg. "Most crimes committed by juveniles are impulsive, stupid acts that occur when they're with their friends, not calculated decisions. To be deterred by the prospect of a long sentence or incarceration or transfer into the adult system, an adolescent needs to think like an adult." (*See sidebar, p. 274.*)

Deterrence, however, isn't the only rationale for keeping extremely severe penalties on the books. "With kids, the deterrent factor is less than with adults," says Minnesota prosecutor Backstrom, accepting a main argument of youth advocates. "A lot of kids don't think before they act."

However, Backstrom says, where violent crime is concerned, "There needs to be accountability," including any punishment short of the death sentence. "Life without parole for a kid would be used in a very limited set of circumstances, but there might be a case where it's warranted. To remove the possibility would be wrong. Juveniles have tied up and tortured elderly people — I don't agree with those who want to argue that people who do that shouldn't be locked up for life."

Some rollback proponents concede that some adolescents should be locked up, even in adult institutions. But focusing on extreme and relatively rare cases obscures a more important question: "The issue is whether the system is smart enough to distinguish high-risk kids from run-of-the-mill delinquents," says Lubow at the Annie E. Casey Foundation.

"About a quarter-million kids whose offenses were committed under the age of 18 are prosecuted annually in the adult system," Lubow says. "These are not, by and large, gang-banging, gun-wielding baby rapists. Are we better off for doing this? Do we deter kids from committing serious crimes?" The Centers for Disease Control study, among others, makes clear that the answer is no, he says.

But Oregon prosecutor Marquis says his contacts with adolescents leave no doubt that they're well-informed about the law change. "I am astounded at how many kids know about this. Over and over I have heard, 'They have a really tough law here in Oregon — you use a gun in a robbery, you get Measure 11.'"

The evidence is conclusive, Marquis says. "Juvenile crime has had a huge drop in Oregon." Statistics on the juvenile crime rate before and after Measure 11 took effect weren't available. But adult crime (which, under the new law, includes serious offenses committed by anyone 15 and older) did drop by 27 percent from 1995 and 1999. By 2006, violent crime in Oregon had decreased to less than 300 crimes per 100,000 persons.[17]

BACKGROUND
Separate System

America's young cities began growing in the early 1800s, largely because of waves of immigration. Given the desperate circumstances in which they arrived, and the long hours they worked, immigrants had little choice but to let their children roam the streets unsupervised. Not surprisingly, some got into trouble.

Alarmed at what they were seeing, early urban reformers established the forerunners of today's juvenile detention institutions. The New York House of Refuge, founded in 1824, was the first. A group of prominent citizens established the Society for the Reformation of Juvenile Delinquents and persuaded the state legislature to create the facility for "boys under a certain age who become subject to the notice of our police, either as vagrants, or homeless, or charged with petty crimes." They would be put to work, and given a basic education, "while at the same time, they are subjected to a course of treatment, that will afford a prompt and energetic corrective of their vicious propensities."[18]

Other cities, including Boston, Philadelphia and Baltimore, followed suit, but hope that "refuges" would put a big dent in juvenile crime proved ill-founded. The explosive growth of poor, often desperate, urban populations far surpassed the institutions' capacities.

Some cities and states concluded they needed another way to house wayward children. The first "reform school" opened in Massachusetts in 1849, but such institutions also proved ineffective.

Meanwhile, civic reformers perceived another problem — children convicted of serious crimes were being imprisoned with adults because adult courts and prisons were the only institutions available. Pressed by concerned citizens who argued that government had a

CHRONOLOGY

1800s *Civic reformers develop private institutions to help youths in trouble.*

1824 Civic leaders found New York House of Refuge to care for young vagrants, petty criminals.

1849 Massachusetts opens "reform school" for young people who hadn't committed crimes but weren't enrolled in school.

1899 Illinois Legislature creates nation's first juvenile court system to handle growing number of youths being tried and sentenced as adults.

1960s-1980s *Youth advocates successfully challenge the constitutionality of juvenile court proceedings nationwide, but liberalization wave ebbs as youth crime skyrockets.*

1966-1970 Supreme Court's *Kent*, *Gault* and *Winship* decisions extend constitutional due-process rights to defendants in juvenile courts.

1974 Congress amends Juvenile Crime and Delinquency Prevention Act of 1968 to ban detention for "status offenses" — curfew-breaking, cigarette purchasing and the like, which only apply to juveniles — for states receiving grants under the law, and to require separation of juveniles from adults in jails and prisons.

1978 Random murders by 15-year-old self-proclaimed "monster" Willie Bosket, spark tough, new juvenile crime law in New York.

1982 Supreme Court's *Eddings* decision limits death penalty for juveniles.

1987 Beginning of crack cocaine epidemic sees youth murder arrests climb to about 10 percent of all arrests, up from 6 percent in 1984.

1988 Supreme Court's *Thompson* decision prohibits capital punishment for juveniles convicted of crimes committed when they were 15 or under.

1990s *Legislatures nationwide respond to juvenile crime wave by toughening laws and easing the transfer of juveniles to adult court, usually in cases involving violent crime.*

1994 Violent juvenile crime nationwide reaches all-time high of 500 arrests per 100,000 under-18s in population . . . . Oregon voters pass Measure 11, requiring long prison sentences for certain serious crimes for offenders 15 and older.

1995 Twenty-one states require juveniles to be tried in adult court for certain serious crimes.

1996 Violent juvenile crime declines 6 percent from previous year. . . . Juvenile courts handle 1.8 million cases, quadruple the 1960 number. . . . Princeton University sociologist John DiIulio and fellow conservatives predict wave of "superpredator" youths.

1997 Forty-five states and Washington, D.C., have made it easier to transfer juvenile defendants to adult court.

2000s *Decline of juvenile violent crime continues; youth advocates seek to roll back hard-line measures.*

2001 DiIulio repudiates "superpredator" thesis.

2004 Violent crime arrests for juveniles fall 22 percent below 1995 level.

2005 Supreme Court in *Roper v. Simmons* abolishes death penalty for defendants under 18 when they committed their crimes; cites brain studies showing adolescents' capacity for judgment not fully developed.

2006 Colorado ends life without parole sentences for juveniles.

2007 Connecticut raises the age at which youths can be tried as adults from 16 to 18. . . . Move to abolish juvenile life without parole in California fails. . . . *American Journal of Preventive Medicine* reports "insufficient evidence" that transferring juveniles to adult court prevents violence.

2008 Gov. Bill Ritter, D-Colo., vetoes legislation to abolish prosecutors' authority to file charges against juveniles directly in adult court. . . . High-profile juvenile justice conference planned on Nov. 6 at Georgetown University to reform harsh laws on youth crime.

special duty to help juveniles mend their ways, the Illinois legislature in 1899 established the nation's first juvenile court in Chicago. Later that year, Colorado lawmakers took the same step in Denver.

Illinois and Colorado also created a category of juvenile offenses seen as gateways to the criminal life, such as "truancy" and "growing up in idleness."[19]

Unlike in adult courts, lawyers and constitutional protections weren't required in juvenile courts since judges would be acting in the juveniles' best interests. Moreover, the courts' stated goal wasn't punishment but rehabilitation.

Judges essentially had unfettered discretion to devise "treatment plans" for juveniles that could leave them confined until they were classified as cured, or they turned 21.

New Standards

By the 1960s, the juvenile court model was coming under growing challenge from liberals, who complained that young offenders not only were being denied legal representation but also other rights that adult defendants enjoyed.[20]

Some of these concerns were addressed in a string of U.S. Supreme Court decisions beginning in the mid-1960s. Starting with the basic questions of young peoples' due-process rights in juvenile courts, the high court eventually found itself grappling with perhaps the weightiest criminal-law issue of all for juveniles — the death penalty.

Before reaching that question, the court in 1966 laid the groundwork for extending adult rights to juveniles. The "essentials of due process" had to be provided to young people, the court said in its landmark *Kent v. United States* ruling. In his majority opinion, Justice Abe Fortas warned that juvenile courts were failing on all fronts: "There may be grounds for concern that the child receives the worst of both worlds: that he gets neither the protections accorded to adults nor the solicitous care and regenerative treatment postulated for children."[21] The following year, the court's *In re Gault* decision laid down specific requirements for juvenile court hearings in which defendants faced commitment to a detention center. In such cases, courts had to grant adequate notice of specific charges, notice of right to a lawyer, the right to confront witnesses and the right against self-incrimination.

Supreme Court decisions found an echo in Congress. The Juvenile Delinquency Prevention and Control Act of 1968 recommended — but did not require — that children charged with "status offenses" be dealt with outside the court system. Status offenses are acts that are illegal only for young people — buying cigarettes, for instance, or violating curfews.

Lawmakers toughened the law in 1974, making states' eligibility for federal grants contingent on removing status offenders from detention, and on physically separating juvenile offenders from adults in jails and prisons. Congress amended the law in 1980 to require that juveniles be removed from all adult jails.

The Supreme Court, meanwhile, continued addressing juvenile justice issues. In its 1970 *In re Winship* decision, justices required states to prove delinquency cases beyond a reasonable doubt — the same standard required in adult criminal convictions. *Breed v. Jones*, in 1975, established that transferring juveniles to adult criminal court after they have been adjudicated in juvenile court constitutes double jeopardy — the unconstitutional practice of trying someone twice for the same crime.

But in 1984, in *Schall v. Martin*, the court approved pretrial, or "preventive," detention. Holding a juvenile defendant thought to pose a risk of committing another crime isn't a punishment, the justices concluded. Procedures were in place, they said, to protect young defendants from improper detention.[22]

A 1985 Supreme Court decision (*New Jersey v. T.L.O.*) loosened Fourth Amendment protections for high school students, allowing school personnel to search students' lockers and belongings if "reasonable grounds" exist to believe that a student has violated school rules or the law. In other circumstances, the search standard is "probable cause."

But the high court began in the 1980s to take up the most morally and emotionally charged juvenile justice issue of all — the death penalty. Finally, following two decisions that limited capital punishment for juveniles, the court in 2005 banned the death penalty for defendants who were under 18 when they committed a capital crime.

Toughening Up

A wave of sensational crimes committed by young offenders — followed by skyrocketing street violence spawned by a crack cocaine boom that began in the 1980s — sparked a new era in juvenile justice in the 1990s.[23]

From 1975 to 1987, the number of juveniles arrested for violent crimes hovered around 300 arrests per 100,000

Should Adolescents Be Treated Like Adults?

Youth advocates and prosecutors square off.

Medical science is playing a key role in the debate over whether juveniles accused of serious crimes should be treated as adults. By scanning the brain in far more detail than ever before, researchers are providing data supporting a key argument by those who advocate rehabilitation rather than jail. They say juveniles shouldn't be treated like adults because the brain scans show they don't think like adults.

That position played a central role in the U.S. Supreme Court's 2005 *Roper v. Simmons* decision banning the death penalty as unconstitutional for juveniles who were under 18 when they committed the crime. Christopher Simmons, a high school junior in Missouri, broke into a house with a friend, with burglary and murder in mind. "Simmons said he wanted to murder somebody," Justice Anthony Kennedy wrote in the majority opinion in the 5-4 decision. Simmons had said they'd get away with the crime because they were minors.[1]

Shirley Crook, 46, was home alone, her husband away on a fishing trip. With duct tape, Simmons and his friend bound her hands and covered her eyes and mouth. They drove her to a railroad trestle at a state park, reinforced her bindings with electrical wire and threw her into the Meramec River.

Notwithstanding the facts of the case, Kennedy accepted evidence that adolescents' capacity for judgment remains immature. "The reality that adolescents still struggle to define their identity means it is less supportable to conclude that even a heinous crime committed by a juvenile is evidence of irretrievably depraved character," Kennedy wrote.[2]

For youth advocates, the decision represented a major breakthrough in their effort to gain acceptance for evidence that teenagers' brains haven't developed sufficiently for them to deserve the full weight of judicial penalties.

Outside the capital-punishment realm, however, youth advocates are still trying to make that case. Earlier this year, the American Bar Association adopted a resolution urging that judges take adolescent maturity levels into account.

"Youth are developmentally different from adults," the resolution says, "and these developmental differences need to be taken into account at all stages and in all aspects of the adult criminal justice system."[3]

The National District Attorneys Association is fighting to overturn the resolution. "We do not believe it is appropriate to take language articulated in a U.S. Supreme Court decision concerning whether or not to impose the death penalty on juvenile murders and apply the same logic in a completely different conceptual framework," the prosecutors said.[4]

In any event, the prosecutors remain skeptical of the adolescent brain studies. "Some of that is hocus pocus,"

---

youths ages 10 to 18 in the population. But in the following seven-year period, 1987-1994, the rate rose by more than 60 percent, to about 500 arrests per 100,000.[24]

Juvenile crime fell again, starting in 1994. By 2004, the juvenile arrest rate for violent crimes had dropped to 271 per 100,000. However, the new hardline laws remained in place. In 1996, juvenile courts handled about 1.8 million delinquency cases — more than four times the 400,000 cases in 1960.[25]

The rapid adoption of the tougher approach reflected not only rising juvenile crime but the fear that far worse was coming. By the mid-1990s, some politically conservative academics attracted considerable publicity and political influence by declaring that a new breed of young "superpredators" was developing. John DiIulio, then a political science professor at Princeton University, coined the term, which soon gained currency.

"Based on all that we have witnessed, researched and heard from people who are close to the action," DiIulio and two co-authors wrote in 1996, "here is what we believe: America is now home to thickening ranks of juvenile 'superpredators' — radically impulsive, brutally remorseless youngsters, including ever more preteenage boys, who murder, assault, rape, rob, burglarize, deal deadly drugs, join gun-toting gangs and create serious communal disorders." DiIulio's co-authors were John P. Walters, now director of the Bush administration's Office of National Drug Control Policy, and William J. Bennett, a prominent conservative who was Education secretary in the Reagan administration,

says Joshua Marquis, a district attorney in Oregon. "Some people mature early."

For his part, James C. Backstrom, the prosecutor in Dakota County, Minn., accepts the brain studies and even their possible relevance at sentencing. But, he adds, "There is a complete disconnect if you say that is a basis why they shouldn't be prosecuting kids as adults. I think a 16- or 17-year-old youth is fully capable of understanding right from wrong, and understanding that it's wrong to murder, rape or torture someone."

Backstrom was echoing an argument by Supreme Court Justice Antonin Scalia. In his dissent in *Roper*, Scalia quoted from an earlier decision in which the court wrote that it was "absurd to think that one must be mature enough to drive carefully, to drink responsibly, or to vote intelligently, in order to be mature enough to understand that murdering another human being is profoundly wrong, and to conform one's conduct to that most minimal of all civilized standards."[5]

Some medical professionals on the liberal side argue, in effect, that biology can trump morality. In the often chaotic circumstances in which most crimes take place, doctors say that the state of adolescent brains is highly relevant to the issue of how to hold juveniles accountable for actions that they don't control in the same ways that adults are capable.

"When children find themselves in emotionally charged situations, the parts of the brain that regulate emotion, rather than reasoning, are more likely to be engaged," said Physicians for Human Rights, a Cambridge, Mass.-based advocacy group, in a brief in the *Roper* case, "[6]

According to the group, brains scanned using magnetic image resonance — the same technology used to detect tumors and other abnormalities — show that adolescent

behaviors are largely controlled by parts of the brain's limbic system, which is part of the so-called primitive part of the brain. It includes the amygdala, specifically important in adolescents' actions, which regulates fear, aggression and impulse. Only when the prefrontal cortex matures — usually when a person is in his 20s — do reasoning and understanding of consequences develop fully, the group said.[7]

The Supreme Court didn't delve into brain structure, but Justice Kennedy did explore the psychological dividing line between adolescence and adulthood, noting that parents are well aware of the differences. "Retribution is not proportional," he wrote, "if the law's most severe penalty is imposed on one whose culpability or blameworthiness is diminished, to a substantial degree, by reason of youth and immaturity."[8]

---

[1] *Roper v. Simmons*, 543 U.S. 551 (2005), www.oyez.org/cases/2000-2009/2004/2004_03_633/.

[2] *Ibid.*

[3] "American Bar Association — 105C," Adopted by the House of Delegates, Feb. 11, 2008, p. 5, www.abanet.org/leadership/2008/midyear/updated_reports/hundredfivec.doc.

[4] "State of the National District Attorneys Association in Response to the Proposed ABA Resolution Concerning Sentence Mitigation for Youthful Offenders," May 4, 2007, www.ndaa.org/ndaa/capital/capital_perspective_july_aug_2007.html.

[5] "Scalia, J. dissenting," *Roper v. Simmons, op. cit.*

[6] "Adolescent Brain Development, a Critical Factor in Juvenile Justice Reform," Physicians for Human Rights, undated, http://physiciansforhumanrights.org/juvenile-justice/factsheets/braindev.pdf.

[7] *Ibid.* See also, "Teenage Risk-taking: Teenage Brains Really Are Different From Child or Adult Brains," *Science Daily*, March 30, 2008, www.sciencedaily.com/releases/2008/03/080328112127.htm.

[8] *Roper v. Simmons, op. cit.*

---

and White House drug policy director under President George H. W. Bush.[26] Five years later, however, DiIulio retracted the entire thesis, prompted by a downturn in juvenile crime — exactly the opposite of what he had predicted. DiIulio, who was then director of the White House Office of Faith-Based and Community Initiatives, said that he had a moment of revelation on the issue in 1996.

"I knew that for the rest of my life I would work on prevention, on helping bring caring, responsible adults to wrap their arms around these kids."[27]

Liberal youth advocates held DiIulio and his collaborators greatly responsible for the get-tough approach that prevailed in the '90s. But it had been foreshadowed in the late 1970s in New York City by a teenager who seemed to fit the "superpredator" archetype.

In 1978, 15-year-old Willie Bosket robbed and murdered two subway passengers. Under state laws at the time, he was sentenced to five years in detention — the maximum he could receive in Family Court, where all defendants under age 16 were automatically sent.

State lawmakers quickly enacted the Juvenile Offender Law, which gave the state Supreme Court (equivalent to district courts in other states) original jurisdiction over 13-, 14- and 15-year-olds charged with violent crimes, with no exceptions.

Bosket, who called himself a "monster" created by the criminal justice system, was released from juvenile detention and later returned to prison for assault. There, he earned two life sentences for crimes committed behind bars, including the stabbing of a prison guard.

Minnesota teenagers Matthew Niedere, left, and Clayton Keister, both 17, were convicted of shooting and killing Matthew's parents. Supporters of tough laws for juveniles say the boys' sentences support their argument that such laws are applied sparingly. Prosecutor James C. Backstrom resisted heavy pressure to seek life without parole for the pair, giving them a chance to apply for release after 30 years in prison. "I just did not feel that locking them up for the rest of their natural lives was the right thing to do," he says.

New York's Juvenile Offender Law, enacted years before other states toughened their juvenile crime laws, remains the nation's toughest, according to Jeffrey Fagan, a professor of law and public health at Columbia University. "The new law signaled a broad attack on the structure and independence of the juvenile court," he wrote this year, "a major restructuring of the border between juvenile and criminal court that was repeated across the nation in recurring cycles for more than two decades."[28]

Other states that revamped their "transfer" laws took a variety of approaches. Fourteen states and Washington, D.C., allowed prosecutors to file charges directly in adult court, without judicial approval, for serious felonies, typically including murder and other "person" crimes — in which a human being, rather than an institution, is the victim.

Studying the Fallout

Virtually as soon as tougher laws took effect, academics and policy makers began researching how effective they were. Focusing on the expanded use of adult court for juveniles, nearly all the researchers concluded that the laws were counterproductive.[29]

Fagan, now co-director of Columbia University's Crime, Community and Law Center, conducted a study published in 1995 that compared re-arrest statistics of youths picked up for robbery and burglary in New York and New Jersey, where adult-court jurisdiction laws differed. He

concluded that the New Yorkers, who had been transferred to adult court, were 39 percent more likely to be re-arrested for a violent crime than the New Jersey juveniles, who had been handled in juvenile court.

And among the New Yorkers who'd been sentenced to prison for more than a year, their recidivism rate for violent crime was twice that of the juvenile-court comparison group from New Jersey.

A study published in 2002 by the Florida Juvenile Justice Department found similar results when comparing youths transferred to adult court and those retained in the state's juvenile system. The transferred juveniles showed a 34 percent higher recidivism rate.

Most other studies yielded similar data. But there were exceptions. Another Florida study published in 1997 found that youths transferred to adult court on property crime charges showed lower recidivism than counterparts arrested for similar crimes and kept in the juvenile system.

Overall, however, the Task Force on Community Preventive Services, appointed by the U.S. Centers for Disease Control, concluded: "The weight of evidence shows greater rates of violence among transferred than among retained juveniles; transferred juveniles were approximately 33.7 percent more likely to be re-arrested for a violent or other crime than were juveniles retained in the juvenile justice system."[30]

The various studies form a key part of rollback advocates' argument that emphasizing adult-court prosecution is counterproductive.

Skeptical prosecutors have faulted the studies, or at least questioned the relevance of a New York-New Jersey study to, say, Washington, D.C. "Are you going to use these statistics as a guide for your jurisdiction?" asks Patricia A. Riley, special counsel to the U.S. Attorney's Office in Washington, which is fighting a rollback proposal. (See "Current Situation," p. 277.)

And she questions the validity of studies conducted within one state, because juveniles who are transferred are — by definition — more serious offenders, hence more likely to recidivate.

Researchers did try to adjust for that factor. But Butts of the University of Chicago, who specializes in juvenile crime statistics, acknowledges that it's impossible to completely control for differences between juvenile defendants.

But he adds that on further reflection, he thinks the realities of juvenile justice can produce counterintuitive results. Property crime, for instance, can be dealt with more leniently by adult court judges and juries, which are used to older and tougher defendants who've done worse. "A jury doesn't want to send a 14-year-old to prison," he says. "But in the juvenile system a 14-year-old defendant can look like a serious case. There are a lot of things about this business that don't hold up when you start looking at them."

CURRENT SITUATION

State Campaigns

Advocates for change in the juvenile justice system are taking the fight to state legislatures. So far, efforts have been mounted in only a handful of states, but the campaigners are planning to expand their efforts.

Major targets are laws authorizing life-without-parole (LWOP) sentences for defendants who were under 18 when they committed their crimes.

In California, state Sen. Leland Yee, D-San Francisco, is planning to renew his efforts to prohibit juvenile LWOP, says a spokesman, Adam Keigwin. California approved juvenile LWOP as part of a sweeping 1990 tough-on-crime ballot initiative. Yee introduced an anti-juvenile LWOP bill this year but didn't take it to the full Senate because he lacked the two-thirds majority necessary to overturn a measure passed by referendum.

Alison Parker, deputy director of U.S. programs for Human Rights Watch, a New York-based global human-rights monitoring organization, says anti-juvenile LWOP campaigns may be mounted next year in Florida, Louisiana, Michigan, Washington state and Nebraska. Colorado in 2006 became the first state in the nation to abolish juvenile LWOP.

Human Rights Watch has made juvenile LWOP one of its major U.S. issues. Parker, who wrote exhaustive reports in 2005 and 2008 on the issue, reported this year — citing the Center for Global Law and Justice

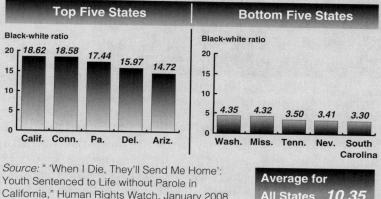

Black Youths Get Most No-Parole Life Sentences

Ten black youths in the United States were serving life-without-parole sentences in 2005 for every white juvenile. In California the ratio is nearly 19-to-1, while South Carolina has the lowest rate, about 3-to-1.

Ratio of Black to White Juveniles Serving Life-Without-Parole Sentences
(among youths under age 18)

Top Five States	Bottom Five States
Calif. 18.62, Conn. 18.58, Pa. 17.44, Del. 15.97, Ariz. 14.72	Wash. 4.35, Miss. 4.32, Tenn. 3.50, Nev. 3.41, South Carolina 3.30

Source: " 'When I Die, They'll Send Me Home': Youth Sentenced to Life without Parole in California," Human Rights Watch, January 2008

Average for All States 10.35

at the University of San Francisco — that the United States is the only country in the world in which juveniles are serving LWOP sentences. According to the center, at least 135 countries have explicitly prohibited juvenile LWOP.[31]

Nevertheless, prosecutors and other hard-liners argue that LWOP for juveniles should stay on the books. "We just believe that some crimes, even if committed by juveniles, deserve a life sentence," says Nina Salarno-Ashford, the California victims' rights advocate and ex-prosecutor. "Heinous murder, kidnap, torture — these are the kinds that get the life sentence. There are those who, in our view, will never be candidates for rehabilitation."

In 2005, Parker's research of juvenile LWOP cases found that most defendants were convicted of murder but that 26 percent "were convicted of felony murder where the teen participated in a robbery or burglary during which a co-participant committed murder, without the knowledge or intent of the teen." Sixteen percent of the juvenile LWOP convicts were 13-15 at the time of their crimes.[32]

AP Photo/Andy Carpenean

Raine Lowry was a juvenile when she was held in an adult facility in Wyoming for running away from home. Wyoming is the only state that has not complied with the federal Juvenile Justice and Delinquency Prevention Act, in part because the state still holds juveniles in adult jails.

Even so, says District Attorney Josh Marquis of Clatsop County, Ore., juveniles sentenced to LWOP make up a "tiny fraction" of youths tried as adults. "They tend to tug at the heartstrings of people who can't accept that there are people so damaged that they're never going to be safe to release."

Those seeking to change juvenile laws also are focusing on laws that require juveniles to be prosecuted as adults for certain crimes, or that make it easy to transfer juveniles into adult court.

"We know so much more about what works, and what doesn't work," says Liz Ryan, president and CEO of Campaign for Youth Justice, a Washington-based advocacy organization. She cites projects such as the Annie E. Casey Foundation's Juvenile Detention Alternatives Initiative, in which authorities in about 100 localities are cooperating. Except in exceptional cases, she says, "Kids can be safely supervised in the community. We don't have to lock up all those kids in detention facilities."

Many youth advocates cite a study earlier this year that compares the Missouri and Maryland juvenile justice systems. Missouri, which has embarked on a rehabilitation approach known as the "Missouri model," complete with dorm-style rooms in treatment centers and an emphasis on education, registered an 8 percent

recidivism rate over three years. Maryland showed a 30 percent recidivism rate for youths coming out of its detention-oriented juvenile program.[33]

Elsewhere, Illinois repealed a law in 2005 that automatically transferred to adult court any juvenile charged with drug offenses in or near public schools or housing projects. Data showed that two-thirds of those transferred were low-level lawbreakers. Judges now decide on transfers.[34]

North Carolina and Illinois are also considering following Connecticut's 2007 move to raise the age of juvenile court jurisdiction to 17 — meaning that 16-year-olds could no longer be sent to adult court.[35]

Conflict in Washington

A debate on whether to limit the transfer of juveniles to adult court in Washington, D.C., shows the intensity of opinions, emotions and racial tensions that juvenile crime issues can arouse.

"I was still bleeding from the scrapes as they pushed my body back and forth, going through my pockets and telling me they were going to kill me," Chandler Goule, who lives on Capitol Hill, told a D.C. City Council hearing in October. The 31-year-old staff director of a House Agriculture subcommittee said one of his three attackers held a gun to his head. The 17-year-old gunman — the only one of the three to be arrested — pleaded guilty in adult court and was sentenced to 48 months in prison. If he successfully completes five years of "supervised release" after prison, his conviction will be expunged from his record.

The bill that Goule was testifying against would not have prevented that transfer to adult court, but it would have shifted the authority to order the transfer from the U.S. attorney (who prosecutes serious local crimes in Washington) to a judge. Youth advocates have argued for some time that the transfer decision should rest with a neutral overseer rather than with one side in a criminal proceeding.

"Why would you, the D.C. Council, give an unelected prosecutor appointed by the president the right to exclusively decide what happens to D.C.'s youth without even so much as a hearing?" Ryan of the Campaign for Youth Justice asked the two council members who held the hearing.

The U.S. Attorney's Office says its standards keep the number of juvenile adult-court cases low. "We do not take every eligible 16- or 17-year-old who commits these offenses,

Should state laws that facilitate prosecuting juveniles in adult court be changed?

YES
Liz Ryan
President and CEO,
Campaign for Youth Justice

Written for *CQ Researcher*, November 2008

When criminologists predicted in the 1990s that a new generation of youthful "superpredators" was on the horizon, state officials responded by passing laws to make it easier to try youths as adults. They intended, understandably, to make their communities safer. However, the latest research suggests the opposite effect: Youths prosecuted as adults are much more likely to re-offend.

For example, in August, the Department of Justice's Office of Juvenile Justice and Delinquency Prevention released a report — "Juvenile Transfer Laws: An Effective Deterrent to Delinquency?" — which found that prosecuting youths as adults has little or no deterrent effect on juvenile crime. In fact, it said, youths prosecuted as adults are more likely to re-offend than youths handled in the juvenile justice system. And late last year the federal Centers for Disease Control and Prevention released a report that showed virtually identical results.

Polling shows the public rejects punitive approaches and supports rehabilitation and treatment for youth, which is virtually non-existent in the adult criminal justice system. Two recent MacArthur Foundation polls showed that 89 percent of Americans agreed or strongly agreed that rehabilitative services and treatment would help reduce crime.

States have different laws for transferring youth under age 18 to adult court. In 44 states and the District of Columbia juvenile court judges under certain circumstances may send a youth to adult court, and 14 states and D.C. allow the prosecutor to directly file cases in adult court. Thirteen states prosecute 16- or 17-year-olds in adult court for any offense. Connecticut, New York and North Carolina try all 16- and 17-year-olds as adults. Connecticut recently raised the age of juvenile court jurisdiction to 18, beginning in 2010.

Many of the youth prosecuted as adults are placed in adult jails pretrial, where they are at risk of harm, abuse and suicide. A November 2007 report released by the Campaign for Youth Justice showed that up to 7,500 youths are in adult jails on any given day, but half to a third of these youngsters are ultimately sent back to the juvenile justice system or not convicted at all, suggesting that their offenses did not merit placement in the adult criminal justice system.

Based on this new research, several states have begun to re-examine and even reverse these harmful statutes. Other state officials should seriously re-examine their state policies governing the trial, sentencing and incarceration of youths as adults.

NO
James C. Backstrom
Dakota County Attorney, Hastings, Minn.;
Member, Board of Directors, National
District Attorneys Association

Written for *CQ Researcher*, November 2008

Prosecuting juvenile offenders in adult court is appropriate and necessary in certain cases to protect public safety and hold youths appropriately accountable for their crimes. Contrary to opponents' claims, this sanction is not being overused by prosecutors. Few jurisdictions prosecute more than 1-2 percent of juvenile offenders as adults. This is a tool reserved for the most serious, violent and chronic offenders, who should face more serious consequences for their crimes than those available in juvenile court.

Don't be misled by claims that large numbers of youths are being prosecuted as adults for low-level offenses, because these statistics come from the 13 states where laws classify 16- or 17-year-olds as adults for purposes of any prosecution. This has nothing to do with transferring juveniles to adult court.

Recent scientific studies have shown that the brain is not fully developed until the early to mid-20's and that the last portion of the brain to reach full maturity is the frontal lobe governing impulse control. While this may explain why some youths lack the reasoning ability to fully appreciate the consequences of their actions, it does not mean they should not be held accountable for their crimes. The vast majority of teenagers understand the difference between right and wrong and know it is wrong to torture or kill someone. This is why our laws rightfully allow adult prosecution for these and other violent crimes.

Juveniles who commit serious and violent crimes, particularly older youths, should face potential adult court sanctions. So, too, must this remedy be available for youngsters who have a long history of convictions for less serious felonies for which juvenile court disposition has not been effective. About one-third of our states also utilize "blended sentencing" models that combine both juvenile and adult sanctions for serious, violent or habitual juvenile offenders whose crimes have been determined not to warrant prosecution in adult court.

Prosecutors and judges thoughtfully and professionally enforce juvenile codes with fairness and impartiality every day, taking into consideration both mitigating factors — such as a juvenile offender's age, maturity and amenability to treatment and probation — and aggravating factors, such as the severity of the crime, the threat to public safety, the impact upon the victim and the offender's criminal history. After properly weighing these factors, the difficult decision to prosecute a juvenile offender as an adult is warranted in some cases.

even though we are importuned to take more of them than we do," Patricia A. Riley, a special counsel to the U.S. attorney, told the hearing.

From 1999 to Oct. 15 of this year, the office prosecuted 428 16- and 17-year-olds in adult court. During most of those years, prosecutors tried about 2,200 to 3,000 juvenile cases a year. "A juvenile who has an extensive and/or violent juvenile record is more likely to be prosecuted as an adult than one who does not," Riley said.

But the juvenile crime debate opens up issues that go far deeper than statistics and prosecutorial versus judicial authority. Witnesses at the hearing also included Jauhar Abraham and Ronald Moten, co-founders of Peaceoholics, a nonprofit that works to quell violence in the District.[36] They questioned why interest in crime prevention seems to be lower when African-American teenagers are the predominant victims, as well as offenders.

"We've been coming here for years complaining about the same type of homicides, and there ain't been an issue," Moten said, suggesting that the upsurge in interest came about because "all of a sudden our city is changing," as traditionally African-American neighborhoods gentrify.

In one horrific case, a 56-year-old man, Mark Kenneth Blank, died from a "severe head trauma" after being robbed and beaten by a trio of youths — ages 13, 14 and 15.

"I guarantee you a couple of those robbers were on drugs," Moten said. "For the last 20 years this government has done nothing to address the youth drug problem. There's more young'uns on PCP committing heinous acts in this city than ever before. We have turned our back on these children, and now the chickens are coming home to roost!"

Moments later, he was pounding the witness table in frustration.

OUTLOOK

Watching the Trends

Experts and activists on both sides of the juvenile justice divide have been analyzing the near-term possibilities as they awaited a change of administration in Washington. State governments are where the action is for juvenile crime laws. But veterans in the field say the federal government can play a big part in influencing policy nationwide.

"Given the problems that any new administration is going to have to take on, none of us would be naïve enough to think that juvenile justice is going to rise above the economy or Iraq or a lot of other things," says Lubow of the Annie E. Casey Foundation. Yet simply appointing the right OJJDP administrator, he says, could make a big difference, because administrators can use federal grant-awarding power to steer states toward a more rehabilitation-oriented approach.

In fact, says Bilchik of Georgetown University's Center for Juvenile Justice Reform, who once held the OJJDP job, the past several years represent a missed opportunity. Money saved as a result of fewer juveniles being arrested could have gone "to known, proven prevention programs and lessened even more the flow of kids into the juvenile justice system. It's mind-numbing to think about the lost opportunity."

The unknowns of a new administration aside, juvenile justice experts are unsure how long the declining crime rate will endure. If the country's ever-worsening economic conditions bring a significant increase in adult and juvenile crime, interest in rolling back some of the get-tough laws could wane.

"Periods of crisis are the absolute worst times in which to discuss crime policy," said Steinberg of Temple University, speaking to the Brookings-Princeton conference. "Panic trumps prudence, and policy gets made on the basis of fear rather than foresight."

On that score, Denver District Attorney Morrissey, whose views on crime and punishment tend to differ from Steinberg's, agrees. "If we continue to take the approaches we take, I hope we'll continue to see violent juvenile crime decline in Denver," Morrissey says. But, he adds, "If it goes up 1 percent, I know we're going to see a headline." And headlines will bring cries for crackdowns, he says.

A counterweight, points out Butts of the University of Chicago's Chapin Hall Center for Children, is the vastly expanded state of knowledge about juveniles and effective programs.

"It has been helpful to show decision makers that there is verifiable, quantitative evidence that young people make decisions differently from adults. You can't estimate someone's level of cognitive development based on his birthday."

Another factor that could moderate any increase in juvenile crime is demographic reality, says Minnesota prosecutor Backstrom. "We are seeing a declining population of juveniles across America overall," he says. "As the population declines, I think the crime rate is going to continue to go down."

That overall trend may not apply everywhere, but it's certainly the case in Backstrom's suburban-and-rural territory. "Our population is dropping in every category, except that when you get to 50 and older it skyrockets. They're not too prone to commit crimes."

But Backstrom says lawmakers shouldn't respond to the downturn the way the youth advocates are proposing. "When serious crimes are committed by youths, they need to be dealt with appropriately, with significant consequences. If we roll back the clock, we're going to be harming our society."

In Washington, at least one recent crime victim is optimistic. "In 10 years, it will be better," he says. "Property values in D.C. are going to continue to rise." Street criminals, he says, will "move on."

As for the social inequalities that underlie at least some D.C. crime, Caspari hopes better schools will provide improved opportunities for young people whose hopes for the future are stunted.

But one major characteristic of juvenile justice is unlikely to change — whatever the future holds. The field is complicated by so many issues of adolescent physiology, family values and local politics that laws and policies don't always determine what happens to individuals caught up in the system.

"It's much easier to articulate sound policy," says Lubow of the Annie E. Casey Foundation, "than it is to implement it."

NOTES

1. See Jeffrey A. Butts, "Juvenile Arrest Rates 1982-2007," Sept. 15, 2008, www.jbutts.com/onlinepps/ucr2007.ppt.

2. See Angela McGowan, *et al.*, "Effects on Violence of Laws and Policies Facilitating the Transfer of Juveniles from the Juvenile Justice System to the Adult Justice System," *American Journal of Preventive Medicine*, Dec. 3, 2006.

3. See Richard E. Redding, "Juvenile Transfer Laws: An Effective Deterrent to Delinquency?" *Juvenile Justice Bulletin*, Office of Juvenile Justice and Delinquency Prevention, U.S. Department of Justice, August 2008, www.ncjrs.gov/pdffiles1/ojjdp/220595.pdf. Also see Michelle Leighton and Connie de la Vega, "Sentencing Our Children to Die in Prison: Global Law and Practice," Center for Global Law and Justice, University of San Francisco School of Law, November 2007, www.usfca.edu/law/home/CenterforLawandGlobalJustice/LWOP_Final_Nov_30_Web.pdf; and "The Rest of Their Lives: Life Without Parole for Youth Offenders in the United States in 2008," www.hrw.org/backgrounder/2008/us1005/us1005execsum.pdf.

4. See Sharon Cohen, "States Rethink Charging Kids as Adults," The Associated Press, Dec. 2, 2007, www.washingtonpost.com/wp-dyn/content/article/2007/12/01/AR2007120100792_pf.html.

5. See "Gov. Ritter Veto Message on HB 08-1208," Colorado Governor's office, May 22, 2008, www.colorado.gov/cs/Satellite/GovRitter/GOVR/1211447629095.

6. "Easy Access to Juvenile Court Statistics: 1985-2005," Office of Juvenile Justice and Delinquency Prevention, U.S. Department of Justice, http://ojjdp.ncjrs.gov/ojstatbb/ezajcs/asp/display.asp.

7. *The Congressional Record — Senate*, Jan. 21, 1997, Sen. John Ashcroft, http://bulk.resource.org/gpo.gov/record/1997/1997_S00145.pdf.

8. For background, see Peter Katel, "Prison Reform," *CQ Researcher*, April 6, 2007, pp. 289-312. Also, Elizabeth Hernandez, "Perry signs TYC reform bill into law," *TheMonitor.com* (McAllen, Texas), June 8, 2007, www.themonitor.com/onset?id=2955&template=article.html; Michael Rothfeld, Jurist orders state to beef up monitoring of youth facilities, ending lawsuit," *Los Angeles Times*, March 13, 2008, p. B5.

9. See "Oregon Violent Crime and Measure 11," Crime Victims United, updated June, 2008, www.crimevictimsunited.org/measure11/presentation/index.htm.

10. See Nancy Merritt, Terry Fain and Susan Turner, "Oregon's Measure 11 Sentencing Reform: Implementation and System Impact," RAND Corp., December 2003, pp. 87-88, www.ncjrs.gov/pdffiles1/nij/grants/205507.pdf.

11. See Howard N. Snyder, "Juvenile Arrests, 1996," Office of Juvenile Justice and Delinquency Prevention, U.S. Department of Justice, November 1997, www.ncjrs.gov/pdffiles/arrest96.pdf.

12. *Ibid.*

13. See Jeffrey A. Butts and Howard N. Snyder, "Too Soon to Tell: Deciphering Recent Trends in Youth Violence," Chapin Hall Center for Children, University of Chicago, November 2006, p. 4, www.jbutts.com/pdfs/toosoon.pdf.

14. Snyder, *op. cit.*

15. For background see Patrick Marshall, "Three-Strikes Laws," *CQ Researcher*, May 10, 2002, pp. 417-432.

16. "Juvenile Justice at the Crossroads," conference proceedings, Dec. 12-14, 1996, reported in "Juvenile Justice," Office of Juvenile Justice and Delinquency Prevention, U.S. Department of Justice, May 1998, http://ojjdp.ncjrs.org/conference/plenarybig.html.

17. See Merritt, Fain and Turner, *op. cit.*, p. 84. Also, "Oregon Violent Crime and Measure 11," *op. cit.*

18. Quoted in Randall G. Shelden, "Delinquency and Juvenile Justice in American Society," 2006, pp. 21-22.

19. *Ibid.*, pp. 30-31.

20. Unless otherwise indicated, material in this subsection is drawn from "Juvenile Justice: A Century of Change," Office of Juvenile Justice and Delinquency Prevention, U.S. Department of Justice, December, 1999, www.ncjrs.gov/pdffiles1/ojjdp/178995.pdf; and Howard N. Snyder and Melissa Sickmund, "Juvenile Offenders and Victims: 1999 National Report, Office of Juvenile justice and Delinquency Prevention, U.S. Department of Justice, September 1999, Chapter 7, www.ncjrs.gov/html/ojjdp/nationalreport99/toc.html.

21. Quoted in Robert E. Shepherd Jr., "The Juvenile Court at 100 Years: A Look Back," Juvenile Justice, National Criminal Justice Reference Service, December, 1999, www.ncjrs.gov/html/ojjdp/jjjournal1299/2.html. The case is *Kent v. U.S.*, 383 U.S. 541 (1966).

22. The case is *Schall v. Martin*, 467 U.S. 253 (1984).

23. Unless otherwise indicated, material in this subsection is drawn from Jeffrey Fagan, "Juvenile Crime and Criminal Justice: Resolving Border Disputes," in "Juvenile Justice" issue of "The Future of Children," Woodrow Wilson School of Public and International Affairs, Princeton University, and the Brookings Institution, Fall, 2008, pp. 81-118; Jan Hoffman, "Quirks in Juvenile Offender Law Stir Calls for Change," *The New York Times*, July 12, 1994, p. B1. For background, see Craig Donegan, "Preventing Juvenile Crime," *CQ Researcher*, March 15, 1996, pp. 217-240.

24. Butts and Snyder, *op. cit.*, p. 4.

25. *Ibid.*, and Snyder and Sickmund, *op. cit.*, p. 141.

26. Quoted in Elizabeth Becker, "As ex-Theorist on Young 'Superpredators,' Bush Aide Has Regrets," *The New York Times*, Feb. 9, 2001, http://query.nytimes.com/gst/fullpage.html?res=9A03EED91531F93AA35751C0A9679C8B63. Also, "Biography, William J. Bennett," BennettMornings.com, undated, www.bennettmornings.com/agnosticchart?charttype=minichart&chartID=22&formatID=1&useMiniChartID=true&destinationpage=/pg/jsp/general/biography.jsp.

27. *Ibid.*

28. See Fagan, *op. cit.*

29. Unless otherwise indicated, material in this subsection, including summaries of other scholars' studies, is drawn from McGowan, *et al.*, *op. cit.*, and Jeffrey A. Butts and Ojmarrh Mitchell, "Brick by Brick: Dismantling the Border Between Juvenile and Adult Justice," Urban Institute, July 1, 2000, www.urban.org/UploadedPDF/1000234_brick-by-brick.pdf.

30. *Ibid.*, McGowan, p. S14.

31. See "The Rest of Their Lives," *op. cit.*, and Leighton and de la Vega, *op. cit.*, p. 4.

32. Ibid., "The Rest of Their Lives," pp. 1-2.

33. See Nancy Cambria, "Teen Offenders Get Help," *St. Louis Post-Dispatch*, Sept. 14, 2008, p. C1.

34. See Douglas W. Nelson, "A Road Map for Juvenile Justice Reform," Annie E. Casey Foundation, 2008, p. 15, www.kidscount.org/datacenter/db_08pdf/2008_essay.pdf.

35. *Ibid.*

36. See Peter Katel, "Fighting Crime," *CQ Researcher*, Feb. 8, 2008, pp. 121-144.

BIBLIOGRAPHY

Books

Bennett, William J., John J. DiIulio and John P. Walters, *Body Count: Moral Poverty . . . And How to Win America's War Against Crime and Drugs, Simon & Schuster*, 1996.
In a revealing look at the political atmosphere surrounding crime during the get-tough years, three leading conservatives sound a warning over a continued increase in juvenile crime — an increase that was ending as their book was published.

Butterfield, Fox, *All God's Children: The Bosket Family and the American Tradition of Violence, Vintage Books*, 2008 (originally published in 1996).
A veteran *New York Times* reporter traces the troubled history of one of New York's most notorious young criminals.

Shelden, Randall G., *Delinquency and Juvenile Justice in American Society, Waveland Press*, 2006.
A veteran criminologist at the University of Nevada examines the history and practice of juvenile justice, with a critical eye toward bias and unfairness that he finds pervasive.

Zimring, Franklin E., *American Juvenile Justice, Oxford University Press*, 2005.
In a series of long essays, a University of California-Berkeley law professor and a leading criminologist makes an extended argument for treating juveniles differently than adult offenders.

Articles

Cambria, Nancy, "Teen Offenders Get Help," *St. Louis Post-Dispatch*, Sept. 14, 2008, p. C1.
Cambria reports on a sophisticated rehabilitation program for juvenile offenders.

Casillas, Ofelia, "A teaching moment for troubled youths," *Chicago Tribune*, Aug. 27, 2006, p. C1.
A newly created Juvenile Justice Department in Illinois has been emphasizing education and therapy.

Cohen, Sharon, "Rethink Charging Kids as Adults," *The Associated Press*, Dec. 2, 2007.
An early report on the movement to turn back some of the hard-line policies of the high-crime years.

Hernandez, Raymond, and Christopher Drew, "It's Not Just 'Ayes' and 'Nays': Obama's Votes in Illinois Echo," *The New York Times*, Dec. 20, 2007, p. A1.
When the Illinois legislature expanded criminal court jurisdiction over some juveniles, state Sen. Barack Obama criticized the measure but voted "present," apparently because voting no would have been politically risky.

Weinstein, Henry, "Focus on Youth Sentences; California has sent more juveniles to prison for life than any other state except one, a report says," *Los Angeles Times*, Nov. 19, 2007, p. B4.
Weinstein reports on a major recent study of a controversial aspect of sentencing law.

Zezima, Katie, "Law on Young Offenders Causes Rhode Island Furor," *The New York Times*, Oct. 30, 2007, p. A16.
Rhode Island lawmakers reconsider the wisdom of their move to drop the age of criminal court eligibility from 18 to 17.

Reports and Studies

Juvenile Offenders and Victims: 2006 National Report, Office of Juvenile Justice and Delinquency Prevention, U.S. Department of Justice, March 27, 2006, http://ojjdp.ncjrs.org/ojstatbb/nr2006/.
The most recent national report provides a wealth of statistics and data analysis.

Backstrom, James C., and Gary L. Walker, "The Role of the Prosecutor in Juvenile Justice: Advocacy in the Courtroom and Leadership in the Community," Dec. 20, 2005, www.co.dakota.mn.us/NR/rdonlyres/0000094a/nkahigkxbxrixmxnjvvvlrmismbxwnmr/RoleProsecutorAdvocacy CtRmLeadershipCommunity 122005FinalVersion.pdf.
Two experts in juvenile prosecution urge their counterparts to involve themselves in programs aimed at spotting and helping troubled youth before they get into more trouble.

Butts, Jeffrey A., and Ojmarrh Mitchell, "Brick by Brick: Dismantling the Border Between Juvenile and Adult Justice," *Urban Institute*, July 1, 2000, www.urban.org/UploadedPDF/1000234_brick-by-brick.pdf.
Two juvenile justice experts succinctly analyze the complicated ties between the parallel, age-defined justice systems.

McGowan, Angela, *et al.*, "Effects on Violence of Laws and Policies Facilitating the Transfer of Juveniles from the Juvenile Justice System to the Adult Justice System," *U.S. Centers for Disease Control and Prevention*, Nov. 30, 2007, www.cdc.gov/mmwr/preview/mmwrhtml/rr5609a1.htm.
A study frequently cited in the juvenile justice debate analyzes the evidence on the effects of juvenile transfer to adult court.

Parker, Alison, "The Rest of Their Lives: Life Without Parole for Child Offenders in the United States," *Amnesty International, Human Rights Watch*, 2005, www.hrw.org/reports/2005/us1005/ (updated May, 2008), www.hrw.org/backgrounder/2008/us1005/us1005execsum.pdf.
Human-rights professional offers meticulous documentation on the effects of the harshest sentence that juveniles can receive.

For More Information

Campaign for Youth Justice, 1012 14th St., N.W., Suite 610, Washington, DC 20005; (202) 558-3580; www.campaign4youthjustice.org. Opposes treatment of youths under 18 as adults in trials, sentencing and incarceration.

Center for Juvenile Justice Reform, Georgetown University, 3300 Whitehaven St., N.W., Suite 5000, Washington, DC 20057; (202) 687-0880; http://cjjr.georgetown.edu/index.html. Trains juvenile system administrators and advocates policies and laws that focus on rehabilitation.

Crime Victims United of California, 1346 N. Market Blvd., Sacramento, CA 95834; (916) 928-4797; www.crimevictimsunited.com. A prosecution-oriented victims' rights organizations advocating on issues of national relevance, including life without parole for juveniles.

National Council on Crime and Delinquency, 1970 Broadway, Suite 500, Oakland, CA 94612; (510) 208-0500; www.nccd-crc.org. A nonprofit think tank and advocacy organization promoting alternatives to incarceration.

National Juvenile Justice Prosecution Center, 44 Canal Center Plaza, Suite 110, Alexandria, VA 22314; (703) 549-9222; www.ndaa.org/apri/programs/juvenile/jj_home.htm. A research and training organization of the National District Attorneys Association.

Office of Juvenile Justice and Delinquency Prevention, 810 Seventh St., N.W., Washington, DC 20531; (202) 307-5911; http://ojjdp.ncjrs.org. A Justice Department agency that provides data on all aspects of juvenile justice.

13

Rebuilding New Orleans

Should Flood-Prone Areas Be Redeveloped?

Peter Katel

The working-class Lower Ninth Ward was among the hardest-hit New Orleans neighborhoods. A rebuilding plan proposed by the Bring New Orleans Back Commission in early January would give residents a role in deciding whether heavily flooded neighborhoods would be resettled. An earlier plan by the Urban Land Institute sparked controversy among African-Americans when it proposed abandoning unsafe areas, including parts of the Lower Ninth.

From *CQ Researcher*,
February 3, 2006.

Hurricane Katrina's floodwaters surged through tens of thousands of houses in New Orleans, including Dennis and Linda Scott's tidy, two-story brick home on Farwood Drive. The first floor has since been gutted, the ruined furnishings and appliances discarded.

Five months after floodwaters breached the city's levees and drainage canals, every other house for miles around is in the same deplorable shape.[1]

Like the Scotts, most of the residents who evacuated the sprawling New Orleans East area cannot decide whether to return, uncertain if their solidly middle class, mostly African-American neighborhoods will ever come back to life.

The disaster that began when Katrina's Category 3 winds hit New Orleans on Aug. 29, 2005, grinds on.[2] Yet the Scotts and their neighbors feel lucky to be alive.

"I'm one of the fortunate ones," says Scott, 47, who fled to Houston with his wife before the storm hit.

Linda's teaching job was swept away when the floods closed down the schools, so she's staying in Texas while Dennis works on the house and goes to his job as a communications specialist at Louis Armstrong International Airport. Their next-door neighbors, an elderly couple who stayed home, were drowned. Some three-quarters of Louisiana's 1,070 Katrina deaths occurred in New Orleans, where about 70 percent of the victims were age 60 and older.[3]

But "the east" is not alone. Similar devastation also afflicts some older neighborhoods, where lush gardens and sprawling villas reflect the city's French and Spanish heritage.[4]

Flooding Affected Most of Greater New Orleans

Flood water up to 20 feet deep covered more than three-quarters of New Orleans when storm surges pushed by Hurricane Katrina breached levees in 34 places. The Lower Ninth Ward and the New Orleans East district were among the hardest-hit areas.

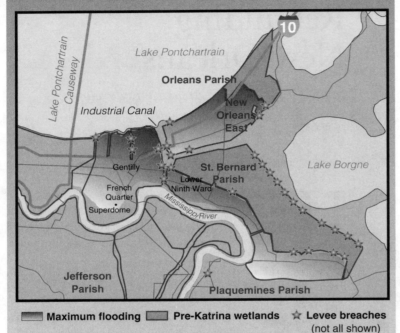

| ▬ Maximum flooding | ▬ Pre-Katrina wetlands | ☆ Levee breaches (not all shown) |

Source: Federal Emergency Management Agency

Losses in destroyed and damaged property, added to losses resulting from the shrinkage of the city's economy, amount roughly to $35 billion, estimates Stan Fulcher, research director of the Louisiana Recovery Authority in Baton Rouge.

Most residents are still gone, largely because most jobs — except those that involve either tearing down houses or fixing them up — have disappeared. Plans are only starting to be made to rebuild the city, and no one knows how much reconstruction money will be available.

Does Scott have a future in New Orleans? "I'm on hold," he replies.

That response comes up a lot among the city's residents and evacuees, often accompanied by a sense that the rest of the country has moved on — or views the French-founded, majority-African-American city as somehow foreign or not worth rebuilding.

"This is America you're talking about," lawyer Walter I. Willard says in frustration.

So American, in fact, that jazz was born there — amid a culture formed by the peculiarities of the city's slavery and segregation traditions.[9] "The West Africans [slaves] were allowed to play their music in Congo Square on Sundays. That happened nowhere else in the United States," famed New Orleans-born trumpeter Wynton Marsalis says.[10]

Slavery's legacy of racial and class divide has been part of the Katrina story from the beginning. New Orleans is two-thirds African-American, and the thousands of impoverished residents who were without cars to flee the approaching hurricane were overwhelmingly black.[11] "As all of us saw on television," President Bush acknowledged, "there's . . . some deep, persistent poverty in this region. That poverty has roots in a history of racial discrimination, which cut off generations from the opportunity of America."[12]

The continuing devastation mocks President Bush's stirring promise two weeks after the storm to mount "one of the largest reconstruction efforts the world has ever seen."[5]

Indeed, when the Senate Homeland Security and Governmental Affairs Committee toured the city four months later, members were "stunned" to see that "so much hasn't been done," said Chairwoman Susan Collins, R-Maine.[6]

Floodwaters up to 20 feet deep covered about 80 percent of the city and didn't recede until late September.[7] Fully half the city's homes — 108,731 dwellings — suffered flooding at least four feet deep, according to the Bring New Orleans Back Commission (BNOBC) formed by Mayor Ray Nagin. In some neighborhoods, Hurricane Rita, which struck later in September, brought additional flooding.[8]

Bush also conceded that the federal response to Katrina amounted to less than what its victims were entitled to — a point reinforced in early 2006, when Sen. Collins' committee released a strikingly accurate prediction of Katrina's likely effects, prepared for the White House two days *before* Katrina hit.[13]

But in a sense, New Orleans was crumbling from within even before the floods washed over the city. "The city had a lot of economic and social problems before — economics, race, poverty, crime, drugs," says musician and Xavier University Prof. Michael White. "Our failure to deal with harsh realities has sometimes been the problem."

In 2004, for example, the city's homicide rate hit 59 per 100,000 — the nation's highest.[14]

In that post-Katrina climate —

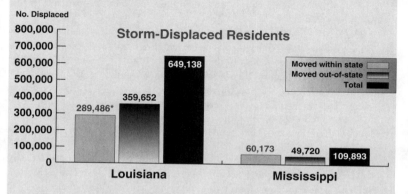

More Than 400,000 Residents Left Home States

Six times more Louisiana residents are still displaced from their homes than Mississippians. Of the more than 750,000 residents from both states displaced by Katrina, more than half are still living outside their home states.

No. Displaced

Storm-Displaced Residents

Legend: Moved within state / Moved out-of-state / Total

Louisiana: 289,486* / 359,652 / 649,138
Mississippi: 60,173 / 49,720 / 109,893

* Based on the number of FEMA aid applicants who have not returned to their pre-Katrina addresses.

Source: Louisiana Recovery Authority

fed by bitter memories of institutional racism — the African-American community is concerned that developers are planning to reduce the black portion of the city's population. U.S. Housing and Urban Development Secretary Alphonso Jackson, who is African-American, intensified those fears when he said, "New Orleans is not going to be as black as it was for a long time, if ever again."[15]

The concern remained an issue into early 2006, when Mayor Nagin, also African-American, declared on Jan. 16 that the city "should be a chocolate New Orleans . . . a majority-African-American city. It's the way God wants it to be."[16] The following day, after furious reactions from both the white and black communities, Nagin apologized.[17]

Nagin's provocative language aside, fears of a demographic shift seem well-founded. In late January, sociologist John R. Logan of Brown University said he had conducted a study that showed about 80 percent of New Orleans' black residents were unlikely to come back, in part because their neighborhoods wouldn't be rebuilt.[18]

The BNOBC sparked the most recent chapter of the race and redevelopment debate. The commission's rebuilding plan, unveiled in early January, would give residents of the most heavily flooded neighborhoods four months to help figure out if their districts could be resettled. Homeowners in neighborhoods that can't be revived could sell their houses to a government-financed corporation for 100 percent of the pre-Katrina values, minus insurance payouts and mortgage obligations. The overall plan would cost more than $18 billion.[19] Federal, state and city approval is needed.[20]

Nowhere did the commission say that the poorest and most heavily damaged African-American neighborhoods should be abandoned. But the Washington-based Urban Land Institute (ULI), flatly recommended against extensive rebuilding in the most flood-prone areas, by implication including much of the working-class, largely African-American Lower Ninth Ward.[21]

Under the Jim Crow segregation system that lasted into the 1960s, residents point out, the Lower Ninth was the only place where African-Americans could buy property. "These people struggled to buy a little bit of land they could call home," says contractor Algy Irvin, 60, standing in the wrecked living room of his mother's house on Egania Street.

Can New Orleans' Musical Culture Be Saved?

Sunpie and the Louisiana Sunspots have the crowd at the House of Blues rocking as the group pounds out "Iko-Iko," a New Orleans standard with Creole lyrics and an irresistible beat.

The first night of Carnival is under way in the French Quarter, and the club is filling up for a long evening of music, with three more acts to follow. In the less touristy Marigny neighborhood, jazz pianist Ellis Marsalis is starting a slightly more sedate set at popular Snug Harbor.

Four months after Katrina hit, New Orleans is making music again. "So far, it's gone better than I would have thought, given the total lack of tourism," says Barry Smith, proprietor of the Louisiana Music Factory, where CDs and vinyl records of New Orleans artists account for some three-quarters of the stock of jazz, blues and gospel artists — both world-renowned and known only to locals. "I've definitely experienced a big increase in the number of local customers coming to the store, and a lot of the people who came here to work — from construction workers to Red Cross volunteers."

Few if any places in the United States come even close to New Orleans as an incubator of musical style and talent. As far back as 1819, a visitor wrote about the African music being played at Congo Square. And by the early 20th century, a musical tradition had formed in which Louis Armstrong — arguably the century's most influential musician — came of age.[1]

"All American music in the 20th century was profoundly shaped and influenced by New Orleans music," Tom Piazza writes in *Why New Orleans Matters.*[2]

The career of famed musician/producer Allen Toussaint illustrates the city's musical power. Toussaint wrote such 1960s hits as "Mother in Law" and produced and arranged the 1973 hit "Right Place, Wrong Time" for fellow New Orleans resident "Dr. John," as well as the disco standard "Lady Marmelade."

"He helped invent things we take as everyday in music — certain beats, certain arrangements," his partner in a record label said recently.[3]

Toussaint fled New Orleans after Katrina and has spoken optimistically of the city's future prospects.[4] But away from the club scene and music stores, the future looks less bright.

That's because the city's music springs from the very streets that Katrina emptied — the fabled "social aid and pleasure clubs," fraternal organizations that sponsor the Mardi Gras "Indian tribes," as well as the brass-band funeral processions that nourished jazz. All these influential institutions are maintained by people who mostly live paycheck to paycheck, says Michael White, a clarinetist and music scholar who holds an endowed chair in arts and humanities at New Orleans' Xavier University.[5]

Irvin recalls earning $35 a week mopping hospital floors and paying $18 a week for his own $1,200 lot on nearby Tupelo Street — now also a ruin. "You can see why people don't want a fat-cat developer coming in, making millions," he says, giving voice to a common suspicion that declaring the neighborhood unsafe is merely a cheap means of clearing out its present inhabitants to make way for lucrative development. But Irvin adds, "If people are compensated, that's another story."

Post-Katrina television coverage also gave the impression that New Orleans' African-American population was uniformly poor. In fact, the city had a substantial black middle class. "I had no clue that people couldn't get out of here," says Anne LaBranche, an African-American from New Orleans East, who returned to the city in January after staying with friends in Birmingham, Ala. "I do not know a person who doesn't own a car."

The LaBranches are moving into a house owned by her father-in-law. Her physician husband Emile, whose family practice was destroyed by Katrina along with all the patients' records, has been looking for work. But other medical offices say they aren't hiring until they know how many people are coming back.

Across town, Cory Matthews, 30, a medical-technology salesman, also wonders whether he still has a place in the city. He is rebuilding the flood-damaged Uptown house he shares with his girlfriend, but as he puts up new Sheetrock and rewires, he worries that his physician customer base has shrunk. "I'm hoping we're making the right move," he says.

Certainly, nobody is expecting redevelopment to bring speedy population growth. An estimated 135,000 people remain in New Orleans — less than a third of the 462,000 pre-Katrina population. Nagin's commission projects

The New Orleans establishment recognizes the problem. "Financial losses for social aid and pleasure clubs, Mardi Gras Indian tribes and [brass band] second-line companies are conservatively estimated at over $3 million," the Bring New Orleans Back Commission reports.[6]

"These were poor people, but people who spent a lot of money on these events," says White, a New Orleans native who comes from a long line of musicians. "The thing of money is serious. If people don't have jobs, they're not going to be able to participate."

White himself suffered another kind of loss — his vast collection of vintage instruments and memorabilia that included a trumpet mouthpiece from jazz saint Sidney Bechet; 4,000 rare CDs and even rarer vinyl recordings; photographs of New Orleans musical legends and notes and tapes of interviews with musicians who have since died. All were stored at his house — and it's all gone.

Is resurrecting an entire popular culture any more possible than restoring White's collection? "It's not like there's a central entity that can be rebuilt," says Piazza. "What steps can be taken to repatriate as many members of the African-American community and other communities — people who don't have the same kinds of resources as others to come back and rebuild, or who lived in areas where logistical challenges to rebuilding are all but insurmountable? That is the most difficult question about cultural renewal."

Legendary jazz pianist Ellis Marsalis is a popular performer in Old New Orleans, which was largely spared by the flooding.

Courtesy www.ellismarsalis.com

[1] For background, see Geoffrey C. Ward and Ken Burns, *Jazz: A History of America's Music* (2000), pp. 7-16; 40-46.

[2] Tom Piazza, *Why New Orleans Matters* (2005), p. 37.

[3] Quoted in Deborah Sontag, "Heat, and Piano, Back in New Orleans," *The New York Times*, Sept. 20, 2005, p. E1; for additional background see, "Inductees: Allen Toussaint," Rock+Roll Hall of Fame and Museum, undated, http://rockhall.com/hof/inductee .asp?id=200.

[4] *Ibid.*

[5] Ward and Burns, *op. cit.*, pp. 7-16.

[6] "Report of the Cultural Committee, Mayor's Bring New Orleans Back Commission," Jan. 17, 2006, pp. 8-9, www.bringneworleansback.org.

247,000 residents by September 2008, while a more optimistic consultant projects 252,000 by early 2007.[22] The totals, however, don't specify whether the residents will be laboring at construction sites or behind desks.

Jay LaPeyre, president of the Business Council of New Orleans and the River Region, says laborers are desperately needed "for every type of manual labor — from skilled electricians and plumbers to low-skilled apprentices and trainees to service jobs at Burger King."

That kind of talk makes white-collar New Orleanians nervous. Tulane University, one of the city's major high-end employers, laid off 230 of its 2,500 professors.[23] Nearly all 7,500 public school employees were laid off as well, though some were rehired by the handful of charter schools that have sprung up.[24]

"It's become a blue-collar market," says Daniel Perez, who lost his night-manager job at the swanky Royal

Sonesta Hotel after business dropped off. Perez applied in vain for dozens of professional or managerial jobs. He had almost decided to leave New Orleans before finally landing a position as a sales manager for *USA Today.*

For now, at least, even the service-industry job market is thinning, though the profusion of help-wanted signs in the functioning parts of the city convey a different impression. A planned Feb. 17 reopening of Harrah's Casino, for example, will take place with only half the pre-Katrina payroll of 2,500, says Carla Major, vice president for human resources.

On his Jan. 11 visit, President Bush touted New Orleans as still "a great place to visit." But his motorcade had skirted most of the devastation, going nowhere near, for instance, the Scotts' deserted neighborhood.[25]

"We can't move forward until we have positive information on what's happening," Scott says. "There are no

Katrina Costs Dwarf Previous Disasters

Hurricane Katrina cost the Federal Emergency Management Agency $25 billion in the Gulf Coast — nearly three times more than the 2001 terrorist attacks on the World Trade Center and eight times more than Hurricane Rita, which followed on the heels of Katrina. The money pays for such services as temporary housing, unemployment assistance, crisis counseling and legal aid.

Disaster	FEMA Cost Estimate* ($ in billions)
Hurricane Katrina (2005)	$24.6
World Trade Center (2001)	$8.8
Hurricane Rita (2005)	$3.4
Hurricane Ivan (2004)	$2.6
Hurricane Wilma (2005)	$2.5
Hurricane Georges (1998)	$2.3
Hurricane Andrew (1992)	$1.8
Hurricane Hugo (1989)	$1.3
Loma Prieta Earthquake (1989)	$0.87
Hurricane Alberto (2000)	$0.6

* Flood-insurance reimbursements not included

Source: FEMA, December 2005

[about 23 feet] below water level during a Category 3 hurricane."[27]

Even so, the extensive levee system was designed to defend the entire metropolitan area from floods. So the Katrina disaster didn't grow out of the development of flood-prone lands that never should have been urbanized, say opponents of shrinking the footprint. Instead, they argue, the catastrophe grew out of human failure in engineering, construction or maintenance — or in all three.

"If we can build levees in Iraq, we can build levees on the Gulf Coast," says Sen. Mary Landrieu, D-La. "And if we can build hospitals in Baghdad and Fallujah, we can most certainly rebuild our hospitals in this metropolitan area."[28]

But congressional power brokers aren't in the mood to redevelop flood-prone areas. "We are committed to helping the people of Louisiana rebuild," said House Appropriations Committee member Rep. Ray LaHood, R-Ill. But, "we are not going to rebuild homes that are going to be destroyed in two years by another flood. We are not just going to throw money at it."[29]

Some who call the flood a man-made failure don't oppose redesigning the city in a more environmentally sensible way — even if it means abandoning their own neighborhoods. "It's not what I want, but I could live with it," says LaBranche, who with her husband owns a home, an office building and rental properties in New Orleans East. "I don't want to go through this again."

But who should decide? "The idea that everybody gets to have what they want" is not practical, says business leader LaPeyre. He wants the government to use its power of eminent domain — the right to condemn private property and compensate the owner — to prevent redevelopment of areas unsuitable for residential and business use.[30]

Private companies, such as utility and insurance companies — will also influence decisions about where development will occur. "The market will do better than

banks, no schools, no electricity. We just want to be home."

As officials plan the city's future, here are some of the questions being debated:

Should some neighborhoods not be rebuilt?

The buzzword summing up the single toughest question about New Orleans' future is "footprint." That's urban-planner jargon for a city's shape and the amount of space it occupies. In New Orleans, the term has become code for the idea that flood-prone districts are best turned back into open-space "sponges" to absorb nature's future onslaughts.

But would that help? New Orleans and the entire Gulf Coast are sinking. New Orleans was built on sandy soil to begin with, but oil and gas extraction and upriver levee construction — which reduces the delta area's natural landfill process, called silting — have exacerbated the problem. And sea levels are rising due to global warming.[26] As a result, writes Virginia R. Burkett of the U.S. Geological Survey's National Wetlands Research Center in Louisiana, by 2100 parts of New Orleans "could lie

most people claim," he says. "If you're not going to have good services, most people will say, 'I don't want to live there.' "

Others argue that a neighborhood's residents should have a big voice. The Bring New Orleans Back Commission proposed letting residents of heavily damaged neighborhoods work with urban and financial planners to determine if their districts could be revived. The "neighborhood planning teams" would have until May to decide. The procedure grew out of opposition to the Urban Land Institute's recommendation against rebuilding in flood-prone areas.

"In an arbitrary and capricious manner to say that these areas — which were populated by black people because they were directed there — should now be turned into green space deepens the wound," says Councilwoman Cynthia Willard-Lewis, who represents several of the city's eastern neighborhoods. Many of the houses can be repaired and the communities brought back, she says, adding that she suspects the plan "was not based on what was safe but on whom they wanted to return."

William Hudnut, a former mayor of Indianapolis who holds the Urban Land Institute's public policy chair, says city leaders do not have the courage to tell residents what they don't want to hear. "The footprint has to be smaller and development more compact," he says. "An honest, tough-minded approach to rebuilding is part of what leadership is all about. It may be that some people would lose their political base or lose their jobs. But if a thing is worth doing, it's worth doing well and worth standing up for."

Ari Kelman, an environmental historian at the University of California at Davis, concedes that some neighborhoods should be abandoned. At the same time,

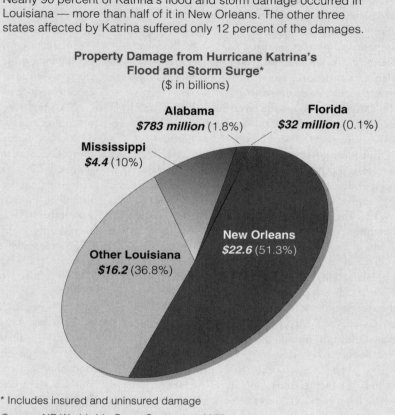

Katrina Saved Worst for New Orleans

Nearly 90 percent of Katrina's flood and storm damage occurred in Louisiana — more than half of it in New Orleans. The other three states affected by Katrina suffered only 12 percent of the damages.

Property Damage from Hurricane Katrina's Flood and Storm Surge*
($ in billions)

Alabama *$783 million* (1.8%)

Florida *$32 million* (0.1%)

Mississippi *$4.4* (10%)

New Orleans *$22.6* (51.3%)

Other Louisiana *$16.2* (36.8%)

* Includes insured and uninsured damage

Source: AIR Worldwide Corp., September 2005

he says, low-income African-American residents have well-founded fears that any planning and decision system will be stacked against them.

"People who don't have money also don't have power," says Kelman, author of a 2003 book on the interplay between human design and nature in New Orleans. "When politics get cooking in New Orleans, it's likely that the poor are going to get screwed."

Should the levee system be upgraded to guard against a Category 5 storm?

The levee system surrounding New Orleans was designed to withstand a Category 3 hurricane. Katrina had weakened to Category 3 by the time it made landfall, but the wall of water it sent ashore — the hurricane's "surge" — was born when the storm was still offshore and raging at Category 4 and 5 strength.[31]

So far, official attention has focused on possible errors in design, construction or maintenance of the levees. But many are also asking whether the system should be upgraded to protect against a Category 5 hurricane — the most powerful. Among Louisianans in general and New Orleanians in particular, support for a Category 5 system seems nearly universal.

"I would like to see the levees brought to Category 5 for my safety and that of my family and properties," says LaBranche, the displaced New Orleans homeowner.

But some government experts say a Category 5 levee system is a pipe dream. They point out Category 5 is open-ended, taking in all hurricanes whose winds exceed 155 mph and create storm surges greater than 18 feet. "What's the top end for a Cat 5 hurricane?" asked Dan Hitchings, director of Hurricane Katrina recovery for the Corps of Engineers. "There isn't one."[32]

That argument carries little weight in Louisiana. Some Louisiana lawmakers say that if the below-sea-level Netherlands can protect itself from floods, New Orleans shouldn't settle for less. "They built once-in-1,000-years flood protection," Sen. Landrieu told a rally of some 75 displaced New Orleanians outside the White House last December. "We don't even have once-in-100-years [protection]."

No hurricanes strike in the North Sea, which surrounds the Netherlands. But the tiny country, some of which lies more than 20 feet below sea level, is vulnerable to powerful storms with winds that can reach 60 mph. Following a 1953 storm that killed more than 1,800 people, the country redesigned its protective system in ways that many New Orleanians say should serve as a model.[33]

The Netherlands system was designed to withstand a once-in-10,000-years storm. New Orleans' levees were designed for a once-in-200-300-years storm, says the Corps of Engineers.[34]

In addition, say Louisiana officials, a Category 5 system would probably cost about $32 billion.[35]

"It will probably be a pretty staggering price tag," acknowledges Craig E. Colten, a geography professor at Louisiana State University in Baton Rouge and author of a recent book on New Orleans' flood-protection history. But the long-term value of property protection would make an upgraded system a wise investment, he added,

citing the prosperous Netherlands, an international shipping center.

But Rep. Richard Baker, R-La., cautions that debating a Category 5 upgrade now could distract from the immediate and urgent tasks facing New Orleans. "Construction toward a Category 5 standard would be a decade-long project," he says. "The statistical probabilities of a Category 5 hitting New Orleans are fairly small, especially since we just got hit. I think we have time."

No one denies the need for fixing the existing flood-protection system. The Bush administration has proposed $1.6 billion to restore the system to a Category 3 level of protection, and another $1.5 billion for further improvements.[36] Thus far, however, it has stayed out of the Category 5 argument. At a White House briefing on the new flood-protection plan, Donald E. Powell, the administration's coordinator of post-hurricane recovery projects, would only say that after the proposed improvements, "The levee system will be better, much better, and stronger than it ever has been in the history of New Orleans."[37]

The White House plan also includes a study, backed by Mayor Nagin, of whether a substantial upgrade to the system is needed. A preliminary report is due in May.

However, city officials argue that a system capable of protecting against a Category 5 storm is well within the range of engineering possibilities and would be good for both the city's and the country's economy. "We need to build toward Category 5 to provide . . . assurance to [potential] investors," says Gary P. LaGrange, director of the Port of New Orleans, and to protect the port, an essential part of the nation's trade system.

Should the nation pay for New Orleans to be rebuilt?

So far, Congress has committed $98.9 billion for post-hurricane recovery and rebuilding programs throughout the Gulf Coast, the Senate Budget Committee calculated. The funding came in two emergency appropriations in September totaling $62.3 billion, followed by several smaller spending authorizations. In December, expanding on a request by Bush, Congress redirected $23.4 billion in funds previously appropriated.[38]

It's uncertain, however, how much will go to New Orleans.

In any event, money appropriated so far includes $6.2 billion in Community Development Block Grants intended for Louisiana, and $22 billion for reimbursements to Gulf Coast homeowners from the federal flood-insurance program. But even with the emergency injections of cash, the flood-insurance program is "bankrupt," Senate Banking Committee Chairman Richard Shelby, R-Ala., said on Jan. 25. The acting director of the Federal Emergency Management Agency (FEMA) insurance division said the agency has paid out $13.5 billion in claims arising from the 2005 hurricane season — nearly as much as the agency has paid out in its 37-year existence. And 30 percent of the 239,000 claims have yet to be resolved.[39]

Federal hurricane-recovery coordinator Powell has been advocating directing much of the block grant money to the estimated 20,000 New Orleans homeowners who didn't have flood insurance because their neighborhoods weren't designated as flood plains.[40]

Given the huge costs involved and all the unknowns, lawmakers from other parts of the country are not exactly champing at the bit to pay for rebuilding New Orleans. The city's tenuous hydrological situation and the likelihood that it will be flooded again by other hurricanes lead some Americans to question whether the rest of the country should have to pay to rebuild the city in such a precarious location.

"There is a lot of — I suppose you can call it 'Katrina fatigue' — that people are dealing with out in the heartland," Rep. Henry Bonilla, R-Texas, told Louisiana Gov. Kathleen Babineaux Blanco, a Democrat, at a House Select Katrina Response Investigation Committee hearing last Dec. 14.

The situation has rekindled a long-simmering debate about whether Americans in the heartland should pay to constantly bail out people — usually those living on the coasts — who choose to live in areas prone to floods, hurricanes, landslides and earthquakes. "Is it fair to make people living in Pennsylvania or Ohio pay billions for massive engineering projects so that some of the people of New Orleans can go back to the way things were and avoid the hard choices that nature presents them?" asks economist Adrian Moore of the libertarian Reason Foundation of Los Angeles.[41]

Some lawmakers agree, although only a few have spoken out. "It looks like a lot of that place could be bulldozed," House Speaker Dennis Hastert, R-Ill. said shortly after the hurricane, raising hackles. He later explained he only meant that danger zones shouldn't be resettled.[42]

Skip LaGrange takes a break from cleaning out his flooded home in the Mid-City section of New Orleans, on Oct. 5, 2005. An estimated 135,000 people are now living in New Orleans — less than a third of the 462,000 pre-Katrina population.

Getty Images/Robyn Beck

New Orleanians respond that other disaster-prone areas, including hurricane-exposed Florida coastal cities, get rebuilt with few questions asked about viability. "People build on mountainsides in California that fall in the ocean," notes Perez, the newspaper sales manager. "We're not the only vulnerable area in the country."

In addition, points out Republican Louisiana Sen. David Vitter, 25 percent of the nation's energy and most of the Midwest's grain exports are shipped through the port of New Orleans. "If people don't think there's a national stake in rebuilding New Orleans, that's fine. But they should get used to much higher gasoline prices," he said. "And people can forget about getting crops to foreign markets. You need a major city as the hub of all that activity."

But if federal funds are forthcoming to rebuild the city, they should have some serious accountability strings attached, given the city's long history of corruption and dysfunction, some argue. "A lot of . . . our constituents now are telling us that they [don't] want us to support funding for the Gulf region at this point without strong plans of accountability," Bonilla said.

Recognizing those sentiments — as well as the reality that the country is at war and its debt and deficits are

CHRONOLOGY

1700s-1800s *From the time of its founding, New Orleans' vulnerability to nature is seen as the price of its incomparably strategic location.*

1718 New Orleans is founded on a natural levee along a bend in the Mississippi River.

1892 Adolph Plessy of New Orleans is arrested after testing segregation laws by riding in a "white" train car. U.S. Supreme Court later upholds his conviction in landmark *Plessy v. Ferguson* decision.

1900-1947 *A catastrophic flood reminds the city of its dangerous location.*

1927 Massive Mississippi floods see many African-Americans forced into levee-reinforcement work; two rural parishes are deliberately flooded to save New Orleans.

1929 The U.S. Army Corps of Engineers begins building a spillway on the Mississippi to channel floodwater away from New Orleans.

1930s *Expansion of city drainage systems allows urban expansion, but new neighborhoods are strictly segregated.*

Sept. 17-19, 1947 A Category 4 hurricane overwhelms levees, causing flooding over nine square miles of the city.

1950s-1970s *The city expands into drained wetlands, increasing its vulnerability to floods.*

1950 Land drained for suburban expansion reaches 49,000 acres.

Sept. 7, 1965 Hurricane Betsy slams the city with Category 3 winds, pushing a 10-foot storm surge through some levees.

Oct. 27, 1965 President Lyndon B. Johnson signs the Flood Control Act, which includes funding for a hurricane-protection system in New Orleans.

Aug. 17, 1969 Category 5 Hurricane Camille devastates Mississippi and Alabama, but reinforced protective systems keep most of New Orleans safe.

May 3, 1978 Heavy rainstorm flooding damages more than 70,000 homes.

1980s-1990s *Attempts by the city to guard against rainstorm floods prove inadequate, as fears of vulnerability to hurricanes begin to grow.*

April 1982 Rainstorm-caused floods damage 1,400 homes and other buildings.

1983 City expands pumping and drainage systems.

May 8-10, 1995 Flooding damages thousands of homes, causes six deaths.

2000-Present *Fears of hurricane vulnerability grow, as journalists and government officials warn about the weakness of the city's defenses.*

June 23-June 26, 2002 *Times-Picayune* warns of New Orleans' hurricane vulnerability.

Sept. 26, 2002 Hurricane Isidore hits Louisiana after weakening to a tropical storm, but still causes major flooding.

July 2004 FEMA officials conduct a drill featuring Category 3 "Hurricane Pam" hitting New Orleans and predict serious flooding, massive evacuation.

Aug. 29, 2005 Hurricane Katrina makes landfall east of New Orleans.

Sept. 15, 2005 President Bush visits New Orleans and pledges a massive disaster-recovery effort.

Jan. 11, 2006 Bring New Orleans Back Commission releases an "Action Plan" for re-creating the city.

Jan 17, 2006 Senators of both parties visit New Orleans and criticize slow progress on recovery.

Jan. 26, 2006 President Bush explains why he refused to support the creation of a public corporation to buy flood-damaged homes.

June 1, 2006 Hurricane season begins; repairs and improvements to levee system due for completion.

rising — Louisiana politicians have proposed two major plans that they say would lower the federal spending burden for rebuilding New Orleans and the rest of the state.

But the White House has already refused to back one of these plans. Its author, Rep. Baker, proposed establishing a public corporation to buy or finance repairs on storm-damaged property. Homeowners who sold their houses to the corporation would get 60 percent of the pre-Katrina value of their holdings. The corporation would then resell the homes, if possible, and turn the proceeds back to the Treasury. Nagin's BNOBC adopted the idea, which some of its members called crucial to reviving the city.

"We were concerned about creating additional federal bureaucracies, which might make it harder to get money to the people," Bush said, explaining his rejection of Baker's idea.[43]

On Feb. 1, according to Baker's office, three former Republican governors of Louisiana — Murphy J. "Mike" Foster, Charles E. "Buddy" Roemer III, and David Treen — urged Bush to change his mind concerning Baker's bill, which they called the only practical method of disposing of thousands of ruined residential and business properties.

The congressman has been vowing to press ahead with his proposal, sponsored in the Senate by Sen. Landrieu. Bush's negative response would "constrict the opportunities for rapid redevelopment, and that's tough," said Reed Kroloff, architecture dean at Tulane University and a BNOBC member.[44]

But developer Joseph Canizaro, who helped put together the commission's plan, said block grant money and other unspecified funds could be found for a property buyback.[45]

The other plan to lower direct federal spending is a longstanding proposal to boost the state's share of money that the federal government earns from petroleum leases on the Outer Continental Shelf in the Gulf of Mexico off Louisiana's coast. One-quarter of U.S. crude oil production comes from Louisiana's offshore waters.[46]

The cost of repairing the state's hurricane-protection system "can be paid for simply by giving Louisiana our fair share of oil and gas revenues from the Outer Continental Shelf," Gov. Blanco told Bonilla at the House Select Committee hearing.

Coastal states like Louisiana receive 27 percent of the revenues from oil and gas leases from waters within their three-mile jurisdictions (federal waters extend another 197 miles). By contrast, states with oil and gas production on public lands receive 50 percent of the federal revenues, leading coastal states to feel they are entitled to a larger share of offshore revenues.[47]

Sen. Landrieu last year pushed a bill to grant coastal states 50 percent of the take from oil and gas leases in the areas off their shores. The bill died at year's end, but she is planning to revive it this year (*see p. 302*).

Rather than creating a new revenue source, however, the proposal would merely divert money to the state before the funds reach federal coffers, which bothered Bonilla. "It is wise when states and local governments come before us to show what they are doing to help themselves in terms of raising whatever revenue dollars you can," Bonilla told Blanco. "People would want to know . . . what is Louisiana doing in terms of everything you possibly can do to help yourself and not just look at the federal government and say, 'We need you to help us pay for these things.' "

But, he added, Americans would not "turn their back on those who want to help themselves."

BACKGROUND

Island City

New Orleans has been battling with nature ever since explorer Jean-Baptiste Le Moyne de Bienville founded the city in 1718. Its original name, in fact, reflected the city's relationship to the four bodies of water surrounding it — the Mississippi River, Lake Pontchartrain, Lake Borgne and the Gulf of Mexico. He called it L'Isle de la Nouvelle Orléans — the Island of New Orleans.[48]

"His enthusiasm for the river's commercial benefits blinded him to many of the challenges of building a city in the delta," environmental historian Kelman writes. These included: epidemics; "terrible to nonexistent" drainage; dampness; and "the threat of catastrophic flooding."

Still, Bienville's insight into the river's economic importance was on the money. The Mississippi was unrivalled as a highway deep into the North American continent, and remains so today. Some 500 million tons of goods — including about 60 percent of U.S. grain exports — are

Experts Blame Levees, Not Storm

The newspaper headlines blamed "Killer Storm Katrina" for devastating New Orleans. But engineers largely blame the levees designed and built by the U.S. Army Corps of Engineers.

A team of experts who examined the protective system found no fewer than 34 storm-induced levee breaches, indicating that the engineering failures were far wider than initial reports indicated.[1]

"The performance of many of the levees and floodwalls could have been significantly improved, and some of the failures likely prevented, with relatively inexpensive modification," the team concluded. The simple addition of concrete "splash slabs," for instance, might have prevented soil levee tops from eroding.

In fact, even a task force assembled by the Corps of Engineers itself concluded "integral parts of the . . . hurricane-protection system failed."[2]

With the June 1 start of the 2006 hurricane season approaching, the Corps is trying to patch the immediate problems. Engineers and lawmakers, meanwhile, are evaluating the system's performance. So far, a lethal combination of design, construction and maintenance errors appears to underlie the disaster.

Blame extends from state-appointed "levee boards" responsible for inspection and maintenance to the Corps of Engineers, Sen. George Voinovich, R-Ohio, told the Senate Homeland Security and Governmental Affairs Committee on Dec. 14. And Congress deserved blame too, he said: "We have been penny-wise and pound-foolish" on funding upkeep and completion of the New Orleans levee system.

The Lake Pontchartrain and Vicinity Hurricane Protection Project includes 125 miles of levees, floodwalls and other structures. The system was supposed to bar storm surges from Lake Pontchartrain and channel any flooding out of the city via a series of canals.[3]

Though Congress approved the project in 1965, it was unfinished when Katrina struck. In the city itself, construction was 90 percent complete, but the lack of completion has not been blamed for the system's failure.[4] Rather, the devastation was intensified by the environmental changes in southern Louisiana since the system was first designed, the *Times-Picayune* reported as early as 2002.[5]

As the oil and gas industry expanded, the Corps of Engineers built or approved the necessary navigation channels in southern Louisiana and the Gulf of Mexico. And the industry expansion swallowed one-third to one-half of the wetlands — which have been disappearing at a rate of at least 25 square miles a year. Experts now know wetlands play a critical role during hurricanes, slowing storms as they make landfall.[6]

shipped downriver to the southern Louisiana port complex, which includes New Orleans.[49]

The first of New Orleans' protective barriers — called levees from the French verb "to lift" — were natural. In fact, New Orleans exists in the first place because the Mississippi's waters helped create a high section of riverbank along the section of the river that forms a crescent embracing old New Orleans — known today as the French Quarter. The sloping, natural levee was only 12 feet above sea level.

Settlers soon began adding to nature's work. Throughout the 18th and early 19th centuries — during the first period of French rule, the Spanish colonial period that followed in 1768-1801 and the French restoration in 1801-1803 — levees were built far upstream, and raised continually after flooding.

The levee work continued after the United States bought the city and vast swaths of the new nation's interior in 1803 for $15 million, or about 3 cents an acre. Nine years after the so-called Louisiana Purchase, Louisiana became a state.

From the beginning, many people realized that building ever-higher levees up and down the river prevented its energy from being dissipated naturally in periodic floods. By the time the Mississippi reached New Orleans, it would be dangerously high and flowing at maximum force.

"We are every year confining this immense river closer and closer to its own bed — forgetting that it is fed by over 1,500 streams — and regardless of a danger becoming every year more and more impending," State Engineer P. O. Herbert warned in 1846. He argued for flood outlets along the river, but landowners resisted, not wanting their plantations flooded.[50]

In 1849, the river broke through several upstream levees, one of them 17 miles above New Orleans. The resultant flooding in the lowest section of New Orleans forced 12,000 mostly poor residents to abandon their dwellings or try to coexist with the water.

Afterward, the city raised the levees higher still. But A. D. Wooldridge, the state engineer who succeeded

So when Katrina made landfall across the region's depleted wetlands, the poorly designed and built levees and floodwalls couldn't withstand the full force of the storm surge.

A section of floodwall along the London Avenue Canal was so weakened that it likely would have been breached by the floodwaters — if the barrier on the opposite side of the canal hadn't failed first, an engineer told the Senate Environment and Public Works Committee on Nov. 17. "Multiple, concurrent failure mechanisms" were present, said Larry Roth, deputy executive director of the American Society of Civil Engineers. "The wall was badly out of alignment and tilting landward; as a result of the tilt, there were gaps between the wall and the supporting soil."

Additional pressure on the flood barriers came from the Mississippi River Gulf Outlet (MRGO), a 76-mile long canal built to give ships a shortcut from the Gulf to the Port of New Orleans. Instead, it gave Katrina a straight shot into the city — a "hurricane alley" — said Sen. David Vitter, R-La., who has called, along with others, for the canal's closure. The Corps says it will not conduct its annual dredging of the waterway, and hurricane experts say it may become less dangerous as it becomes shallower.[7]

Meanwhile, engineers have suggested that some residential areas be abandoned — to provide a flood-absorbing floodplain — and building codes amended to require that houses be elevated.

But the levee system also must be dealt with, Roth said. "If we are to rebuild the city," he said, "we must also rebuild its protections."[8]

---

[1] The team was assembled by the National Science Foundation (NSF), the American Society of Civil Engineers (ASCE) and the University of California at Berkeley. See R. B. Seed, *et al.*, "Preliminary Report on the Performance of the New Orleans Levee Systems in Hurricane Katrina on Aug. 29, 2005," Nov. 2, 2005, Figure 1.4, p. 1-10, www.ce.berkeley.edu/~inkabi/KRTF/CCRM/levee-rpt.pdf.

[2] "Performance Evaluation Plan and Interim Status, Report 1 of a Series: Performance Evaluation of the New Orleans and Southeast Louisiana Hurricane Protection System," Interagency Performance Evaluation Task Force, Jan. 10, 2006, Appendix A, p. 2, https://ipet.wes.army.mil.

[3] *Ibid*, pp. 1.2-1.3; Seed, *et al.*, *op. cit.*, p. A-2.

[4] "Performance Evaluation Plan," *op. cit.*, Appendix A, p. 2.

[5] John McQuaid and Mark Schleifstein, "Evolving Danger; experts know we face a greater threat from hurricanes than previously suspected," *The Times-Picayune* (New Orleans), June 23, 2002, p. A1.

[6] John McQuaid and Mark Schleifstein, "Shifting Tides," *The Times-Picayune* (New Orleans), June 26, 2002, p. A1.

[7] John Schwartz, "New Orleans Wonders What to Do With Open Wounds, Its Canals," *The New York Times*, Dec. 231, 2005, p. A26; Seed, *et al.*, *op. cit.*, p. 3.1; Matthew Brown, "Corps suspends plans to dredge MRGO," *The Times-Picayune* (New Orleans), Breaking News Weblog, Nov. 21, 2005, www.nola.com/t-p/.

[8] For background, see Larry Roth statement to Senate Committee on Environment and Public Works, Nov. 17, 2005, http://epw.senate.gov/hearing_statements.cfm?id=249000.

Herbert, declared in 1850 that reliance on levees "will be destructive to those who come after us." By then, some rose 15 feet.

Dynamiting the Levee

The engineers' warnings came to pass in early 1927. A series of rainstorms, coupled with unusually heavy spring runoff, swelled the huge river and overwhelmed the levees. Floodwater inundated 28,545 square miles of the Mississippi Valley as far north as Illinois, killing 423 people. By mid-April, more than 50,000 people had fled their homes.[51]

In New Orleans, powerful pumps kept floodwaters at bay — until a bolt of lightning disabled the power plant that kept the pumps humming.

A group of city leaders, who had formed the Citizens Flood Relief Committee, began campaigning to stop the flooding of the city by blowing a hole in the levee some 12 miles downstream.

Residents of the two thinly populated wetland parishes downstream, St. Bernard and Plaquemines, largely made their living fishing and trapping muskrats for their fur. The New Orleans political class persuaded Louisiana Gov. Oramel Simpson that those rural activities were worth sacrificing to protect New Orleans. Simpson gave the "river parish" residents three days to clear out. Muskrat trapping took years to recover.

For the poor African-Americans living along the river's southern reaches, the 1927 flood left bitter memories of racial oppression and death. Especially in Mississippi, thousands of black men were conscripted into labor gangs that shored up the levee, often working at gunpoint. Some drowned as they worked, and a community leader who refused a summons because he'd been working all night was shot on the spot.

The race-hatred exacerbated by the flood triggered a vast expansion of the "great migration" of African-Americans from South to North.[52] For the black community, 1927 established a connection between natural disaster, racism

Getty Images/Robyn Beck

Engineers inspect a Katrina-damaged section of the London Avenue Canal on Sept. 21, 2005, three days before Hurricane Rita hit and reopened some levees that had been partially repaired. The city's flood-control system may not be completely repaired by June 1, the start of the 2006 hurricane season.

and black exodus — a chain of events that many would later see repeating itself with Katrina.

New Expansion

The 1927 disaster led to improved federal flood-control systems and also launched a continuing debate over whether all Americans should have to pay to protect people living in disaster-prone places.[53]

In New Orleans, the 1927 flood also undermined the total dependence on levees for protection. Two years later, the Corps of Engineers began building a spillway at Bonnet Carré that could release river water into Lake Pontchartrain if the Mississippi rose to 20 feet in New Orleans. The spillway was completed in 1936.

By then, New Orleans residents had other reasons to feel safer. Electric and gasoline-powered motors had relieved

major drainage problems. In a city sitting below sea level in a swampy area, difficulties in disposing of human and other waste had long endangered health and lowered the quality of life. Mosquito-borne yellow fever alone killed about 41,000 people between 1817 and 1905.

Draining surrounding swampland also allowed the opening up of new lands for settlement. From the 1930s to the post-World War II years, acreage to the north, east and west of the original city were transformed from wetlands into tract-housing territory. The suburbanization expanded into Jefferson Parish, just outside of the city.

Within New Orleans itself, the amount of land that had been drained for settlement expanded from 12,349 acres in 1895 to more than 90,000 by 1983.

On a dry day, the newly drained territory appeared suitable for housing. But after Katrina, the local paper, the *Times-Picayune*, published an 1878 map showing that nearly every part of the city that flooded in 2005 had been uninhabited in the years before the land was drained. Early residents understood exactly where not to live, the paper concluded.[54]

Hurricanes and Floods

Since the city's founding, the protective levee system had aimed mainly at holding back Mississippi flooding. But beginning in the mid-20th century, a series of powerful hurricanes changed the perception of where danger lay.

In 1947, a 112-mile-an-hour hurricane (they didn't have names yet) brought two-foot floods in a nine-square-mile area. Hurricanes Flossy (1956) and Hilda (1964) caused some damage but were dwarfed by the 160-mile-an-hour winds of Hurricane Betsy in 1965. Floodwaters reached eight feet in parts of the city; 75 people died, and 7,000 homes suffered damage.

In response, Congress passed the Flood Control Act of 1965, which funded expansion of the levee and canal system in and around New Orleans to protect against what today would be classified as a Category 3 hurricane.[55]

In 1969, just as construction of the expanded system began, Hurricane Camille slammed the Gulf Coast. Mississippi was hit hardest, but a section of the New Orleans levee complex also failed, flooding part of the city.

In succeeding years, even rainstorms became problematic. Nine inches of rain during a 1978 storm caused flooding of up to 3.5 feet in low-lying sections, damaging

71,500 homes. A series of heavy rainstorms between 1979 and 1995 also caused widespread damage, and in 1998, Hurricane Georges, a Category 2 storm that barely touched New Orleans, brought a water surge to within a foot of topping the levees.[56]

Waiting for the Big One

The steady growth in the number and intensity of hurricanes during the 1990s fed unease in New Orleans and prompted the *Times-Picayune* to publish — at the beginning of the 2002 hurricane season — a series of articles unflinchingly examining the risks New Orleans faced. "Officials at the local, state and national level are convinced the risk is genuine and are devising plans for alleviating the aftermath of a disaster that could leave the city uninhabitable for six months or more," the authors presciently wrote.[57]

In January 2005, Ivor van Heerden, deputy director of the Louisiana State University (LSU) Hurricane Center, told a conference on "coastal challenges" that a Category 3 or above storm striking New Orleans or any other coastal Louisiana city would be a "disaster of cataclysmic proportion."[58]

By then, the city's new-and-improved flood-protection system consisted of about 125 miles of levees, floodwalls and flood-proofed bridges and other barriers. In Orleans Parish, the renovation work was 90 percent complete.[59]

On Aug. 28, as Hurricane Katrina was rolling through the Gulf and heading for New Orleans, the National Weather Service called it "a most powerful hurricane with unprecedented strength." After landfall, "Most of the area will be uninhabitable for weeks, perhaps longer."[60]

Mayor Nagin, a newcomer to politics, had ordered the city evacuated. But buses for the tens of thousands of elderly and poor residents who didn't own cars were never dispatched.

Even before Katrina touched down near New Orleans on the morning of Aug. 29, a storm surge breached the levees along the Inner Harbor Navigation Canal (the "Industrial Canal"). At about the same time, an 18-foot surge from Lake Borgne pushed through a wall along the Mississippi River Gulf Outlet east of St. Bernard Parish and the Lower Ninth Ward. The resulting flooding soon reached the Lower Ninth.[61]

Over the next few hours, additional surges over the Industrial Canal sent even more floodwater into the Lower Ninth. Then, with Katrina moving westward near Lake Pontchartrain, another section of levee along the Industrial Canal gave way, followed by a breach of the 17th Street Canal floodwall, flooding the western end of the parish.[62]

In the days following the storm, New Orleans became an international symbol of government dysfunction. Tens of thousands of residents unable to evacuate clung to rooftops or flocked to the New Orleans Superdome, which was unequipped to receive them. By Sept. 12, FEMA Director Michael Brown had resigned under pressure — only days after being congratulated by President Bush. Belatedly, federal officials organized bus convoys and flights out of the city.[63]

Then, on Sept. 24, Hurricane Rita, a Category 3 storm, hit the Gulf Coast. New Orleans didn't lie directly in the storm's path, but the hurricane reopened some partly repaired levee breaches. As a result, the Lower Ninth Ward and the Gentilly neighborhood flooded again. Elsewhere in Louisiana and East Texas, the damage was far worse, with tens of thousands left homeless.[64]

By early December, only 10 percent of the city's businesses were up and running, and 135,000 residents, at most, had stayed or returned. They had a name for the only fully functioning part of the city — a strip of high ground that includes the French Quarter and other sections of old New Orleans: Like explorer Bienville, they called it "the island."[65]

CURRENT SITUATION
Redevelopment Plans

On Jan. 11, the Bring New Orleans Back Commission released its "Action Plan" for rebuilding the city, but action doesn't seem to be on the near horizon.

The plan recommended the formation of 13 neighborhood-planning committees, with work on recommendations to start on Feb. 20, finish by May 20 and be submitted to the city for approval by June 20. Reconstruction would begin by Aug. 20.[66]

Within days of the plan's release, however, a FEMA official said updated floodplain maps of the city wouldn't be available until the summer, depriving crucial information to homeowners considering rebuilding.

"If I were putting my lifetime savings in the single, biggest investment I'll ever make, I'd want to make sure I had minimized every possible risk," said Tulane's Kroloff, chairman of the commission's urban-design subcommittee.[67]

The delay in obtaining the updated flood-zone information would slow the reconstruction timetable, Kroloff said, but wouldn't prevent the neighborhood committees from canvassing past and present residents. "There are some people who are going to return no matter what, and some who aren't," he said.[68]

Other obstacles could further slow the plan's execution. Congress may not approve Rep. Baker's proposal to create a public corporation to buy and sell distressed properties, although BNOB Commissioner Canizaro hopes funds can be rounded up from FEMA and elsewhere.[69] The Louisiana legislature would have to create the nonprofit entity, provisionally entitled the Crescent City Recovery Corp., and New Orleans voters would have to OK changes to the city charter to authorize it.[70]

That vote could come as soon as April. But it remains to be seen how receptive voters will be to measures recommended by Nagin and the commission, especially in light of the criticism that greeted the plan when it was unveiled. Some property owners attacked the proposal as a land grab.

"If you come to take our property, you'd better come ready," homeowner Rodney Craft of the Lower Ninth Ward told the commissioners.[71]

"I hear the politicians talk, and nothing is being said — nothing," says Gail Miller, a retired New Orleans police officer who has returned to her home in New Orleans East, living upstairs but cooking in a motor home she and her husband park in the driveway. "The political situation worries me — the levees don't worry me a bit."

Another widespread worry is education. Since the state took over 102 of the city's 117 schools by designating them as a "recovery district," only about 8,000 of 60,000 pre-Katrina students are attending the handful of public and parochial schools that are operating.[72]

"One of the barriers to families returning is that the state took over the schools and is not opening them," says Councilwoman Willard-Lewis.

Many residents say, however, that reopening the schools as they were wouldn't be much help. The Urban Land Institute reported that before Katrina the public school system had an "educational quotient" ranking of 1 out of 100 — the nation's lowest.[73]

"Everybody knew that public education was broken before the storm," says Heather Thompson, a New Orleans native and Harvard Business School student. A graduate of the public schools' only secondary-level crown jewel, Benjamin Franklin High School, Thompson helped organize a consulting project by four dozen of her fellow business students to recommend recovery ideas for schools and other elements of civic life.[74]

Meanwhile, the shortage of school space seems likely to continue. "I want to get the very best leaders and the very best teachers for every child in Orleans Parish," said State Education Superintendent Cecil Picard, adding that he expects 15,000 public-school students when classes reopen in August.[75]

Port Bounces Back

Giant cranes are swinging containers off and on ships, warehouses are filled with bundles of rubber and coils of steel, and trucks headed inland are filling up with coffee beans. The Port of New Orleans is back up and running, though only months ago a quick comeback seemed improbable.

"On Aug. 30, somebody told me it would be six months before we got the first ship back," port Director LaGrange says. "I said our goal was to be at 70 percent of pre-Katrina activity by March 1 — the six-month anniversary [of Katrina]. We're pushing 65 percent now."

Immediately after Katrina struck, while Americans watched thousands of human tragedies unfolding in real time on television, shippers and merchants focused on the southern Louisiana port complex — the country's fourth-largest.[76] "The longer the ports remained closed, the greater the risk that we'd all be paying higher prices for coffee, cocoa, lumber, steel, zinc, aluminum and any number of other things," said Mark M. Zandi, chief executive of the Economy.com research firm.[77]

In a seeming paradox, Katrina largely spared the riverfront port area. Like the old French Quarter, most of the port sits atop the natural levee on which Bienville founded the city. However, a major container terminal and a new cold-storage warehouse in eastern New Orleans were both destroyed.

Louisiana politicians frequently cite the port's importance to the economy as an argument for rebuilding New Orleans to its pre-Katrina scale. When she heard that the port might be able to function at full strength with a city somewhat smaller than pre-Katrina New Orleans, Sen. Landrieu, responded: "Where are the workers going to

AT ISSUE

Should New Orleans be completely rebuilt on its old footprint?

YES

Sen. Mary Landrieu, D-La.
Member, Senate Appropriations Committee

Written for the *CQ Researcher*, January 2006

More than five months ago, Hurricane Katrina and the subsequent breaks in numerous flood-control levees decimated one of our nation's greatest cities, my hometown, New Orleans.

Some have since questioned whether or not we should rebuild New Orleans, saying that we should abandon a city that has contributed so much to our great nation.

New Orleans is the capitol of our nation's energy coast. It was put there for a reason. We did not go there to sunbathe. We went there to set up the Mississippi River, to tame that river, to create channels for this country to grow and prosper. New Orleans was established so the cities and communities along the Mississippi River would have a port to trade with the world.

The indispensible Higgins boats that saved us during World War II were built in New Orleans. Forty-three thousand people built those boats and headed them out to Normandy. We're going to rebuild our shipping industry. We're going to rebuild our maritime industry, we will maintain our great port and we will continue to provide the energy that keeps our lights on across the nation.

Just because parts of New Orleans are below sea level is no reason to allow this great city to die. The Netherlands is a nation that is 21 feet below sea level at its economic heart, yet they still operate Europe's largest port — just as we operate America's largest port system.

The Dutch have proved that you can live below sea level and still keep your feet dry. They believe in an integrated system of water management. After a flood destroyed their nation in 1953, the Dutch said "Never again," and today they have created the world's most advanced storm-protection and flood-control system. If a nation half the size of Louisiana can do it, then surely the United States of America can.

We can and should rebuild every neighborhood — but maybe not exactly the way we did it the first time. This time we can build better, smarter, stronger neighborhoods.

One fact is certain: Every, single American citizen who calls New Orleans home has a right to come back and rebuild their neighborhoods, and the federal government should generously support that right.

New Orleans helped build America, and now America must help rebuild New Orleans, because America needs that great city — right where it is.

NO

William Hudnut
Joseph C. Canizaro Chair in Public Policy, Urban Land Institute

Written for the *CQ Researcher*, January 2006

There are those who understandably feel that New Orleans should be rebuilt in its entirety, and that blocks and neighborhoods throughout the pre-Katrina city should be rebuilt house by house as resources permit.

The emotional tug of going back to one's "roots" is strong. One cannot blame the City Council and others for demanding that all areas of the city, especially East New Orleans and the Lower Ninth, as well as Lakeview and Gentilly, be rebuilt simultaneously. But we need to ask: Is such a plan realistic? Does it make sense?

The city will not have the resources to take care of a widely dispersed population, and not all the evacuees will be returning. Critics of a smaller city dismiss such plans and ideas as "arrogant," "elitist" and "racist," because the low-lying areas are where mostly black and low-income residents lived before Katrina. But the questions persist.

I can think of two compelling reasons to envision a smaller New Orleans in the future. It will have a smaller population, and it will be safer.

As is often said, "Demography is destiny." If New Orleans once had 465,000 people, that was once and no more. The city was losing population before Katrina and has shrunk to a little over 100,000 today, with prospects of that number climbing to perhaps 250,000 by the time Katrina's third anniversary rolls around.

Is it prudent to think that this smaller number of people should occupy all the territory that almost twice that number did before August 2005, especially when the city will not have the financial resources, police, fire and EMS services and the like to care for such a scattered population? Two keys to a successful, vibrant city are diversity and density, which a sprawled-out land base does not provide.

Katrina has given New Orleans a chance to reinvent itself as a more compact, connected city on a smaller footprint. The city's recovering economy built on restored building blocks — culture, food, music, art, entertainment, tourism, bioscience and medical research, the port, energy production — will attract people back into mixed-use, mixed-income, racially balanced, pedestrian-friendly neighborhoods carefully planned by citizens, with parks, open space, new wetlands and light-rail transit added to the mix. All of that can be accomplished on less space than the city occupied heretofore.

Who was it that said, "Small is beautiful?"

come from? You can't have a port without New Orleans."

LaGrange takes a more nuanced view. "You've got to have the work force here," he says, and they will need "the support services that a city provides — transit, schools, places to worship, grocery stores, gasoline stations. But if the city, for some reason, is smaller, I don't think that would be a tremendous effect on the output of the port."

Politics and Legislation

New Orleans' future lies in many hands, but federal lawmakers may be the most important, because they control the biggest money source.

"We are at your mercy," Gov. Blanco told Senate Homeland Security Committee members as they toured the disaster zone on Jan. 17. "We are begging you to stay with us."[78]

Landrieu plans to revive her proposal to channel 50 percent of offshore petroleum-lease revenues to the state. The money would be earmarked for post-Katrina reconstruction, says her spokesman, Adam Sharp.

Besides the Landrieu and Baker proposals, Louisiana politicians will continue to push for $2.1 billion in supplemental Medicaid funds to help pay for health care for Katrina victims — many suddenly homeless and unemployed — who had to enter the federally subsidized medical insurance program for low-income people. Congress adjourned at year's end without passing the Medicaid bill, but Landrieu says she'll also continue to push for that.

The fact that none of these proposals passed while Katrina's devastation was fresh would seem to show that the state's politicians "have some work to do" to get Congress' attention, said one of Baker's aides. Blanco, meanwhile, is preparing to call a special 12-day legislative session, beginning on Feb. 6. She wants state lawmakers to make the "levee boards" that supervise maintenance more accountable. The boards were widely criticized — even ridiculed — for laxity, following Katrina.[79]

Getting the schools going again remains a priority, and Blanco must hammer together by May a plan to reorganize the city's school system, now largely under state control. The state Board of Elementary and Secondary Education would have to rule on the plan. The BNOBC in January proposed a leaner administrative office — one superintendent and four or five assistants — and expanded authority for principals, who would be able to hire and fire their own staffs. Differences between "have" and "have-not" schools would be eliminated under the plan, and early-education programs would be initiated.[80]

Meanwhile, the often-criticized Blanco tangled with the City Council over what she called its resistance to installing FEMA-supplied trailers for needy families. The council was responding, in part, to complaints from some residents who objected to trailer villages in their neighborhoods.

"Disagreements over housing must end — and must end now," she told the council on Jan. 5. Council members denied that they had obstructed trailer installation. After a subsequent meeting between the governor and council members, sites for a total of 40,000 trailers were identified.[81]

Even the demolition of unsafe houses stirred controversy. When it appeared the city was about to bulldoze some Lower Ninth Ward houses deemed unsafe, residents and some council members sought a court order to stop it. U.S. District Judge Martin L. C. Feldman then OK'd a deal between the Nagin administration and Lower Ninth Ward residents requiring at least seven-days' notice before demolition.[82]

The court-approved settlement apparently resolved the demolition issues, but political conflicts between Nagin and the council remain. The beleaguered mayor is among the candidates up for re-election on April 22.

OUTLOOK
Pessimism and Paralysis

Optimism is in short supply in New Orleans, notwithstanding the brave talk of Louisiana politicians. The failure of the flood-protection system, the tragedy and chaos of the early days of the disaster and the devastated conditions that remain in much of the city five months after Katrina have not provided grounds for much hope.

President Bush, in his State of the Union address on Jan. 31, devoted 162 of the speech's 5,432 words to New Orleans, proposing no specific, new remedies. "As we meet . . . immediate needs, we must also address deeper challenges that existed before the storm arrived," Bush said, citing a need for better schools and economic opportunity. Among Louisiana politicians, even the president's fellow Republicans felt left out. "I was very disappointed at how small a part those national challenges — and I think are national challenges — were given in the speech," Sen. Vitter told the *Times-Picayune*.

"There's no sense of urgency from the city government, the state government or the federal government," says Dennis Scott, looking out on his devastated New Orleans East neighborhood.

Indeed, as of late January, the U.S. Army Corps of Engineers had completed only 16 percent of the levee repairs scheduled for completion by June 1, when the 2006 hurricane season begins.[83]

An outsider draws essentially the same conclusion as Scott. "The lack of unity in the political establishment is the paralyzing factor," says the Urban Land Institute's Hudnut. "There's almost a political stand-off between the governor's office, the mayor's office, the City Council and the Bring New Orleans Back Commission; but this is also partially a Washington issue. I don't see a lot of leadership coming from the White House team."

Republican Hudnut is one of many politicians and ordinary citizens to question the high cost of the war in Iraq with the needs of New Orleans. The war's direct cash cost alone through November 2005 was calculated at $251 billion, according to a study released in January by two former Clinton administration officials.[84] Thompson, the Harvard Business School student working on redevelopment plans, observes that the government ought to be able to "make money appear" for New Orleans in the same way as deficit financing is arranged for the war.

If talking openly about race relations holds promise for making them better, the New Orleans disaster might have served some purpose. Some black New Orleanians wonder aloud, though, if the color of the majority of the city's residents hasn't also slowed down the pace of recovery. Anne LaBranche, the doctor's wife from New Orleans East, can't think of any other reason.

"This was a man-made problem," she says, referring to the failure of the flood-protection system. And yet, previous hurricane damage in Florida and other Gulf Coast states has been paid for without debate on whether people should be living in such potentially risky areas, she says. "President Bush says he resents it when people say 'racism,' so tell me what it is," she says quietly. "Why the different treatment?"

If New Orleans has one advantage concerning race, it may be that the city's geography tends to throw people of different colors together more than in other locales. Another point in the city's favor is New Orleanians' loyalty to their city. It remains to be seen whether that's enough to overcome the economic, political and environmental obstacles.

Piano technician David Doremus has lived in New Orleans most of the past 30 years. He and his wife live in the unflooded Algiers neighborhood on the Mississippi's west bank, and they are committed to remaining in town with their daughters.

While he's unsure about how much piano tuning and rebuilding work he'll have in the near future, he can't imagine anywhere else that offers the pace of life, the social graces and the fishing that he enjoys in New Orleans — as well as the musical variety. "I work for a recording studio, and one of the first sessions I worked on after the storm was with Allen Toussaint and Elvis Costello," he says.

So Doremus is ready to commute 40 miles to work at a friend's piano business in Covington, La., for a year, if he has to, or even work at Home Depot. "My family back in Virginia thinks I'm nuts," he adds. "And my wife's family in Pittsburgh thinks she's nuts."

If the Doremuses are crazy, New Orleans needs all the nuts it can muster.

NOTES

1. Gary Rivlin, "Anger Meets New Orleans Renewal Plan," *The New York Times*, Jan. 12, 2006, p. A18.

2. When Hurricane Katrina made landfall at Buras, La., 35 miles east of New Orleans at about 6 a.m., it was originally rated at Category 4, the classification for storms with wind speeds of 131-155 mph. The National Hurricane Center later revised that classification down to Category 3, with winds of 111-130 mph. Some 24 hours before reaching Louisiana, Katrina varied between categories 4 and 5. For further detail, see Peter Whoriskey and Joby Warrick, "Report Revises Katrina's Force," *The Washington Post*, Dec. 22, 2005, p. A3; Richard D. Knabb, *et al.*, "Tropical Cyclone Report: Hurricane Katrina, 22-30 August, 2005," National Hurricane Center, Dec. 20, 2005, p. 3, www.nhc.noaa.gov/pdf/TCR-AL122005_Katrina. pdf; and National Aeronautics and Space Administration, "Hurricane Season 2005:

Katrina," www.nasa.gov/vision/earth/looking atearth/h2005_katrina.html.

3. Nicholas Riccardi, "Most of Louisiana's Identified Storm Victims Over 60," *Los Angeles Times*, Nov. 5, 2005, p. A11; Nicholas Riccardi, Doug Smith and David Zucchino, "Katrina Killed Along Class Lines," *Los Angeles Times*, Dec. 18, 2005, p. A1.

4. While Katrina had weakened to Category 3 upon reaching Louisiana, the surges it created began when the storm was at categories 4 and 5 strength. For further detail, see "Tropical Cyclone Report," *op. cit.*, p. 9.

5. "President Discusses Hurricane Relief in Address to the Nation," White House, Sept. 15, 2005, www .whitehouse.gov/news/releases/2005/09/ print/20050915-8.html.

6. Bill Walsh, "Senators say recovery moving at snail's pace," *The Times-Picayune* (New Orleans), Jan. 18, 2006, p. A1.

7. Ralph Vartabedian, "New Orleans Should be Dry by End of Week," *Los Angeles Times*, Sept. 19, 2005, p. A8; "Performance Evaluation Plan and Interim Status, Report 1 of a Series: Performance Evaluation of the New Orleans and Southeast Louisiana Hurricane Protection System," Interagency Performance Evaluation Task Force, Jan. 10, 2006, p. 1, https://ipet.wes.army.mil.

8. "Action Plan for New Orleans: The New American City," Bring New Orleans Back Commission, Urban Planning Committee, Jan. 11, 2006, Introduction, www.bringneworleansback.org.

9. "It was not unusual for slaves to gather on street corners at night, for example, where they challenged whites to attempt to pass. . . ," historian Joseph G. Tregle is quoted in Eugene D. Genovese, *Roll, Jordan, Roll: The World the Slaves Made* (1972), pp. 412-413.

10. Quoted in Reed Johnson, "New Orleans: Before and After," *Los Angeles Times*, Sept. 5, 2005, p. E1. For more background on Congo Square, see Craig E. Colten, *An Unnatural Metropolis: Wresting New Orleans From Nature* (2005), p. 72; and Gerald Early, "Slavery," on Web site for "Jazz," PBS documentary, www.pbs.org/jazz/time/time_slavery.htm.

11. "A Strategy for Rebuilding New Orleans, Louisiana," Urban Land Institute, Nov. 12-18, 2005, p. 17,

www.uli.org/Content/NavigationMenu/ ProgramsServices/AdvisoryServices/KatrinaPanel/ ULI_Draft_New_Orleans%20Report.pdf.

12. "President Discusses Hurricane Relief," *op. cit.*

13. Joby Warrick, "White House Got Early Warning on Katrina," *The Washington Post*, Jan. 24, 2005, p. A2.

14. Steve Ritea and Tara Young, "Cycle of Death: Violence Thrives on Lack of Jobs, Wealth of Drugs," *The Times-Picayune* (New Orleans), p. A1; Adam Nossiter, "New Orleans Crime Swept Away, With Most of the People," *The New York Times*, Nov. 10, 2005, p. A1. Dan Baum, "Deluged, When Katrina hit, where were the police?" *The New Yorker*, Jan. 9, 2006, p. 59.

15. Quoted in, Joel Havemann, "New Orleans' Racial Future Hotly Argued," *Los Angeles Times*, Oct. 1, 2005, p. A14.

16. Brett Martel, The Associated Press, "Storms Payback From God, Nagin Says," *The Washington Post*, Jan. 17, 2006, p. A4.

17. Manuel Rog-Franzia, "New Orleans Mayor Apologizes for Remarks About God's Wrath," *The Washington Post*, Jan. 18, 2006, p. A2.

18. James Dao, "Study Says 80% of New Orleans Blacks May Not Return," *The New York Times*, Jan. 27, 2006, p. A16.

19. *Ibid.*; see also "Action Plan," (pages unnumbered); Frank Donze and Gordon Russell, "Rebuilding proposal gets mixed reception," *The Times-Picayune* (New Orleans), Jan. 12, 2006, p. A1.

20. Donze and Russell, *ibid.*; Rivlin, *op. cit.*

21. "A Strategy for Rebuilding," *op. cit.*; Frank Donze, "Don't write us off, residents warn," *The Times-Picayune* (New Orleans), Nov. 29, 2005, p. A1.

22. "Action Plan," Introduction, *op. cit.*; Gordon Russell, "Comeback in Progress," *The Times-Picayune* (New Orleans), Jan. 1, 2006, p. A1.

23. "Battered by Katrina, Tulane University forced into layoffs, cutbacks," The Associated Press, Dec. 9, 2005.

24. Susan Saulny, "Students Return to Big Changes in New Orleans," *The New York Times*, Jan. 4, 2006, p. 13; Steven Ritea, "School board considers limited

role," *The Times-Picayune* (New Orleans), Dec. 7, 2005, p. A1.

25. Elizabeth Bumiller, "In New Orleans, Bush Speaks With Optimism But Sees Little of Ruin," *The New York Times*, Jan. 13, 2006, p. A12.

26. For background, see Marcia Clemmitt, "Climate Change," *CQ Researcher*, Jan. 27, 2006, pp. 73-96.

27. Virginia R. Burkett, "Potential Impacts of Climate Change and Variability on Transportation in the Gulf Coast/Mississippi Delta Region," Center for Climate Change and Environmental Forecasting, Oct. 1-2, 2002, p. 7, http://climate.volpe.dot.gov/workshop1002/burkett.pdf. Burkett is chief of the Forest Ecology Branch of the U.S. Geological Survey's National Wetlands Research Center, in Lafayette, La.

28. In 2006, the Bush administration does not plan to seek new funds for reconstruction in Iraq. See, Ellen Knickmeyer, "U.S. Has End in Sight on Iraq Rebuilding," *The Washington Post*, Jan. 2, 2006, p. A1.

29. Michael Oneal, "GOP Cools to Katrina Aid," *Chicago Tribune*, Nov. 12, 2005, p. A7.

30. For background, see Kenneth Jost, "Property Rights," *CQ Researcher*, March 4, 2005, pp. 197-220.

31. R. B. Seed, *et al.*, "Preliminary Report on the Performance of the New Orleans Levee Systems on August. 29, 2005," University of California at Berkeley, American Society of Civil Engineers, Nov. 2, 2005, pp. 1.2-1.4.

32. Schwartz, *op. cit.*

33. For details, see John McQuaid, "The Dutch Swore It Would Never Happen Again," "Dutch Defense, Dutch Masters," "Bigger, Better, Bolder," *The Times-Picayune* (New Orleans), Nov. 13-14, 2005, p. A1.

34. "Performance Evaluation Plan," *op. cit.*, appendix A-2. John Schwartz, "Category 5: Levees are Piece of $32 Billion Pie," *The New York Times*, Nov. 29, 2005, p. A1.

35. *Ibid.*

36. Richard W. Stevenson and James Dao, "White House to Double Spending on New Orleans Flood Protection," *The New York Times*, Dec. 16, 2005, p. A1.

37. *Ibid.*

38. President Bush said on Jan. 26 the congressional appropriations amounted to $85 billion. For background and detail, see Joseph J. Schatz, "End-of-Session Gift for the Gulf Coast," *CQ Weekly*, Dec. 26, 2005, p. 3401; "Cost of Katrina Nearing $100 Billion, Senate Budget Says," *CQ Budget Tracker News*, Jan. 18, 2006; "Senate Budget Committee Releases Current Tally of Hurricane-Related Spending," Budget Committee, Jan. 18, 2006, http://budget.senate.gov/republican. "Press Conference of the President," [transcript] Jan. 26, 2006, www.whitehouse.gov/news/releases/2006/01/20060126.htm..

39. Quoted in Jacob Freedman, "Additional Flood Funds Needed to Cover Extensive Gulf Coast Damage," *CQ Today*, Jan. 25, 2006; Statement of David I. Maurstad, Acting Director/Federal Insurance Administrator, Mitigation Division, Federal Emergency Management Agency, Committee on Senate Banking Housing and Urban Affairs, Jan. 25, 2006, http://banking.senate.gov/_files/ACF43B7.pdf.

40. Frank Donze, Gordon Russell and Lauri Maggi, "Buyouts torpedoed, not sunk," *The Times-Picayune* (New Orleans), Jan. 26, 2006, p. A1.

41. Adrian Moore, "Rebuild New Orleans Smarter, Not Harder," Reason Foundation, Jan. 11, 2006, /www.reason.org/commentaries/moore_20060111.shtml.

42. David Greising, *et al.*, "How Do They Rebuild a City?" *Chicago Tribune*, Sept. 4, 2005, p. A1.

43. "Press Conference of the President," *op. cit.*

44. Donze, Russell and Maggi, *op. cit.*

45. *Ibid.*

46. Robert L. Bamberger and Lawrence Kumins, "Oil and Gas: Supply Issues After Katrina," Congressional Research Service, updated Sept. 6, 2005, p. 1, www.fas.org/sgp/crs/misc/RS22233.pdf. For background on offshore leases, see Jennifer Weeks, "Domestic Energy Development," *CQ Researcher*, Sept. 30, 2005, pp. 809-832.

47. Marc Humphries, "Outer Continental Shelf: Debate Over Oil and Gas Leasing and Revenue Sharing," Congressional Research Service, Uupdated Oct. 27, 2005, pp. 1-4. http://fpc.state.gov/documents/organization/56096.pdf.

48. Unless otherwise indicated, all material in this section comes from Colten, *op. cit.*; and Ari Kelman, *A River and Its City: The Nature of Landscape in New Orleans* (2003).

49 Caroline E. Mayer and Amy Joyce, "Troubles Travel Upstream," *The Washington Post*, Sept. 5, 2005, p. A23.

50. Colten, *op. cit.*, pp. 25-26.

51. For background, see C. Perkins, "Mississippi River Flood Relief and Control," *Editorial Research Reports*, 1927, Vol. 2; and M. Packman, "Disaster Insurance," *Editorial Research Reports 1956*, Vol. I.

52. John M. Barry, *Rising Tide: The Great Mississippi Flood of 1927 and How it Changed America* (1998), pp. 311-317; p. 332.

53. For background, see "Economic Effects of the Mississippi Flood," *Editorial Research Reports, 1928*, Vol. I.

54. Gordon Russell, "An 1878 Map Reveals that Maybe Our Ancestors Were Right to Build on Higher Ground," *The Times-Picayune* (New Orleans), Nov. 3, 2005, p. A1.

55. "Performance Evaluation Plan," *op. cit.*, Appendix A, p. 1; Willie Drye, " 'Category Five': How a Hurricane Yardstick Came To Be," *National Geographic News*, Dec. 20, 2005, http://news.nationalgeographic.com/news/2005/12/1220_051220_saffirsimpson.html.

56. John McQuaid and Mark Schleifstein, "The Big One," *The Times-Picayune* (New Orleans), June 24, 2002, p. A1.

57. *Ibid.*

58. Ivor van Heerden, "Using Technology to Illustrate the Realities of Hurricane Vulnerability," Jan. 25, 2005, www.laseagrant.org/forum/01-25-2005.htm.

59. "Performance Evaluation Plan," *op. cit.*, Appendix A, pp. 2-3.

60. "Urgent Warning Proved Prescient," *The New York Times*, Sept. 7, 2005, p. A21.

61. "How New Orleans Flooded," in "The Storm That Drowned a City," NOVA, WGBH-TV, October 2005, www.pbs.org/wgbh/nova/orleans/how-nf.html.

62. *Ibid.*

63. See Pamela Prah, "Disaster Preparedness," *CQ Researcher*, Nov. 18, 2005, pp. 981-1004.

64. "Rita's Aftermath," *Los Angeles Times*, Sept. 28, 2005, p. A1; Shaila Dewan and Jere Longman, "Hurricane Slams Into Gulf Coast; Flooding Spreads," *The New York Times*, Sept. 25, 2005, p. A1.

65. Anne Rochell Konigsmark, "Amid ruins, 'island' of normalcy in the Big Easy," *USA Today*, Dec. 19, 2005, p. A1; Gordon Russell, "Comeback in Progress," *The Times-Picayune* (New Orleans), Jan. 1, 2006, p. A1.

66. "Action Plan," *op. cit.*, Sec. 4, (pages unnumbered).

67. Gordon Russell and James Varney, "New flood maps will likely steer rebuilding," *The Times-Picayune* (New Orleans), Jan. 15, 2006, p. A1.

68. *Ibid.*

69. *Ibid.*

70. *Ibid.*

71. Russell and Donze, *op. cit.*, Jan. 12, 2006.

72. Ritea and Saulny, *op. cit.*

73. "A Strategy for Rebuilding New Orleans," *op. cit.*, p. 19.

74. For background, see, George Anders, "How a Principal in New Orleans Saved Her School," *The Wall Street Journal*, Jan. 13, 2006, p. A1.

75. Steve Ritea, "La. won't run N.O. schools by itself," *The Times-Picayune* (New Orleans), Jan. 3, 2006, p. B1.

76. Vanessa Cieslak, "Ports in Louisiana: New Orleans, South Louisiana, and Baton Rouge," Congressional Research Service, Oct. 14, 2005, p. 1, http://fpc.state.gov/documents/organization/57872.pdf.

77. Keith L. Alexander and Neil Irwin, "Port Comes Back Early, Surprisingly," *The Washington Post*, Sept. 14, 2005, p. D1.

78. Bill Walsh, "Senators say recovery moving at a snail's pace," *The Times-Picayune* (New Orleans), Jan. 18, 2006, p. A1.

79. Ed Anderson, "Special session set to begin Feb. 6," *The Times-Picayune* (New Orleans), Jan. 12, 2006, p. A2.

80. Steve Ritea, "Nagin's schools panel issues reforms," *The Times-Picayune* (New Orleans), Jan. 18, 2006, p. A1; "Rebuilding and Transforming: A Plan for

World-Class Public Education in New Orleans," Bring New Orleans Back Commission, Jan. 17, 2006, pp. 10, 48.

81. Ed Anderson, "N.O. needs 7,000 more trailer sites, Blanco says," *The Times-Picayune* (New Orleans), Jan. 9, p. A1.

82. Adam Nossiter, "New Orleans Agrees to Give Notice on Home Demolitions," *The New York Times*, Jan. 18, 2006, p. A10.

83. Spencer S. Hsu, "Bush's Post-Katrina Pledges," *The Washington Post*, Jan. 28, 2006, p. A12.

84. Linda Bilmes and Joseph Stiglitz, "The Economic Costs of the Iraq War: An Appraisal Three Years After the Beginning of the Conflict," http://ksghome.harvard.edu/~lbilmes/paper/iraqnew.pdf. Former Deputy Assistant Commerce Secretary Bilmes is now at the Kennedy School of Government at Harvard; Stiglitz, a Nobel laureate economist, teaches at Columbia University.

BIBLIOGRAPHY

Books

Colten, Craig E., *An Unnatural Metropolis: Wresting New Orleans from Nature*, Louisiana State University Press, 2005.
A Louisiana State University, Baton Rouge, geographer chronicles the city's ongoing efforts to tame its watery environment.

Dyson, Michael Eric, *Come Hell or High Water: Hurricane Katrina and the Color of Disaster*, Basic Civitas Books, 2006.
A professor of humanities at the University of Pennsylvania — and a prolific author and commentator on issues of race and culture — dissects what he views as structural racism, government incompetence and class warfare against the poor in the Katrina disaster.

Kelman, Ari, *A River and its City: The Nature of Landscape in New Orleans*, University of California Press, 2003.
Using New Orleans' long and complicated relationship with the Mississippi River as a framework, an environmental historian at the University of California, Davis, examines why New Orleans developed as it did.

Piazza, Tom, *Why New Orleans Matters*, HarperCollins, 2005.
A jazz historian, novelist and New Orleans resident who evacuated the city during Katrina argues that American culture will be poorer if the working people who keep the city's traditions alive are permanently uprooted from the city.

Ward, Geoffrey C., and Ken Burns, *Jazz: A History of America's Music*, Alfred A. Knopf, 2000.
An author of popular history (Ward) and a renowned documentary filmmaker provide — with contributions by jazz scholars — a one-volume history of America's major cultural creation, with much attention to New Orleans' role.

Articles

Baum, Dan, "Deluged: When Katrina hit, where were the police?" *The New Yorker*, Jan. 9, 2006, p. 50.
A writer recounts how police and city government coped — or failed to — in the post-hurricane disaster.

Cooper, Christopher, "Old-Line Families Escape Worst of Flood and Plot the Future," *The Wall Street Journal*, Sept. 8. 2005, p. A1.
A profile of one of New Orleans' aristocrats brings the city's social inequalities to light in dispassionate fashion.

McQuaid, John, and Mark Schleifstein, "In Harm's Way," "Evolving Danger," "Left Behind," "The Big One," "Exposure's Cost," "Building Better," "Model Solutions," "Tempting Fate," "Shifting Tides," [series] *The Times-Picayune*, June 23-June 26, 2002.
Three years before Katrina, two reporters spell out the city's growing vulnerability to a massive hurricane, virtually telling the Katrina story.

Sontag, Deborah, "Delrey Street," *The New York Times*, Oct. 12, 2005, p. A1; Oct. 24, 2005, p. A1; Nov. 12, 2005, p. A9; Nov. 14, 2005, p. A1; Dec. 2, 2005, p. A20; Jan. 9, 2006, p. A1.
In a series of detailed profiles, a *New York Times* reporter examines how the lives of families from New Orleans' Lower Ninth Ward have been upended by Katrina.

Tizon, Alex Tomas, and Doug Smith, "Evacuees of Hurricane Katrina Resettle Along a Racial Divide," *Los Angeles Times*, Dec. 12, 2005, p. A1.
Two reporters analyzed change-of-address data to draw early conclusions on the racial effects of the disaster.

Reports and Studies

"Action Plan for New Orleans: The New American City," *Bring New Orleans Back Commission, Urban Planning Committee*, Jan. 11, 2006, www.bringneworleansback.org. Civic leaders and officials provided the first detailed plan for redevelopment of New Orleans.

"An Unnatural Disaster: The Aftermath of Hurricane Katrina," *Scholars for Progressive Reform*, Sept. 2005, www.progressivereform.org/Unnatural_Disaster_512.pdf. A liberal organization analyzes the disaster as a failure of unrestrained energy development and inadequate government regulation.

Katz, Bruce, *et al.*, "Katrina Index: Tracking Variables of Post-Katrina Reconstruction," updated Dec. 6, 2005, *The Brookings Institution*, www.brookings.edu/metro/pubs/200512_katrinaindex.htm. To be updated periodically, this report compiles and organizes statistics in order to show economic and social trends as New Orleans recovers.

Seed, R. B., *et al.*, "Preliminary Report on the Performance of the New Orleans Levee Systems in Hurricane Katrina on August 29, 2005," *University of California at Berkeley, American Society of Civil Engineers, National Science Foundation*, Nov. 2, 2005, www.berkeley.edu/news/media/releases/2005/11/leveereport_prelim.pdf. Engineering experts provide an early look at the failures of the levee system that led to disaster.

For More Information

Bring New Orleans Back Commission, www.bringneworleansback.org. The commission has been issuing detailed redevelopment plans.

The Brookings Institution, Katrina Issues and the Aftermath Project, Metropolitan Policy Program, 1775 Massachusetts Ave., N.W., Washington, DC 20036; (202) 797-6139; www.brookings.edu/metro/katrina.htm. The think tank provides policy proposals, commentary and statistics.

Center for the Study of Public Health Impacts of Hurricanes, CEBA Building, Suite 3221, Louisiana State University, Baton Rouge, LA 70803; (225) 578-4813; www.publichealth.hurricane.lsu.edu. A research center focusing on disaster prevention and mitigation.

Federal Emergency Management Agency, 500 C St., S.W., Washington, DC 20472; (202) 566-1600; www.fema.gov. The lead federal agency on disaster recovery; provides information on relief program requirements and application deadlines.

Greater New Orleans Community Data Center, www.gnocdc.org. A virtual organization that provides links to the city's most recent social, economic and demographic statistics.

Louisiana Recovery Authority, 525 Florida St., 2nd Floor, Baton Rouge, LA 70801; (225) 382-5502; http://lra.louisiana.gov. The state government's post-disaster reconstruction agency; provides information on the aid flowing to New Orleans.

New Orleans Area Habitat for Humanity, P.O. Box 15052, New Orleans, LA 70175; (504) 861-2077, www.habitat-nola.org. A self-help housing organization building new homes in the city and nearby suburbs.

Savenolamusic, www.savenolamusic.com/index.php. An exhaustive listing of performance bookings and other resources (including medical assistance) for New Orleans musicians, including those forced out of the city.

Urban Land Institute, 1025 Thomas Jefferson St., N.W., Suite 500 West, Washington, DC 20007; (202) 624-7000; www.uli.org. The nonprofit organization for land-use and development professionals is the New Orleans city government's disaster-recovery consultant.

14

Regulating
Toxic Chemicals

Do We Know Enough About Chemical Risks?

Jennifer Weeks

Concern about exposure to the chemical bisphenol A (BPA), widely used in hundreds of products, is prompting consumers and retailers to switch to BPA-free products. The National Toxicology Program warned recently that BPA poses some concern for "effects on the brain, behavior and prostate gland in fetuses, infants and children." Wal-Mart, Target, and other large companies have stopped selling products containing BPA.

From *CQ Researcher*,
January 23, 2009.

I n October 2007, the Eastman Chemical Co. of Kingsport, Tenn., introduced Tritan, a new plastic boasting "faster molding cycles compared to many other types of transparent polymers," plus enhanced durability and high gloss.[1]

But Tritan had another feature that made the plastics market take special notice: The new resin did not contain bisphenol A (BPA), a chemical widely found in rigid plastic products like food containers and baby bottles.

BPA has been used in consumer products for decades, although researchers have known since the 1930s that in mammals the chemical mimics estrogen, the natural hormone that regulates female sexual development and reproductive cycles. Endocrine disruption, as the effect is known, has been linked to developmental, reproductive and other problems in wildlife and laboratory animals, and some researchers believe it has a similar impact in humans.[2]

Until the late 1990s scientists thought BPA was only harmful at high doses, but then some studies showed that quantities as low as a few parts per billion could have toxic effects. They also demonstrated that BPA could leach from bottles and can linings into infant formula and food.[3] Then in 2008 the federally funded National Toxicology Program warned that current exposure levels to BPA posed some concern for "effects on the brain, behavior and prostate gland in fetuses, infants and children."[4] In contrast, the Food and Drug Administration (FDA), which regulates exposure to BPA from food packaging, maintained it was safe.

Consumers and retailers opted to be safe rather than sorry, especially after Canada banned BPA from baby bottles in April. Wal-Mart,

A sign warns that dangerous industrial chemicals were dumped in a lake near Gary, Ind. In 1979 the Environmental Protection Agency banned production of polychlorinated biphenyls (PCBs), which cause cancer and birth defects in laboratory animals, and set timetables for phasing out their use in various industries.

Target, REI, Costco and other large companies pulled products containing BPA from their shelves or found substitutes. Many of these stores now sell hard plastic water bottles made with Tritan copolyester, prominently marked as BPA-free.

"They're selling fantastically," says Carolyn Beem, public affairs manager for L.L. Bean, the Maine outdoor retailer. "We're not experts in science, but we are experts in listening and responding to our customers. With all of the reports out there, it seemed like a good time to start again." In March 2008 Eastman expanded Tritan production to keep up with demand.[5]

In addition to BPA, environmentalists and consumer advocates warn that many other materials in commercial products may be harmful to human health, including:

- polyvinyl chloride (PVC) plastic, used in items such as shower curtains and water pipes;
- phthalates, a group of chemicals used to make plastics soft and pliable; and
- polybrominated diphenyl-ethers (PBDEs), chemicals added to foams and fabrics as flame retardants.

In addition, some consumer goods contain materials widely known to be toxic, such as lead in popular brands of lipstick.[6]

"Consumers assume when they buy a product that someone has vetted it to make sure it's safe, but that doesn't always happen," says Sarah Janssen, a physician and environmental health expert at the Natural Resources Defense Council (NRDC), an advocacy group. "They're at a disadvantage because most of the important information isn't even on the label."

Humans are exposed to many potentially harmful substances in their daily lives, from air and water pollutants to household contaminants like mold and dust. Over time some of these exposures may cause cancer or other serious problems, such as birth defects or organ damage. Some of these illnesses result from lifestyle choices: for example, smoking, inactivity and obesity are major causes of cancer in the United States.[7] But workers and consumers also can be exposed unknowingly to risky materials that are legally used in commercial products.

Human exposure to toxic chemicals is controlled by several different agencies, depending on how the chemical is used and where people come in contact with it. The Environmental Protection Agency (EPA) regulates industrial chemicals and pesticides, while the FDA controls food additives, drugs and cosmetics and the Consumer Product Safety Commission (CPSC) oversees thousands of other consumer goods, from personal care products to toys. Workplace exposure to chemicals is regulated by the Occupational Safety and Health Administration (OSHA). Some materials must be tested for toxicity before marketing, but in other cases manufacturers merely have to notify regulators that they are going to start producing them.

Many experts think federal policy should be more consistent. "The agencies have very different approaches because they are covered by laws that find wildly varying levels of risk acceptable," says David Michaels, a professor of environmental and occupational health at George

Washington University and a former assistant secretary of Energy. "We should be thinking about ways to harmonize standards across these agencies, because their actions affect each other. They allow different levels of exposure for many of the same chemicals."

Chemical manufacturers say the Toxic Substances Control Act (TSCA) — the core law that regulates industrial chemicals — is working. "TSCA has protected human health and the environment," says Michael Walls, managing director for regulatory and technical affairs at the American Chemistry Council (ACC), a chemical industry trade group. "There are areas where we can reform it, and we're encouraged that proposals have been offered to amend the law, not to replace it." The chemical industry is working with EPA to make more data available on hazards from widely used chemicals and assess how chemical exposures affect children's health.

Many critics worry that it is too easy for new materials to enter commerce before their effects have been well studied. They are especially concerned about the growing field of nanotechnology, which uses microscopic particles to enhance products ranging from sunscreen to medications. Reducing materials to the nano scale makes it easier to apply them precisely: for example, chemotherapy drugs can be targeted directly at tumors. But materials acquire new properties at this scale, and scientists are still analyzing the toxicity of many nanomaterials.

"Nanotechnology is taking our understanding of what makes something harmful and how we deal with that, and turning it upside down," said Andrew Maynard, chief science advisor to the Project on Emerging Nanotechnologies, in April 2008 congressional testimony. "New, engineered nanomaterials are prized for their unconventional properties. But these same properties may also lead to new ways of causing harm to people and the environment."[8]

Concerns Linger Over Exposure to Bisphenol A

Scientists say they have "negligible concern" to "some concern" about the health effects of exposure to bisphenol A, a chemical commonly used in the production of plastics.

The National Toxicology Program uses the following five-level scale of concern for adverse effects from exposure to BPA:

- Serious concern
- Concern
- Developmental toxicity for fetuses, infants and children (effects on the brain, behavior and prostate gland) → Some concern
- Developmental toxicity for fetuses, infants and children (effects on the mammary gland and early puberty in females, and reproductive toxicity in workers) → Minimal concern
- Reproductive toxicity in adult men and women and malformations in newborns → Negligible concern

How to Reduce Your Exposure to Bisphenol A:

Don't microwave polycarbonate plastic food containers. Bisphenol A may break down from repeated use at high temperatures.

Avoid plastic containers with the number 7 on the bottom. (www.recyclenow.org/r_plastics.html)

Don't wash polycarbonate plastic containers in the dishwasher with harsh detergents.

Reduce your use of canned foods.

When possible, opt for glass, porcelain or stainless steel containers, especially for hot foods or liquids.

Use infant formula bottles and toys that are bisphenol A-free.

Source: National Toxicology Program

In 2007 the European Union (EU) launched a new system for regulating chemicals that differs markedly from the U.S. approach. Under the REACH (Registration, Evaluation, Authorization and Restriction of Chemicals) policy, companies that produce or import chemicals in large volumes have to register their products with the EU and provide data on their properties and uses. Chemicals must be shown to be safe before they can enter commerce. U.S. companies doing business in Europe have to comply with the directive.[9] (*See sidebar, p. 324.*)

REACH is based on the so-called precautionary principle, which can be traced back through history but was articulated as a basis for environmental regulation at an international conference in 1998: "When an activity raises threats of harm to the environment or human health, precautionary measures should be taken even if some cause-and-effect relationships are not fully established scientifically."[10]

In contrast, many U.S. laws require regulators to produce scientific evidence that a substance is harmful before it can be removed from the market. Environmental and health advocates want the United States to adopt a more precautionary approach to regulation. But critics say the precautionary principle is too vague to be a viable basis for regulation and fails to balance risks and benefits. (*See "At Issue," p. 326.*)

For example, EPA banned use of the insecticide DDT in the U.S. in 1972 because it harmed the environment, but in 2006 the World Health Organization endorsed DDT for controlling mosquito-borne malaria in developing countries.[11] Some environmental groups want DDT banned worldwide, along with other persistent organic pollutants, but other advocates — including health experts — say it should remain in use until safer alternatives are developed.[12]

As Congress, regulators, businesses and advocates debate how to protect consumers from harmful exposures, here are some issues they are considering:

Do we know enough about chemical risks?

Chemicals are central to the economy and to many products that Americans associate with modern living. They underpin a $637 billion industry in the United States and generated over $135 billion in export revenues as of 2006.[13] Innovations in chemistry have contributed to technical advances such as composite materials for vehicles, stronger adhesives, faster microprocessors for computers and recyclable plastics.

Core responsibility for regulating the massive chemical industry falls to the EPA, which is authorized under the Toxic Substances Control Act of 1976 (TSCA) to collect information about industrial chemicals from manufacturers and to limit or ban those that pose unreasonable risks.[14] Today EPA has some 82,000 chemicals in its TSCA inventory, of which about 62,000 were already in use when the law was passed. On average, more than 700 new chemicals are introduced each year.[15]

Although TSCA gives EPA the power to review chemicals already in commerce, the testing burden falls mainly on the agency rather than on manufacturers. As a result, EPA has required testing for fewer than 200 of the 62,000 chemicals that were in commerce in the 1970s. TSCA also requires the agency to show substantial evidence that a substance already in use poses an unreasonable risk in order to limit its use. EPA has banned only five chemicals or classes of chemicals under TSCA, and one of these efforts was overruled by a federal court in 1991.[16]

For new chemicals, manufacturers have to notify EPA before they start production and provide information on production volumes, expected uses and any test data that they have. However, most companies do not voluntarily test their products. Instead of testing new chemicals directly, EPA uses scientific models to compare their properties to similar existing chemicals and identify potential hazards. According to the Government Accountability Office (GAO), these reviews have led to actions that reduced risks from over 3,600 new chemicals.[17]

Critics say that the U.S. needs a broader and more proactive policy for regulating chemicals. "Our approach is barbaric and out of date. We used to be the leader decades ago, but now we're behind," says Lois Gibbs, founder and director of the Center for Health, Environment & Justice. In the late 1970s Gibbs organized homeowners in Niagara Falls, N.Y., after learning that their neighborhood had been built on top of a leaking toxic waste dump called Love Canal; after two years, the federal government relocated the families.

"The U.S. is much more science-bound than other countries. There's a presumption that we understand all of the harmful interactions from exposure to toxics, but we don't," Gibbs argues. "Industry doesn't want anything changed until there's proof beyond the shadow of a doubt that it will cause harm, but we're just not that smart."

Manufacturers say that the U.S. regulatory system is fundamentally sound. "TSCA gives EPA broad authority to collect information, order testing, prohibit new uses of a substance and label or ban substances," says Walls at the American Chemistry Council. "We can enhance it to promote more systematic review and give the public more information about what chemicals are being produced." Under a program called the High Production Volume (HPV) Challenge, launched in 1998, chemical companies are voluntarily testing about 2,800 chemicals that are produced or imported in quantities of at least 1 million pounds per year and providing the information to EPA. About 1,400 data sets have been completed to date.

But GAO, while calling the HPV Challenge "laudable," has concluded that TSCA makes it too expensive and time-consuming for EPA to review chemical hazards.[18] In order to force companies to do testing EPA has to issue a regulation, a process that can take several years. "Given the difficulties involved in requiring testing, EPA officials do not believe that TSCA provides an effective means for testing a large number of existing chemicals," GAO reported in 2006. As a solution, it recommended empowering EPA to require companies to do chemical testing and provide the data to regulators.

Both EPA and FDA also need better testing methods in order to regulate toxic substances effectively. Scientists agree that current approaches, which rely heavily on animal testing, are too slow and expensive to cover hundreds of new chemicals each year and are not well-suited to predict harm from very low doses. "We need to bring our methodologies into the 21st century by making them less animal-intensive and getting higher throughputs," or testing many substances quickly, says John Bucher, associate director of the federally funded National Toxicology Program (NTP), which studies the impact of chemicals on human health.

Current test methods typically give rats or mice large doses of chemicals, look for end points like cancer or organ damage and then extrapolate those responses from animals to humans — a complex and often controversial process. A 2007 report by the National Research Council called for a new approach focused on "toxicity pathways" — changes that occur in networks of cells due to chemical exposure and which eventually may lead to adverse health

Getty Images/China Photos

Scientists agree that animal testing is generally too slow and expensive to cover hundreds of new chemicals each year and is not well-suited to predict harm from very low doses. The National Research Council has called for testing chemicals in cell cultures, a shift it predicted would greatly reduce the need for animal testing.

effects. For example, exposure might initially cause hormone levels to change or tissues to become inflamed. The study recommended developing rapid systems for testing chemicals in cell cultures to identify toxicity pathways — a shift that it predicted would greatly reduce the need for animal testing and focus more attention on human biology and exposures.[19]

The NTP shares this vision, says Bucher. "These would be short-term assays [tests] with very simple readouts that could be run 24/7 just by punching buttons and would give a signature of biological interactions that a particular chemical would have," he says. "We hope that certain structures will be related to particular chemical classes and that that will let us make judgments about which chemicals should go through more sophisticated studies or should not be authorized for significant human exposures."

Are we commercializing nanotechnologies too quickly?

Many nanoscale materials (particles as small as 1/100,000th of the width of a human hair) have unique chemical, physical or biological characteristics that are different from larger particles of the same materials. Because they have distinctive properties such as high electrical conductivity, nanomaterials have special uses and are showing up in hundreds of consumer products, from kitchenware with anti-bacterial silver coatings to paints impregnated with silica particles that repel graffiti.

Consumer advocates worry that some of these applications could pose health risks, and that government agencies do not know enough about nanomaterials to regulate them effectively. An EPA fact sheet states the challenge bluntly: "At this early stage of the development of nanotechnology, there are few detailed studies on the effects of nanoscale materials in the body or the environment . . . it is not yet possible to make broad conclusions about which nanoscale substances may pose risks."[20]

Twenty-six federal agencies, including EPA, FDA and the CPSC, participate in the National Nanotechnology Initiative, a federal program that supports research on promising applications of nanotechnology and on environmental health and safety (EHS) issues. From fiscal 2005 through 2008, these agencies spent an estimated $180 million on research to address EHS questions.[21]

But keeping up with this fast-growing field is challenging for regulators. "I do not pretend to understand nanotechnology, and our agency does not pretend to have a grasp on this complicated subject either," CPSC Commissioner Thomas H. Moore told a Senate subcommittee in March 2007. "For fiscal year 2007, we were only able to devote $20,000 in funds to do a literature review on nanotechnology. Other agencies are asking for, and getting, millions of dollars for research in this area."[22]

Four months later an FDA task force report on regulating nanomaterials pointed out that because of their unique properties, the agency might need new testing equipment and methods to predict how they will react in body tissues.[23] Other agencies studying nanotechnologies confirm they often behave in surprising ways. "It is a daily occurrence in our labs that one of our standard assays doesn't work because of the unusual properties of these nanomaterials," said Scott E. McNeil, director of the National Cancer Institute's Nanotechnology Characterization Laboratory, at a conference last March.[24]

Some watchdog groups want to stop the marketing of nanoproducts until they are proven safe. Last May a coalition of health, environmental, and consumer groups petitioned EPA to control products containing nano-silver, which is highly effective at killing bacteria, fungi and other microorganisms. Because of this property, nano-silver has been added to garments (to kill odor), food storage containers, soaps, air purifiers and dozens of other products.

The petitioners argued that nano-silver in the environment could kill plants, benign microbes, fish and other aquatic species and might also threaten human health. They called on EPA to regulate the material as a pesticide and require comprehensive safety testing before any products containing it could be marketed.[25]

At a minimum, critics say, manufacturers should be required to label products containing nanoparticles so that consumers can choose whether or not to buy them. A study by Consumers Union found that four out of five sunscreens that claimed to be nano-free actually contained nanoparticles of titanium dioxide and/or zinc oxide, two compounds that help protect against ultraviolet radiation.[26]

"Size matters. Materials at the nanoscale should be considered new particles and have to go through new safety assessments," says Michael Hansen, senior staff scientist at Consumers Union. "Right now, it's assumed that if a material has been tested for bulk applications, it's safe. But when you reduce things to such small sizes, their behavior and surface area can change drastically. You can't assume that something safe at the macro scale is safe at the nano scale."

Some experts say that health concerns may be exaggerated. "When we started looking at them, we found that the properties of nanomaterials in products, such as particle size, often were different from what manufacturers said they were. People didn't always know what they were studying," says the NTP's Bucher. "We completely characterized the materials we were working with and then administered them to animals in ways that might mimic human exposures. Our studies suggest that some risks are lower than reports in the literature have suggested."

For example, according to Bucher, many reports predicted that titanium dioxide in sunscreen would penetrate skin readily, but the NTP concluded that won't happen unless the skin is cut or scraped. However, he cautions, this does not prove that all nanomaterials are harmless. "Every product is going to be different," he says.

Even if many nanomaterials are harmless, weak and underfunded regulatory agencies may have trouble distinguishing benign products from risky ones. Marla Felcher, an expert on marketing and consumer issues at Harvard University's Kennedy School of Government, says the Consumer Product Safety Commission is unprepared for the challenge. "CPSC is playing catch-up," she says. "More than half of the nanotechnology goods on the market come under its jurisdiction, and the funding it has to work with is a drop in the bucket."

But Felcher says the CPSC needs more than additional staffing and funding to ensure that nanomaterials in consumer products are safe. It also needs new authority to make manufacturers identify products that contain these substances and to impose mandatory safety standards for products based on new technologies, she says. (Today the agency relies on industry to develop and comply with voluntary safety standards).[27]

Hansen is hopeful the EPA will regulate nano-silver as a pesticide, but he says the FDA has so far refused to agree that nanomaterials are categorically different from their conventional counterparts. "The biggest exposures come from items that you put on or in your body and that contain free [non-bound] nanoparticles, like food ingredients and personal care products," says Hansen. "Scientific studies are saying that these materials need to be regulated."

Use of Nanotechnology Is Increasing

More than 800 consumer products containing nanomaterials were in the marketplace as of August 2008 — nearly quadruple the amount from just two years earlier. Some 60 percent of the products were related to health and fitness.

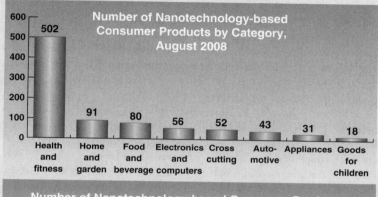

Number of Nanotechnology-based Consumer Products by Category, August 2008

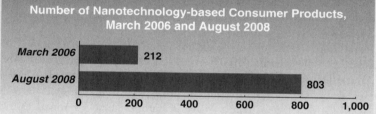

Number of Nanotechnology-based Consumer Products, March 2006 and August 2008

Source: The Project on Emerging Nanotechnologies, Woodrow Wilson International Center for Scholars

Would stricter regulations hurt manufacturers and consumers?

Regulating chemicals in consumer products more stringently would affect chemical companies and manufacturers that use those chemicals to make retail goods. Chemical toxicity testing is expensive, and many business leaders say that substitutes for widely used materials like BPA will be more expensive and could produce inferior products. Health, environmental and consumer groups respond that safer products don't always cost much more, and in any case are worth a cost premium.

Even if a chemical poses risks, only some uses may require substitutes. "Health and safety concerns about a chemical like BPA are dictated as much by how it's used as by its chemistry. I'm more concerned about using it in baby bottles than in auto parts or compact discs," says

Terry Collins, a professor of chemistry and director of the Institute for Green Science at Carnegie Mellon University.

Industry representatives say the risks of BPA have been widely studied and that removing it from food containers, as many critics urge, will be difficult. "BPA has been part of epoxy resin can linings for more than 50 years, and that's why we have canned goods with long shelf lives," says Steven Hentges, executive director of the American Chemistry Council's Polycarbonate/BPA Global Group. These linings prevent metallic flavors from migrating into food and keep acidic foods like tomatoes from corroding the cans.

"Substitutes have to be safer than what's being replaced, and no alternatives have been as thoroughly tested as BPA" says Hentges. The European Food Safety Authority concluded last July that BPA was safe in food packages. Canada banned BPA in baby bottles as a precautionary measure and declared it a toxic substance, but has not removed the material from food packaging for adults. Canadian regulators are funding studies to see whether further steps are needed to limit how much BPA is released into the environment.[28]

But other options exist. Many Americans buy juices, soups, sauces and chopped tomatoes in brick-shaped cartons, which were originally introduced in Europe. These containers, which are about 75 percent paper, 20 percent polyethylene plastic and 5 percent aluminum, typically cost more than canned goods, but foods packaged this way retain more color and flavor than canned foods because the food and the container are sterilized separately. The boxes can be recycled with milk and juice cartons.[29]

"There also are ways to can food so that it doesn't contain such high levels of BPA," says Janssen at the Natural Resources Defense Council. "Japan set voluntary standards for reducing BPA use in canned food, and now a lot of Japanese canned goods have a polyethylene layer inside that seals the epoxy resin lining so that BPA doesn't migrate into food. Industry should be thinking about more sustainable techniques instead of fighting to maintain the status quo."

Chemical companies and consumer advocates are also debating risks associated with phthalates, especially in children's toys. In 1998 a coalition of environmental and consumer groups petitioned the CPSC to ban toys that contained phthalates and were designed for children under age 6, citing studies suggesting that these materials could be toxic and the fact that young children commonly chewed on soft plastic toys. An expert panel convened by the commission found some risk from the phthalate DINP and asked toy makers to remove phthalates from toys voluntarily. According to the American Chemistry Council most companies removed phthalates from teethers, rattlers and pacifiers.[30]

The European Union also studied phthalates and imposed a temporary ban on six forms of the chemicals in toys and teething items in 1999. In 2005 it made the ban permanent, requiring manufacturers to eliminate three phthalates (DEHP, DBP and BBP) from all toys and to remove three others (DINP, DIDP and DNOP) from toys and child-care items that could be mouthed by children.[31]

EU regulators acted in response to studies that suggested, but did not prove, that exposure to phthalates could have toxic effects or cause abnormal reproductive development, especially in boys. A 2005 study in the United States reached a similar conclusion, but a U.S. government review panel found that animal studies that showed a connection were not necessarily applicable to people.[32] However, California, Washington and Vermont passed state-level bans. In 2008 Congress permanently banned DEHP, DBP and BBP from toys and set an interim ban on the other three types pending a safety review by the CPSC.[33]

Many businesses opposed the measure. "[M]anufacturers would be forced to use more expensive alternatives that may subject them to additional safety and legal liability concerns, and consumers would be exposed to products containing alternatives that have not been approved for use in children's products by any federal agency," the U.S. Chamber of Commerce wrote to Senate members early in 2008.[34]

Chemical industry representatives maintain that scientific evidence shows phthalates to be safe. "The pro-regulatory side considers phthalates guilty until proven innocent. They want to act even though the data is not conclusive," says Allen Blakey, vice president of the Vinyl Institute, a trade group for companies that manufacture vinyl and vinyl products (many of which contain phthalates to make them soft and flexible).

But toy manufacturers seem to be adapting to phthalate restrictions. Observers predicted the U.S. ban could drive some of China's small and uncompetitive toy man-ufacturers out of business, but other Chinese companies already make toys with phthalates for U.S. markets and without them for sale in the EU.[35] BASF, a major German chemical company, still produces DEHP in the United States but developed a new plasticized version called Hexamoll DINCH, which it markets to toy makers as a product "whose health safety is beyond all question" and "an ideal solution to adapting their products to the requirements of the new EU regulation."[36]

BACKGROUND

Reactive Regulation

Through the 19th century, as the United States grew from a nation of small-scale farmers into an industrial powerhouse, few standards protected people from hazard-ous materials. And even when government began to regu-late dangerous products and substances in the early 20th century, controls were almost always put in place belat-edly after scandals or disasters.

Muckraking journalist Upton Sinclair spurred passage of early consumer-protection laws with his 1906 novel *The Jungle*, which described filthy conditions in Chi-cago's meatpacking industry. Simultaneously, a series of articles in *Collier's* magazine spotlighted false claims and unsafe ingredients in so-called patent (non-prescription) medicines. Many of these concoctions were sold as cure-alls for numerous diseases but contained addictive substances like cocaine, heroin or alcohol. "[F]raud, exploited by the skillfulest [sic] of advertising bunco men, is the basis of the trade," author Samuel Hopkins Adams charged.[37]

In response Congress passed the Meatpacking Act and the first Food and Drug Act, which authorized govern-ment regulators to inspect meat processing plants and to seize products that were mislabeled or contained harmful or spoiled ingredients. However, manufacturers were not required to list all of the ingredients in foods or medicines or submit any information to the government before marketing them.

In 1933 the Food and Drug Administration proposed a complete revision of the Food and Drug Act, but Congress failed to act until 107 people in 15 states died in 1937 after taking elixir of sulfanilamide for strep infections.

A chemist had created a liquid form of sulfanilamide, a new and effective prescription medicine, by dissolving the powdered medication in diethylene glycol, a chem-ical normally used as antifreeze, which he failed to real-ize was poisonous.[38] This tragedy spurred passage of the Federal Food, Drug, and Cosmetic Act, which required new drugs to be tested for safety before marketing.

As Congress debated safety standards, the fast-growing chemical industry was inventing myriad new materials. Important chemical products developed in the 1920s and '30s included polychlorinated biphenyls (PCBs), used as coolants and lubricants; synthetic estrogens (female hormones); and organic pesticides like the mosquito-killer DDT. Nineteenth-century inventors had already discovered many basic types of plastic, and by World War II these materials were widely used in applications including cellophane, vinyl, nylon and Teflon coatings.

During this period the labor movement gained strength as workers formed unions and won the right to collective bargaining. One of their priorities was making workplaces safer. Government agencies started to regulate safety, ini-tially focusing on industries like mining and manufacturing, where workers were frequently injured by machinery, fires and explosions. Toxic exposure was also emerging as a seri-ous hazard. For example, by the 1930s manufacturers knew that workers who inhaled silica dust or asbestos had high rates of lung disease, and medical researchers were starting to connect asbestos inhalation with cancer.

Under President Franklin D. Roosevelt (1933-1945), the Labor Department worked with industry and unions to improve workplace safety, mainly through voluntary safety codes and better training programs. During the economic boom of the 1950s occupational safety became a more established professional field, but it focused on traumatic injuries such as falls or machine accidents rather than exposure to dangerous materials. In one industrial hygienist's words, it was hard to draw attention to safety issues "unless people saw the blood drip."[39]

New Guardians

As corporations shifted from wartime manufacturing to civilian products, new materials streamed into commer-cial use, including vaccines, food additives, pesticides and herbicides and an avalanche of consumer goods. Most Americans welcomed these products, but research soon showed that some were unsafe.

By the late 1950s government regulators had banned more than a dozen food additives because they caused cancer, organ damage or other toxic effects in animals. [40] In 1958 Congress adopted the Delaney Clause, which barred all food additives that had been shown to cause cancer in laboratory animals. Six years later Surgeon General Luther Terry released a report stating that smoking caused cancer, and Congress passed a law requiring cigarette packs to carry health warning labels. [41] However, the tobacco industry — which had created its own scientific arm, the Tobacco Industry Research Committee, to refute incriminating studies — argued that smoking was a personal choice, and successfully lobbied against any limits on cigarette advertising or marketing.

By this time other toxic exposures were in the news. Rachel Carson's 1962 bestseller *Silent Spring* warned that persistent organic pesticides like DDT were accumulating in the environment, harming fish and birds and contaminating food supplies. In the same year it was disclosed that thousands of babies in Asia, Africa and Europe had been born with deformed or missing limbs after their mothers took thalidomide, a new sedative that the FDA was then close to approving for sale in the United States.

In 1965 consumer activist Ralph Nader amplified pressure on the government to regulate dangerous products with his book *Unsafe at Any Speed*, which attacked U.S. automakers for refusing to include safety devices like seat belts and filling cars with confusing and distracting features. [42] The book and subsequent congressional hearings generated new safety requirements and oversight agencies for passenger cars.

The Nixon administration created other agencies to regulate industry more tightly and protect consumers, including the Environmental Protection Agency (EPA) in 1970, the Occupational Safety and Health Administration (OSHA) in 1971 and the Consumer Product Safety Commission (CPSC) in 1972. Along with the FDA, each of the new agencies had some responsibility for protecting workers and the public from toxic threats, although the scope of their powers varied.

During the 1970s federal regulators passed some important protective measures. EPA banned DDT use in 1972 because of its harmful environmental impacts. In 1979 the agency banned production of PCBs, which had been shown to cause cancer and birth defects in laboratory animals, and set timetables for phasing out their use in various industries. In 1977 the CPSC banned lead paint and its use on toys and furniture. OSHA set occupational exposure standards for many hazardous substances and developed requirements (finalized in the early 1980s) for businesses to identify and label hazardous chemicals in the workplace and tell employees how to use them safely.

Business Pushes Back

Many controls adopted in the 1970s made the environment cleaner and improved public health. But industry and conservative politicians argued that government regulators were becoming high-handed and that excessive regulations slowed economic growth. President Ronald Reagan (1981-1989) made reducing government's power a centerpiece of his administration, cutting budgets at EPA, OSHA and CPSC and appointing officials who were hostile to regulation.

Reagan also required the White House Office of Management and Budget to review proposed new rules — a policy that his successors continued — and signed an executive order directing agencies not to issue new ones unless their potential benefits to society were greater than their costs. [43] Many policy experts agreed that cost-benefit analysis was a useful tool for setting priorities, but others worried that health and environmental benefits were hard to quantify and would be undervalued.

Critics charged the Reagan administration with leading a "retreat from safety." "The agencies no longer respond to the needs of unorganized victims of technological hazards. Instead, they service the business executives and stockholders who are responsible for the hazards," wrote Joan Claybrook, president of the Nader-founded activist group Public Citizen and former head of the National Highway Traffic Safety Administration, in 1984. [44]

In this climate some of the most significant new steps were so-called right-to-know policies, which did not limit the use of risky materials but gave people more information about potential exposures. After methyl isocyanate, a deadly industrial gas, leaked from a chemical plant in Bhopal, India, in 1984 (killing some 4,000 people) and a plant in West Virginia the next year (with no deaths), Congress passed the Emergency Planning and Community Right to Know Act in 1986. The law required companies to tell EPA and state officials what hazardous chemicals were used in significant quantities at their plants and to notify emergency responders about any chemical releases.

CHRONOLOGY

1900-1960 *Government begins regulating consumer goods to protect buyers; fast-growing chemical industry produces thousands of materials that are quickly put to use.*

1906 Congress authorizes federal inspections of meatpacking plants and outlaws adulterated or mislabeled foods and drugs.

1929 Chemical companies start making polychlorinated biphenyls (PCBs).

1930 Food and Drug Administration (FDA) is established . . . . Studies show asbestos can cause cancer.

1938 After tainted medicine kills 105 people, Congress passes Food, Drug and Cosmetic Act, requiring food additives and drugs to be proven safe. . . . British scientists produce diethylstilbestrol (DES), a synthetic estrogen, which is approved to treat gynecological ailments.

1939 Swiss chemist Paul Müller discovers that the synthetic chemical DDT is an effective insect killer. Müller later wins Nobel Prize after DDT-B is widely used to protect troops from typhus and malaria during World War II.

1954 Cigarette manufacturers create Tobacco Industry Research Council in response to scientific findings of health threats from smoking.

1958 Delaney Clause to Food, Drug and Cosmetic Act bans food additives that cause cancer in animals.

1960-1980 *New agencies protect consumers and workers from hazardous substances. Studies find popular chemicals can harm health.*

1962 Rachel Carson's bestseller *Silent Spring* warns of environmental and health threats from DDT.

1964 Surgeon general declares smoking hazardous to health; warning labels are required on cigarettes.

1970 Environmental Protection Agency (EPA) is created.

1971 Occupational Safety and Health Administration (OSHA) is established. . . . DES is linked to vaginal cancer.

1972 Consumer Product Safety Commission (CPSC) is established. . . . EPA bans DDT.

1976 Toxic Substances Control Act (TSCA) authorizes EPA to regulate chemicals but exempts 62,000 substances.

1977 CPSC bans nearly all uses of lead paint, including on toys.

1980-2000 *Anti-regulatory forces challenge new, protective standards.*

1980 Supreme Court's "benzene decision" says OSHA must show significant risks before limiting a chemical's use.

1983 OSHA requires employers to show workers how to use toxic chemicals.

1986 California mandates warning labels for products with chemicals that cause cancer or birth defects.

1990 Congress requires leaded gasoline to be phased out by 1996.

1996 Food Quality Protection Act tightens standard for pesticide residues in food and requires special protection for infants and children.

1997 Study finds bisphenol A (BPA) alters reproductive development in mice.

2000-Present *Support grows for natural and organic products.*

2000 National Nanotechnology Initiative is launched.

2003 Congress approves $3.7 billion over four years for nanotech research, but only a small amount is earmarked for studying health impacts.

2007 European Union's REACH chemical regulation enters into force.

2008 Congress strengthens CPSC and bans lead and six phthalates from children's toys. . . . Canada bans BPA from baby bottles. . . . FDA committee concludes BPA is not harmful in food packaging, but the agency's review panel faults the study's methods.

Americans' Bodies Contain Over 100 Chemicals

Some Cause Cancer and Other Health Problems.

Studies indicate that all human beings alive today carry traces of many industrial chemicals in their bodies. Some of these substances enter during fetal development or infanthood, carried by maternal blood and breast milk. In addition, we inhale airborne pollutants, ingest pesticide residues and chemical additives with our food and drinking water and absorb others through our skin. Exposure can happen in the workplace, outdoors or inside homes and schools.

Some so-called "chemical body burdens" in humans are harmless, but others can cause cancer, birth defects, developmental problems and other serious health impacts. Many are still being studied. The presence of a chemical in the body does not necessarily mean it will cause harm, but scientists say chemical exposure is pervasive in modern society, and they underscore the importance of testing widely used chemicals for toxic effects.

"I find it remarkable that in this day and age one of the primary ways by which the toxic effects of chemicals are discovered is still the 'body in the morgue' method," writes epidemiologist and former Assistant Secretary of Energy David Michaels. "An industrial worker dies from some very unusual condition, and we ask why. Well, some of us ask."

For example, Michaels notes, chemical companies that make diacetyl (the main ingredient in artificial butter flavor) did not know that breathing the compound could cause lung damage until workers in popcorn factories became ill. Manufacturers had been required to test diacetyl as a food ingredient, but not as an airborne contaminant in the workplace.[1]

According to a 2005 report from the Centers for Disease Control and Prevention (CDC), well over 100 chemicals are present in Americans at detectable levels, including heavy metals like cadmium and mercury, phthalates and many pesticides. Levels of some chemicals have fallen in recent years, notably lead (which has been banned from gasoline and house paint) and substances found in secondhand cigarette smoke.[2]

Others are more worrisome. For example, the CDC found that almost 6 percent of women of childbearing age had blood levels of mercury that were borderline dangerous. Mercury is a potent neurotoxin that can cause birth defects, nervous system damage and other harmful effects. It is emitted into the air from sources including coal-burning power plants and incinerators, then falls back to the surface and concentrates in the food chain. Humans are exposed mainly by eating fish that contain high amounts of mercury.

Another major step in 1986 was the passage in California of Proposition 65, a ballot initiative that directed the state to publish an annual list of chemicals used in California that were known to cause cancer, birth defects or other reproductive harms. Businesses had to warn people before exposing them to significant risks from listed chemicals — for example, by putting warning labels on processed food or signs in workplaces where listed substances were used.[45]

The Clinton administration (1993-2001) was more receptive to new health and safety regulations than its predecessors. FDA Commissioner David Kessler declared cigarettes to be "drug delivery devices," an acknowledgment that nicotine was addictive, and called for limiting marketing and sales to young people. In 1996 Clinton signed the Food Quality Protection Act, which tightened standards for pesticide residues in foods and required EPA to consider children's higher sensitivity to these chemicals when it set tolerance levels.

In 1998 the administration called for chemical companies to perform voluntary toxicity testing on chemicals in use that had not been tested — and threatened to require it if industry did not comply. Subsequently the EPA, the chemical industry and the advocacy group Environmental Defense announced the High Production Volume (HPV) Challenge, which aimed to complete toxicity testing by 2004 on about 2,800 industrial chemicals made or imported into the U.S. in large quantities.

Along with cancer and birth defects, Americans started to hear in the 1990s about so-called endocrine disruptors — chemicals that interfered with hormones responsible for regulating biological processes throughout the body, such as brain growth and sexual development. Scientists were finding evidence that endocrine disruptors were causing reproductive abnormalities, population declines and other negative impacts in

Another 2005 study commissioned by two advocacy organizations, Commonweal and the Environmental Working Group, tested umbilical cord blood from 10 babies born in U.S. hospitals during the previous year. Researchers found an average of 200 industrial chemicals and pollutants in the samples, including mercury, environmental pollutants known as dioxins and pesticides. This study showed that pollutants cross the placenta from mother to fetus as infants grow in utero, exposing the gestating infants to a complex mixture of chemicals during critical months of development.[3]

Living far from industrial sources does not necessarily make people safer from exposures. Indigenous peoples in the Arctic have some of the highest body concentrations of mercury, PCBs and other pollutants of any region on the planet, thanks to global wind patterns and ocean currents that carry pollutants to the poles. Inuit, Aleut, and other native people in Greenland, Alaska and Canada eat large quantities of locally caught meat and fish, which contain high concentrations of chemicals.[4]

Recent studies show that concentrations of some toxins in Arctic food animals are stabilizing, thanks to international agreements limiting use of some of the most hazardous chemicals.[5] However, toxic chemicals remain a major threat to Arctic indigenous peoples' traditional way of life — an ironic fate for people who neither produce nor use most of these products.

An Inuit woman in Iqaluit, Nunavut, Canada, dries a caribou skin. Toxic chemicals remain a major threat to traditional ways of life in the Arctic.

---

[1] David Michaels, *Doubt Is Their Product: How Industry's Assault on Science Threatens Your Health* (Oxford University Press, 2008), p. 247.

[2] "Third National Report on Human Exposure to Environmental Chemicals," 2005, www.cdc.gov/exposurereport/report.htm, and "Spotlight on Mercury," both Centers for Disease Control and Prevention, www.cdc.gov/ExposureReport/pdf/factsheet_mercury .pdf.

[3] "Body Burden — The Pollution in Newborns," *Environmental Working Group*, July 14, 2005, http://archive.ewg.org/reports/ bodyburden2/contentindex.php.

[4] Marla Cone, *Silent Snow: The Slow Poisoning of the Arctic* (2006).

[5] "Toxic Chemical Levels Finally Dropping in Arctic Food Animals, New Study Shows," The Canadian Press news agency, July 14, 2008.

---

wildlife. Some studies linked pesticides that mimicked estrogen, the female sex hormone, with increased risk of breast cancer.

Other researchers were alarmed by falling human sperm counts. "Every man sitting in this room today is half the man his grandfather was," University of Florida zoologist Louis Guillette told a Senate committee in 1993. "Are our children going to be half the men we are?" Three years later, the best-selling book *Our Stolen Future* argued that endocrine disruptors posed pervasive health risks but that federal controls on toxic chemicals were overly focused on detecting and controlling cancer risks. "The assumptions about toxicity and disease that have framed our thinking for the past three decades are inappropriate and act as obstacles to understanding a different kind of damage," the authors contended.[46]

New Worries

Under President George W. Bush (2001-2009), momentum once again swung from strong regulation to voluntary compliance strategies in which companies agreed to police themselves. With pro-business officials in charge at many regulatory agencies and limited budgets, the pace of federal regulation dropped sharply. Rulemaking fell by more than 50 percent at FDA and 57 percent at EPA between 2001 and 2008 compared with those agencies' records during the Clinton administration. OSHA withdrew more than a dozen regulations that had been proposed under Clinton and delayed taking action on silica dust after identifying it as a workplace health threat.[47]

Conservative advocates generally supported the shift to deregulation, arguing that excessive health and safety regulations were burdens on the economy and often were not the most effective way to protect public health or the

The Polluted Arctic

Atmospheric and ocean currents carry persistent organic pollutants around the world far from their sources and concentrate them in some regions, notably the Arctic, where they are a threat to indigenous peoples and wildlife. Even relatively "clean" air from non-industrial areas contains low levels of pesticides and other chemicals.

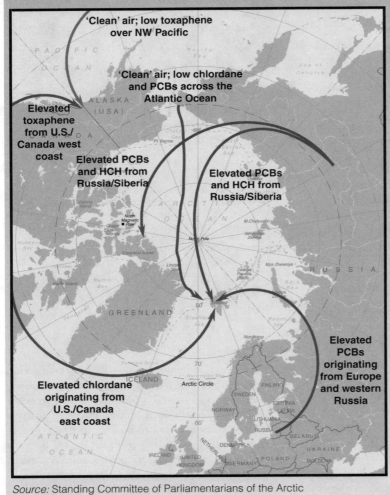

Source: Standing Committee of Parliamentarians of the Arctic

is decreasingly relevant to the local environmental problems that remain to be tackled."[48]

At the same time, however, consumers and even some large industries were asking federal agencies for more regulation. From 2007 through mid-2008 a string of product scares made headlines, including U.S.-grown spinach carrying hazardous bacteria, imported pet food and seafood adulterated with chemicals, and recalls of toys found to contain lead paint.[49] Many of these products, including the tainted pet food and toys, came from China, while contaminated fish was shipped from China and other countries in Asia and from Latin America. In May 2008 FDA Commissioner Andrew von Eschenbach asked Congress for $275 million in immediate funding to improve oversight of drugs, medical products and imported food.[50]

Two months later Congress passed the Consumer Product Safety Act of 2008, which overhauled the CPSC and increased its staffing, required toys and other children's products to be tested for safety before they entered the market and banned lead and several types of phthalates from children's products. "This reform is much needed, long overdue and necessary to ensure that CPSC can successfully ensure the safety of consumer products," said Rachel Weintraub, director of product safety and senior counsel at the Consumer Federation of America.

environment. "Regulations unquestionably force the issue, but usually at a very high cost to the economy and to property rights," wrote American Enterprise Institute analyst Steven Hayward in 2008. "This kind of bureaucratic environmentalism has about played itself out, and

CURRENT SITUATION
FDA and BPA

As debate continues over potential health risks from BPA, the FDA is at the center of controversy. Last August the agency released a draft assessment concluding that BPA in food packaging did not pose a health risk. But an advisory panel that reviewed the draft report found a number of flaws, such as omitting studies suggesting BPA could have harmful effects, using too few infant formula samples and not considering cumulative exposures. The reviewers concluded that "the Margins of Safety defined by the FDA as 'adequate' are, in fact, inadequate."[51] (A margin of safety is the gap between the lowest dose of BPA expected to cause harm and the actual exposure that scientists expect to occur.)

The FDA is reviewing these arguments and has pledged to provide a response by this February. "FDA agrees that, due to the uncertainties raised in some studies relating to the Potential effects of low-dose exposure to bisphenol A, additional research would be valuable," says agency spokesperson Michael Herndon. "[The agency] is already moving forward with planned research to address the potential low-dose effects of bisphenol A, and we will carefully evaluate the findings of these studies."

Critics argue the FDA has deliberately downplayed low-dose exposures to avoid having to issue new regulations. "We're replaying what happened with lead regulation," says Carnegie Mellon chemistry Professor Terry Collins. "Trade associations fought against banning lead from house paint and gasoline for 70 years by beating up doctors who said lead was bad for children and funding studies that only looked at high doses. EPA chose for years not to look at risks from ultra-low doses, and FDA is doing the same thing now. It's very confusing to the public, and these impacts are showing up across the population."

The National Toxicology Program's Bucher agrees that the FDA needs new methods to evaluate BPA. "The academic studies that found effects at low doses assessed exposures to very fine degrees," he says. "FDA's guidelines for industry studies don't require such detail, and they're just not adequate to pick up subtle changes that can occur from low-dose exposures, such as behavior differences between male and female mouse pups."

The NTP is still trying to answer important questions about BPA, says Bucher: "We know what doses animals receive in studies, but we don't know much about where it goes and how much of it reaches different tissues, or how quickly it's eliminated from the body. It's not eliminated as quickly in young animals as in older ones, and we think that's true in humans as well." He expects that the NTP will soon initiate a study to see whether prenatal exposure to BPA can lead to cancer. "Earlier studies started dosing in young adults, but clearly the most sensitive periods are earlier than that," Bucher adds.

Activist Congress

Although research is ongoing, some members of Congress have already called for new limits on chemicals in consumer products, starting with a ban on BPA in food and beverage containers. Several legislators cited a November 2008 study by the *Milwaukee Journal Sentinel* that found plastic products labeled as "microwave safe" leached potentially harmful doses of BPA when they were heated. "Parents always err on the side of caution when it comes to their kids' health. We think the law should do the same," said Sen. Charles E. Schumer, D-NY.[52] He introduced legislation in 2008 that would have banned BPA from products designed for children ages 7 and under, while Rep. Edward J. Markey, D-Mass., introduced a House bill that would have eliminated BPA from all food and beverage packaging.[53]

At least 13 states are also considering BPA bans. However, one such proposal failed in California in August 2008. Food processors, chemical manufacturers and packaging companies opposed the bill, which would have banned use of BPA in products for children ages 3 and under. "California's legislators made the right decision for consumers," said the American Chemistry Council's Hentges.

Another 2008 congressional bill that is likely to be reintroduced, the Kid-Safe Chemicals Act, would require more sweeping reforms to the Toxic Substances Control Act and the chemical-testing process.[54] The measure seeks to "eliminate the exposure of all children, workers, consumers and sensitive subgroups to harmful chemicals distributed in commerce by calendar year 2020." The measure would:

- require industry to demonstrate that chemicals in use are safe;
- authorize EPA to require additional testing for health effects at low doses and for nanomaterials;
- expand analysis by the Centers for Disease Control and Prevention (CDC) of chemical residues in humans; and
- provide new funds to promote safer alternatives.

"It is critical that we modernize our nation's chemical safety laws," said Rep. Henry A. Waxman, D-Calif., a sponsor of the House bill and the new chair of the Energy

European Regulators Take 'Precautionary' Approach

Chemical companies must show products are safe.

Chemicals are big business in Europe as well as in the United States, but the European Union (EU) has taken a sharply different approach to regulating chemical risks. In 2007 the EU's new REACH policy (Registration, Evaluation, Authorization, and Restriction of Chemicals) went into effect. In the United States, regulators must show that chemicals pose risks to human health or the environment before they can limit their production or use. But REACH takes essentially the opposite approach: Companies must show that chemicals will not harm human health or the environment before they can be marketed.

During an 11-year phase-in period, businesses that produce or import any chemical into the EU in quantities greater than one metric ton per year will have to register it with the new European Chemicals Agency and submit information about its physical and chemical properties, how it will be made, how to use it safely and how it affects human health and the environment. More detailed information is required for chemicals that are produced in larger volumes. EU officials estimate that about 30,000 chemicals now in use will be subject to REACH.[1]

Manufacturers of chemicals deemed to pose especially high risks — such as those that cause cancer, birth defects or endocrine disruption or that persist and are toxic in the environment — will have to apply to the European Commission for authorization. They will have to show that it is not technically or economically feasible to use safer substitutes, and that the risks from using the chemical can be controlled. REACH allows regulators to ban or restrict the use of chemicals that pose unacceptable risks to human health or the environment and limits the amount of health-related data that manufacturers can shield as proprietary information.

Many U.S. health and environmental advocates say REACH is a better model for regulating hazardous substances than the Toxic Substances Control Act (TSCA), and that the U.S. should emulate Europe by moving in a more precautionary direction. "TSCA is really ineffective and needs to be updated," says Sarah Janssen, a scientist at the Natural Resources Defense Council. "It limits EPA's ability to request toxicity information from manufacturers; there are thousands of chemicals on the market now without

and Commerce Committee. "The Kid-Safe Chemicals Act will deliver what its name implies — a non-toxic environment for our children."

Another chemical issue on Congress's agenda is reauthorization of the National Nanotechnology Initiative (NNI), which coordinates nanotechnology research by federal agencies. The House passed a reauthorization bill with little controversy in 2008, but nanotechnology may face a bumpier ride in the Senate. In December 2008 the National Research Council released a review of NNI's research plan for studying potential health and environmental risks of nanotechnologies. While the study did not address whether current uses of nanomaterials posed risks to the public, it found that NNI did not have an adequate strategy for answering that question.

NNI's plan "does not describe a clear strategy for nano-risk research. It lacks input from a diverse stakeholder group,

and it lacks essential elements, such as a vision and a clear set of objectives, a comprehensive assessment of the state of the science, a plan or road map that describes how research progress will be measured, and the estimated resources required to conduct such research," the NRC review stated.[55]

Making Exceptions

Banning products does not always end debate over them. Bans on phthalates in children's products under the 2008 Consumer Product Safety Improvement Act were scheduled to start on Feb. 10, 2009, but lawyers representing toy wholesalers and retailers wrote to the CPSC in late 2008 that the ban would impose "significant financial hardship" on their clients — especially if they were left with useless products after the deadline passed.

In response CPSC General Counsel Cheryl Falvey held that the law did not contain a "clear statement of

toxicity information; and there's no requirement for companies to notify EPA if they increase production or start using chemicals in new ways. REACH isn't perfect, but it's definitely a lot better than what we have, which is basically a free-for-all."

U.S. chemical companies and the Bush administration lobbied hard against REACH, arguing that it was too complex and expensive, posed a barrier to foreign exporters outside of Europe and could cause American workers to lose their jobs. C. Boyden Gray, the U.S. ambassador to the EU, said REACH would "be hell for American multinationals. . . . Our position is if we don't stop it, it will multiply like kudzu."[2] Now, however, U.S. manufacturers are reformulating their products for sale in Europe and preparing to register them.

Michael Walls, managing director at the American Chemistry Council, the main U.S. chemical industry trade group, acknowledges that REACH breaks some valuable ground. "It's raised the issue of how we assure safe use, and it's promoted dialogue about how certain chemicals are used in sectors like electronics, automobiles and aerospace," he says. But, Walls argues, REACH does not pay enough attention to how chemicals are used, which is one determinant of how risky they are. "There are some opportunities to consider specific uses, but chemicals are identified for regulation specifically based on hazardous characteristics, and we don't think that's the way to prioritize," he says.

No regulatory decisions have been made under REACH yet. A preregistration phase for existing chemicals ended last November, and regulators now are considering which substances should require authorization before they can be used. By December 2010 companies must submit data on high-volume chemicals (those produced in quantities over 1,000 metric tons per year) and highly toxic chemicals produced in smaller quantities. "REACH is still untested and unproven, and we have concerns about whether some of its provisions are workable," says Walls.

But some activists already would like to make REACH even more stringent. For example, Janssen argues the system does not pay enough attention to endocrine-disrupting chemicals. "Some chemicals aren't produced in very big volumes, but they have serious impacts at very low volumes," she says. "Hormones work in the parts-per-billion to parts-per-trillion range in your body — very small doses have really big impacts." And REACH does not explicitly cover nanomaterials, although manufacturers who want to use an existing chemical substance at the nano level will have to supply additional information on the nanoform's specific properties and describe measures to minimize risks from them.[3]

[1] "Chemical Regulation: Comparison of U.S. and Recently Enacted European Union Approaches to Protect Against the Risks of Toxic Chemicals," U.S. Government Accountability Office, August 2007.

[2] Mark Schapiro, *Exposed: The Toxic Chemistry of Everyday Products and What's at Stake for American Power* (2007), p. 253.

[3] "REACH and Nanomaterials," European Commission, http://ec.europa.eu/enterprise/reach/reach/more_info/nanomaterials/index_en.htm.

unambiguous intent" to apply the ban to existing toys, so manufacturers could keep selling items in their inventories that contained the proscribed materials.[56] Two advocacy groups, the Natural Resources Defense Council and Public Citizen, filed suit against the agency, arguing that all items containing the phthalates in question should be removed from shelves by the February 2009 deadline. "The CPSC decision will generate and prolong exposure to known hormone-disrupting chemicals. . . . There is no way for [consumers] to know whether products on store shelves after the ban date contain phthalates or not," the groups argued.[57]

Many toy vendors and manufacturers also say the law's Feb. 10 deadline for applying tough, new lead levels could cost them heavily. By that date toys may contain no more than 600 parts per million by weight of lead, a trace amount that will ratchet further down over time. Falvey ruled in

November that unlike the phthalate ban, the new lead ban (which was worded differently in the law) did apply to existing toys. But some toy company owners said that testing their entire inventories for lead would be extremely expensive, and that retailers might send entire shipments back if there were worries about whether some items met the standard.[58] According to the CDC, only certified laboratories can test toys accurately for lead.[59]

Another proposed ban, on polyvinyl chloride (PVC) plastics, passed through the California Assembly and two Senate committees last year but then stalled in the Senate Appropriations Committee. PVC is used for many applications, including water pipes, medical tubing and numerous types of packaging. But critics like the Center for Health, Environment, and Justice (CHEJ) call PVC "poison plastic" because it can release chemicals such as phthalates and dioxins (a family of persistent, toxic, chlorinated hydrocarbon

Does the precautionary principle make us safer?

YES
Wendy E. Wagner
Professor of Law, University of Texas

Written for *CQ Researcher*, January 2009

The regulation of chemicals in the United States epitomizes what can go wrong when a legal system adopts a non-precautionary approach. Under the Toxic Substances Control Act (TSCA), manufacturers are not required to do any pre- or post-market testing on their chemicals unless mandated by the Environmental Protection Agency. At the same time, there are few to no rewards under the act for producing safer or better-tested chemicals, at least with regard to latent hazards.

In fact, chemical manufacturers that do voluntarily test their chemicals may put themselves at a competitive disadvantage: They not only produce evidence that can be used against them by regulators and plaintiffs' attorneys but also dedicate resources to testing that are unlikely to be recouped in sales — either because the testing reveals unwelcome risks or because the positive results cannot be validated readily by consumers or investors.

The TSCA's non-precautionary approach is partly to blame for the resulting ignorance about the long-term safety of most chemicals and for the lack of incentives to develop safer, "greener" chemicals. Over the 30-year-plus history of the legislation, EPA has required testing for fewer than 200 chemicals. Most of the remaining 75,000 chemicals produced during that period are essentially unrestricted and unreviewed with regard to their health and environmental impacts. While such a counterproductive regulatory scheme would seem at first blush a perfect candidate for public-spirited reform, the highest-stakes participants in toxics policy are the chemical manufacturers, who not surprisingly have become well-organized and steadfast in their opposition to reform.

Fortunately, the European Union's REACH directive will produce valuable toxicity information on chemicals, whether U.S. manufacturers want it or not. Through its mandatory testing requirements, REACH (registration, evaluation, authorization and restriction of chemicals) may also generate incentives for safer chemical substitutes.

In the United States, the precautionary features of REACH could be supplemented by creating additional rewards for producing safer chemicals. For example, EPA could preside over petitions filed by manufacturers seeking regulatory certification of a chemical's superiority relative to its competitors. Pitting manufacturers against one another through such adjudication will help draw out information on the toxicity of chemicals and reward greener chemical companies, while at the same time undermining the unified resistance of chemical manufacturers to modifications in TSCA's non-precautionary approach.

NO
Gary Marchant
Professor of Law, Arizona State University

Written for *CQ Researcher*, January 2009

The precautionary principle (PP) attempts to address a serious problem: How should we deal with uncertain risks? Bisphenol A, Teflon, thimerosal in vaccines, melamine in baby formula and phthalates in fire retardants are just some of the uncertain risks on the front pages of newspapers today. Which ones should we restrict now, and which should we just study more before taking action?

Unfortunately, the PP fails to provide a coherent or useful answer to this critical question. The problem, as H. L. Mencken once noted: "[t]here is always an easy solution to every human problem — neat, plausible, and wrong."

Since originating in Europe approximately 40 years ago, the PP is now binding law in Europe, Canada, Australia and several Asian nations, has been incorporated in over 60 international treaties and has been adopted by several U.S. cities. Yet, the PP is problematic, especially when enacted as a binding legal rule. First, there is no standard or official definition of "the" precautionary principle, and dozens of unofficial versions exist. Which version applies will make a huge difference in many decisions.

Second, available interpretations of the PP offer no clear guidance on key questions, such as what manufacturers must do to satisfy the PP and how costs are factored in. Without answering these fundamental questions, the PP opens the door to arbitrary decisions motivated by political bias, protectionism and other inappropriate motives, rather than objective scientific evidence of risk.

Thus, relying on the PP, Norway banned Kellogg's Corn Flakes because the added vitamins could theoretically harm some ultra-susceptible person. France banned Red Bull energy drinks because the caffeine might harm pregnant women (but did not ban coffee or wine) and Denmark banned cranberry fruit drinks because vitamin C might harm some people.

More tragically, Zambia cited the PP to deny U.S. food aid to its starving population because of the possible presence of genetically modified corn (which Americans routinely eat with no apparent consequences). The European Union even used the PP to justify governmental subsidization of the coal industry, even though coal is not generally perceived as the most environmentally friendly energy source. With the PP, however, no further explanation is needed.

Finally, the PP fails to consider that many new technologies, such as biotechnology and nanotechnology, offer the promise of enormous benefits, including health and environmental gains. By failing to consider these effects, the PP fails its own test for seeking to prohibit dangerous innovations.

chemicals) during its life cycle, and its production exposes workers to other hazardous materials.

Debate over the California bill showed the difficulty of making up-or-down decisions about substances that have many uses but also pose risks. As the bill moved through various committees, legislators exempted a number of products from the ban, including medical devices, packaging for medications and containers for petroleum products. "It's easy for attackers to dismiss PVC, but not so easy for the marketplace," says the Vinyl Institute's Blakey.

Many large manufacturers and retailers have adopted policies to phase out PVC in products or packaging, including Mattel, Nike, Sony, Target, Wal-Mart, K-Mart and Sears. But Blakey calls these steps responses to political pressure and argues that PVC products are safe. Retailers, he says, "are misinformed and pressured. They don't have a lot of staff to verify critiques, and they want the issue to go away."

Activists don't deny that they're pushing companies to drop PVC, but they say safer alternatives are available. "There are some substances that don't have substitutes, so we have to use them carefully. But there are all kinds of substitutes for PVC," says CHEJ President Lois Gibbs. The center published a guide in 2008 that lists dozens of sources for toys, clothing, mattresses and other goods made without PVC. (However, as the guide notes, the center does not endorse any of the listed substitute products, manufacturers, or retailers.)[60]

The Obama Administration

Many environmentalists are optimistic about what the newly inaugurated President Barack Obama will do about toxic chemicals. Obama has embraced green issues during his campaign and since his election. Although the economic meltdown undoubtedly will force Obama to pare down his campaign wish list, his transition team has been examining new environmental policies that could be adopted quickly, including some Clinton-era initiatives that could be resurrected.

During his inaugural speech on Jan. 20, Obama said he would "restore science to its rightful place" and has vowed to listen more closely to scientific advisers and environmental experts, whose advice the Bush administration often ignored or overruled. "I think we are in store for something new," said William Reilly, who led the Environmental Protection Agency under President George

H. W. Bush. "His pledge to follow the science will be reassuring to a lot of people, including those who fear the regulators are going to run amok."[61]

Within hours after Obama's inauguration, his Chief of Staff Rahm Emmanuel ordered a halt on all work on unfinished Bush administration regulations until they can be reviewed by the new team. Bush issued 100 new rules after Obama was elected in November, including one that President Obama strenuously opposes, which would make it much harder for the government to regulate toxic substances and hazardous chemicals in the workplace.[62]

Earlier, Obama and four other senators had proposed a measure to block the new rule and wrote a letter urging the department to scrap it, saying it would "create serious obstacles to protecting workers from health hazards on the job."[63]

The administration probably will also reconsider a Jan. 15 EPA health advisory urging Americans not to drink water with more than 0.4 parts per billion (ppb) of perfluorooctanoic acid (PFOA) — a toxic chemical linked to cancer, liver damage and birth defects that is used to make Teflon and other non-stick coatings.[64]

Some scientists have urged limits as low as 0.02 parts per billion of PFOA, and, in fact, his pick to lead the EPA, New Jersey Environmental Protection Commissioner Lisa Jackson, recommended a level of 0.04 parts per billion in her state — 10 times stricter than the new federal limit.

Richard Wiles, executive director of the Environmental Working Group — a nonprofit organization that has pushed for stricter regulation of PFOA — said the EPA's new advisory was "essentially legalizing unsafe exposure levels. Nobody should have to drink a cancer-causing Teflon chemical in their water."[65]

OUTLOOK

Green Chemistry

The task of regulating the chemical industry's constant stream of new products for health and safety risks can seem hopelessly daunting. But some experts see a way: green chemistry, which seeks to design chemicals and chemical processes with reduced environmental impacts.[66]

Since the mid-1990s, green chemistry has developed into an active research field. The EPA provides grants, awards and fellowships for green chemistry achievements, and the American Chemical Society's Green Chemistry Institute works to advance green principles across all fields of chemical research. About a dozen U.S. universities offer green chemistry programs, and major corporations like GE and BASF are investing billions of dollars in green applications, such as alternative energy systems.

Winners of the EPA's green chemistry awards for 2008 included Battelle, which developed bio-based resins and toners for office copiers and printers. Made from soy and corn feedstocks instead of petroleum products, the inks are easier to remove from paper than conventional toner, which reduces the amount of energy needed to recycle waste paper. Another winner, Nalco, designed technology to monitor the water that circulates through many building cooling systems. The Nalco system adds chemicals to keep cooling water clean only when needed, saving water and energy and reducing the quantity of chemicals in discharged cooling water.[67]

Although the field is growing rapidly, Carnegie Mellon Professor Collins says government leadership is needed. "Federal investment in green chemistry is almost nonexistent, and we desperately need it," he says. "We need to prioritize hazards and figure out how to design against them." Collins recently invented an environmentally friendly catalyst that can break down harmful pollutants into less-toxic substances.[68]

The Green Chemistry Research and Development Act, which was passed by the House in 2007 and introduced in the Senate, would provide $188 million over three years for agencies to support research, development, education and training in green chemistry.

"Modern science keeps giving us new warnings about many of the chemicals we use every day, from home cleaning products to the food we put on our family's table," said Sen. John Kerry, D-Mass., a cosponsor of the Senate bill. "It's time for Washington to respond by helping to build a whole, new chemistry industry that's on a mission to make America greener."

Reducing serious risks is key, says Collins. "Green chemistry could exist without focusing on hazardous products, and it would probably do all kinds of nice little things. But to be authentic, it has to deal with hazards."

NOTES

1. "All About Eastman Tritan Copolyester," www.eastman.com/company/news_center/News_archive/2007.

2. "Endocrine Disruptors," National Institute of Environmental Health Sciences, February 2007.

3. "Timeline: BPA from Invention to Phase-Out," Environmental Working Group, April 22, 2008, www.ewg.org/node/26291/print.

4. "Bisphenol A (BPA)," National Toxicology Program, September 2008, www.niehs.nih.gov/health/docs/bpa-factsheet.pdf.

5. "Eastman Expanding Tritan Copolyester Capacity," Reuters, March 13, 2008. For background, see Jennifer Weeks, "Buying Green," *CQ Researcher*, Feb. 29, 2008, pp. 193-216.

6. "A Poison Kiss: The Problem of Lead in Lipstick," Campaign for Safe Cosmetics, October 2007, www.safecosmetics.org/docUp-loads/A%20Poison%20Kiss.pdf.

7. For background, see Marcia Clemmitt, "Preventing Cancer," *CQ Researcher*, Jan. 9, 2009, pp. 25-48.

8. Testimony of Andrew D. Maynard before Committee on Science and Technology, U.S. House of Representatives, April 16, 2008, p. 5.

9. For background, see Brian Beary, "The New Europe," *CQ Global Researcher*, August 2007, pp. 181-210, and Kenneth Jost, "Future of the European Union," *CQ Researcher*, Oct. 28, 2005, pp. 909-932.

10. "Wingspread Statement on the Precautionary Principle," www.sehn.org/ppfaqs.html.

11. "WHO gives indoor use of DDT a clean bill of health for controlling malaria," World Health Organization, Sept. 15, 2006.

12. "Alternatives to DDT on International Radar," United Nations Environment Programme, November 2008.

13. "The Business of Chemistry," American Chemistry Council, August 2007.

14. Exceptions include pesticides, which EPA regulates under a separate law, and food additives, drugs, and cosmetics, which are controlled by the Food and Drug Administration.

15. "Chemical Regulation: Actions Are Needed to Improve the Effectiveness of EPA's Chemical Review Program," U.S. Government Accountability Office, Aug. 2, 2006, p. 1.

16. The five chemicals are PCBs, chlorofluorocarbons, dioxin, asbestos, and hexavalent chromium for use as a water treatment chemical. EPA's decision banning asbestos was reversed in *Corrosion Proof Fittings v. EPA*, 947 F. 2d 1201 (1991).

17. GAO, *op. cit.*, p. 3.

18. "Toxic Substances Control Act: Legislative Changes Could Make the Act More Effective," Sept. 26, 1994; "Chemical Regulation: Options Exist to Improve EPA's Ability to Assess Health Risks and Manage Its Chemical Review Program," June 1, 2005; and "Chemical Regulation: Actions Are Needed to Improve the Effectiveness of EPA's Chemical Review Program," Aug. 2, 2006, all U.S. Government Accountability Office.

19. "Toxicity Testing in the 21st Century: A Vision and a Strategy," National Research Council (2007), pp. 48-52.

20. "Fact Sheet for Nanotechnology Under the Toxic Substances Control Act," U.S. Environmental Protection Agency, www.epa.gov/oppt/nano/nano-facts.htm.

21. E. Clayton Teague, Director, National Nanotechnology Coordination Office, testimony before House Subcommittee on Research and Science Education, Oct. 31, 2007, pp. 1-4.

22. Thomas H. Moore, Commissioner, Consumer Product Safety Commission, testimony before Senate Commerce Subcommittee on Consumer Affairs, Insurance, and Automotive Safety, March 21, 2007, p. 7.

23. "Nanotechnology: A Report of the U.S. Food and Drug Administration Nanotechnology Task Force," July 25, 2007, pp. 12-15.

24. David J. Hanson, "FDA Confronts Nanotechnology," *Chemical & Engineering News*, March 17, 2008.

25. Online at www.nanoaction.org/nanoaction/doc/CTA_nano-silver%20petition__final_ 5_1_08 .pdf.

26. "No-Nano Sunscreens?" *Consumer Reports*, December 2008.

27. E. Marla Felcher, "The Consumer Product Safety Commission and Nanotechnology," *PEN 14*, Project on Emerging Nanotechnologies, August 2008.

28. "Baby Bottle Chemical Levels Safe, EU Agency Says," Reuters, July 23, 2008; "Health Canada Responds to Concerns Raised About Bisphenol A in Canned Food," Health Canada, May 29, 2008; "Canada Declares BPA a Health Hazard," *USA Today*, Oct. 18, 2008.

29. Kate Murphy, "Business: Thinking Outside the Can," *The New York Times*, March 14, 2004; "Frequently Asked Questions," Hain Celestial Canada, www.hain-celestial.ca/index.php/faq/.

30. For a chronology see "Phthalates and Children's Toys," American Chemistry Council, Phthalate Information Center, www.phthalates.org/yourhealth/childrens_toys.asp.

31. "New EU Phthalates Directive Finalised," *Intertek Labtest*, July 2005.

32. Jocelyn Kaiser, "Panel Finds No Proof That Phthalates Harm Infant Reproductive Systems," *Science*, Oct. 21, 2005.

33. "Congress Passes Consumer Product Safety Improvement Act," *Beveridge & Diamond*, July 31, 2008.

34. Letter online at www.uschamber.com/issues/letters/2008/080304_phthalate_ban.htm.

35. Bohan Loh and Judith Wang, "U.S. Ban To Shake up China Toy Sector," *ICIS News*, July 31, 2008; Mark Schapiro, *Exposed: The Toxic Chemistry of Everyday Products and What's at Stake for American Power* (2007), pp. 56-57.

36. "A Plasticizer for Sensitive Applications," *Science Around Us*, BASF, June 2007.

37. Samuel Hopkins Adams, "The Great American Fraud: Articles on the Nostrum Evil and Quacks," Reprinted from *Collier's Weekly* (Collier, 1905), p. 3.

38. "Taste of Raspberries, Taste of Death: The 1937 Elixir Sulfanilamide Incident," *FDA Consumer Magazine*, U.S. Food and Drug Administration, June 1981.

39. Gregg LaBar, "Seven Decades of Safety: Good Times Take Their Toll," *EHS Today*, Oct. 1, 2008.

40. "Food Additives," Center for Science in the Public Interest, www.cspinet.org/reports/chemcuisine .htm#Food%20additive.

41. "The Reports of the Surgeon General," National Library of Medicine, http://profiles.nlm.nih.gov/ NN/Views/Exhibit/narrative/smoking. html.

42. Ralph Nader, *Unsafe at Any Speed: The Designed-In Dangers of the American Automobile* (1965).

43. Philip Shabecoff, "Reagan Order on Cost-Benefit Analysis Stirs Economic and Political Debate," *The New York Times*, Nov. 7, 1981.

44. Joan Claybrook *et al.*, *Retreat From Safety: Reagan's Attack on America's Health* (1984), p. xi.

45. "Proposition 65 in Plain Language," California Office of Environmental Health Hazard Assessment, www .oehha.org/prop65/background/p65plain.html.

46. Theo Colborn, Dianne Dumanoski and John Peterson Myers, *Our Stolen Future: Are We Threatening Our Fertility, Intelligence, and Survival?* (1996).

47. Stephen Labaton, "OSHA Leaves Worker Safety in Hands of Industry," *The New York Times*, April 25, 2007.

48. Steven Hayward, "Happy Earth Day," *Human Events Online*, April 22, 2008.

49. For background see Jennifer Weeks, "Fish Farming," *CQ Researcher*, July 27, 2007, pp. 625-648, and Peter Katel, "Consumer Safety," *CQ Researcher*, Oct. 12, 2007, pp. 841-864.

50. Gardiner Harris, "F.D.A. Chief Writes Congress for Money," *New York*, May 14, 2008.

51. "Scientific Peer-Review of the Draft Assessment of Bisphenol A for Use in Food Contact Applications," U.S. Food and Drug Administration Science Board Subcommittee on Bisphenol A, Oct. 31, 2008, p. 4.

52. Meg Kissinger, "Lawmakers to Seek Ban on BPA," *Milwaukee Journal Sentinel*, Nov. 17, 2008.

53. S. 2928, introduced April 29, 2008, and H.R. 6228, introduced June 10, 2008.

54. S. 3040 and H.R. 6100, both introduced May 20, 2008.

55. National Research Council, *Review of Federal Strategy for Nanotechnology-Related Environmental, Health, and Safety Research* (2008), prepublication version, p. 6.

56. The letter and CPSC advisory opinion are online at www.cpsc.gov/LIBRARY/FOIA/advisory/320.pdf.

57. The complaint is online at http://docs.nrdc.org/ health/files/hea_08120401a.pdf.

58. Melanie Trottman, "Vendors Urge Relaxed Lead-Safety Rule," *The Wall Street Journal*, Nov. 18, 2008.

59. "Toys and Childhood Lead Exposure," Centers for Disease Control and Prevention, www.cdc.gov/ nceh/lead/faq/toys.htm.

60. "Pass Up the Poison Plastic," Center for Health, Environment and Justice, November 2008, www .besafenet.com/pvc/documents/PVC-Guide-1.pdf.

61. Michael Hawthorne, "Change gets green light; His plans for environmental legislation may have big impact," *Chicago Tribune*, Nov. 19, 2008, p. C4.

62. Robert Pear, "Bush Aides Rush to Enact a Rule Obama Opposes," *The New York Times*, Nov. 29, 2008, www.nytimes.com/2008/11/30/ washington/30labor.html?ref=us.

63. Quoted in *ibid*.

64. See Michael Hawthorne, "U.S. warns of Teflon chemical in water," *Chicago Tribune*, Jan. 16, 2009, p. C18.

65. *Ibid*.

66. "Introduction to the Concept of Green Chemistry," U.S. Environmental Protection Agency, www.epa .gov/greenchemistry/pubs/about_gc.html.

67. "Award Winners," U.S. Environmental Protection Agency, www.epa.gov/greenchemistry/pubs/pgcc/ past.html.

68. "Green Catalysts Provide Promise for Cleaning Toxins and Pollutants," *Science Daily*, Aug. 20, 2008.

BIBLIOGRAPHY

Books

Hilts, Philip J., *Protecting America's Health: The FDA, Business, and One Hundred Years of Regulation*, Knopf, 2003.
A health and science reporter traces the history of the Food and Drug Administration and business resistance to regulation.

Michaels, David, *Doubt Is Their Product: How Industry's Assault on Science Threatens Your Health*, Oxford University Press, 2008.
An epidemiologist and former assistant secretary of Energy criticizes what he calls the "product defense industry" for promoting doubt and uncertainty about whether unsafe products should be regulated.

Schapiro, Mark, *Exposed: The Toxic Chemistry of Everyday Products and What's at Stake for American Power, Chelsea Green*, 2007.
An investigative journalist argues that Europe is replacing the United States as a commercial leader by setting high standards that require manufacturers to develop safer products.

Shabecoff, Philip, and Alice Shabecoff, *Poisoned Profits: The Toxic Assault on Our Children, Random House*, 2008.
Two journalists link rising levels of childhood illness and death to toxic exposures in children's homes, schools and neighborhoods.

Articles

Cone, Marla, "A Greener Future," *Los Angeles Times*, Sept. 14 and 19, 2008.
Once an obscure subfield, green chemistry is slowly changing the chemical industry, but more funding and training are needed before it becomes the mainstream approach.

Henig, Robin Marantz, "Our Silver-Coated Future," *On Earth*, fall 2007.
Nano-silver, the most widely used nanomaterial, illustrates the need for safety testing and new regulations for nanotechnologies.

Hogue, Cheryl, "The Future of U.S. Chemical Regulation," *Chemical & Engineering News*, Jan. 8, 2007.
American Chemistry Council Managing Director Michael Walls and University of Massachusetts-Lowell Professor Joel Ticknor debate whether U.S. law regulating commercial chemicals is stringent enough.

Pereira, Joseph, "Protests Spur Stores to Seek Substitute for Vinyl in Toys," *The Wall Street Journal*, Feb. 12, 2008.
Under pressure from consumers and advocacy groups, toy makers are exploring substitute materials without vinyl or phthalates.

Rosenberg, Tina, "What the World Needs Now is DDT," *The New York Times Magazine*, April 11, 2004.
DDT is a cheap way to kill mosquitoes that carry malaria, but the pesticide's toxic reputation and the challenging logistics of effective spraying campaigns have made it hard for the countries that most need help to use it.

Spivak, Cary, Susanne Rust and Meg Kissinger, "Are Your Products Safe? You Can't Tell," *Milwaukee Journal Sentinel*, Nov. 25, 2007.
Shampoo, carpets, skin lotions, clothing and many other consumer products contain endocrine-disrupting chemicals that cause cancer and other health problems in laboratory animals. Critics call U.S. government efforts to regulate these substances "an abject failure."

Reports and Studies

"Chemical Regulation: Comparison of U.S. and Recently Enacted European Union Approaches to Protect Against the Risks of Toxic Chemicals," *U.S. Government Accountability Office*, Aug. 17, 2007.
The report compares U.S. chemical regulation under the Toxic Substances Control Act (TSCA) and the European Union's REACH directive.

"Third National Report on Human Exposure to Environmental Chemicals," *Centers for Disease Control and Prevention*, 2005, www.cdc.gov/exposurereport/report.htm.
This ongoing assessment of human exposure to environmental chemicals, based on human specimens such as blood and urine, finds that levels of some substances such as blood lead and secondhand cigarette smoke have fallen, but that many other chemicals are widely present throughout the U.S. population, including known hazardous substances.

"Toxicity Testing in the 21st Century: A Vision and a Strategy," *National Research Council*, 2008.
The council charts a course for making chemical toxicity testing faster, more affordable and more accurate while reducing reliance on animal studies.

Felcher, E. Marla, "The Consumer Product Safety Commission and Nanotechnology," *Project on Emerging Nanotechnologies*, August 2008, www.nanotechproject.org/process/assets/filed/7033/pen14.pdf.
An expert on business and consumer protection argues that the commission is ill-prepared to regulate nanomaterials in consumer products.

For More Information

American Chemistry Council, 1300 Wilson Blvd., Arlington, VA 22209; (703) 741-5000; www.american-chemistry.com. The main trade organization for the U.S. chemical industry.

Center for Health, Environment and Justice, P.O. Box 6806, Falls Church, VA 22040; (703) 237-2249; www.chej.org. A grassroots advocacy group that works to protect communities from exposure to dangerous environmental chemicals.

Consumer Product Safety Commission, 4330 East West Highway, Bethesda, MD 20814; (301) 504-7921; www.cpsc.gov. The federal agency charged with protecting the public from unreasonable risks from products.

Consumers Union, 101 Truman Ave., Yonkers, NY 10703; (914) 378-2000; www.consumersunion.org. A nonprofit group that tests products.

National Nanotechnology Coordination Office, 4201 Wilson Blvd., Stafford II Room 405, Arlington, VA 22230; (703) 292-8626; www.nano.gov. Provides information about federal research and development of nanotechnologies.

National Toxicology Program, 111 T.W. Alexander Dr., Research Triangle Park, NC 27709; (919) 541-3665; http://ntp.niehs.nih.gov. A Department of Health and Human Services agency that studies the impact of chemicals on human health.

Project on Emerging Nanotechnologies, One Woodrow Wilson Plaza, 1300 Pennsylvania Ave., N.W., Washington, DC 20004; (202) 691-4282; www.nanotechproject.org. Provides independent, objective analysis of nanotechnology.

Project on Scientific Knowledge and Public Policy, 2100 M St., N.W., Suite 203, Washington, DC 20052; (202) 994-0774; www.defendingscience.org. Examines how science is used and misused in government decision-making.

15

Cost of the Iraq War

Are Economic Woes a Casualty of
Unexpectedly High Costs?

Peter Katel

Combat boots at the Ohio Statehouse in Columbus in June 2006 memorialize U.S. soldiers killed in Iraq as part of the exhibit "Eyes Wide Open: The Human Cost of War in Iraq." More than 4,000 U.S. troops have been killed so far, and some 30,000 wounded. Experts differ on the eventual cost of the fighting in Iraq and Afghanistan, but several projections — including care for wounded veterans, reconstruction of Iraqi cities and towns and interest on foreign debt — approach or exceed $2 trillion.

From *CQ Researcher*,
April 25, 2008.

Two major events in American life intersected in March 2008. A major Wall Street investment bank collapsed. And the country marked the five-year anniversary of the U.S.-led invasion of Iraq.

The demise of Bear Stearns came amid a national mortgage crisis that has helped precipitate an economic slowdown and rising joblessness. And the war's anniversary prompted a grim accounting: more than 4,000 Americans killed, tens of thousands wounded (plus millions of Iraqis killed or forced to flee their homes) and some $700 billion in taxpayer money spent so far.[1]

Experts differ on the eventual total cost of the conflict, but several projections approach or exceed $2 trillion.

As both parties gear up for the November presidential election, foes of the George W. Bush administration are insisting on a direct linkage between the big issues of the political season. "There are not two concerns in this coming election. There is one," says economist Joseph E. Stiglitz of Columbia University in a conference call with reporters. "The war is very much related to the weakness of the economy."

In a best-selling new book, Stiglitz, winner of the 2001 Nobel Prize in economics, lays out the most detailed and sustained economic case against the Iraq intervention, which he and co-author Linda J. Bilmes calculate will cost the United States upwards of $3 trillion.[2]

President Bush summarily rejects the war-economy link. "I think the economy is down because we've built too many houses," he told the NBC "Today Show."[3]

War Cost Could Exceed $3 Trillion

The total budgetary cost of the Iraq and Afghanistan wars could exceed $3 trillion, according to a best-selling new book by Nobel Prize-winning economist Joseph E. Stiglitz and federal budget expert Linda J. Bilmes, former chief financial officer of the U.S. Department of Commerce. Under a "best-case" scenario, they say the costs still would exceed $2 trillion.

Budgetary Costs of the Iraq and Afghanistan Wars

Cost factors	COST SCENARIOS (in $ billions)	
	Best-Case*	Realistic-Moderate**
Total Operations to Date	$646	$646
Future Operations	521	913
Future Veterans' Costs	422	717
Other Military Costs	132	404
Total (without interest payments)	**$1,721**	**$2,680**
Plus interest		
Interest costs On past, present and future debt	$613	$816
TOTAL (with interest)	**$2,334**	**$3,496**

* Assumes most optimistic speed of U.S. withdrawal, casualty levels and veterans' needs. Troops decline to 180,000 in 2008, then fall to 75,000 by 2010 and by 2012 to a non-combat force of 55,000. Overall, the number of unique troops deployed to the conflict by 2017 will total 1.8 million, which is "critical in determining future veterans' medical and disability costs."

** Assumes longer deployment timeframe for active-duty troops, higher demand for medical needs and more comprehensive tally of costs to the government and country. Assumes troop levels decline more slowly as number approaches 75,000 in 2012; these troops continue in primarily military function, including offensive operations against al Qaeda. The number of troops needed will total 2.1 million by 2017.

Source: Joseph E. Stiglitz and Linda J. Bilmes, The Three Trillion Dollar War: The True Cost of the Iraq Conflict, 2008

Even some Bush administration critics share that opinion. The war "didn't have much effect on the housing market or on the willingness or unwillingness of banks or others to provide credit," says Robert D. Hormats, vice chairman of Goldman, Sachs (International), a Wall Street firm.

Still, the Democratic contenders for the presidential nomination, New York Sen. Hillary Rodham Clinton and Illinois Sen. Barack Obama, are starting to echo some of the Stiglitz-Bilmes critique. And some of their fellow lawmakers, Republicans included, are taking up the simpler argument that the United States is spending money that the Iraqi government — a major oil producer — ought to be paying for defense and rebuilding.

"Isn't it time for the Iraqis to start bearing more of those expenses, particularly in light of the windfall in revenues due to the high price of oil?" Sen. Susan Collins, R-Maine, asked Ambassador Ryan Crocker, the U.S. envoy to Iraq, during an April 8 hearing of the Senate Armed Services Committee.[4]

"Senator, it is," Crocker replied. He and Gen. David H. Petraeus, the top military commander in Iraq, said the Iraqi government has agreed to channel $300 million to U.S. authorities for reconstruction projects.[5]

The presumptive Republican presidential nominee, Sen. John McCain, R-Ariz., a vocal supporter of the Iraq intervention, endorses that approach. "The Iraqis . . . need to move a portion of their growing budget surpluses into job-creation programs," he said at the same hearing, "and look for other ways to take on more of the financial burdens currently borne by American taxpayers.[6]

President Bush had already signaled a shift toward insisting that the Iraqi government lessen financial dependence on the United States.

"The Iraqi government is stepping up on reconstruction projects," Bush said in a March 27 speech at the National Museum of the United States Air Force in Dayton, Ohio. "Soon we expect the Iraqis will cover 100 percent of those expenses. The same is true when it comes to security spending. Initially, the United States paid for most of the costs of training and equipping the Iraqi security forces. Now Iraq's budget covers three-quarters of the cost of its security forces, which is a total of more than $9 billion in 2008."[7]

But Stiglitz and Bilmes calculate that the United States spends more than that — $12 billion — in just one month on Iraq operations. Their overall estimate of $3 trillion includes interest payments on the entirely borrowed funds for the war, and takes in the cost of Iraq (and Afghanistan) operations since 2001 — when the Global War on Terrorism was launched, the Afghanistan intervention began and pre-invasion planning for the Iraq conflict started up — through 2017.[8] The Democratic staff of the Congressional Joint Economic Committee produced a nearly identical estimate of $2.8 trillion. And the nonpartisan Congressional Budget Office (CBO) came up with an estimate of $1.2 trillion to $1.7 trillion. The CBO's total could rise to as much as $2.4 trillion if future interest payments on borrowed money are added.[9] Scott Wallsten, a former economist at the World Bank and the American Enterprise Institute who has been tracking Iraq costs for years, told the congressional Joint Economic Committee that Iraq expenses would reach close to $2 trillion.[10]

None of these estimates are easily compared with one another because the underlying calculations were based on different methodologies and time horizons; some also do not account for oil-price fluctuations, debt interest payments and the effects of inflation. Some of these contrasts are apparent in projections on the costs of veterans' care. (*See sidebar, p. 344.*)

In any event, however much the United States spends in the future, there's no question that it already has spent far more than the administration ever projected. The closest thing to an official cost estimate ran to $60 billion tops, by Mitchell E. Daniels, then the head of the Office of Management and Budget, in December 2002. And a White House adviser, Lawrence B. Lindsey, then director of the administration's National

Economic Council, who in 2002 gave an unofficial projection of up to $200 billion, was fired shortly after that.

The administration's projections "presupposed a relatively short conflict that would have had us out of there in a matter of months," says Dov Zakheim, who was assistant secretary of Defense and the Pentagon's budget chief in 2001-2004. Instead, "The war became a lot more intense than people anticipated, and the thing has gone on a lot longer than people anticipated." Zakheim is now a vice president of Booz Allen Hamilton, a McLean, Va.-based consulting firm.

Cost figures and economic theories notwithstanding, the Iraq-costs debate ultimately turns on issues of national security policy.

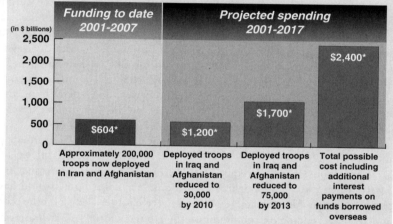

Cost Reductions Hinge on Troop Reductions

The United States has spent more than $600 billion on the Bush administration's "War on Terrorism" in the six years following the Sept. 11, 2001, terrorist attacks, according to the Congressional Budget Office. The CBO calculates the cost at $1.7 trillion if current troop levels in Iraq and Afghanistan — about 200,000 — are reduced to 75,000 by 2013. The total cost could be as much as $2.4 trillion if interest on foreign debt is added to the CBO numbers.

U.S. Costs for War on Terrorism

(in $ billions)

Funding to date 2001-2007	Projected spending 2001-2017		
$604*	$1,200*	$1,700*	$2,400*
Approximately 200,000 troops now deployed in Iran and Afghanistan	Deployed troops in Iraq and Afghanistan reduced to 30,000 by 2010	Deployed troops in Iraq and Afghanistan reduced to 75,000 by 2013	Total possible cost including additional interest payments on funds borrowed overseas

* Includes funding for U.S. and Iraqi military operations, diplomatic operations and reconstruction and veterans' benefits and services

Sources: "Estimated Costs of U.S. Operations in Iraq and Afghanistan and of Other Activities Related to the War on Terrorism," Congressional Budget Office, Oct. 24, 2007; Senate Budget Committee

"We're supporting a vital national interest, which is a stable Middle East," says James Jay Carafano, a retired U.S. Army lieutenant-colonel and a military affairs and foreign policy specialist at the conservative Heritage Foundation. Of the Stiglitz-Bilmes thesis, he says, "These are political arguments, not economic arguments. It's not even saying 'I've done this economic analysis and it has political implications.' That's prostitution, as far as I'm concerned. [It amounts to] 'I'm going to prostitute my craft for politics.' "

In an interview later, Stiglitz says, "We've obviously hit a raw nerve." He adds that he and Bilmes make clear they oppose the war. But they began laying out their thesis in a 2006 paper published by the National Bureau of Economic Research, a nonpartisan, nonprofit forum.[11] "It was open for people to give us comments, which is in the nature of an academic process. We were very careful in responding to issues that were raised in open debate. The remarkable thing, from the Heritage Foundation or from the administration, is that they won't come up with their own numbers."

White House press secretary Dana Perino said in March, when asked about Stiglitz's calculations: "I'm not going to dispute his estimates. . . . But it's very hard to anticipate, depending on conditions on the ground and circumstances, how much the war is going to cost."[12]

In any case, war critics aren't the only economists studying Iraq. Even before the invasion, a trio of economists at the University of Chicago began examining projected costs of the war against the alternative — maintaining military operations to enforce United Nations sanctions against Iraq (referred to as "containment"). The analysts calculated that containment would have cost $297 billion while military action would cost $414 billion.[13]

"The cost of the war is certainly far in excess of the baseline cost that we estimated for containment," the study's leader, Steven J. Davis, concedes. But Davis, a professor at the University of Chicago's Graduate School of Business, disputes the idea that the underestimate by him and his colleagues strengthens the Stiglitz-Bilmes argument. "Their whole premise is that, in the absence of war, things would have been fine and dandy in the Middle East," something he calls a "questionable assumption."

A military-spending specialist agrees decisions on going to war can't be reduced entirely to dollars and cents. "It's not like investing in real estate," says Steven M. Kosiak, vice president for budget studies at the Center for Strategic and Budgetary Assessments, a nonpartisan think tank. "You have to take into account all kinds of things that are not budgetary or economic; there may be times when it is worth going to war even if there is a high budgetary cost."

Still, President Bush has not been relying on that argument alone in recent comments about the costs and economic effects of the war. "I think, actually, the spending on the war might help with jobs, because we're buying equipment and people are working," he told the NBC "Today Show."[14]

But the classic World War II argument that military spending benefits the entire economy is finding little resonance among lawmakers managing a $239 billion federal budget deficit, whose constituents fear recent economic developments and hear that no plans exist to end the war.

As Hormats points out, "War spending is a highly inefficient way of boosting U.S. jobs and growth; spending on roads, bridges, energy research and education at home would have a far more beneficial and enduring effect on the economy than artillery and tanks."

"We must ask ourselves," said Sen. Charles Schumer, D-N.Y., said at a February hearing of the Joint Economic Committee, "is it worth spending trillions of dollars on such an uncertain and unpredictable outcome?"[15]

As Congress, the presidential candidates and the public debate the cost of the Iraq War, here are some of the questions being asked:

Are war costs contributing to the current economic downturn?

Shortly before the April 9 anniversary of the fall of Baghdad, the federal Bureau of Labor Statistics announced that the economy had shed 232,000 jobs since the beginning of the year — 80,000 in March alone. With that, the national jobless rate rose to 5.1 percent, the highest since the post-Hurricane Katrina month of September 2005. But the latest losses aren't confined to any one region or industry.[16]

And the statistics all but ended debate on whether the country has entered a recession. Ian C. Shepherdson, chief domestic economist for High Frequency Economics, an analytical firm for institutional investors, told *The*

New York Times: "We are in for a much longer recession than Wall Street thinks. This particular downturn is driven by a rare contraction in consumer spending, and that is starting to hurt a broader range of people than those hurt by the mortgage crisis."[17]

In general, economists say an oversupply of houses started the dominoes falling. Previously, a construction boom had enabled a big uptick in homeownership, particularily among first-time buyers whose incomes had been too low to qualify for mortgages. These customers were steered into high-interest "subprime" house loans. When property values dropped, subprime mortgage-holders couldn't borrow against their houses to keep making payments, which had risen when their variable-interest-rate loans "reset" to higher rates. The result: foreclosure proceedings against 1.5 million homeowners last year alone.[18]

Foreclosures rocked Wall Street investment banks that had invested heavily in packages of home loans that included subprime mortgages. One major firm, Bear Stearns, went broke and was rescued only by a competitor's acquisition, made possible by a $29 billion line of credit from the Federal Reserve. Other banks slowed their lending.[19]

In this climate, it's not surprising that politicians are reciting lists of government services that could have been funded with money being spent in Iraq. The Democratic staff of the Joint Economic Committee, for instance, picked a few examples of what the Iraq spending could have paid for: enrollment of 57,500 more children in the Head Start preschool program, 153,000 Pell Grant higher-education scholarships and 9,000 more police officers nationwide.[20]

Stiglitz and Bilmes argue that such investments in domestic health, education and infrastructure would have provided more benefits to the country at large than the money spent in Iraq. For instance, they say, a Nepalese employee of a military contractor in Iraq doesn't use his paycheck to buy goods and services in the United States (unless he happens to buy a U.S.-made product).

Going a step further, Stiglitz and Bilmes claim that the combined effect of Iraq War investments was to suppress demand in the United States — because of the goods and services not bought here.

The Federal Reserve, whose mission includes helping keep the economy running, responded by increasing liquidity, Stiglitz explains, in order to ensure an easy supply of money to lenders. "They kept interest rates lower than they otherwise would have been. The more they had to do that, the more they encouraged people to take out mortgages. If the economy were going gangbusters, they would have raised interest rates, and at higher interest rates there's less refinancing of mortgages."

But other economists including Hormats of Goldman, Sachs — while emphasizing that funds spent in Iraq would been much more beneficial to the U.S. economy if spent on infrastructure, worker training, medical research and homeland security in this country — discount the role of war spending in the foreclosure-sparked financial emergency that helped slow the entire economy. "This downturn was the result of two major factors that converged — overbuilding of homes and disruption in the credit markets," Hormats says. "People lent too much money in the housing sector, and the assets that backed up those loans deteriorated. It's hard to see that war spending itself played much of a role."

Bilmes and Stiglitz also point to the war as a force pushing up world oil prices. Crude oil in late April was fetching about $116 a barrel, creating higher costs for producers, retailers and — ultimately — consumers, throughout the economy. In their book, Stiglitz and Bilmes said the war contributed, conservatively, $10 to $15 of that price increase.

In the week that the war started, the average world oil price was about $27 a barrel. "The futures markets expected that the price would remain for the next decade at that kind of price," Bilmes says during a March 18 conference call with reporters. "They anticipated that demand was rising considerably from India and China but that supply would rise to meet the demand. So the question is what changed that equation; certainly the main thing that changed was the invasion of Iraq."[21]

Still, many economists reject their view of the war's effect on the oil market. "If you press me and ask, 'Did the war cause less Iraqi oil in the market than would otherwise be the case?' I'd say 'yes,'" Hormats says. "But you can't make the case that that's the main factor in pushing oil prices to $116 a barrel."

Some other economists echo that view. But they add that war costs are now playing a major role in how the economic drama unfolds. "You can absolutely blame the war for restricting our ability to take countermeasures,"

Many Weapons Given to Iraq Are Missing

U.S. military officials cannot account for hundreds of thousands of rifles, pistols and other equipment issued to Iraqi security forces by the Multinational Security Transition Command-Iraq (MNSTC-I). Some analysts say the unaccounted-for weapons fell into the hands of insurgents or were sold on the black market. MNSTC-I officials blame a lack of adequate staffing for the inconsistencies.

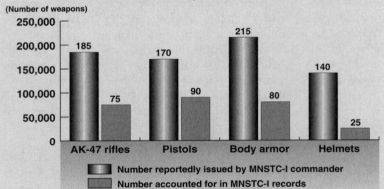

Missing Equipment Issued to Iraqi Security Forces, June 2004-September 2005

(Number of weapons)

	AK-47 rifles	Pistols	Body armor	Helmets
Number reportedly issued by MNSTC-I commander	185	170	215	140
Number accounted for in MNSTC-I records	75	90	80	25

Source: "Stabilizing and Rebuilding Iraq: Actions Needed to Address Inadequate Accountability Over U.S. Efforts and Investments," Government Accountability Office, March 2008

says Jared Bernstein, living standards program director at the Economic Policy Institute, a liberal think tank. "We've added so much to our national debt. If you believe, and many people do, that we'll ultimately need another stimulus package, it's going to be tougher to add to the debt because of the war. Those $600 billion you wouldn't have spent otherwise limit your future options."

But Davis, the University of Chicago economist, says that thesis amounts to "picking on the Iraq War to make a political argument." In an environment in which the government is already borrowing money to finance its activities, anything and everything on which the government spends money drives up the debt, Davis argues.

Did the Bush administration low-ball projections of war costs?

The costs of fighting wars have challenged American leaders as far back as the American Revolution, before the

United States had even been founded. During the Vietnam War, the most recent lengthy conflict before Iraq and Afghanistan, President Lyndon B. Johnson's struggle to finance escalating costs without sacrificing domestic social programs led to political standoffs that prompted him to abandon politics.

Today's war-spending conflict centers on the administration's pre-war cost projections. The disparity between those estimates and actual spending lead some to argue that the administration deliberately kept projections low.

These numbers ranged no higher than White House economic adviser Lindsey's high-end estimate of $200 billion. In a newly published memoir, he wrote that his Iraq estimate surely contributed to his ouster, though the administration never said that. "Putting out only a best-cases scenario without preparing the public for some worse eventuality was the wrong strategy," Lindsey wrote this year. "It is too bad for the country that his [Bush's] credibility was squandered by the White House for not being upfront about what the war might cost."[22]

The first non-governmental expert to question the validity of official estimates was William D. Nordhaus, a professor of economics at Yale University. In a paper published six months before the war began, he wrote: "The Bush administration has made no serious public estimate of the costs of the coming war. . . . Perhaps, the administration is fearful that a candid discussion of wartime economics will give ammunition to skeptics of the war."[23]

Shortly after the conflict began, skepticism about administration numbers emerged in the mass media. Stiglitz and Bilmes quote an exchange between Ted Koppel, then-host of ABC News' "Nightline," and Andrew Natsios, then the administrator of the U.S. Agency for International Development (USAID). Clearly in a state of disbelief, Koppel challenged his guest's claim that U.S. expenses for rebuilding Iraq would total $1.7 billion.

"I want to be sure that I understood you correctly," Koppel said at one point, before asking Natsios to spell out his estimate all over again. The USAID official said that Iraq's projected $20 billion a year in oil revenues would pay for the bulk of the rebuilding.[24]

Lawrence J. Korb, a former assistant Defense secretary in the Reagan administration and a U.S. Navy veteran, points to the Natsios episode as evidence that the administration was deliberately keeping estimates low. "They had to be," he says. "I can't imagine anybody who knew anything saying it was going to cost $1.7 billion, or that Iraqi oil would pay for it. They would have known that the oil infrastructure was not in very good shape and that the people who did not want us to come there were not going to allow the oil to be used."

Bush administration officials had plenty of reason to downplay costs, Korb says. If the American people had been given an idea of the real price they'd be paying to topple a tyrant and sponsor elections in his country, the administration would have been "laughed out of the ballpark." Korb, a frequent critic of the administration's foreign policy, is now a fellow at the Center for American Policy, a liberal think tank directed by former staff members in Democratic administrations and congressional offices.

From the other side of the political divide, former Pentagon financial officer Zakheim rebutted the low-balling claim. Administration estimates might have proved accurate, Zakheim argues, if top U.S. administrator in Iraq L. Paul Bremer III had not dissolved the Iraqi armed forces. That May 23, 2003, order effectively turned 350,000 trained fighters onto the street in a state of rage at the United States.

"If anything drove the cost of the war up, it was the decision to demobilize the Iraqi army," Zakheim says, adding that he opposed the move at the time (as did

Iraq War Cost Ranks Near the Top

If Congress approves a pending request for additional 2008 funds, the wars in Iraq and Afghanistan will have cost $756 billion by the end of this fiscal year (not including costs for veterans' care) — second only to World War II. A recent best-selling book contends that when veterans' costs and other factors are added, the war will end up costing $3 trillion, about the same as World War II.

Costs of Major U.S. Wars
(in $ billions, in 2007 dollars)

($ billions)

World War II	$3,200
Iraq/Afghanistan	$655*
Vietnam War	$670
World War I	$364
Korean War	$295
Persian Gulf War	$94
Civil War	$81**
American Revolution	$4

* Total for fiscal years 2001-2008. Another $102.9 billion supplemental request is expected to be approved for this fiscal year.

** Total Union and Confederate costs

Sources: Congressional Research Service and Office of Management and Budget data, available on Web site of the Center for Arms Control and Non-Proliferation, www.armscontrolcenter.org/policy/securityspending/articles/historical_war_costs/.

Colin L. Powell, then the secretary of State, and others). "If they had stuck to the original plan, I think it would have been a very different situation."[25]

Zakheim's argument echoes the impression at the time of Army Col. Alan King, then a civil affairs officer in Iraq. "When Bremer did that, the insurgency went crazy," he told *Washington Post* correspondent Thomas Ricks.[26]

Still, some longtime government-watchers remain skeptical of the intentions behind the prewar estimates. "If you had sat down and tried to realistically tally expected costs, you would have come up with a number well above $50 billion," says Bernstein at the Economic Policy Institute. "You would have been low-balling, but you wouldn't have known that at the time."

All in all, Bernstein says, the prewar estimates were an exercise in "fuzzy math." He defined the term as production of "a number intentionally biased to achieve support for something instead of a more realistic, higher, number that might cost you that support. It's easier to sell something if it's cheaper."

But Davis, the University of Chicago economist, argues that intentional underestimating wouldn't explain the gap between prewar projections and actual costs. Officials under-projected, he argues, out of belief in U.S. power to speedily alter the reality on the ground in Iraq.

"There was active resistance in the administration, by [Defense Secretary] Donald Rumsfeld, and I don't know how deeply in the Defense Department itself, to costing out the longer-term consequences of the war, to careful planning for the post-major combat phase," he says. "That meant that the American electorate wasn't sufficiently prepared for what would come after."

Is the Iraq war worth the expense?

Debates over Iraq War costs and their consequences are only partly economic. The question of what the United States has obtained for the billions of dollars it spent underlies the entire subject. Even before the war began, some of the debates over whether to support what was clearly a looming invasion touched on the issue of costs vs. benefits.

Three days before Bush ordered the invasion, a reporter for *The Washington Post* summed up the dollars-and-cents side of the debate this way: "In the longer term, many experts expect that the cost of reconstructing Iraq, including care of refugees, troops to keep the peace and repairs to bombed infrastructure, would add scores — even hundreds — of billions to a federal deficit already spiraling out of control," David Von Drehle wrote.[27]

Five years later, President Bush focused on the other side of the argument during one of a trio of speeches timed to the war anniversary season. The costs of withdrawing at this point — "retreating," as he put it, "would be a propaganda victory of colossal proportions for the global terrorist movement, which would gain new funds, and find new recruits, and conclude that the way to defeat America is to bleed us into submission."[28]

Critics of that view include foreign-policy veterans. Rand Beers, who resigned from the State Department in 2003 in disagreement over the imminent Iraq War, argues that it is distracting the United States from growing dangers in Afghanistan and Pakistan. Osama bin Laden and his network have "reconstituted their capabilities to operate in Afghanistan and around the world,

to be able to become much more of a threat to the United States," Beers says, speaking on a conference call with reporters in February. "They now represent a danger to the homeland of the United States."

Given the scale of U.S. operations in Iraq, "all the other challenges we face are being unattended or are only modestly dealt with," Beers says. When he quit the State Department, he was assistant secretary for international narcotics and law enforcement affairs. Earlier, he had served Republican and Democratic administrations on the National Security Council.

But Zakheim, the former Pentagon finance chief, argues that some benefits of the U.S. invasion and its aftermath are already evident. "Having an Iraq that doesn't attack its neighbors and whose neighbors don't attack it is already a factor in the region," he says. "If the outcome of this exercise is not just that we got rid of Saddam Hussein but got a country that's no longer busily attacking others and being attacked, you've made a fundamental change in the region, in a situation that's gone on for 60 years."

By contrast, Bernstein, a liberal critic of the war, makes an effort to distinguish between his "gut reaction" that the war clearly isn't worth the cost and a more objective view. "I can imagine making an argument that in the absence of this action we'd be faced with a whole set of difficult and deadly situations that we're not facing," he says. "At this point it's awfully hard to make a positive cost-benefit analysis. But it's probably also too soon to do that in any believable way."

Carafano at the Heritage Foundation argues against any suggestion that the costs overwhelm benefits. "We've already turned the corner," he says, arguing that U.S. forces are stabilizing the country. "I can spin numbers that show that the American Revolution was a mistake. To argue that the United States, with a $13 trillion economy, is even remotely near spending so much on security that it's undermining the ability to be a viable state is laughable. You can hate the war and think all the money has been squandered, and even if that's true, in terms of the long-term competitiveness of the nation, it's irrelevant."

At the same time, Carafano argues against withdrawing U.S. forces too quickly. "What has to happen is that there has to be trust and confidence among the Shia, the Sunni and the Kurds such that nobody can steal a march on anyone else," he says. "Now, the guarantor that they can't do that is the U.S. government. You don't want to whittle that military presence below the point at which people don't think it's influential any more."

Korb of the Center for American Progress holds that the five-year results of the Iraq war make clear that the American investment yielded negative results. "Say this was early 2003 and we did a cost-benefit analysis of getting rid of Saddam Hussein — you would not have gone in," he argues. "Are you more secure than you were five years ago? No."

Other pro-war arguments vanished soon after Saddam was toppled, Korb says, referring to the administration's claims that Saddam's regime was storing weapons of mass destruction and developing new ones. In addition, he argues, speaking as if to an administration official, "The costs you projected were way too low. That's where you are now."

President Bush, in an April 10 speech, responded to cost-benefit criticism by noting that today's military budget amounts to about 4 percent of the nation's gross domestic product (GDP). "During other major conflicts in our history, the relative cost has been even higher," he said from the White House. At the height of the Cold War, during the Truman and Eisenhower administrations, the defense budget ran to 13 percent of GDP, Bush said.[29]

"We should be able to agree that this is a burden worth bearing," he said, predicting that "violent extremists" would be encouraged if America were to "fail" in Iraq. "This would . . . increase the threat of another terrorist attack on our homeland."[30]

But Wallsten, formerly of the World Bank and now vice president for research at the Technology Policy Institute, a think tank (formerly iGrowthGlobal), responds that that benefit is far from clear. "Nobody has shown that there's a link between Iraq and discouraging another 9/11," he says.

Stiglitz and other economists note that pointing to the relative size of the defense budget leaves out the Bush administration's reliance on borrowing to pay for the war — hence adding billions in interest payments ($800 billion alone for Iraq and Afghanistan by 2017, Stiglitz and Bilmes calculate).[31] "This war was financed, for the first time for any major war in American history, entirely by borrowing — entirely," Goldman, Sachs' Hormats

says. "The next generation of taxpayers will have to foot the entire bill." The lenders are the same buyers of U.S. bonds who finance all of the federal government's deficit spending. Forty percent of that money, Hormats says, comes from abroad.

BACKGROUND

Paying for War

How the United States should pay for war is an issue that goes back to the founding of the nation. When the Continental Congress couldn't come up with all the money George Washington needed to feed and supply his troops, independence movement leaders assigned Benjamin Franklin and John Adams to negotiate loans from France and from backers in the Netherlands.[32]

Those loans precipitated one of the first political crises in U.S. history. After independence, several states separately owed a total of $25 million in war debts. Alexander Hamilton, the first Treasury secretary, wanted the federal government to assume the loans. Otherwise, he argued, any state stalling on repayments would affect the entire country's creditworthiness.

Southern politicians led by James Madison of Virginia, then a member of the House of Representatives, feared "assumption" as an erosion of states' rights. The standoff was resolved when Hamilton and his allies gave way on the Southerners' proposal to establish the national capital in the South, further from the Northern financial and commercial interests that Madison and his allies distrusted.

From then until the present conflicts in Iraq and Afghanistan, the federal government never had to borrow money from abroad to finance wars. But from the War of 1812 onward, war spending did force the central government to go into debt to domestic creditors.

Washington and Thomas Jefferson, among other early members of the U.S. political class, discouraged long-term indebtedness. The first president, in his celebrated farewell address, urged future lawmakers and presidents to "discharge the debts which unavoidable wars may have occasioned, not ungenerously throwing upon posterity the burdens we ourselves ought to bear."[33]

His successors maintained that doctrine during the nation's first 150 years. Military spending caused major budget deficits during the Civil War and both world wars.

But postwar spending limits, as well as prosperity that raised tax revenues, brought down indebtedness following all those conflicts.

Of all U.S. military conflicts, World War II involved by far the biggest U.S. commitment of troops, industrial production and money. With all males ages 20-45 eligible for the draft, 5.4 million men were serving in the U.S. Army alone by the end of 1942.[34]

Hormats, in his history of war financing, emphasizes that the military mobilization was accompanied by rationing and wage controls in the civilian sector. President Franklin D. Roosevelt also fought for tax policies that didn't unfairly benefit the well-to-do. The Roosevelt administration also raised millions by selling war bonds to ordinary citizens, a policy that Hormats credits with "binding Americans together in a commitment to the success of the war effort."

But the end of World War II in 1945 saw the opening of a new kind of conflict, the Cold War, which lasted until 1991. The rivalry between the United States and the Soviet Union, both nuclear powers (from 1949 on, in the case of the Soviets), saw the establishment of a large standing military during peacetime for the first time in U.S. history. Conscription — the draft — for all males above the age of 18 made that possible, though a system of deferments for college enrollment, disabilities and hardship made avoiding military service relatively simple for many potential draftees.

Military preparedness during the Cold War also saw active development and manufacture of weapons of all kinds, including nuclear arms. As military spending grew, along with new social programs, massive government borrowing became commonplace.

Korea and Vietnam

The Cold War was punctuated by the wars in Korea and in Vietnam. Both involved fierce combat by draftees. But neither war measured up to World War II in scale, and neither conflict was fought under a congressional declaration of war.

Even so, the 1950-53 Korean conflict provoked a major debate on how to pay its costs. "During World War II, taxes were not high enough, and the government was forced to borrow too much," President Harry S. Truman told Congress in 1951, arguing for a $10 billion tax increase.[35] Congressional Republicans fought the proposal. But they didn't scrap it, instead cutting it to $7.2 billion.

During the next two decades, fights over funding the Vietnam War far exceeded the Korea debates in intensity. (*See graph, p. 339.*) These debates also spanned a greater period of time, as did the conflict itself. It began in 1964 (though U.S. military aid began earlier), when Congress passed a resolution authorizing the president to take whatever action he saw fit to defend Southeast Asia, and ended in 1973, when the United States formally ended its combat role. In 1975, communist North Vietnam conquered U.S. ally South Vietnam, finally concluding the war.

The scale of U.S. military participation can be measured by the size of the American contingent — from 22,000 at the end of 1964, to 184,000 at the end of 1965, to 536,000 by the end of 1968.

President Lyndon B. Johnson, who escalated the war during his time in office (1963-1969), was determined not to let military spending erode the "Great Society" social programs he had created. But reality dictated otherwise. As the war intensified, its costs pushed defense spending from $50 billion in 1965 to $78 billion in 1968. As a result, economists projected that the budget deficit would quadruple from $4.5 billion in 1967 to $19 billion in 1968. Top presidential advisers pressed Johnson to propose a tax hike to meet skyrocketing revenue demands.

But Johnson resisted, fearing that conservatives would counter by threatening to slash Great Society programs. Finally, in 1967, Johnson proposed a temporary 6 percent surcharge on income taxes for individuals and corporations. As he had feared, Congress' leading tax legislator demanded some spending cuts as the price for supporting a tax hike.

With the financial picture worsening, Johnson upped his proposal to a 10 percent tax surcharge. But he made no headway in Congress. Then, in January 1968 the North Vietnamese mounted a major offensive to coincide with the lunar New Year, known as Tet. American and South Vietnamese forces came under intense and sudden attacks that reached into the U.S. Embassy itself.

The Tet Offensive failed militarily but put a major dent in public support for the war. Stalemated on the financing and political sides, Johnson made a surprise announcement that he was pulling out of the 1968 presidential nomination race, in effect, retiring from public office.

Following that move, congressional leaders pushed through the president's 10 percent surcharge, at the cost of $6 billion in cuts from Great Society programs. In the end,

CHRONOLOGY

1790-1941 *From the American Revolution until World War II, presidents and lawmakers avoided long-term indebtedness to pay for military conflict.*

1790 Alexander Hamilton and James Madison forge a compromise in which the federal government assumes the states' Revolutionary War debt.

1796 George Washington warns against saddling "future generations" with war debts.

1866 Federal government's interest obligations on its Civil War borrowing amount to twice the size of the entire budget before the war started.

1919 Major increase on personal and business income taxes covers about one-third of U.S. costs for World War I.

1941-1968 *An era of massive, long-term military spending begins with the biggest military conflict in U.S. history and continues into the long Cold War.*

1942 A year after the United States enters World War II, 5.4 million men are serving in the U.S. Army.

1945 The fifth of a series of war bond drives among the public raises $20.6 billion — for an overall total of $87.3 billion. . . . Wage and price controls and rationing hold down civilian spending. . . . Wartime tax-code changes dramatically boost the number of taxpayers to 42 million, from 4 million before the war.

1951 With the Korean War under way, President Harry S Truman proposes a $10 billion tax increase, settles for $7.2 billion.

1968 Tet Offensive by North Vietnamese military fails but shocks U.S. public and politicians with its sweep and intensity. . . . President Lyndon B. Johnson proposes 10 percent tax surcharge to help finance war, accepting $6 billion in cuts to social programs. . . . U.S. forces in Vietnam expand to 536,000.

1990-2002 *Iraq becomes a major U.S. military adversary under dictator Saddam Hussein.*

1990 Saddam orders his military to seize neighboring Kuwait. . . . President George H. W. Bush begins assembling an international military coalition to free Kuwait.

1991 After weeks of air bombardment, U.S.-led land forces expel Iraq in 100 hours of fighting. . . . The United States provides the majority of troops and thus pays only $13 billion of the cost, while allies pay $48 billion.

2000 U.S. contingent enforcing United Nations sanctions against Iraq includes 30 Navy vessels, 175 airplanes and as many as 25,000 soldiers at a total cost of $16.4 billion.

2002 Following the Sept. 11, 2001, terrorist attacks, President George W. Bush begins laying the groundwork for "regime change" in Iraq. . . . President's chief economic adviser estimates war costs at $100-$200 billion and is replaced. . . . Office of Management and Budget director estimates cost at up to $60 billion.

2003-2008 *U.S. intervention in Iraq is far longer, more complicated and expensive than the administration projected.*

2003 Assistant Defense Secretary Paul D. Wolfowitz predicts Iraq's oil revenues will finance reconstruction. . . . Bush administration proposes $20.3 billion for Iraq reconstruction. . . . Senate proposes money be lent instead of granted but backs down.

2006 Iraq spends only 23 percent of its annual reconstruction budget.

2007 Iraq's oil revenues rise to $41 billion, up from $31 billion in 2006.

2008 War costs rise to an expected $160 billion for the year, following year-by-year increases from $53 billion in 2003. . . . Increase in foreclosures, collapse of Bear, Stearns and rise in unemployment prompt growing concerns over economy. . . . Number of American military personnel killed exceeds 4,000. . . . Economist Joseph E. Stiglitz and budget expert Linda J. Bilmes intensify war-cost debate by calculating total eventual cost at $3 trillion.

Veterans' Care Cost Estimates May Be Too Low

Authors of new book challenge government calculations.

The authors of the *Three Trillion War: The True Cost of the Iraq Conflict* say the future costs of caring for wounded Iraq War veterans may be vastly more than government estimates.[1]

"Some of the issues in Iraq and in the Middle East are not entirely under our control," says co-author Linda J. Bilmes. "But the way we take care of the veterans is entirely under our control. It's a matter of national attention and priorities."

Bilmes, a lecturer in public policy at Harvard University's Kennedy School of Government, did the research underlying veterans'-care cost projections in the new book, co-authored with Columbia University economist Joseph E. Stiglitz, winner of the 2001 Nobel Prize in economics.

Stiglitz and Bilmes base their projections on lifetime costs for all care and government compensation for veterans who served in Iraq from 2003 to 2017, including Social Security payments. The projections range from a minimum of $371 billion under a reduced combat scenario to a more "realistic" sum of $630 billion.[2]

The Congressional Budget Office (CBO), using the same military scenarios utilized by Stiglitz and Bilmes, calculated its own projections, which differ enormously from the Stiglitz-Bilmes estimates, in part because the CBO didn't try to project lifetime care costs. As a result, the CBO estimates that veterans' benefits and services in 2001-2017 would cost $12 billion under a more favorable reduced-combat scenario, or $13 billion under a scenario that envisions more fighting.

The "best-case" scenario (the Stiglitz-Bilmes term) involves a gradual drawdown of U.S. forces in Iraq and Afghanistan by 2012 to 55,000 stationed there long-term (or 30,000 by 2010, in one of the CBO scenarios) and assigned to non-combat duty. A more realistic scenario, according to Stigliltz and Bilmes, would involve continued combat and drawing troops down to about 75,000 by 2013.

"The funding needs for veterans' benefits comprise an additional major entitlement program along with Medicare and Social Security," they write.[3]

Numerous other assumptions about federally funded care for vets are built into the Stiglitz-Bilmes projections, including:

- Forty-eight percent of Iraq-Afghanistan veterans will seek treatment from the Department of Veterans Affairs (VA). That projection is based on an average of five disabling conditions claimed by Iraq-Afghanistan veterans, compared to the average of three disabling conditions claimed by Persian Gulf War veterans and on the 45 percent claim rate by Gulf War veterans and the 37 percent rate by Iraq-Afghanistan veterans thus far, of which 88 percent have been granted, or partially granted.

- The number of post-traumatic stress disorder (PTSD) claims will "rapidly" pass the 19,000 filed when Stiglitz and Bilmes did their research in 2005-2007; the average payment per PTSD claim will remain at $7,109, the 2007 amount. They base that projection in part on a Government Accountability Office (GAO) finding last year that processing a PTSD claim can take up to a year. An earlier GAO report notes that the "intense and prolonged combat" that characterizes service in Iraq and Afghanistan puts veterans of those conflicts at high risk for PTSD.[4]

Bilmes and Stiglitz based their calculations on official data. But they were charting new territory. Neither the executive branch, nor the GAO nor the Congressional Research Service has made public any long-range projections of the costs of Iraq War veterans' care, says Bilmes, who has made a specialty of calculating the cost of veterans' care. She adds that a VA economist's request for a professional conference last year on her methodology was canceled by the agency but that she then met with then Veterans Affairs Secretary R. James Nicholson, at his request, to discuss her calculations.

The longest-range projections of the costs of caring for veterans are those by the CBO. CBO Director Peter Orszag wrote a detailed blog posting laying out his agency's differences with the Stiglitz-Bilmes numbers. The differences go beyond the lifetime versus year-to-year costs.

"It may appear surprising to some readers, but veterans of the recent conflicts, on average, require less medical care from VA than veterans of other conflicts," Orszag wrote, noting also that he is a friend of Stiglitz, with whom he has co-authored papers. For instance, he wrote, the VA in 2006 actually spent only an average of $2,610 per veteran — an average that reflects a relatively low incidence, so far, of more expensive PTSD treatment.[5]

And Stiglitz and Bilmes seem to have failed to account for disability claims by veterans who didn't serve in Iraq or Afghanistan, Orszag said. Those projected costs should be subtracted from war-related projections, he wrote.[6]

And he concludes that because of the differing spending estimates, and related disagreements, Stiglitz and Bilmes over-estimated VA medical costs by at least $100 billion.[7]

Stiglitz notes, for his part, that CBO didn't take into account data showing that PTSD claims escalate with repeated deployments. "We have not yet gotten back those who've been deployed, two, three, four times. Our number is an underestimate." CBO may have relied too heavily, he says, on initial health-care costs for Iraq veterans, which arise more from diagnosis than from treatment, which is more expensive.

Paul Sullivan, executive director of Veterans for Common Sense, a Washington-based nonprofit that advocates more efficient treatment, says of the Stiglitz-Bilmes calculations: "I know with absolute certainty that their estimates are low." For instance, he points out, veterans with other options — such as private insurance or a free county clinic — may avoid the VA after hearing of long waits for treatment.

Sullivan, a veteran of the 1991 Gulf War and a former VA project manager, adds, "The VA is the best place to go for mental health care, traumatic brain injury and prosthetics, because they have real combat injury experience of 60-plus years."

But veterans' needs are going unmet because the agency is underfunded, say advocates. Veterans for Common Sense and a California-based organization, Veterans for Truth, have filed a class-action lawsuit against the VA claiming its system for processing PTSD claims has essentially collapsed under the weight of growing demand. "Delays have become an insurmountable barrier preventing many veterans from obtaining health care and benefits," the organizations said in the 2007 complaint launching the suit.[8]

The Bush administration takes the position that "wounded warriors" — the term now in favor — are getting all the service they require. "I believe that we are getting adequate funding," Michael J. Kussman, the VA's undersecretary for health, told the Senate Veterans Affairs Committee last year. "And with your support and the administration's support, we've been very appreciative of the very significant increases in the budget over the last couple years."[9]

Appropriations to the VA for all operations have risen from $47.95 billion in fiscal 2001 to $87.6 billion in the present fiscal year. The latter amount includes about $3 billion in emergency supplemental funding.[10]

For Stiglitz and Bilmes, the VA's reliance on supplementals shows that the agency is pushed past its limits. "The pattern of underfunding . . . has repeated itself every year of the war," they write about the stream of supplemental funds. "The VA has told Congress that it can cope with the surge in demand, despite overwhelming evidence to the contrary."[11]

---

[1] For background, see Peter Katel, "Wounded Veterans," *CQ Researcher*, Aug. 31, 2007, pp. 697-720.

[2] See Joseph E. Stiglitz and Linda J. Bilmes, *The Three Trillion Dollar War: The True Cost of the Iraq Conflict* (2008), pp. 81-90.

[3] *Ibid.*, p. 89.

[4] See "Post Traumatic Stress Disorder: DOD Needs to Identify the Factors Its Providers Use to Make Mental Health Evaluation Referrals for Servicemembers," General Accountability Office, May 2006, p. 20, www.gao.gov/new.items/d06397.pdf. See also "GAO Findings and Recommendations Regarding DOD and VA Disability Systems," Government Accountability Office, May 25, 2007, p. 3, www.gao.gov/new.items/d07906r.pdf.

[5] See Peter Orszag, "Director's Blog," April 8, 2008, http://cboblog.cbo.gov/.

[6] *Ibid.*

[7] *Ibid.*

[8] See *Veterans for Common Sense and Veterans United for Truth, Inc., v. R. James Nicholson, et al.*, July 23, 2007, www.veteransptsdclassaction.org/pdf/courtfiled/veteranscomplaint.pdf. Other documents in the case are available at www.veteransptsdclassaction.org/.

[9] See "Senate Veterans' Affairs Committee Holds Hearing on Veterans' Affairs Health Care Funding," *Congressional Transcripts*, July 25, 2007.

[10] See Daniel H. Else, "Military Construction, Veterans Affairs, and Related Agencies: FY 2008 Appropriations," Congressional Research Service, June 12, 2007, p. 7, www.fas.org/sgp/crs/natsec/RL34038.pdf.

[11] Stiglitz and Bilmes, *op. cit.*, p. 85.

Soldier's Death Reveals Spending Abuses, Corruption

Billions of reconstruction dollars may have been lost.

Staff Sgt. Ryan Maseth died in Iraq, but not by enemy fire. "When Staff Sergeant Maseth stepped into the shower and turned on the water," House Oversight and Government Reform Committee Chairman Henry A. Waxman, D-Calif., wrote in a March 19 letter to Defense Secretary Robert M. Gates, "an electrical short in the pump sent an electrical current through the water pipes to the metal shower hose, and then through Staff Sergeant Maseth's arm to his heart."[1] Contractor KBR was responsible for maintenance on the living quarters at the Radwaniyah Palace Complex in Baghdad, where Maseth died.

In looking into Maseth's death, Waxman's committee staff found reports of the electrocution deaths of at least 12 military personnel since 2003. Waxman's letter prompted a Pentagon investigation, which is still under way. Whether contractor construction or maintenance may have played a part in the other deaths isn't clear, a committee staff member says.

A lawyer for Maseth's family found documents showing that KBR had spotted the hazard that killed him, *The New York Times* reported. But the company's contract may not have required repair. KBR said in a statement that "safety and security . . . remains KBR's priority, and we remain committed to pledging our full cooperation with the agencies involved in investigating this matter."[2]

News of Maseth's death surfaced amid renewed attention to Iraq War spending and concern that billions of dollars in U.S. funds may have been stolen, wasted or otherwise misspent. "The United States is entering its fifth year of efforts to rebuild and stabilize Iraq, but these efforts have neither consistently achieved their desired outcomes nor done so in an economic and efficient manner," U.S. Comptroller General David M. Walker told the Senate Appropriations Committee on March 11.[3]

Walker said waste is the biggest problem for U.S. agencies working in Iraq.

On the Iraqi side, the issue is criminal conduct, Walker said. "With regard to fraud and corruption, that is a major problem with regard to the Iraqi government and Iraqi funds," Walker said. Those funds include, he said, the $45 billion that the United States has spent on Iraq reconstruction.[4]

A subsequent witness spoke in graphic terms about the predominant outcome of that rebuilding work. "Sir, the infrastructure in Iraq is equal to zero," said Radhi Hamza al-Radhi, an Iraqi judge who headed Iraq's Commission on Public Integrity in 2004-2007, which was established to root out corruption in government contracting. "If you visited Baghdad, you would see for yourself that there is no water, no electricity, no sewer systems, no streets. Everything is destroyed."[5]

Al-Radhi resigned his post after death threats and what he termed political pressure from the Iraqi government. Thirty-one of his former staff members and 12 of their

Vietnam turned into an American defeat, and little of the Great Society package survived the decades that followed.

Desert Storm

After Vietnam, U.S. military actions in Grenada, Panama, Somalia, Haiti and the Balkans were all of relatively short duration and low cost. The next major military theater would be the Middle East, specifically, the region bordering the Persian Gulf. The United States had long maintained a political presence there through an alliance with Shah Mohammad Reza Pahlavi. He was overthrown in 1979 and replaced by a Shiite Muslim and fiercely anti-American government. The U.S. government cut relations with Tehran after radical students stormed the American Embassy there and took 52 embassy personnel hostage for more than 14 months.

In 1980, Saddam launched a war against the new government in neighboring Iran. During that eight-year conflict, in which Iraq openly used chemical weapons, the Reagan administration tilted toward Iraq — the enemy of America's new Iranian enemy. Reagan ordered his top diplomatic, military and intelligence officials in 1984 to plan on how to "avert an Iraqi collapse." Paradoxically, during the Iran-Iraq War members of the Reagan administration set up a scheme — later known as "Iran-Contra" — to illegally sell weapons to Iran, using the proceeds to fund

family members have been assassinated, he told the committee. *Portfolio* magazine reported a total of 28 of al-Radhi's former colleagues have been killed.[6]

The former anti-corruption boss fled to the United States, where he is seeking asylum. Some officials who worked with him praise his past efforts. Stuart Bowen Jr., the special inspector general for Iraq reconstruction who had frequently praised al-Radhi, told the Appropriations Committee that he stood by a past statement that the Iraqi is "an honorable man and an effective crime fighter in Iraq."[7]

Bowen and Walker also generally agreed with al-Radhi's explosive allegation that about one-third of U.S. grants and contracts to the Iraqi government wind up in the hands of Shiite and Sunni militias. "I can't attest to that percentage, but it is a significant problem," Bowen said.[8]

Lack of accountability has plagued U.S.-controlled projects as well — apart from heavily reported past episodes such as the disappearance of some of the 363 tons of cash — $12 billion worth — flown into Iraq in 2003-2004.[9]

Defense Department Inspector General Claude Kicklighter told the Senate Appropriations Committee that his agency's investigations of fraud and corruption have yielded $840 million in questionable spending and 34 federal criminal charges, 25 felony convictions and recovery or forfeiture of $11.1 million.

Republicans on the committee challenged none of the expert testimony. "Over the past five years, U.S. programs to bolster anti-corruption institutions in Iraq have been inconsistent, suffering from poor coordination, weak planning and limited resources," Bowen said.[10]

Other U.S. officials have used even stronger language. "Challenges to the rule of law, especially corruption, are enormous," Ryan Crocker, U.S. ambassador to Iraq, told the Senate Armed Services and Foreign Relations committees on April 8. The next day he told the House Foreign Affairs Committee, "We're engaged in doing everything we can to assist on this. We've recently reorganized our own anti-corruption effort within the embassy."[11]

---

[1] See letter, Rep. Henry A. Waxman, chairman, House Oversight and Government Reform Committee, to Secretary of Defense Robert M. Gates, March, 19, 2008, http://oversight.house.gov/documents/20080319091300.pdf. Also, see James Risen, "G.I.'s Death Prompts 2 Investigations of Iraq Electrocutions," *The New York Times*, March 20, 2008, p. A15.

[2] Quoted in Risen, *ibid.*

[3] See David M. Walker, "Stabilizing and Rebuilding Iraq: Actions Needed to Address Inadequate Accountability over U.S. Efforts and Investments," Government Accountability Office, March 11, 2008, www.gao.gov/new.items/d08568t.pdf.

[4] "Senate Appropriations Committee Holds Hearing on Iraq Funding Waste, Fraud, and Abuse," *Congressional Transcripts*, March 11, 2008.

[5] *Ibid.*

[6] See Christopher S. Stewart, "The Betrayal of Judge Radhi," *Portfolio.com*, April, 2008, www.portfolio.com/news-markets/international-news/portfolio/2008/03/17/Iraq-Top-Fraud-Cop-Judge-Radhi.

[7] *Ibid.* Quoted in Alissa J. Rubin, "Blaming Politics, Iraqi Antigraft Official Vows to Quit," *The New York Times*, Sept. 7, 2007, p. A12. Also, see Matt Kelley, "Iraqi's resignation hurts fight against corruption," *USA Today*, Sept. 10, 2007, p. A4.

[8] *Ibid.*

[9] For background, see Peter Katel, "New Strategy in Iraq," *CQ Researcher*, Feb. 23, 2007, pp. 169-192.

[10] See "Senate Appropriations Committee Holds Hearing on Iraq Funding Waste, Fraud and Abuse," *op. cit.*

[11] See "Senate Armed Services Committee Holds Hearing on Iraq," April 8, 2008; "Senate Foreign Relations Committee Holds Hearing on Iraq," *Congressional Transcripts*, April 8, 2008; "House Foreign Affairs Committee Holds Hearing on the Crocker/Petraeus Iraq Report," *Congressional Transcripts*, April 9, 2008.

---

the "contra" guerrillas fighting the left-wing government of Nicaragua.[36]

U.S. policy shifted in 1990, when Saddam ordered his forces to invade tiny Kuwait, declaring the oil-producing nation a province of Iraq. The favorable U.S. attitude to Iraq during the war with Iran may have persuaded Saddam that he could count on President George H. W. Bush to turn a blind eye to the invasion. But Kuwait's long and close alliance with the United States weighed far more heavily. And neighboring Saudi Arabia, America's most important Arab friend in the region — and at the time the world's single biggest oil exporter — feared it might be next on Saddam's list.

President Bush assembled a military coalition that would drive Iraq out of Kuwait in 1991 in a massive military operation dubbed Operation Desert Storm. Victory came after about 100 hours of ground fighting.

The American military contingent comprised more than 540,000 troops; coalition members contributed another 270,000 personnel. When it came to paying for the conflict, though, the roles were reversed. In cash terms, the allies paid most of the money — $48 billion paid by coalition members, in addition to $6 billion in fuel and other material. The U.S. share was $13 billion.

The war seemed to end Saddam's territorial ambitions. But it opened a new chapter in efforts by the United

States and other nations to limit Saddam's attempts to repress rebellious sectors of the Iraqi population — members of the Shiite majority in southern Iraq, and Kurds, a Muslim but non-Arab people in the north. The United States and other U.N. members were also involved in trying to pressure Hussein into ending what was believed to be his development and production of weapons of mass destruction (WMD).

U.S. forces protected what became a Kurdish enclave in the northern mountains. And U.S. warplanes patrolled over both southern and northern Iraq, enforcing U.N. Security Council resolutions prohibiting Saddam from using his air power to bomb and strafe the Kurds and Shiites (though the latter did suffer Saddam's retaliation before the protective measures took effect). Naval vessels from the United States and seven other countries also enforced economic sanctions designed to force Iraqi compliance with U.N. weapons inspections.[37]

The forces used in these operations reached nowhere near the strength of current U.S. military contingents in Iraq and Afghanistan, but they were sizable. Davis and his trio of economists from the University of Chicago calculated that about 28,000 troops, 30 ships, and 300 aircraft participated in the various "containment" efforts. The economists estimated the total cost at about $14.5 billion a year.[38]

Estimates and Reality

Only about seven months after President George W. Bush was sworn into office, suicide terrorists from al Qaeda — the Afghanistan-based terrorist network headed by Osama bin Laden — carried out the attacks of Sept. 11, 2001.

In their wake, the Bush administration launched an intensive search for possible connections between al Qaeda and Saddam's dictatorship. Meanwhile, U.S. forces spearheaded the invasion of Afghanistan in October, 2001, an operation that toppled the fundamentalist Taliban, who had allowed bin Laden to set up his headquarters in their country.

And on a third track, the Bush administration ratcheted up the pressure on Saddam to dismantle the WMD development and manufacturing operations that — according to U.S. intelligence agencies — Iraq still maintained.

By 2002, Bush was signaling that he planned to topple Saddam by military means. Indications were strong

enough that Sen. Kent Conrad, D-N.D., chairman of the Senate Budget Committee, and Rep. John M. Spratt Jr., D-S.C., top Democratic member of the House Budget Committee, asked the Congressional Budget Office (CBO) to project the costs of war in Iraq.

In September of that year, the CBO estimated the war would cost $6 billion to $9 billion a month and that the monthly cost of an occupation would run from $1 billion to $4 billion. The budget specialists said they couldn't estimate the costs of reconstruction.[39]

That same month, Lindsey, then the president's chief economic adviser, estimated war costs at $100 billion to $200 billion. (That projection led to Lindsey's replacement three months later.)[40]

In October, Congress authorized "the use of armed forces against Iraq." By the end of 2002, the director of the Office of Management and Budget (a White House position), put war costs at $50 billion to $60 billion.[41]

By 2003, a combination of supposed intelligence reports and Saddam's on-again, off-again handling of U.N. weapons-inspection demands persuaded the administration that the Iraqi dictator (and his two sons, who held top Iraqi government posts) represented a global menace that could not be contained by peaceful means.

On March 17, 2003, Bush delivered a public ultimatum. In a televised speech, he warned Saddam and his sons to leave Iraq "within 48 hours" or face "military conflict."[42]

Two days later, Bush launched the invasion with a bomb and missile strike on a complex of buildings where Saddam had been reported to be hiding (he was in the same compound, its owner said, but not in a targeted building). A fast-moving ground assault followed. By April 9, 2003, U.S. troops had started taking over Baghdad, even as looting broke out there and in other cities where Iraqi soldiers fled or surrendered.[43]

In Baghdad, the looting — carried out in full view of U.S. troops as well as numerous journalists — dominated news coverage and raised a barrage of questions about the near-term stability of a country over which the United States was assuming control. Defense Secretary Rumsfeld responded furiously at a Pentagon news conference. "Stuff happens," he said. "Freedom's untidy. And free people are free to make mistakes and commit crimes and do bad things. They're also free to live their lives and do wonderful things. And that's what's going to happen here."[44]

Rumsfeld's optimism was shared by his colleagues. "We're dealing with a country that can really finance its own reconstruction, and relatively soon," Deputy Defense Secretary Paul D. Wolfowitz had told the House Appropriations Subcommittee on Defense five days after the war began, citing Iraq's potential oil revenues.[45]

But by October of that year, the administration proposed spending $20.3 billion on Iraq reconstruction projects. The Senate trimmed that figure to $18.4 billion and also voted to lend rather than grant the money. But key senators backed down on the loan proposal after Bush threatened a veto.[46]

As for combined military and reconstruction costs, these soon outstripped administration projections. Costs for the first two fiscal years of Iraq operations alone (not including Afghanistan spending) totaled nearly $129 billion. By early 2008, spending had risen to almost $526 billion.[47]

In the intervening years, Iraq costs rose from $53 billion in fiscal 2003 to $76.4 billion so far in fiscal 2008, with $82.3 billion more expected to be requested.[48]

CURRENT SITUATION

Iraq in Politics

Vying for the Democratic presidential nomination, Sens. Clinton and Obama are trying to tie the United States' economic slide to the war in Iraq. In doing so, they're trying to reawaken an apparently fading public interest in the conflict.

"When you're spending over $50 to fill up your car because the price of oil is four times what it was before Iraq, you're paying a price for this war," Obama, an Illinois Democrat, told a rally at the University of Charleston, in West Virginia. "When Iraq is costing each household about $100 a month, you're paying a price for this war."[49]

Obama was broadcasting the thesis developed by Stiglitz and Bilmes — and taking advantage of the fact that he opposed the war from the beginning. Clinton, too, though she voted for the 2002 "war powers" resolution that authorized Bush to order the invasion, is linking the economy to the war.

"We spend $12 billion a month in Iraq, and that does affect the economy," New York Democrat Clinton said, also echoing Stiglitz and Bilmes. "That's one of the reasons

we've gone into more and more debt. We've got to begin not only to withdraw our troops but bring that money back home."

No data have emerged publicly on whether the candidates, along with antiwar critics in general, have persuaded citizens that Iraq spending helped create today's economic climate. But in late February and early March, a survey by the nonpartisan Pew Research Center for the People and the Press found that only 28 percent of adults questioned knew that about 4,000 U.S. military personnel had died in Iraq. Less than a year earlier, more than half of respondents accurately estimated the number of deaths. The survey correlated the drop in awareness to a decline in media coverage of the war.[50]

On the Republican side, Sen. McCain hasn't shown any inclination to get drawn into the dollars-and-cents debate on war costs. But he isn't shying away entirely. "We must increase levels of reconstruction assistance so that Iraq's political and economic development can proceed in the security that our forces and Iraqi security forces provide," he said in a speech to the Veterans of Foreign Wars in Kansas City, Mo., on April 7. McCain didn't mention amounts. And a statement he gave at a Senate Armed Services Committee hearing swung more towards Iraqi government funding. In that statement, McCain called on the Iraqi government to contribute to the Commander's Emergency Relief Program, a U.S. fund that pays for reconstruction and jobs programs. Other rebuilding projects should be entirely Iraqi-funded, he said.[51]

But two of his key supporters have responded to the war-costs criticism from the Democratic contenders and other war critics. "Today's antiwar politicians have effectively turned John F. Kennedy's inaugural address on its head, urging Americans to refuse to pay any price, or bear any burden, to assure the survival of liberty," Sens. Joseph I. Lieberman, I-Conn., and Lindsey Graham, R-S.C., wrote in *The Wall Street Journal* in early April. "There is no question the war in Iraq — like the Cold War, World War II and every other conflict we have fought in our history — costs money. But as great as the costs of this struggle have been, so too are the dividends to our national security from a successful outcome, with a functioning, representative Iraqi government and a stabilized Middle East."[52]

Lieberman and Graham published their op-ed on the eve of congressional testimony by Gen. Petraeus and U.S. Ambassador to Iraq Crocker. Arguably, their appearance,

Are the Iraq War's results worth the price paid?

YES James J. Carafano
Senior Research Fellow
Heritage Foundation

Written for *CQ Researcher*, April 2008

Critics of the Iraq War repeatedly focus on its financial costs, as if dollar signs alone explain why the United States should leave precipitously, whether or not the Iraqi government can provide security and services for its people.

Unfortunately, most attempts to estimate the actual costs of the war are driven by political considerations and ultimately end up as convenient overestimations. The recent book by Joseph Stiglitz and Linda Bilmes, *The Three Trillion Dollar War*, is one such example.

The reality is that the overall U.S. defense budget for both Iraq and Afghanistan still makes up only a modest proportion of the federal budget. Indeed, spending on Social Security, Medicare and Medicaid far exceeds what taxpayers spend to provide for defense.

As a percentage of the U.S. economy, defense spending remains historically low. Over the last 40 years, America has spent an average of 5.7 percent of gross domestic product (GDP) annually on defense. Yet Iraq War costs have ranged from 0.5 percent of GDP in fiscal 2003 to 1.3 percent in 2008.

By comparison, it cost almost 9 percent of GDP to pay for the Vietnam War and 14 percent for the Korean War. Also, the U.S. economy is significantly larger than during any previous conflict and can more easily absorb the costs today.

Moreover, our resources are not just being invested in bullets and bombs. We're also rebuilding schools, courthouses and other infrastructure, as well as Iraq's security forces, so the country can thrive long after American forces depart.

There's no denying the Iraq War is costing more than anticipated. However, spending alone doesn't tell the whole story. We must also consider what the war means to the wider war on terror and America's security.

Iraq, unlike Afghanistan, is a potentially wealthy country at the heart of the Middle East. Currently both al Qaeda (through its subsidiary al Qaeda in Iraq) and Iran (using its Qods Force unit) have committed their resources to America's defeat.

An early pullout would cede the battlefield to al Qaeda and Iran and allow Iraq to fall into chaos. Eventually, Americans could well be required to intervene again.

Meanwhile, America's enemies would interpret our retreat as a testament to their strength and our weakness. American credibility among Islamic moderates, our true allies in the region, would suffer. We must prevent this outcome by spending the reasonable amount needed to defeat al Qaeda and Iranian proxy forces while upholding our moral commitment to Iraqis. That's a wise investment in a more secure future.

NO Brian Katulis
Senior Fellow
Center for American Progress

Written for *CQ Researcher*, April 2008

Five years into the Iraq War, the results do not justify the substantial costs to date; nor do the potential results from maintaining the current policy of strategic confusion justify staying in Iraq with no end in sight for decades to come — the policy that most conservatives propose today.

More than 4,000 families have lost loved ones serving in the military in Iraq. All too often forgotten — particularly in the Bush administration's mishandling of veterans' affairs — are the tens of thousands, scarred both physically and mentally by the war. Beyond the human costs, U.S. taxpayers are now spending about $12 billion a month for the war — all for a policy that still has no clear and realistic end.

In addition, overall military readiness has suffered as a result of the extended troop deployments. Now the vast majority of active-duty units are suffering from critical shortages in personnel and equipment, resulting in low readiness levels.

Beyond these immediate costs, the Iraq War has resulted in broader strategic failures. When historians look back on 2003 to 2008, they will see a period when America's ability to shape and influence events around the world became severely constrained. Preventing advances in nuclear weapons programs in North Korea and Iran, completing the mission left unaccomplished in Afghanistan and eradicating the threat posed by the global al Qaeda movement have all fallen in the list of national security priorities because of the fixation on Iraq.

For what benefit? Saddam Hussein was a brutal dictator, but he was contained, and his regional and global influence was weakening. It is good that he is no longer in power, but the sad truth is that the living situation and human-rights conditions for most Iraqis is not much better — and for some groups like Christians and women — conditions have deteriorated.

America's leaders must recognize that it is past time to begin drawing down the troops — that the strategic costs outweigh the benefits of staying with some open-ended commitment, whether it is wrapped up in a banner of "strategic patience" or "conditional engagement."

Until America's leaders demonstrate the courage to take back control of U.S. national security by setting a course of redeployment of U.S. forces, the United States will remain trapped in this quagmire, paying rising costs for a policy with no end in sight.

along with a continuation of recent combat between the U.S.-backed Iraqi government and a militia commanded by Moqtada al-Sadr, a nationalist, anti-American Iraqi Shiite leader, would help return the war to public awareness. Thirty-four Americans were killed in March, the month that the Iraq government launched an offensive against Sadr, though not all those deaths could be attributed to those clashes. During the first week of April, however, five of 11 U.S. military deaths did result from combat with Sadr's militia.[53]

In the days leading up to the Petraeus-Crocker appearance, attempts to link Iraq and the economy weren't noticeable in campaigns for congressional seats — all 435 House seats and 33 Senate posts are up for election in November. "They do talk all the time about spending for the war — that those dollars haven't been used for other things, like health care," says Stuart Rothenberg, a specialist in congressional elections and publisher of the nonpartisan *Rothenberg Political Report*. "I haven't heard anyone say there's a causal relationship between [Iraq] spending and the recession." He adds, "I could see how they could."

Funding Fight

Even as presidential candidates, economists, administration officials and others joust over the big picture of war spending, lawmakers and administration officials have started skirmishing over the newest appropriation proposals for military and reconstruction operations in Iraq.

Democratic critics demand to know why Iraq, flush with oil revenue, isn't spending more of its own money on reconstruction. The Bush administration insists that Iraq's contribution is growing.

The fight is coming to a head over the Bush administration's latest "emergency supplemental" request for funds to maintain operations in Iraq. The administration is expected to seek $82.3 billion to cover operations for the remainder of the current fiscal year, which will end Sept. 30.[54] Supplementals — money requests that aren't part of the regular budget — have been the major source of Iraq funds since the beginning of the conflict. Reliance on supplementals is controversial in itself. But whether to wean the Iraqi government from its reliance on U.S. reconstruction funds presents a more immediate issue — and one with potential to arouse taxpayer interest in a time of economic insecurity marked by higher gasoline prices.

Iraq saw its petroleum revenues increase from $31 billion in 2006 to $41 billion in 2007, the Government Accountability Office (GAO) reported. Sen. Ben Nelson, D-Neb., told *USA Today* that he expected Iraq to earn $60 billion from oil this year.[55]

Opening a Senate Armed Services Committee hearing where Petraeus and Crocker would testify, panel Chairman Carl Levin, D-Mich., said he heard the following story from an unnamed senior U.S. military officer: "He asked an Iraqi official, 'Why is it that we're using our U.S. dollars to pay your people to clean up your towns instead of you using your funds?' The Iraqi replied, 'As long as you are willing to pay for the cleanup, why should we do it?' "[56]

To illustrate the point, Levin cited a January 2008 report by the Special Inspector General for Iraq Reconstruction showing the Iraqi government spent only 23 percent of its $6.2 billion capital budget for 2006.[57]

Matters apparently didn't improve in 2007. Levin noted that the GAO reported it could not confirm an administration claim that the Iraqi government had spent 24 percent of its capital budget as of July 15, 2007. The GAO said Iraqi government data showed that only 4.4 percent of the budget had been spent as of August 2007.[58]

McCain, the senior Republican on Armed Services, showed some sensitivity to the issue of the Iraqi government's use of U.S. funds. Acknowledging that the Iraqi government "continues to take in revenues it finds difficult to disburse," McCain proposed that Iraq put some of its money in the main U.S. fund that pays for reconstruction projects — the Commanders Emergency Response Program.[59]

The administration is expected to ask for $500 million for that program. How much an Iraqi contribution — if it's made — might reduce that request, wasn't immediately clear.

As for major rebuilding — of power stations, oil production facilities and the like — Ambassador Crocker told Armed Services that the Bush administration has turned a page. "The era of U.S.-funded major infrastructure projects is over," the envoy said. "We are seeking to ensure that our assistance, in partnership with the Iraqis, leverages Iraq's own resources."[60]

However the fight over the next big Iraq appropriation is resolved, the conflict over reliance on supplementals remains. "That is unusual," says Kosiak of the Center for Strategic and Budgetary Assessments. "In the case of Vietnam and Korea, the last wars of any significant duration, in the first years they did use supplementals but then attached war funding to regular budget requests."

Supplementals are taken up on a faster track and come with less in the way of supporting material, Kosiak says. The administration did use the conventional process for the 2008 fiscal year (which will end on Sept. 30), but returned to the supplemental route in the present budget cycle for fiscal 2009.

"I haven't met one person who believes this is a good way to do business," says Bernstein of the Economic Policy Institute, saying that attitude crosses party lines. "It's obviously a tactic to keep these expenditures off-budget."

But Zakheim, on whose watch the first supplementals went to Capitol Hill, says the rapidly changing nature of the conflict left no other option. "When you put together a budget, you're putting it together roughly 18 months before the money is actually to be available on the ground," he says. "The nature of this kind of war is so highly unpredictable that to risk through the normal budget process defies logic. You can't have the resources in the field immediately."

OUTLOOK

Continued Presence

Critics and supporters alike of the decision to invade Iraq tend to agree that the United States will stay in Iraq or its neighborhood for the foreseeable future — a presence that will keep costing billions of dollars.

Davis of the University of Chicago says that whatever happens to the military presence in Iraq there's no question that troops will be stationed nearby. "It's a politically and militarily unstable part of the world that sits on most of the world's low-cost oil reserves. Whether we have forces in Iraq or not is going to be influenced by elections and political decision-making. But I have no doubt that there will be a major U.S. military presence in the region; we have had one for 50 years or so."

To the extent that conditions today set the stage for conditions 10 years from now, the official short-range forecast from the top U.S. military commander in Iraq offers little in the way of conclusive judgments. During

two days of hearings before Senate and House committees Petraeus refused to be pinned down even concerning the pace at which the U.S. would keep drawing down forces in the wake of last year's "surge" of troops.

Explaining his recommendation to Bush for a 45-day halt in withdrawals now under way, Petraeus said that his proposal "does not allow establishment of a set withdrawal timetable." He added, "It does provide the flexibility those of us on the ground need to preserve the still-fragile security gains our troopers have fought so hard and sacrificed so much to achieve." By July, the total U.S. military contingent in Iraq is expected to number about 140,000 personnel.[61]

In a recent interview, former Assistant Defense Secretary Zakheim says he's advocated for some time that the United States keep 70,000 troops stationed on Iraq's borders. That move could put troops facing Iran, Syria, Saudi Arabia, and Kuwait. "If the goal is to protect stability in the region, that is what it's going to take."

And Zakheim says he's optimistic that "at a minimum" the U.S. intervention will have brought that stability. "Petraeus seems to have turned the corner," Zakheim adds. "I'm reasonably optimistic. He's had a huge impact."

It would be hard to find a more opposite view than that of Paul Sullivan, executive director of Veterans for Common Sense. "The war is lost — irretrievably," says Sullivan, a veteran of the 1991 Persian Gulf War and a former Veterans Administration data project manager. "We can temporarily hold violence down, but until the Iraqis have a legitimate, functioning government and the electricity, water, schools and economic infrastructure are rebuilt, Iraq will remain in a state of violent anarchy."

On the home front, Sullivan says only major funding increases for the Department of Veterans Affairs will head off catastrophe in veteran care. "On the current course we see disaster on the horizon — an increase in the number of broken families, divorce, increased unemployment, drug and alcohol [abuse], homelessness, preventable suicides," he warns.

Hormats of Goldman, Sachs agrees with Sullivan — and with *Three Trillion Dollar War* authors Stiglitz and Bilmes — that the demand for veterans health care will be far higher than the Bush administration projected. And that demand will have to be met. "Our nation has a moral obligation to provide the best care possible," Hormats says.

And he points to another expense category often overlooked outside of specialist circles. It's called "reset" in Pentagon jargon — the reconditioning of military equipment worn down by the grind of performing in a combat zone. "A lot of weapons have been destroyed or deteriorated to the point that they're unusable," Hormats says. "The Defense Department has to repair or replace them. That's expensive."

Bilmes sees little chance of Iraq intervention costs coming down in the next two years. The big question, she says, centers on what the next president does to start paying the bills for the operation — "whether we continue to borrow all this money or ask people to tighten their belts or raise taxes or buy war bonds."

How those decisions turn out, she says, "will determine not just the cost of the war, but how much of the war we're asking our children to shoulder and how much interest we have to pay on all the money we're borrowing."

NOTES

1. See Peter Orszag, "Estimated Costs of U.S. Operations in Iraq and Afghanistan and of Other Activities Related to the War on Terrorism," Congressional Budget Office, Oct. 24, 2007, p. 3, www.cbo.gov/ftpdocs/86xx/doc8690/10-24-Cost OfWar_Testimony.pdf.

2. Joseph E. Stiglitz and Linda J. Bilmes, *The Three Trillion Dollar War: The True Cost of the Iraq Conflict* (2008).

3. "Bush Dismisses Iraq Recession," "The Today Show," NBC, March 18, 2008, http://youtube .com/watch?v=lIbdnM8Ts88&feature=related.

4. See "Senate Armed Services Committee Holds Hearing on Iraq," *Congressional Transcripts*, April 8, 2008.

5. *Ibid.*

6. *Ibid.*

7. "President Bush Visits Dayton, Ohio, Discusses Global War on Terror," The White House, March 27, 2008, www.whitehouse.gov/news/releases/2008/03/20080327-2.html.

8. *Ibid.* See also Stiglitz and Bilmes, *op. cit.*, pp. 32-60, 114-131.

9. See Orszag, *op. cit.*; "House Budget Committee Holds Hearing on the Costs of the Iraq War," House Budget Committee, Oct. 24, 2007, *Congressional Transcripts*; Ken Dilanian, "Wars may cost $2.4T," *USA Today*, Oct. 24, 2007, p. A1; "War At Any Price?: The Total Economic Costs of the War Beyond the Federal Budget," Joint Economic Committee majority staff, November 2007, http://jec.senate.gov/Documents/Reports/11.13.07IraqEconomicCostsReport.pdf.

10. See Orszag, *op. cit.*; and "House Budget Committee Holds Hearing on the Costs of the Iraq War," *op. cit.*; Dilanian, *op. cit.*; Joint Economic Committee majority staff, *op. cit.*

11. See Linda Bilmes and Joseph E. Stiglitz, "The Economic Costs of the Iraq War: An Appraisal Three Years After the Beginning of the Conflict," February 2006, National Bureau of Economic Research, www2 .gsb.columbia.edu/faculty/jstiglitz/download/2006_ Cost_of_War_in_ Iraq_NBER.pdf.

12. "Press Briefing by Dana Perino," The White House, March 10, 2008, www.whitehouse.gov/news/releases/2008/03/20080310-4.html.

13. Steven J. Davis, *et al.*, "War in Iraq versus Containment," National Bureau of Economic Research, March 2006, www.nber.org/papers/w12092.pdf.

14. "Bush Dismisses Iraq Recession," Feb. 18, 2008, http://video.google.com/videoplay?docid=-774425 8843658177585.

15. See Joint Economic Committee, *op. cit.* For monthly Iraq operations cost, see Orszag, *op. cit.*, p. 2.

16. For statistics, see Louis Uchitelle, "80,000 Jobs Lost; Democrats Urge New Aid Package," *The New York Times*, April 5, 2008.

17. Quoted in *ibid.*

18. For foreclosure proceedings statistic, see "Chairman Ben S. Bernanke at the Independent Bankers of America Annual Convention," Federal Reserve, March 4, 2008, www.federalreserve.gov/newsevents/speech/bernanke20080304a.htm. Extensive coverage of the ongoing events includes Julie Creswell, "A Nervous Wall St. Seems Unsure What's Next," *The New York Times*, March 31, 2008, p. C1; David Leonhardt, "Can't Grasp the Credit Crisis? Join the

Club," *The New York Times*, March 19, 2008, p. A1; Carrick Mollenkamp and Mark Whitehouse, "Banks Fear a Deepening of Turmoil," *The Wall Street Journal*, March 17, 2007, p. A1, and Marcia Clemmitt, "Mortgage Crisis," *CQ Researcher*, Nov. 2, 2007, pp. 913-936.

19. *Ibid.*

20. Joint Economic Committee, *op. cit.*, p. 22.

21. For 2003 oil prices see, "All Countries Spot Price FOB Weighted by Estimated Export Volume (Dollars per Barrel)," Energy Information Administration, http://tonto.eia.doe.gov/dnav/pet/hist/wtotworldw.htm.

22. See Lawrence B. Lindsey, "What the Iraq war will cost the U.S.," *Fortune*, Jan. 11, 2008, http://money.cnn.com/2008/01/10/news/economy/costofwar.fortune/index.htm?postversion=2008011112.

23. See William D. Nordhaus, "The Economic Consequences of a War With Iraq," Yale University, Oct. 29, 2002, p. 41, www.econ.yale.edu/~nordhaus/iraq.pdf. A shorter version was published in the *New York Review of Books*, Dec. 5, 2002, www.nybooks.com/articles/15850.

24. Quoted in Stiglitz and Bilmes, *op. cit.*, pp. 7-8.

25. Concerning Powell, see Michael Gordon, "Fateful Choice on Iraq Army Bypassed Debate," *The New York Times*, March 17, 2008, p. A1.

26. Quoted in Thomas E. Ricks, *Fiasco* (2006), p. 164.

27. See David Von Drehle, "Economic Costs Could Weaken Bush Politically," *The Washington Post*, March 16, 2003, p. A13.

28. See "President Bush Visits Dayton, Ohio, Discusses Global War on Terror," The White House, March 27, 2008, www.whitehouse.gov/news/releases/2008/03/20080327-2.html.

29. See "President Bush Discusses Iraq," The White house, April 10, 2008, www.whitehouse.gov/news/releases/2008/04/print/20080410-2.html.

30. *Ibid.*

31. Stiglitz and Bilmes, *op. cit.*, pp. 54-55.

32. Unless otherwise indicated, material in this subsection is drawn from Robert D. Hormats, *The Price of Liberty: Paying for America's Wars* (2007).

33. Quoted in *ibid.*, p. 26.

34. See "Mobilization: The U.S. Army in World War II," undated, U.S. Army, www.history.army.mil/documents/mobpam.htm.

35. Quoted in *ibid.*

36. See "National Security Decision Directive 139," April 5, 1984, www.gwu.edu/~nsarchiv/NSAEBB/NSAEBB82/iraq53.pdf. The document is part of a collection, "Shaking Hands With Saddam Hussein: The U.S. Tilts Toward Iraq, 1980-1984," The National Security Archive, www.gwu.edu/~nsarchiv/NSAEBB/NSAEBB82/.

37. See Davis, *et al.*, *op. cit.*

38. *Ibid.*, pp. 8-18.

39. See "Estimated Costs of a Potential Conflict With Iraq," Congressional Budget Office, September 2002, p. 4, www.cbo.gov/ftpdoc.cfm?index=3822&type=0.

40. See Elizabeth Bumiller, "White House Cuts Estimate of Cost of War With Iraq," *The New York Times*, Dec. 31, 2002, p. A1; and Lindsey, *op. cit.*

41. *Ibid.*

42. See "Bush: 'Leave Iraq within 48 hours,'" CNN, March 17, 2003, www.cnn.com/2003/WORLD/meast/03/17/sprj.irq.bush.transcript.

43. For Saddam's whereabouts during raid, see Robert F. Worth, "Advice of Iraqi, Now in Beirut Cell, Finally Heeded," *The New York Times*, April 2, 2008.

44. See "DOD News Briefing — Secretary Rumsfeld and Gen. [Richard B.] Myers," U.S. Department of Defense, April 11, 2003, www.defenselink.mil/transcripts/transcript.aspx?transcriptid=2367.

45. See "House Appropriations Subcommittee on Defense Holds Hearing on FY2004 Appropriations," *Congressional Transcripts*, March 27, 2003.

46. See Juliet Eilperin, "Senators Overturn Vote on Aid to Iraq," *The Washington Post*, Oct. 30, 2003, p. A6; "Splitting Hairs Over Iraq's Reconstruction," *The New York Times*, Nov. 9, 2003, Sect. 4, p. 2.

47. See Amy Belasco, "The Costs of Iraq, Afghanistan, and Other Global War on Terror Operations Since 9/11," Congressional Research Service, updated Feb. 8, 2008, pp. 11-12, www.fas.org/sgp/crs/natsec/RL33110.pdf.

48. *Ibid.*, p. 8.

49. Quoted in Jeff Zeleny and Michael Cooper, "Obama Links Effects of War Costs to Fragility in the Economy," *The New York Times*, March 21, 2008, p. A19.

50. For survey data see "Awareness of Iraq War Fatalities Plummets," Pew Research Center for the People and the Press, March 12, 2008, http://people-press.org/reports/display.php3?ReportID=401.

51. "Remarks by John McCain to the Members of the Veterans of Foreign Wars," McCain campaign Web site, April 7, 2008, http://johnmccain.com/Informing/News/Speeches/3d837545-5ac8-4124-929c-33c3f0ee9fe5.htm. See also "Text of Sen. John McCain's Opening Statement at Senate Armed Services Committee," *USA Today*, April 8, 2008, www.usatoday.com/news/mmemmottpdf/McCain-Iraq-Hearing-4-8-2008.pdf.

52. Joe Lieberman and Lindsey Graham, "Iraq and Its Costs," *The Wall Street Journal*, April 7, 2008, p. A13, http://online.wsj.com/article_print/SB120752308688293493.html.

53. See "Iraq Index," Brookings Institution, updated March 31, 2008, p. 18, www.brookings.edu/saban/~/media/Files/Centers/Saban/Iraq%20Index/index.pdf. Also see Leila Fadel, "Sadr cancels million-man march; 11 U.S. dead since Sunday," McClatchy Newspapers, April 8, 2008, www.mcclatchydc.com/homepage/story/32986.html.

54. See Belasco, *op. cit.*

55. See Matt Kelley, "Dems take aim at Iraq reconstruction," *USA Today*, April 8, 2008, p. A8. For past oil revenues, see "Stabilizing and Rebuilding Iraq: Actions Needed to Address Inadequate Accountability over U.S. Efforts and Investments," Government Accountability Office, March 11, 2008, p. 22, www.gao.gov/new.items/d08568t.pdf.

56. "Opening Statement of Sen. Carl Levin, Senate Armed Services Committee Hearing on the Situation in Iraq with Ambassador Crocker and Gen. Petraeus," press release, April 8, 2008, http://levin.senate.gov/newsroom/release.cfm?id=295684.

57. See "Update on Iraq Reconstruction — Section 2," Special Inspector General for Iraq Reconstruction, Jan. 30, 2008, p. 123, www.sigir.mil/reports/Quarterly Reports/Jan08/pdf/Section2_-_January_2008.pdf.

58. "Iraq Reconstruction: Better Data Needed to Assess Iraq's Budget Execution," Government Accountability Office, January 2008, p. 1, www.gao.gov/new.items/d08153.pdf.

59. McCain opening statement, *op. cit.*, http://tpmmuckraker.talkingpointsmemo.com/2008/04/mccain_on_iraq_we_should_choos.php.

60. "Testimony of Ambassador Ryan C. Crocker Before the Senate Armed Services Committee," April 8, 2008, http://armed-services.senate.gov/statemnt/2008/April/Crocker%2004-08-08.pdf.

61. Quoted in Thom Shanker and Steven Lee Myers, "General Resists Timetable for Withdrawal of Troops in Iraq," *The New York Times*, April 8, 2008, www.nytimes.com/2008/04/09/world/middleeast/08cnd-petraeus.html?hp.

BIBLIOGRAPHY

Books

Feith, Douglas J., *War and Decision: Inside the Pentagon at the Dawn of the War on Terrorism*, Harper, 2008.
The undersecretary of Defense for policy when the war was planned and begun explains how errors were made and defends himself against charges that pitfalls weren't foreseen.

Hormats, Robert D., *The Price of Liberty: Paying for America's Wars*, Times Books, 2007.
A Wall Street banker with lengthy government experience examines the financing of major U.S. military conflicts and the political and economic issues involved.

Ricks, Thomas E., *Fiasco: The American Military Adventure in Iraq*, Penguin, 2006.
A Pulitzer Prize-winning reporter provides one of the best accounts of the war on the ground and how little it corresponded to pre-invasion plans.

Stiglitz, Joseph E., and Linda J. Bilmes, *The Three Trillion Dollar War: The True Cost of the Iraq Conflict*, W.W. Norton, 2008.
A Nobel Prize-winning economist (Stiglitz) and a federal budget expert make a detailed case that the Iraq war is damaging the U.S. economy.

Articles

Kaplan, Fred, "This Is Not an Emergency: Supplemental war funds are a backdoor way to boost the defense budget," *Slate*, Feb. 20, 2008, www.slate.com/id/2184804.
The online magazine's military-affairs correspondent argues against funding war via "supplemental" budget requests.

Krueger, Alan B., "The Cost of Invading Iraq: Imponderables Meet Uncertainties," *The New York Times*, March 30, 2006, p. C1.
The *Times*' then-economics columnist, a Princeton professor, examines the complexities of running a cost-benefit analysis of a war.

Shapiro, Robert, "The Cost of Toppling Saddam: Will an Iraq war hurt the economy?," *Slate*, Oct. 2, 2002, www.slate.com/id/2071811.
Before the Iraq War, a Clinton administration economist predicts it will not hurt the economy.

Stevenson, Richard W., "War Budget Request More Realistic But Still Uncertain," *The New York Times*, Sept. 10, 2003, p. A1.
The Bush administration acknowledges for the first time that Iraq War costs will be far higher than projected.

Weisman, Jonathan, and Juliet Eilperin, "In GOP, Concern Over Iraq Price Tag," *The Washington Post*, Sept. 26, 2003, p. A1.
Some early uneasiness over Iraq War costs came from President George W. Bush's own party.

White, Josh, " 'Hidden Costs' Double Price of Two Wars, Democrats Say," *The Washington Post*, Nov. 13, 2007, p. A14.
Democratic congressional staffers come up with Capitol Hill's first trillion-dollar-plus forecast of war costs.

Zorpette, Glenn, "Keeping Iraq in the Dark," *The New York Times*, March 11, 2008, p. A23.
The executive editor of *IEEE Spectrum*, the magazine of the Institute of Electrical and Electronics Engineers, examines Iraq's continuing electricity crisis.

Reports and Studies

"Integrated Strategic Plan Needed to Help Restore Iraq's Oil and Electricity Sectors," *Government Accountability Office*, May 2007, www.gao.gov/new.items/d07677.pdf.
Congress' auditors examine the extent of damage to the two most critical elements of Iraq's infrastructure, concluding that billions of dollars more are required for rebuilding — even after the U.S.' expenditure of $8.9 billion.

Davis, Steven J., et al., "War in Iraq versus Containment," *National Bureau of Economic Research*, March 2006, www.nber.org/papers/w12092.pdf.
A trio of economists from the University of Chicago analyzes the projected costs of the Iraq intervention against the costs of the pre-invasion "containment" operations.

Grasso, Valerie Bailey, "Defense Contracting in Iraq: Issues and Options for Congress," *Congressional Research Service*, updated Jan. 29, 2008, www.fas.org/sgp/crs/natsec/RL33834.pdf.
The nonpartisan agency lays out the extent of military contracting and the choices lawmakers face.

Walker, David M., "Stabilizing and Rebuilding Iraq: Actions Needed to Address Inadequate Accountability over U.S. Efforts and Investments," *Government Accountability Office*, March 11, 2008, www.gao.gov/new.items/d08568t.pdf.
In one of his last projects before resigning to head a new foundation, the GAO director examines the state of spending controls on U.S.-funded projects and concludes the Defense Department still lags in overseeing contractors' work.

For More Information

American Enterprise Institute, 1150 17th St., N.W., Washington, DC 20036; (202) 862-5800; www.aei.org. Conservative think tank has strong links to the Bush administration and the "surge" strategy.

Brookings Institution, 1775 Massachusetts Ave., N.W., Washington, DC 20036; (202)-797-6000; www.brookings.edu. The liberal-leaning think tank's Iraq research projects include a regularly updated statistical summary of conditions in Iraq.

Center for American Progress, 1333 H St., N.W., 10th Floor, Washington, DC 20005; (202) 682-1611; www.americanprogress.org. Liberal think tank Web-publishes analyses of the Iraq conflict and U.S. options.

Heritage Foundation, 214 Massachusetts Ave., N.E., Washington, DC 20002; (202) 546-4400; www.heritage.org. Conservative think tank maintains a program of military and political analyses of the Iraq conflict.

Special Inspector General for Iraq Reconstruction, 400 Army Navy Dr., Arlington, VA 22202; (703) 428-1100; www.sigir.mil. The congressionally created agency investigates and audits rebuilding projects in Iraq.

Veterans for Common Sense, P.O. Box 15514, Washington, DC 20003; www.veteransforcommonsense.org. Publicizes deficiencies in veterans' care.

The Obama Presidency

Can Barack Obama Deliver the Change He Promises?

Kenneth Jost and the *CQ Researcher* Staff

16

The largest crowd in Washington history cheers President Barack Obama after his swearing in on Jan. 20, 2009. An estimated 1.8 million high-spirited, flag-waving people gathered at the Capitol and National Mall, but thousands more were turned away by police due to overcrowding.

From *CQ Researcher*,
January 30, 2009.

They came to Washington in numbers unprecedented and with enthusiasm unbounded to bear witness and be a part of history: the inauguration of Barack Hussein Obama on Jan. 20, 2009, as the 44th president of the United States and the first African-American ever to serve as the nation's chief executive.

After taking the oath of office from Chief Justice John G. Roberts Jr., Obama looked out at the estimated 1.8 million people massed at the Capitol and National Mall and delivered an inaugural address nearly as bracing as the subfreezing temperatures.

With hardly the hint of a smile, Obama, 47, outlined the challenges confronting him as the fifth-youngest president in U.S. history. The nation is at war, he noted, the economy "badly weakened" and the public beset with "a sapping of confidence."

"Today I say to you that the challenges we face are real," Obama continued in his 18-minute speech. "They are serious and they are many. They will not be met easily or in a short span of time. But know this, America — they will be met."[1] (*See economy sidebar, p. 368; foreign policy sidebar, p. 374.*)

The crowd received Obama's sobering message with flag-waving exuberance and a unity of spirit unseen in Washington for decades. Despite Democrat Obama's less-than-landslide 7 percentage-point victory over John McCain on Nov. 4, hardly any sign of political dissent or partisan opposition surfaced on Inauguration Day or during the weekend of celebration that preceded it. (*See maps, p. 360; poll, p. 362.*)

"It's life-changing for everyone," said Rhonda Gittens, a University of Florida journalism student, "because of who he is,

359

Obama Victory Changed Electoral Map

Barack Obama won nine traditionally Republican states in the November 2008 election that George W. Bush had won easily in 2004, and his electoral and popular vote totals were significantly higher than Bush's. In 2004, Bush won with 50.7 percent of the vote to John Kerry's 48.3 percent. By comparison Obama garnered 52.9 percent to Sen. John McCain's 45.7. In the nation's new political map, the Democrats dominate the landscape, with the Republicans clustered in the South, the Plains and the Mountain states.

Presidential Election, 2004

Democrat: John Kerry
Republican: George W. Bush

Total electoral vote:		Total popular vote:	
Republican	286	Republican	62,040,610
Democrat	251	Democrat	59,028,444

* One "unfaithful" elector voted for John Edwards.

Presidential Election, 2008

Democrat: Barack Obama
Republican: John McCain

Total electoral vote:		Total popular vote:	
Democrat	365	Democrat	69,456,897
Republican	173	Republican	59,934,814

* Due to the Nebraska's proportional allocation system, McCain received four electoral votes and Obama one.

Source: Federal Election Commission

because of how he represents everyone." Gittens traveled to Washington with some 50 other members of the school's black student union.

The inaugural crowd included tens of thousands clustered on side streets after the U.S. Park Police determined the mall had reached capacity. The crowd was bigger than for any previous inauguration — at least three times larger than when the outgoing president, George W. Bush, had first taken the oath of office eight years earlier. The total number also exceeded independent estimates cited for any of Washington's protest marches or state occasions in the past.*

The spectators came from all over the country and from many foreign lands. "He's bringing change here," said Clayton Preira, a young Brazilian accompanying three fellow students on a two-month visit to the United States. "He's bringing change all over the world." The spectators were of all ages, but overall the crowd seemed disproportionately young. "He really speaks to young people," said Christian McLaren, a white University of Florida student.

Most obviously and most significantly, the crowd was racially and ethnically diverse — just like the new first family. Obama himself is the son of a

---

* Crowd estimates for President Obama's inauguration ranged from 1.2 million to 1.8 million. Commonly cited estimates for other Washington events include: March on Washington for Jobs and Freedom, 1963, 250,000; President John F. Kennedy's funeral, 1963, 800,000; inauguration of President Lyndon B. Johnson, 1965, 1.2 million; Peace Moratorium, 1969, 250,000; Million Man March, 1995, 400,000-800,000; March for Life, 1998, 225,000; March for Women's Lives, 2004, 500,000-800,000.

black Kenyan father and a white Kansan mother. His wife Michelle, he often remarks, carries in her the blood of slaves and of slave owners. Among those behind the first lady on the dais were Obama's half-sister, Maya Soetoro-Ng, whose father was Indonesian, and her husband, Konrad Ng, a Chinese-American. Some of Obama's relatives from Kenya came as well, wearing colorful African garb.

The vast numbers of black Americans often gave the event the air of an old-time church revival. In quieter moments, many struggled to find the words to convey the significance, both historic and personal. "It hasn't sunk in yet," Marcus Collier, a photographer from New York City, remarked several hours later.

David Moses, a health-care supervisor in New York City, carried with him a picture of his late father, who had encouraged him and his brother to join the anti-segregation sit-ins of the early 1960s in their native South Carolina. "It's the culmination of a long struggle," Moses said, "that still has a long way to go."

Shannon Simmons, who had not yet been born when Congress passed major civil rights legislation in the 1960s, brought her 12-year-old daughter from their home in New Orleans. "It's historic," said Simmons, who made monthly contributions to the Obama campaign. "It's about race, but it's more than that. I believe he can bring about change." (*See sidebar, p. 364.*)

For black Americans, old and young alike, the inauguration embodied the lesson that Obama himself had often articulated — that no door need be viewed as closed to any American, regardless of race. For Obama himself, the inauguration climaxed a quest that took him from the Illinois legislature to the White House in only 12 years.

To win the presidency, Obama had to defy political oddsmakers by defeating then-Sen. Hillary Rodham Clinton, the former first lady, for the Democratic nomination and then beating McCain, the veteran Arizona senator and Vietnam War hero. Obama campaigned hard against the Bush administration's record, blaming Bush, among other things, for mismanaging the U.S. economy as well as the wars in Iraq and Afghanistan.

After a nod to Bush's record of service and help during the transition, Obama hinted at some of those criticisms in his address. "The nation cannot prosper long when it favors only the prosperous," he declared, referencing tax cuts enacted in Bush's first year in office that Obama had called for repealing.

On national defense, "we reject the false choice between our safety and our ideals," Obama continued. The Bush administration had come under fierce attack from civil liberties and human rights advocates for aggressive detention and interrogation policies adopted after the Sept. 11, 2001, terrorist attacks on the United States. (*See "At Issue," p. 384.*)

Despite the attacks, Obama also sounded conservative notes throughout the speech, blaming economic woes in part on a "collective failure to make hard choices" and calling for "a new era of responsibility." Republicans in the audience were pleased. "He wasn't pointing fingers just toward Bush," said Rhonda Hamlin, a social worker from Alexandria, Va. "He was pointing fingers toward all of us."

With the inauguration behind him, Obama went quickly to work. Within hours, the administration moved to institute a 120-day moratorium on legal proceedings against the approximately 245 detainees still being held at the Guantánamo Bay Naval Base in Cuba. Obama had repeatedly pledged during the campaign to close the prison; two days later he signed a second decree, ordering that the camp be closed within one year.

Then on his first full day as president, Obama on Jan. 21 issued stringent ethics rules for administration officials and conferred separately with his top economic and military advisers to begin mapping plans to try to lift the U.S. economy out of its yearlong recession and bring successful conclusions to the conflicts in Iraq and Afghanistan.

By then, the Inauguration Day truce in partisan conflict was beginning to break down. House Republicans pointed to a Congressional Budget Office study questioning the likely impact of the Democrats' $825-billion economic stimulus package, weighted toward spending instead of tax cuts. "The money that they're going to throw out the door, at the end of the day, is not going to work," said Rep. Devin Nunes, R-Calif., a member of the tax-writing House Ways and Means Committee. (*See "At Issue," p. 385.*)

The partisan division raised questions whether Democratic leaders could stick to the promised schedule of getting a stimulus plan to Obama's desk for his signature by the time of the Presidents' Day congressional recess in mid-February. More broadly, the Republicans' stance presaged continuing difficulties for Obama as he turned to other ambitious agenda items, including his repeated pledge to overhaul the nation's health-care system. (*See sidebar, p. 378.*)

Public Gives Obama Highest Rating

Barack Obama began his presidency with 79 percent of Americans having a favorable impression of him — higher than the five preceding presidents. George W. Bush entered office with a 62 percent favorability rating; he left with a 33 percent approval rating, lowest of post-World War II presidents except Harry S. Truman and Richard M. Nixon.

Do you have a favorable impression of . . . ?

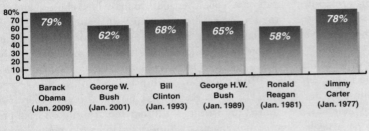

Barack Obama (Jan. 2009)	George W. Bush (Jan. 2001)	Bill Clinton (Jan. 1993)	George H.W. Bush (Jan. 1989)	Ronald Reagan (Jan. 1981)	Jimmy Carter (Jan. 1977)
79%	62%	68%	65%	58%	78%

Source: The Washington Post, Jan. 18, 2009

Obama included health care in his inaugural litany of challenges, along with education, climate change and technology. For now, those initiatives lie in the future. In the immediate days after his euphoric inauguration, here are some of the major questions being debated:

Is President Obama on the right track in fixing the U.S. economy?

As president-elect, Obama spent his first full week in Washington in early January first warning of trillion-dollar federal budget deficits for years to come and then making urgent appeals for public support for a close to trillion-dollar stimulus to get the economy moving.

Members of Congress from both parties and advocates and economic experts of all persuasions agree on the need for a good-sized federal recovery program for the seriously ailing U.S. economy. And most agree on a prescription that combines spending increases and tax cuts. But there is sharp disagreement as to the particulars between tax-cutting conservatives and pump-priming liberals, with deficit hawks worried that both of the prescribed remedies could get out of hand.

With the plan's price tag then being estimated somewhere around $800 billion, Obama made his first sustained appeal for public support in a somber, half-hour address on Jan. 8 at George Mason University in Fairfax,

Va., outside Washington. Any delay, he warned, could risk double-digit unemployment. He outlined plans to "rebuild America" ranging from alternative energy facilities and new school classrooms to computerized medical records, but he insisted the plan would not entail "a slew of new government programs." He reiterated his campaign promise of a "$1,000 tax cut for 95 percent of working-class families" but made no mention of business tax cuts being included as sweeteners for Republican lawmakers.

Within days, Obama's plan was taking flack from left and right in the blogosphere. Writing on the liberal HuffingtonPost.com, Robert Kuttner, co-editor of *American Prospect* magazine, denounced the spending plan as too small and the business tax cuts as "huge concessions" in a misguided effort at "post-partisanship." From the right, columnist Neal Boortz accused Obama on the conservative TownHall.com of using the economic crisis as "cover for increased government spending that he's been promising since the day he announced his candidacy."

Allen Schick, a professor of economics at the University of Maryland in College Park and formerly an economics specialist with the Congressional Research Service, sees weaknesses with both components of the Obama plan. "We really have no model to deal with the question of what's the right number" for the stimulus, he says. "And we're not even sure that the stimulus will do the job, especially if a lot of the spending is wasteful."

As for the tax cuts, Schick calls them "harebrained, more intended to look good and buy support than to actually get the economy moving." In particular, he criticized a proposed $3,000 jobs credit for employers. "We know from the past that employers don't hire people for just a few shekels," he says. Eventually, the jobs credit was dropped, but the package still includes business tax breaks such as a $16 billion provision to allow businesses to use 2008 and 2009 losses to offset profits for the previous five years instead of two.

Conservatives favor tax cuts, but not the middle-class tax cut that Obama is proposing. "A well-designed tax cut is the only effective short-term stimulus," says J. D. Foster, a senior fellow at the Heritage Foundation. But Foster, who worked in the Office of Management and Budget in the Bush administration, calls either for extending or making permanent Bush's across-the-board rate cuts, which primarily benefited upper-income taxpayers.

From the opposite side, Chad Stone, chief economist with the liberal Center on Budget Policy and Priorities, endorses Obama's approach. "Tax cuts should be focused on people of low and moderate means, who are much more likely to spend the extra money they get," he says.

Academic economists, however, caution that tax cuts may not deliver a lot of bang for the buck in terms of short-term stimulus. Studies indicate that taxpayers pocketed at least one-third of the $500 tax rebate the government disbursed to counteract the 2001 recession.

Advocates and observers on both sides warn that the spending side of the package may also be less effective than hoped if political forces play too large a role in shaping it. "If it goes to pork, if it goes to green jobs that may sound good in the short term but may not have a market response or a market for them, then it's a waste," Paul Gigot, editorial page editor of *The Wall Street Journal*, said on NBC's "Meet the Press" on Jan. 11.

"If the stuff that gets added is not very effective as stimulus or the things that are good get pulled out, that would not be good," says Stone.

For its part, the budget-restraint advocacy group Concord Coalition sees political forces as driving up the total cost of the package — in spending and tax cuts alike — with no regard for the long-term impact. "Nothing is ever taken off the table," says Diane Lim Rogers, the coalition's chief economist.

Rogers complains of "political pressure to come up with tax cuts even though economists are having trouble figuring out whether they're going to do any good." At the same time, she says spending has to be designed "as thoughtfully as possible, not in a way that the federal government ends up literally just throwing money out the door."

A range of experts also call for renewed efforts to solve the mortgage and foreclosure crisis, saying that homeowners are not going to start spending again without confidence-restoring steps. Indeed, Federal Reserve Chairman Ben Bernanke pointedly told a conference in December that steps to reduce foreclosures "should be high on the agenda" in any economic recovery plan.[2]

Despite questions and concerns about the details, however, support for strong action is all but universal. "We have no choice," said Mark Zandi, chief economist of Moody's Economy.com and a former adviser to the McCain campaign, also on "Meet the Press." "If we don't do something like this — a stimulus package, a foreclosure mitigation plan — the economy is going to slide away."

Is President Obama on the right track in Iraq and Afghanistan?

At the start of his presidential campaign in February 2007, candidate Obama was unflinchingly calling for withdrawing all U.S. combat forces from Iraq within 16 months after taking office. But his tone began changing as he neared the Democratic nomination in summer 2008. And in his first extended broadcast interview after the election, President-elect Obama said on NBC's "Meet the Press" on Dec. 7 only that he would summon military advisers on his first day in office and direct them to prepare a plan for "a responsible drawdown."

Obama also did nothing to knock down host Tom Brokaw's forecast of a "residual force" of 35,000 to 50,000 U.S. troops in Iraq through the end of his term. "I'm not going to speculate on the numbers," Obama said, but he went on to promise "a large enough force in the region" to protect U.S. personnel and to "ferret out any terrorist activity." In addition, Obama voiced disappointment with developments in Afghanistan and said that "additional troops" and "more effective diplomacy" would be needed to achieve U.S. goals there.

Many foreign policy observers are viewing Obama's late campaign and post-election stances as a salutary shift from ideology to pragmatism. "It seems very clear that he will not fulfill his initial pledge to withdraw all U.S. forces from Iraq in 16 months — which is only wise," says Thomas Donnelly, a resident fellow on defense and national security issues at the American Enterprise Institute (AEI).

"I personally have been very impressed with [Obama's] thinking and his way of assembling a national security team," says Kenneth Pollack, director of the Brookings Institution's Saban Center for Middle East Policy. "This is not a man who plays by the traditional American political rules."

First Black President Made Race a Non-Issue

Obama's personal attributes swept voters' doubts aside.

Barack Obama took the oath of office the day after this year's Martin Luther King holiday, and he accepted the Democratic presidential nomination last August on the 45th anniversary of King's celebrated "I Have a Dream" speech.

For millions of Americans, Obama's election as the nation's first African-American president seemed to fulfill the promise of King's "dream" of a nation in which citizens "will not be judged by the color of their skin, but by the content of their character."

"Obviously, for an African-American to win the presidency, given the history of this country . . . is a remarkable thing," Obama said after the election. "If you think about grandparents who are alive today who grew up under Jim Crow, that's a big leap."[1]

While Obama clearly benefited from the sacrifices of the civil rights generation — to which he has paid homage — his politics are different from the veterans of that movement. Older black politicians such as the Rev. Jesse Jackson seemed to base their candidacies mainly on issues of particular concern to African-Americans. But black politicians of Obama's generation, such as Massachusetts Gov. Deval Patrick and Newark Mayor Cory Booker (both Democrats), have run on issues of broader concern — in Obama's case, first on the war in Iraq and later on the economic meltdown.

"The successful ones start from the outside by appealing to white voters first, and work back toward their base of black voters," said broadcast journalist Gwen Ifill, author of the new book *The Breakthrough: Politics and Race in the Age of Obama*.[2]

Black voters initially were reluctant to support Obama — polls throughout 2007 showed Sen. Hillary Rodham Clinton with a big lead among African-Americans — but he picked up their support as it became clear he was the first black candidate with a realistic hope of winning the White House. Clinton's support among blacks dropped markedly in the wake of remarks by former President Bill Clinton that many found demeaning.

But many white Democratic voters remained reluctant to support Obama, particularly in Appalachia. Exit polling during the Pennsylvania primary, for example, showed that 16 percent of whites had considered race in making their pick, with half of those saying they would not support Obama in the fall.[3]

Obama also was bedeviled by videotaped remarks of his pastor, the Rev. Jeremiah Wright, which were incendiary and deemed unpatriotic. But Obama responded with a widely hailed speech on race in March 2008 in which he acknowledged both the grievances of working-class whites and the continuing legacy of economic disadvantages among blacks. Obama said his own life story "has seared into my genetic makeup the idea that this nation is more than the sum of its parts — that out of many, we are truly one."[4]

As the general election campaign got under way, it was clear that race would continue to be a factor. One June poll showed that 30 percent of Americans admit prejudice.[5] And, despite Obama's lead, there was debate throughout the campaign about the so-called Bradley effect — the suggestion that people will lie to pollsters about their true intentions when it comes to black candidates.*

But neither Obama nor Arizona Sen. John McCain, his Republican rival, made explicit pleas based on race, with McCain refusing to air ads featuring Wright. As the campaign wore on, no one forgot that Obama is black — but most doubters put that fact aside in favor of more pressing concerns.

"For a long time, I couldn't ignore the fact that he was black. I'm not proud of that," Joe Sinitski, a 48-year-old Pennsylvania voter, told *The New York Times*. "I was raised to think that there aren't good black people out there."[6] But Sinitski ended up voting for Obama, along with many other whites won over by Obama's personal attributes or convinced that issues such as the economy trumped race.

Exit polls showed that Obama prevailed among those who considered race a significant factor, 53 to 46 percent.[7] "In difficult economic times, people find the price of prejudice is just

---

* The Bradley effect refers to Tom Bradley, an African-American who lost the 1982 race for governor in California despite being ahead in voter polls going into the election.

a little too high," said outgoing North Carolina Gov. Mike Easley, a Democrat.[8]

"The Bradley effect really was not a significant factor, despite much concern, fear and hyperventilation about it leading up to the election," says Scott Keeter, a pollster with the Pew Research Center. "Race was a consideration to people, but what it wasn't, invariably, was a negative consideration for white voters. It was a positive consideration for many white voters who saw Obama as a candidate who could help the country toward racial reconciliation."

Obama carried more white voters than former Vice President Al Gore or Sen. John Kerry of Massachusetts, the two previous Democratic nominees. Still, he could not have prevailed without black and Hispanic voters, particularly in the three Southern states he carried. In Virginia — a state that had voted Republican since 1964 — Obama lost by 21 points among white voters, according to exit polls.

His victory clearly did not bring racial enmity to its end. In December, Chip Saltsman, a candidate for the Republican Party chairmanship, sent potential supporters a CD containing the song "Barack the Magic Negro," a parody popularized by right-wing talk show host Rush Limbaugh during the campaign. And, when Senate Democrats initially balked in January at seating Roland Burris as Obama's replacement, Rep. Bobby Rush, D-Ill, played the race card, warning them not to "hang or lynch the appointee," comparing the move to Southern governors who sought to block desegregation.[9]

But still polls suggest that most Americans believe Obama's presidency will be a boon for race relations. A *USA Today*/Gallup Poll taken the day after the November election showed that two-thirds predicted black-white relations "will eventually be worked out" — by far the highest total in the poll's history.[10]

In the future, white males may no longer be the default inhabitants of America's most powerful position. The present generation and those in the future are likely to grow up thinking it's a normal state of affairs for the country to be led

Michelle Obama holds the Bible used to swear in President Abraham Lincoln as Barack Obama takes the oath of office from Supreme Court Chief Justice John G. Roberts Jr.

AFP/Getty Images/Tim Sloan

by a black president. "For a lot of African-Americans, it already has made them feel better and more positive about the country and American society," says David Bositis, an expert on black voting at the Joint Center for Political and Economic Studies.

"When you ask my kids what they want to be when they grow up, they always say they want to work at McDonald's or Wal-Mart," said Joslyn Reddick, principal at a predominantly black school in Selma, Ala., a city from which King led an historic march for voting rights in 1965.

"Now they will see that an African-American has achieved the highest station in the United States," Reddick said. "They can see for themselves that dreams can come true."[11]

— Alan Greenblatt,
staff writer, *Governing* magazine

[1] Bryan Monroe, "The Audacity of Victory," *Ebony*, January 2009, p. 16.

[2] Sam Fulwood III, "The New Face of America," *Politico.com*, Jan. 13, 2009.

[3] Alan Greenblatt, "Changing U.S. Electorate," *CQ Researcher*, May 30, 2008, p. 459.

[4] The Obama speech, "A More Perfect Union," is at www.youtube.com/watch?v=pWe7wTVbLUU. The text of the March 18, 2008, speech, "A More Perfect Union," is found in *Change We Can Believe In: Barack Obama's Plan to Renew America's Promise* (2008), pp. 215-232.

[5] Jon Cohen and Jennifer Agiesta, "3 in 10 Americans Admit to Race Bias," *The Washington Post*, June 22, 2008, p. A1.

[6] Michael Sokolove," The Transformation," *The New York Times*, Nov. 9, 2008, p. WK1.

[7] John B. Judis, "Did Race Really Matter?" *Los Angeles Times*, Nov. 9, 2008, p. 34.

[8] Rachel L. Swarns, "Vaulting the Racial Divide, Obama Persuaded Americans to Follow," *The New York Times*, Nov. 5, 2008, p. 7.

[9] Clarence Page, "Hiding Behind Black Voters," *Chicago Tribune*, Jan. 4, 2009, p. 24.

[10] Susan Page, "Hopes Are High for Race Relations," *USA Today*, Nov. 7, 2008, p. 1A.

[11] Dahleen Glanton and Howard Witte, "Many Marvel at a Black President," *Chicago Tribune*, Nov. 5, 2008, p. 6.

Cabinet Includes Stars, Superstars and Surprises

President Obama made his Cabinet selections in record time, and his appointees run the gamut of race, ethnic origin, gender, age and even party affiliation. Those in top posts include Sen. Hillary Rodham Clinton at State and Robert Gates continuing at Defense. Besides Gates, one other Republican was chosen: Transportation's Ray LaHood. New Mexico Gov. Bill Richardson's withdrawal left the Commerce post unfilled along with the director of Drug Control Policy. Cabinet-level appointees include four women, two Asian-Americans, two Hispanics and two African-Americans.

Name, Age Department	Date of Nomination	Date of Confirmation	Previous Positions
Hillary Rodham Clinton, 61, State	Dec. 1	Jan. 21	New York U.S. senator (2001-09); first lady (1993-2001); Arkansas first lady (1979-81, 1983-92)
Timothy Geithner, 47, Treasury	Nov. 24	Jan. 26	President, Federal Reserve Bank of New York (2003-09); under secretary, Treasury (1998-2001)
Robert Gates, 65, Defense*	Dec. 1	Dec. 6, 2006 *	Defense secretary (2006-present); director, CIA (1991-93); deputy national security adviser (1989-91)
Eric Holder, 57, Attorney General	Dec. 1		Deputy attorney general (1997-2001); U.S. attorney (1993-97); judge, D.C. Superior Court (1988-93)
Ken Salazar, 53, Interior	Dec. 17	Jan. 20	Colorado U.S. senator (2005-09); Colorado attorney general (1999-2005)
Tom Vilsack, 58, Agriculture	Dec. 17	Jan. 20	Iowa governor (1999-2007); Iowa state senator (1992-99)
Hilda Solis, 51, Labor	Dec. 19		California U.S. representative (2001-09); California state senator (1995-2001)
Tom Daschle, 61, Health & Human Services	Dec. 11		South Dakota U.S. senator (1987-2005); Senate majority leader (2001, 2001-03); South Dakota U.S. representative (1979-87)
Shaun Donovan, 42, Housing and Urban Development	Dec. 13	Jan. 22	Commissioner, New York City Dept. of Housing Preservation and Development (2004-08); deputy assistant secretary, HUD (2000-01)

Obama invited speculation about a shift toward the center by selecting Clinton and Robert Gates as the two Cabinet members on his national security team along with a retired Marine general, James Jones, as national security adviser. (*See chart, at left.*) Clinton had voted for the Iraq War in late 2002, though she echoed Obama during the campaign in calling for troop withdrawals. As Bush's secretary of Defense, Gates had overseen the "surge" in U.S. forces during 2007.

"This is a group of people who are very sober, very intelligent, fully aware of the importance of Iraq to America's security interests and of the fragility of the situation there," says Pollack.

Some anti-war activists were voicing concern about Obama's seeming shift within days of his election. "Obama has very successfully branded himself as anti-war, but the fact remains that he's willing to keep a residual force in Iraq indefinitely, [and] he wants to escalate in Afghanistan," said Matthis Chiroux of Iraq Veterans Against the War. "My hope is that he starts bringing home the troops from Iraq immediately, but I think those of us in the anti-war movement could find ourselves disappointed."[3]

Since then, however, criticism of Obama's emerging policies has been virtually nonexistent from the anti-war and Democratic Party left. "He seems to be accelerating the withdrawal, which is terrific," says Robert Borosage, co-director of the Campaign for America's Future. Borosage is "concerned" about the residual force in Iraq because of the risk that U.S. troops will become involved in "internecine battles." But he adds, "That's what he's promised, and I think he'll fulfill his promise."

Donnelly and Pollack, however, both view a continuing U.S. role in Iraq as vital. "There's good progress, but a long way to go," says Donnelly. "A huge American role is going to be needed through the four years of the Obama administration." Pollack agrees. "Iraq is far from solved. Whether we like it or not, Iraq is a vital interest for the United States of America."

In his campaign and since, Obama has treated Afghanistan as more important to U.S. interests and harshly criticized the Bush administration for — in his view — ignoring the conflict there. Afghanistan "had had a huge rhetorical place in the Obama campaign," says Donnelly. "The idea being that Afghanistan was the good war, the more important war, and that Iraq was a dead end strategically."

P. J. Crowley, a senior fellow at the liberal think tank Center for American Progress, calls Obama's focus on Afghanistan "correct" but emphasizes the need for a multipronged effort to stabilize and reform the country's U.S.-backed government. "Returning our weight of effort [to Afghanistan] is a right approach," says Crowley, who was spokesman for the National Security Council under President Bill Clinton.

"More troops may help in a narrow sense," Crowley continues, "but I don't think anyone suggests that more troops are the long-term solution in Afghanistan. The insertion of U.S. forces is logical in the short- to mid-term, but it has to be part of a broader strategy."

But Pollack questions the value of any additional U.S. troops at all. "The problems of Afghanistan are not principally military; they are principally political and diplomatic," he says. "Unless this new national security team can create a military mission that is of value to what is ultimately a diplomatic problem, it's going to be tough to justify to the country the commitment of those additional troops."

Name, Age, Department	Date of Nomination	Date of Confirmation	Previous Positions
Ray LaHood, 63, Transportation	Dec. 19	Jan. 22	Illinois U.S. representative (1995-2009); state representative (1982-83)
Steven Chu, 60, Energy	Dec. 15	Jan. 20	Director, Lawrence Berkeley National Laboratory, Dept. of Energy (2004-09); professor, UC-Berkeley (2004-present); Nobel Prize winner, physics (1997)
Arne Duncan, 44, Education	Dec. 16	Jan. 20	C.E.O, Chicago Public Schools (2001-09)
Eric Shinseki, 66, Veterans Affairs	Dec. 7	Jan. 20	Chief of staff, Army (1999-2003)
Janet Napolitano, 51, Homeland Security	Dec. 1	Jan. 20	Arizona governor (2003-09); attorney general (1999-2002)
Rahm Emmanuel, 49, Chief of Staff	Nov. 6	NA	Illinois U.S. representative (2003-09); senior adviser to the president (1993-98)
Lisa Jackson, 46, Environmental Protection Agency	Dec. 15	Jan. 22	Chief of staff, governor of New Jersey (2008-09); commissioner, New Jersey Dept. of Environmental Protection (2006-2008)
Peter Orszag, 40, Office of Management and Budget	Nov. 25	Jan. 20	Director, Congressional Budget Office (2007-08); adviser, National Economic Council (1997-98)
Susan Rice, 44, Ambassador to the United Nations	Dec. 1	Jan. 22	Assistant secretary, State (1997-2001); National Security Council (1993-97)
Ron Kirk, 54, Trade Representative	Dec. 19		Mayor of Dallas (1995-2002)

Department heads are listed in order of succession under Presidential Succession Act; nondepartment heads were given Cabinet-level status.

* Gates was confirmed when first nominated by President George W. Bush and did not have to be re-confirmed.

Compiled by Vyomika Jairam; all photos by Getty Images

Bleak Economy Getting Bleaker

Economists widely agree a stimulus plan is needed.

When Barack Obama took office on Jan. 20, he inherited the most battered U.S. economy since World War II — and one of the shakiest to confront a new president in American history.

And the view from the Oval Office is likely to get bleaker before the gloom begins to lift.

"There are very serious questions on the financial side and apprehension among many parties that there may be more bad news to come," says Kent Hughes, director of the Program on Science, Technology, America and the Global Economy at the Woodrow Wilson Center for Scholars.

Already, Obama has stepped into the worst unemployment picture in 16 years, with the jobless rate at 7.2 percent and 11.1 million people out of work. The economy lost 1.9 million jobs during the last four months of 2008 — 524,000 in December alone.[1]

Economists worry that rising unemployment in manufacturing, construction, retailing and other sectors foreshadows an even more dismal future, at the very least in the short term. Dean Baker, co-director for the Center for Economic and Policy Research, a liberal think tank in Washington, says he expects another million or so jobs to disappear through February, then the pace of job loss to slow if Congress acts to stimulate the economy.

Obama must figure out not only how to get people back to work but also how to restore their confidence in the economy. A punishing credit crisis and cascade of grim news from Wall Street has led consumers to stop spending on everything from restaurant meals to houses and autos.[2]

Home sales have plunged in recent months, foreclosures are hitting record levels and a study by PMI Mortgage Insurance Co. estimates that half of the nation's 50-largest Metropolitan Statistical Areas have an "elevated or high probability" of experiencing lower home prices by the end of the third quarter of 2010 compared to the same quarter of 2008.[3]

Retail sales, a key indicator of consumer confidence, fell in December 2008 for the sixth month in a row, according to the Commerce Department.[4] The International Council of Shopping Centers said chain-store sales in December posted their biggest year-to-year decline since researchers began tracking figures in 1970.[5]

Rebecca Blank, a senior fellow at the Brookings Institution and former member of President Bill Clinton's Council of Economic Advisers, says the unemployment numbers "suggest the economy is still on the way down," and the decline in holiday sales is "surely going to lead to some bankruptcies and belt tightening in the retail sector."

Indeed, such trouble is already occurring. The shopping centers group estimated that 148,000 retail stores closed last year and that more than 73,000 will be shuttered in the first half of 2009.[6] Among the latest examples: Bankrupt electronics chain Circuit City said in January that it was closing its remaining 567 stores, putting some 30,000 employees out of work.

To revive the economy, the new administration — most visibly Obama himself — is urging Congress to quickly approve a stimulus package that could approach $900 billion. Much of the money would likely go toward tax cuts and public infrastructure projects, though how, exactly, the government would allocate it remains a matter of intense political debate.

One thing seems certain, though: The cost of a stimulus package, added to the hundreds of billions of dollars already spent to shore up the nation's flagging financial system, will add to the bulging federal deficit.

"The thing you know for sure is that a stimulus is going to add to the debt, which is [now] quite frightening, and it's going to make it worse," says June O'Neill, an economics professor at the City University of New York's Baruch College

---

Borosage also worries about an increased U.S. military presence in Afghanistan. "A permanent occupation of Afghanistan is a recipe for defeat," he says.

All of the experts stress that U.S. policy in Afghanistan now plays a secondary part in the fight with the al Qaeda terrorist group, which carried out the 9/11 attacks in the United States. "There is no al Qaeda in Afghanistan," says Donnelly. "Al Qaeda has now reconstituted itself in the tribal areas of northwest Pakistan."

Donnelly questions Afghanistan's importance to U.S. interests altogether but ultimately supports continued U.S. involvement. "The only thing worse than being engaged in Afghanistan," he says, "is turning our backs on it."

and a former director of the Congressional Budget Office (CBO) during the Clinton administration.

In January the CBO projected a $1.2 trillion deficit for the fiscal year. A stimulus plan would add even more pressure on Obama to get federal spending under control. "My own economic and budget team projects that, unless we take decisive action, even after our economy pulls out of its slide, trillion-dollar deficits will be a reality for years to come," Obama said.[7]

The battered economy that confronts President Obama includes record foreclosure rates and plummeting home values. Above, a foreclosed home in Nevada, the state with the nation's highest foreclosure rate.

of the financial markets, more closely resembles the Great Depression than any other recession since then.

Most postwar recessions "were the result of the Fed raising rates," says Baker. "That meant we knew how to reverse it. This one, there's not an easy answer to. We're not going to see [another] Great Depression — not double-digit unemployment for a decade." But in terms of the severity of the problem, Baker adds, the Great Depression is the "closest match" to what confronts the new administration.

— Thomas J. Billitteri

Still, a wide spectrum of economists — including conservatives who typically look askance at government spending — agree that a stimulus plan is necessary.

Martin Feldstein, a Harvard University economist and former chair of the Council of Economic Advisers in the Reagan administration, told a House committee in January that stopping the economic slide and restoring "sustainable growth" requires fixing the housing crisis and adopting a "fiscal stimulus of reduced taxes and increased government spending."[8]

Feldstein pointed out that past recessions started after the Federal Reserve raised short-term interest rates to fight inflation. Once inflation was under control, the Fed cut rates, which spurred a recovery. But the current recession is different, Feldstein said: It wasn't caused by the Fed tightening up on fiscal policy, and thus rate cuts haven't succeeded in reviving the economy.

"Because of the dysfunctional credit markets and the collapse of housing demand, monetary policy has had no traction in its attempt to lift the economy," he said.

That poses an especially daunting challenge for Obama.

Baker of the Center for Economic and Policy Research says that the current crisis, occurring amid a broad collapse

[1] Bureau of Labor Statistics, "Employment Situation Summary," Jan. 9, 2009, www.bls.gov/news.release/empsit.nr0.htm.

[2] For coverage of the economic crisis, see the following *CQ Researcher* reports: Thomas J. Billitteri, "Financial Bailout," Oct. 24, 2008, pp. 865-888; Kenneth Jost, "Financial Crisis," May 9, 2008, pp. 409-432; Marcia Clemmitt, "Regulating Credit Cards," Oct. 10, 2008, pp. 817-840; and Marcia Clemmitt, "The National Debt," Nov. 14, 2008, pp. 937-960.

[3] News release, "PMI Winter 2009 Risk Index Indicates Broader Risk Spreading Across Nation's Housing Markets," PMI Mortgage Insurance Co., Jan. 14, 2009.

[4] Bob Willis, "U.S. Economy: Retail Sales Decline for a Sixth Month," Bloomberg, Jan. 14, 2009, www.bloomberg.com.

[5] V. Dion Haynes and Howard Schneider, "A Brutal December for Retailers," *The Washington Post*, Jan. 9, 2009, p. 2D.

[6] *Ibid.*

[7] Quoted in David Stout and Edmund L. Andrews, "$1.2 Trillion Deficit Forecast as Obama Weighs Options," *The New York Times*, Jan. 8, 2009, www.nytimes.com/2009/01/08/business/economy/08deficit.html?scp=2&sq=deficit&st=cse.

[8] Martin Feldstein, "The Economic Stimulus and Sustained Economic Growth," statement to the House Democratic Steering and Policy Committee, Jan. 7, 2009, www.nber.org/feldstein/Economic StimulusandEconomicGrowthStatement.pdf.

Is President Obama on the right track in winning support for his programs in Congress?

As president of Harvard University, Lawrence Summers clashed so often and so sharply with faculty and others that he was forced out after only five years in office. But when Summers went to Capitol Hill as President-elect Obama's

designee to be top White House economic adviser, the normally self-assured economist told lawmakers that he and other administration officials plan to be all ears.

"All of us have been instructed that when it comes to Congress, to listen and not just talk," Summers told House Democrats in a Jan. 9 meeting to discuss Obama's economic recovery plan.[4]

State Department staffers greet new Secretary of State Hillary Rodham Clinton on her first day of work, Jan. 22, 2009.

Within days after the new Congress was sworn in on Jan. 6, however, lawmakers on both sides of the political aisle were, in fact, taking pot shots at Obama's plan. Republicans were calling for hearings after the plan was unveiled — a move seen as jeopardizing Obama's goal of signing a stimulus bill into law before Congress' mid-February recess. Meanwhile, some Democratic lawmakers were questioning the business tax cuts being considered for the package, calling them examples of what they considered the discredited philosophy of "trickle-down economics."

Despite the criticisms, Obama was upbeat about his relations with Congress in an interview broadcast on ABC's "This Weekend" on Jan. 11. "One of the things that we're trying to set a tone of is that, you know, Congress is a co-equal branch of government," Obama told host George Stephanopoulos. "We're not trying to jam anything down people's throats."

Veteran Congress-watchers in Washington are giving Obama high marks in his dealings with Capitol Hill so far, while also praising Congress for asserting its own constitutional prerogatives.

"Obama is off to a very good start with Congress, and, just as importantly, Congress is off to a good start with him," says Thomas Mann, a senior fellow at the Brookings Institution. "No more [status as a] potted plant for the first branch or an inflated sense of presidential authority by the second, but instead a serious engagement between the players at the opposite ends of Pennsylvania Avenue."

Obama is "in good shape," says Stephen Hess, a senior fellow emeritus at Brookings who began his Washington career as a White House staffer under President Dwight D. Eisenhower in the 1950s. Hess credits Obama in particular with seeking to consult with Republican as well as Democratic lawmakers.

"He was very shrewd after talking with Democrats to talk with Republicans," says Hess, who also teaches at George Washington University. "He has given the opposition the sense that he's open, he's listening. He's reached out to them when he doesn't need them — which of course is the right time to reach out to them."

Norman Ornstein, a resident scholar at the American Enterprise Institute, similarly credits Obama with having gone "further in consulting members of the opposition party than any president I can remember." Writing in the Capitol Hill newspaper *Roll Call*, Ornstein also said Obama is well aware of lawmakers' "issues and sensitivities." For example, Ornstein noted the president-elect's personal apology to Senate Intelligence Committee Chair Dianne Feinstein, D-Calif., for failing to give her advance word in early January of the planned nomination of Leon Panetta to head the Central Intelligence Agency.[5]

The lapse of protocol on the Panetta nomination — which Feinstein later promised to support — may well have been the only avoidable misstep by the Obama team in its dealings with Congress. Criticisms of the economic recovery program as it took shape could hardly have been avoided. And Republican senators naturally looked for ways to find fault with some of Obama's Cabinet nominees — such as their criticism of Attorney General-designate Eric Holder for his role in President Clinton's pardon of fugitive financier Marc Rich and Treasury Secretary-designate Timothy Geithner for his late payment of tens of thousands of dollars in federal income taxes.

A prominent, retired GOP congressman, however, says Obama is doing well so far and predicts the economic crisis may give him a longer than usual pass with lawmakers from both parties. "He has the advantage of a honeymoon, and perhaps the second advantage of the economic conditions of the country, which I think will help the Congress to gather around his program," says Bill Frenzel, a guest scholar at the Brookings Institution and a Minnesota congressman for two decades before his retirement in 1991.

AFP/Getty Images/Mark Ralston

Big-Name Policy 'Czars' Head for West Wing

Appointments may signal decline in Cabinet's influence.

President Barack Obama has tapped several high-profile Washington insiders to fill new and existing senior White House positions, indicating the new administration is shifting policy making from the Cabinet to the influential White House West Wing.

The new so-called policy "czars" include former Sen. Tom Daschle, D-S.D., at the Office of Health Reform (he is also Health and Human Services secretary); former assistant Treasury secretary Nancy Killefer, leading efforts to cut government waste as the nation's first chief performance officer; former Environmental Protection Agency Administrator Carol Browner as the new coordinator of energy and climate policy; and former New York City Council member Adolfo Carrion Jr., who is expected to head the Office of Urban Affairs.

"We're going to have so many czars," said Thomas J. Donohue, president of the U.S. Chamber of Commerce. "It's going to be a lot of fun, seeing the czars and the regulators and the czars and the Cabinet secretaries debate."[1]

In another major West Wing appointment, former Treasury secretary and Harvard President Lawrence Summers becomes director of the existing National Economic Council. In the weeks leading up to the inauguration, analysts noted that Summers, and not then-Treasury secretary-designate Timothy Geithner, was leading then-President-elect Obama's efforts to draft a new financial stimulus package.

But Paul Light, an expert on governance at New York University, questions the role the new "czars" will play. "It's a symbolic gesture of the priority assigned to an issue, and I emphasize the word symbolic," he said. "There've been so many czars over the last 50 years, and they've all been failures. Nobody takes them seriously anymore."[2]

— Vyomika Jairam

---

[1] Michael D. Shear and Ceci Connolly, "Obama Assembles Powerful West Wing; Influential Advisers May Compete With Cabinet," *The Washington Post*, Jan. 14, 2009, p. A1.

[2] Laura Meckler " 'Czars' Ascend at White House," *The Wall Street Journal*, Dec. 15, 2005, p. A6.

"We're talking about both Republicans and Democrats," Frenzel continues. "Democrats are going to want to be independent, and Republicans are going to want to take whacks at him when they can. But I think there is a mood of wanting to help the president when they can for a while."

Ornstein and Hess caution, however, that new presidents cannot expect the honeymoon to last very long. Ornstein writes that Obama's hoped-for supermajority support in Congress "may be doable on stimulus" and "perhaps even on health care." But he says an era of "post-partisan politics" will require "some serious steps" by party leaders and rank-and-file members.

For his part, Hess says Obama may eventually begin to disappoint some within his own party — but not yet. "Democrats will for a while cut him a great deal of slack," Hess explains. "Reason No. 1, he's not George W. Bush. Reason No. 2, they're going to get some of what they want. And reason No. 3, some of those folks have become wiser about the way politics is played in this town."

BACKGROUND

'A Mutt, Like Me'

Barack Obama's inauguration as president represents a 21st-century version of the American dream: the election of a native-born citizen, both black and white, with roots in Kansas and Kenya. Abandoned by his father and later living apart from his mother, Obama was nurtured in his formative years by doting white grandparents and educated in elite schools before turning to community organizing in inner-city Chicago and then to a political career that moved from the Illinois statehouse to the White House in barely 12 years.[6]

Barack Hussein Obama was born in Honolulu on Aug. 4, 1961, to parents he later described in his memoir *Dreams from My Father* as a "white as vanilla" American mother and a "black as pitch" Kenyan father. Barack Obama Sr. and Stanley Ann Dunham married, more or less secretly, after having met as students at the University of Hawaii. Stanley Ann's "moderately liberal" parents

CHRONOLOGY: 1961-2006

1960s-1970s *Obama born to biracial, binational couple; begins education in Indonesia after mother's remarriage, then returns to Hawaii.*

1961 Barack Hussein Obama born on Aug. 4, 1961, in Honolulu; parents Stanley Ann Dunham and Barack Obama Sr. meet as students at University of Hawaii; father leaves family behind two years later for graduate studies at Harvard, return to native Kenya.

1967-1971 Obama's mother remarries, family moves to Indonesia; Obama attends a secular public elementary school with a predominantly Muslim student body until mother decides he should return to Hawaii for schooling.

1971-1979 "Barry" Obama lives with grandparents Stanley and Madelyn Dunham; graduates with honors from Punahou School, one of three black students at the elite private school; enrolls in Occidental College in Los Angeles but transfers later to Columbia University in New York City.

1980s-1990s *Works as community organizer in Chicago, gets law degree, enters politics.*

1983 Obama graduates with degree in political science from Columbia University; floods civil rights organizations with job applications.

1985-1988 Works on housing, employment issues as community organizer in Far South Side neighborhood in Chicago.

Summer 1988 Visits Kenya for first time.

1988-1991 Enrolls in Harvard Law School in fall 1988; graduates in 1991 after serving as president of *Harvard Law Review* — the first African-American to hold that position.

1992-1995 Returns to Chicago; marries Michelle Robinson in 1992; runs voter registration project; works as lawyer, lecturer at University of Chicago Law School.

1995 *Dreams from My Father* is published; mother dies just after publication (Nov. 7, 1995).

1996 Elected to Illinois legislature as senator representing Chicago's Hyde Park area; serves for eight years.

2000-2006 *Enters national political stage as U.S. senator, Democratic keynoter.*

2000 Loses badly in Democratic primary for U.S. House seat held by Rep. Bobby Rush.

2002 Opposes then-imminent war in Iraq.

2004 Gains Democratic nomination for U.S. Senate from Illinois. . . . Wins wide praise for keynote address to Democratic National Convention. . . . Elected U.S. senator from Illinois: third African-American to serve in Senate since Reconstruction.

2005-2006 Earns reputation as hard worker in Senate; compiles liberal voting record; manages Democrats' initiative on ethics reform. . . . *Audacity of Hope* is published (October 2006). . . . Deflects intense speculation about possible presidential bid.

accepted the union. In Kenya — where Barack Sr. already had a wife and child — the family did not. The marriage lasted only two years; Barack left his wife and child behind to go to graduate school at Harvard. Stanley Ann filed for divorce, citing standard legal grounds.

His mother's second marriage, to an Indonesian student, Lolo Soetoro, took young Barry, as he was then called, to his Muslim stepfather's native country at the age of 6. Lolo worked as a geologist in post-colonial Indonesia; his mother taught English. They had a child, Obama's half-sister, Maya. (Maya Soetoro-Ng now teaches high school history in Honolulu.) Barry attended a predominantly Muslim school that would be falsely depicted as an Islamist madrassa during the 2008 campaign. His mother, meanwhile, taught her son about the civil rights struggles in America and eventually sent him back to Hawaii for schooling. The marriage ended later, a victim of cultural and personality differences.

Barry returned to live with grandparents Stanley and Madelyn Dunham — "Gramps" and "Toot" (her nickname came from the Hawaiian word for grandmother). They provided him the stable, supportive home life that he had

CHRONOLOGY: 2007 - PRESENT

2007 *Obama enters presidential race as underdog to New York Sen. Hillary Rodham Clinton; nearly matches Clinton in "money primary" in advance of Iowa caucuses.*

Feb. 10, 2007 Obama announces candidacy for Democratic nomination for president at rally in Springfield, Ill., three weeks after Clinton, former first lady, joined race; Democratic field eventually includes eight candidates.

March-December 2007 Democratic candidates engage in 17 debates, with no knockout punches; Obama closes gap with Clinton in polls, fundraising.

2008 *Obama gains Democratic nomination after drawn-out contest with Clinton; beats Republican Sen. John McCain as economic issues take center stage.*

January-February Obama scores upset in Iowa caucuses (Jan. 3); Clinton wins New Hampshire primary (Jan. 9); field narrows to two candidates by end of January.

March-April Clinton wins big-state primaries, including Ohio (March 4) and Pennsylvania (April 22); Obama edges ahead in delegates.

May-June Obama gains irreversible lead after Indiana, North Carolina primaries (May 6); clinches nomination after final primaries (June 2).

July Obama goes to Iraq, reaffirms 16-month pullout timetable; speaks at big rally in Berlin, Germany.

August Obama picks Delaware Sen. Joseph R. Biden as running mate; accepts nomination with speech promising

Iraq withdrawal, domestic initiatives; McCain chooses Alaska Gov. Sarah Palin as running mate.

September-October Obama holds his own in three debates with McCain (Sept. 26, Oct. 7, Oct. 15); McCain challenge to go to Washington to push financial bailout plan ends with advantage to Obama.

Nov. 4 Obama victory is signaled with victories in "red states" in East, Midwest; networks declare him winner as polls close in West (11 p.m., Eastern time).

November-December Obama completes Cabinet selections; works on economic recovery plan; vacations in Hawaii.

2009 *Obama inaugurated before largest crowd in Washington history.*

Jan. 5-19 Obama, in Washington, starts public campaign for economic recovery plan. . . . Congress reconvenes with Democrats holding 256-178 majority in House with one vacancy, 57-41 majority in Senate with two seats vacant. . . . More high-level nominations; Commerce post in limbo after Bill Richardson withdraws because of ethics investigation in New Mexico.

Jan. 20 Obama is inaugurated as 44th president; uses inaugural address to detail "serious" challenges at home, abroad; promises that challenges "will be met." . . . President moves quickly over next week to reverse some Bush administration policies; lobbies Congress on economic stimulus package, but Republicans continue to push for less spending, more tax cuts.

somewhat lacked so far. He gained admission to the prestigious Punahou School as one of only three black students. His father visited once — Barack's only time spent with him after the divorce — and spoke to one of his son's classes about life in Africa. Obama's mother came back to Hawaii for studies in anthropology, but when she returned to Indonesia for field work Barack chose to stay in Hawaii.

At Punahou, Obama excelled as a student and played with the state championship basketball team his senior year. He graduated in 1979 and enrolled at Occidental

College in Los Angeles. Two years later, he transferred to Columbia University in New York. By now, Obama was well aware of racial issues in the United States — and his ambiguous place in the story. "I learned to slip back and forth between my black and white worlds," he wrote in *Dreams from My Father.* More recently, as president-elect, Obama referred self-deprecatingly to his background. In describing the kind of puppy he would have preferred to get for his two young daughters, but for Malia's allergies, Obama said, "A mutt, like me."

Myriad Global Problems Confront Obama

Two wars, the Middle East and terrorism top the list.

President Barack Obama faces immense foreign-policy challenges — two wars and a turbulent global scene that includes continuing conflict in the Middle East — all against the backdrop of a global economic crisis.

Tens of thousands of U.S. troops are at war in Iraq and Afghanistan. Israel, America's closest Mideast ally, has just suspended a devastating military offensive in the Gaza Strip that could restart at any time. And Islamist terrorism remains a constant threat, with al Qaeda leader Osama bin Laden still at large.[1]

Obama divided his early days in office between wartime matters, the latest Mideast crisis and the economic meltdown. By all indications, he will be walking a tightrope between domestic and international affairs for the foreseeable future.

"A president in these circumstances is going to want to do everything possible to ensure that the transformative and ambitious and very difficult projects of domestic policy that have been designated as the priority for this new administration are not inhibited or disrupted by early failures, in counterterrorism or foreign policy," Steve Coll, president and CEO of the New America Foundation, a nonpartisan think tank, told a pre-inauguration conference on security issues.

Obama's inaugural address restated his commitment to withdraw U.S. forces from Iraq, which is more peaceful after more than five years of war but still violent and torn by political intrigue.[2]

In Afghanistan, however, escalating warfare is tied to another source of U.S. worries: Pakistan. Concern escalated in late November following coordinated terrorist attacks on hotels and other sites in Mumbai — India's financial and cultural capital — which were traced to a jihadist group in Pakistan with deep ties to that country's intelligence agency.[3] Some 175 people were killed and 200 wounded.

The group, Lashkar-e-Taiba, also has at least some operational link to al Qaeda and bin Laden, who is believed to be hiding in Pakistan's northern tribal region, bordering Afghanistan. Another al Qaeda ally, the Taliban guerrillas who are fighting the Afghan government and U.S.

and NATO troops in Afghanistan, use Pakistan as a headquarters.[4]

"Moreover," a government commission on weapons of mass destruction and terrorism said in December, "given Pakistan's tense relationship with India, its buildup of nuclear weapons is exacerbating the prospect of a dangerous nuclear arms race in South Asia that could lead to a nuclear conflict."[5]

The other daunting foreign-policy issue facing the new Obama administration — conflict between Israel and the Palestinians — offers slender prospects for peace. "Two states living side by side in peace and security — right now that stands about as much chance as Bozo the Clown becoming president of the United States," says Aaron David Miller, a former Mideast peace adviser to six secretaries of State.

The biggest obstacle, Miller says, is the "broken and dysfunctional" state of the Palestinian national movement. Fatah, the secular party that runs the West Bank, has a negotiating relationship with Israel. Hamas, the elected Islamist party and militia that initially seized power in an anti-Fatah coup in Gaza in 2007, deems Israel illegitimate. Hamas sponsored or tolerated rocket fire into Israel from Gaza but halted rocketing at the beginning of a cease-fire that began in June 2008. But Israel accused Hamas of building up its arsenal and retaliated by limiting the flow of goods into the region. In December, Hamas announced it wouldn't renew the already shaky truce, blaming the Israeli embargo and military moves. From then on, Hamas stepped up rocketing.

Israel's recent 22-day anti-Hamas offensive in Gaza cost some 1,300 Palestinian lives. The Palestinians estimated the civilian death toll at 40 percent to 70 percent of the fatalities; Israel put the toll at about 25 percent of the total. Israeli fatalities totaled 13, including three civilians.[6]

The scale of Israel's Gaza offensive is renewing calls for the U.S. government to change its relationship to Israel. "The days of America's exclusive ties to Israel may be coming to an end," Miller wrote in *Newsweek* in January. Obama, however, reaffirmed his support for Israel in his Jan. 26 interview with the Arabic-language network Al Arabiya.[7]

Those interests also would require devising a response to what the United States believes is a nuclear arms development project by Iran, which supports Hamas politically and financially — a sign, for some, of how all Middle Eastern issues are interconnected.

"One of the great mistakes we have made has been to believe we can compartmentalize these different policies, that we can somehow separate what is happening between Israel and the Palestinians from what's happening in Iraq and what's happening in Iran and what's happening in Egypt and Saudi Arabia and everywhere else in the Middle East," said Kenneth M. Pollack, a senior fellow at the Brookings Institution and former CIA analyst of the region. "Linkage is a reality."[8]

Another set of connections ties past U.S. support for NATO membership by Ukraine and Georgia to chilled U.S. relations with Russia, which views the potential presence of Western military allies — and U.S. missiles — on its borders as hostile.

Despite the Cold War echoes of that dispute, some foreign-affairs experts argue that Obama actually confronts a less perilous international panorama than some of his recent predecessors. "We don't have the Cold War and World War II," says Michael Mandelbaum, director of the foreign policy program at Johns Hopkins University's School of Advanced International Studies. "Those were existential threats. What the incoming president faces are annoying and troublesome, but not existential threats."

That picture could change if jihadist radicals took over nuclear-armed Pakistan. For now, Mandelbaum argues the biggest international and domestic dangers are one and the same — the economic meltdown.

But success for the huge spending package that Obama wants will require participation by China, America's major creditor. "China has been lending us money by buying our bonds," Mandelbaum says. "That huge stimulus package is not going to work unless we get some cooperation from the Chinese."

Palestinians in Gaza search the rubble of their homes for usable items after an Israeli air strike on Jan. 5, 2009.

Getty Images

In short, the American way of life very much depends on China, Mandelbaum says: "For what Americans care about, for what matters in the world, the issue of where and how we borrow money for the stimulus and where and how we rebalance the economy dwarfs Gaza in importance, and is more important than Iraq and Afghanistan."

— Peter Katel

---

[1] For coverage of the Iraq and Afghanistan wars, the Middle East and Islamic fundamentalism, see the following *CQ Researcher* reports: Peter Katel, "Cost of the Iraq War," April 25, 2008, pp. 361-384; Peter Katel, "New Strategy in Iraq," Feb. 23, 2007, pp. 169-192; and Peter Katel, "Middle East Tension," Oct. 27, 2006, pp. 889-912. Also see the following *CQ Global Researcher* reports: Roland Flamini, "Afghanistan on the Brink," June 2007, pp. 125-150; Robert Kiener, "Crisis in Pakistan," December 2008, pp. 321-348; and Sarah Glazer, "Radical Islam in Europe," November 2007, pp. 265-294.

[2] Alissa J. Rubin, "Iraq Unsettled by Political Power Plays," *The New York Times*, Dec. 25, 2008, www.nytimes.com/2008/12/26/world/middleeast/26baghdad.html; and Alissa J. Rubin, "Bombs Kill 5 in Baghdad, but Officials Avoid Harm," *The New York Times*, Jan. 20, 2009, www.nytimes.com/2009/01/21/world/middleeast/21iraq.html.

[3] Jane Perlez and Somini Sengupta, "Mumbai Attack is Test for Pakistan on Curbing Militants," *The New York Times*, Dec. 3, 2008, www.nytimes.com/2008/12/04/world/asia/04pstan.html?scp=5&sq=MumbaiLashkar ISI&st=cse.

[4] For a summary and analysis, see K. Alan Kronstadt and Kenneth Katzman, "Islamist Militancy in the Pakistan-Afghanistan Border Region and U.S. Policy," Congressional Research Service, Nov. 21, 2008, http://fpc.state.gov/documents/organization/113202.pdf.

[5] See "World at Risk," Commission on the Prevention of Weapons of Mass Destruction Proliferation and Terrorism, December 2008, p. xxiii.

[6] See Steven Erlanger, "Weighing Crimes and Ethics in the Fog of Urban Warfare," *The New York Times*, Jan. 16, 2009, www.nytimes.com/2009/01/17/world/middleeast/17israel.html?scp=1&sq=Gaza civilian deathpercent&st=cse; Amy Teibel, "Last Israeli troops leave Gaza, completing pullout," The Associated Press, Jan. 21, 2009, http://news.yahoo.com/s/ap/ml_israel_palestinians.

[7] Aaron David Miller, "If Obama Is Serious, He should get tough with Israel," *Newsweek*, Jan. 3, 2009, www.newsweek.com/id/177716.

[8] Quoted in Adam Graham-Silverman, "Conflict in Gaza Strip Presents Immediate Challenge for New President," *CQ Today*, Jan. 20, 2009.

Barack Obama's riveting, highly personal keynote address at the 2004 Democratic National Convention made him an overnight star and presidential contender.

Graduating from Columbia in 1983 with a degree in political science, Obama decided to take on the so-called Reagan revolution by becoming a community organizer — aiming, as he wrote, to bring about "change . . . from a mobilized grass roots." Obama flooded civil rights organizations to no avail until he was hired in 1985 by Gerald Kellman, a white organizer looking for an African-American to help with community development and mobilization in a Far South Side section of Chicago. Obama's three years in Chicago brought him face to face with the gritty realities of urban life and the disillusionment of the disadvantaged. He later described the time as "the best education I ever had."[7]

Obama enrolled in Harvard Law School in 1988.[8] He wrote nothing about the decision in his memoir and has said little about it elsewhere. Before going, he visited Kenya, where his father had died in an automobile accident six years earlier. Obama described enjoying the meeting with his extended family while acutely conscious of the cultural gap. At Harvard, he excelled as a student, played pick-up basketball and had only a limited social life after meeting his future wife, Michelle Robinson, a lawyer he had met while working for a Chicago law firm as a summer associate. His election in 1990 as president of the *Harvard Law Review* — as a compromise between conservative and liberal factions — marked the first time an African-American had held the prestigious position.

His barrier-breaking gained enough attention to get Obama an invitation from a literary agent, Jane Dystel, to write a book.[9] Obama planned to write about race relations, but in the three years of writing it turned into more of a personal memoir. Obama has said he was unmindful of political consequences in the writing and that he rejected a suggestion from one of his editors to delete references to drug use while in college. The book garnered respectable reviews — and the audio version won a Grammy — but no more than middling sales. Obama's mother read page proofs and lived just long enough to see it published. She died of ovarian cancer in November 1995.[10]

Red, Blue and Purple

Obama needed only 10 years to rise from the back benches of the Illinois legislature to a front seat on the national political stage. His political ambition misled him only once: in a failed run for the U.S. House. But he succeeded in other endeavors on the strength of hard work, personal intelligence, political acumen and earnest efforts to bridge the differences of race, class and partisan affiliation.

Obama entered politics in 1995 as the chosen successor of a one-term state senator, Alice Palmer. But he turned on his mentor when she sought re-election after all, following a losing bid in a special election for a U.S. House seat. Obama successfully challenged signatures on Palmer's nominating petitions and had her disqualified (and the other candidates too) to win the Democratic nomination unopposed and eventual election.

As a Democrat in a Republican-controlled legislature and a liberal with no connection to his party's organization, Obama worked to develop personal ties — some formed in a weekly poker game. Among his accomplishments: ethics legislation, a state earned-income tax credit and a measure, backed by law enforcement, to require videotaped interrogations in all capital cases.[11]

After four years in office, Obama decided in 2000 to mount a primary challenge to the popular and much better known Democratic congressman, Bobby Rush. The race was foolhardy from the outset. But — as Obama recounts in his second book, *The Audacity of Hope* — he suffered a grave embarrassment when he failed to return from a family vacation in Hawaii in time to vote on a major gun control bill in a specially called legislative session. Rush won handily.[12] In the 2008 presidential campaign, Obama's absence on the gun control vote was cited along with many other instances when he voted "present" as evidence of risk-averse gamesmanship on his part — a depiction vigorously disputed by the campaign.

His ambition unquenched, Obama began deciding by fall 2002 to run for the U.S. Senate seat then held by Republican Peter Fitzgerald, a vulnerable incumbent who eventually decided not to seek re-election. In October, at the invitation of a peace activist group, he delivered to an anti-war rally in Chicago his now famous speech opposing the then-imminent U.S. war in Iraq. Obama formally entered the Senate race in 2003 as the underdog to multimillionaire Blair Hull and state Comptroller Dan Hynes. But Hull's candidacy collapsed after allegations of abuse against his ex-wife. Hynes ran a lackluster campaign, while Obama waged a determined, disciplined drive that netted him nearly 53 percent of the vote in a seven-way race.[13]

Obama's debut on the national stage came in July 2004 after the presumptive Democratic presidential nominee, Massachusetts Sen. John Kerry, picked him to deliver the keynote address at the party's convention. Obama drafted the speech himself, according to biographer David Mendell. The night before, he told a friend, "My speech is pretty good." It was better than that. Obama wove his personal story together with verbal images of working-class America to lead up to the passage — rebroadcast thousands of times since — envisioning a unified nation instead of the "pundits' " image of monochromatic "Red States" and "Blue States." The speech "electrified the convention hall," *The Washington Post* reported the next day, and made Obama a rising star to be watched.[14]

By the time of the speech, political fortune had already shone on Obama back in Illinois. Divorce files of his Republican opponent in the Senate race, Jack Ryan, made public in June, showed that Ryan had pressured his wife to go with him to sex clubs and have sex in front of others. Ryan, a multimillionaire businessman, resisted pressure to withdraw for more than a month. Once Ryan bowed out — three days after Obama's speech — GOP leaders had to scramble for an opponent. They eventually lured Alan Keyes, a conservative African-American from Maryland, to be the sacrificial lamb in the race. Obama won with a record-setting 70 percent of the vote to take his seat in January 2005 as only the third African-American to serve in the U.S. Senate since Reconstruction.

Obama entered the Senate with the presidency on his mind but also the recognition that he must succeed first in a club with low tolerance for celebrity without substance. A profile in Congressional Quarterly's *Politics in America* published with his presidential campaign under way in 2007 credited Obama with "a reputation as a hard worker, a good listener and a quick study."[15]

With Democrats in the majority, Obama was designated in 2007 to spearhead the party's work on ethics reform — a role that prompted an icy exchange with his future opponent, Sen. McCain, who had expected to work with Democrats on a bipartisan approach. The eventual package included a ban on senators' discounted trips on corporate jets, but not — as Obama had pushed for — outside enforcement of ethics rules.

Obama had more success working with other Republicans, including Oklahoma's Tom Coburn (Internet access to government databases) and Indiana's Richard Lugar (international destruction of conventional weapons). Overall, however, his voting record was solidly liberal and reliably party-line. In the 2008 race, the McCain campaign repeatedly tried to debunk Obama's image of post-partisanship by challenging him to cite a significant example of departing from Democratic Party positions.

'Yes, We Can'

Obama won the Democratic nomination for president in a come-from-behind victory over frontrunner Hillary Clinton on the strength of fundraising prowess, message control and a pre-convention strategy focused on amassing delegates in caucus as well as primary states. He took an even bigger financial advantage into the general election but pulled away from McCain only after the nation's dire economic news in October drove the undecideds decisively toward the candidate promising "change we can believe in."[16]

Despite intense speculation and Obama's evident interest, he decided to run only after heart-to-heart talks with Michelle while vacationing in Hawaii in December 2006. Michelle's reluctance stemmed from the effects on the family and fear for Obama's personal safety. In the end, she agreed — with one stipulation: Obama had to give up smoking. That promise remains a work in progress. In his post-election appearance on NBC's "Meet the Press" on Dec. 7, Obama promised only that, "you will not see any violations" of the White House's no-smoking rule while he is president.

Daschle Appointment Shows Commitment to Health-Care Reforms

But a vote on a specific plan may be delayed until next year.

"The flaws in our health system are pervasive and corrosive. They threaten our health and economic security," said former Sen. Tom Daschle, D-S.D., President Obama's nominee for secretary of Health and Human Services (HHS), at his initial confirmation hearing before the Senate Health, Education, Labor, and Pensions (HELP) Committee on Jan. 8.[1]

Throughout his campaign, Obama promised to make good-quality health care accessible to all Americans. Many observers see his choice of Daschle — who recently coauthored a book laying out a plan for universal insurance coverage — to lead both HHS and a new White House Office of Health Policy as a sign of the new president's commitment to health-care reform, which he has called the key to economic security.[2] "I talk to hardworking Americans every day who worry about paying their medical bills and getting and keeping health insurance for their families," Obama said.[3]

In the final presidential debate on Oct. 15, 2008, Obama laid out the essence of his health overhaul. "If you've got health insurance through your employer, you can keep your health insurance," he said. "If you don't have health insurance, then what we're going to do is to provide you the option of buying into the same kind of federal pool [of private insurance plans] that [Republican presidential nominee] Sen. McCain and I enjoy as federal employees, which will give you high-quality care, choice of doctors at lower costs, because so many people are part of this insured group," Obama said.[4]

In addition, Obama's plan would:

- require insurance companies to accept all applicants, including those with already diagnosed illnesses — or

"preexisting conditions" — that insurers often decline to cover;
- create a federally regulated national "health insurance exchange" where people could buy coverage from a range of approved private insurers and possibly from a public insurance program as well;
- provide subsidies to help lower-income people buy coverage;
- require all children to have health insurance; and
- require employers except small businesses to either provide "meaningful" coverage to workers or pay a percentage of payroll toward the costs of a public plan.[5]

Points of potential controversy include whether all Americans should be required to buy health coverage.

During the presidential primary campaign, Obama sparred with fellow Democratic candidate Sen. Hillary Rodham Clinton, D-N.Y., who called for a mandate on individuals to buy insurance. Obama disagreed, saying, "my belief is that if we make it affordable, if we provide subsidies to those who can't afford it, they will buy it," and that only children's coverage should be required.[6]

But many analysts, including Daschle, point out that unless coverage is required many people will buy it only after they become sick, making it impossible for health insurance to perform its main task — spreading the costs of care among as many people as possible, not just among those who happen to be sick at a given time.

"The only way we can achieve universal coverage is to require everybody to either purchase private insurance or enroll in a public program," Daschle wrote.[7]

Obama entered the race with a speech to an outdoor rally on a cold Feb. 10, 2007, in Springfield, Ill. After acknowledging the "audacity" of his campaign, Obama laid out a platform of reshaping the economy, tackling the health-care crisis and ending the war in Iraq. He started well behind Clinton in the polls and in organization. In the early debates — with eight candidates in all — Obama himself rated his performance as "uneven," according to *Newsweek*'s post-election account.[17] By December, however, Obama had pulled ahead of Clinton

If Obama ends up authorizing a new government-run insurance plan to compete with private insurers for enrollees, as most Democrats favor, the plan could face tough opposition from Republicans.

"Forcing private plans to compete with federal programs, with their price controls and ability to shift costs to taxpayers, will inevitably doom true competition and could ultimately lead to a single-payer, government-run health-care program," said Sen. Michael Enzi, R-Wyo., the top Republican on the HELP Committee. "Any new insurance coverage must be delivered through private health-insurance plans."[8]

Congressional Democrats stand ready to work with the Obama administration to move health-care reform quickly. Two very influential senators, HELP Committee Chairman Sen. Edward Kennedy, D-Mass., and Finance Committee Chairman Sen. Max Baucus, D-Mont., were already crafting health-reform legislation last year and are expected to begin a strong push for legislation soon. But the press of other business and the time-consuming process of gathering support for a specific plan will put off a vote until the end of this year or the beginning of 2010, predicted Rep. Pete Stark, D-Calif., chairman of the House Ways and Means Health Subcommittee. "I don't think we'll do it in the first 100 days," said Stark.[9]

Ironically, the struggling economy, which leaves many more Americans worried about their jobs and therefore their health coverage, may have opened the door for reform by giving business owners, doctors and others a greater stake in getting more people covered, said Henry Aaron, a senior fellow in economic studies at the centrist Brookings Institution. "Before the economic collapse . . . the odds of national reform were nil," but the nation's economic stress makes it somewhat more likely, especially since Congress has been spending large amounts of money on other industries, Aaron said.[10]

Nevertheless, Aaron and some other analysts say the climate for health-care reform may not be much different from that in 1993 when the tide quickly turned against the Clinton administration's attempt at providing universal health care.

The times are "similar," and despite the desire of many for reform, the details will be painful and will spark pushback, Stuart Butler, vice president of the conservative Heritage Foundation, told PBS' "NewsHour." "When you say, 'We've got to make the system efficient by reducing unnecessary costs' . . . that means people's jobs and . . . doctors are going to rebel against that."[11]

— Marcia Clemmitt

[1] Quoted in "Daschle: Health Care Flaws Threaten Economic Security," CNNPolitics.com, Jan. 8, 2009, www.cnn.com/2009/POLITICS/01/08/daschle.confirmation.

[2] For background see the following *CQ Researcher* reports by Marcia Clemmitt: "Universal Coverage," March 30, 2007, pp. 265-288, and "Rising Health Costs," April 7, 2006, pp. 289-312.

[3] Barack Obama, "Modern Health Care for All Americans," *The New England Journal of Medicine*, Oct. 9, 2008, p. 1537.

[4] Quoted in "In Weak Economy, Obama May Face Obstacles to Health Care Reform," PBS "NewsHour," Nov. 20, 2008, www.pbs.org.

[5] "2008 Presidential Candidate Health Care Proposals: Side-by-Side Summary," health08.org, Kaiser Family Foundation, www.health08.org.

[6] Quoted in Jacob Goldstein, "Clinton and Obama Spar Over Insurance Mandates," *The Wall Street Journal* Health Blog, Feb. 1, 2008, http://blogs.wsj.com.

[7] Quoted in Teddy Davis, "Obama and Daschle at Odds on Individual Mandates," ABC News blogs, Dec. 11, 2008, http://blogs.abcnews.com.

[8] "Enzi Asks Obama Health Cabinet Nominee Daschle Not to Doom Health-Care Competition," press statement, office of Sen. Mike Enzi, Jan. 8, 2009, http://enzi.senate.gov.

[9] Quoted in Jeffrey Young, "Rep. Stark: No Health Reform Vote in Early '09," *The Hill*, Dec. 17, 2008, http://thehill.com.

[10] Quoted in Ben Weyl, "Experts Predict a Health Overhaul Despite Troubled Economy," *CQ Healthbeat*, Dec. 9, 2008.

[11] "In Weak Economy, Obama May Face Obstacles to Health Care Reform," *op. cit.*

in some New Hampshire polling and was in a virtual dead-heat in the all-important "money primary."

The Iowa caucuses on Jan. 3, 2007, gave Obama an unexpected win with about 38 percent of the vote and left only two other viable candidates standing: former North Carolina Sen. John Edwards, who came in second; and Clinton, who finished a disappointing third. Five days later, however, Clinton regained her stride with a 3-percentage-point victory over Obama in the first-in-the-nation New Hampshire primary. Edwards' third-place

Vice President Biden Brings Foreign-Policy Savvy

"I want to be the last guy in the room on every important decision."

The inauguration of Joseph R. Biden Jr. as the 47th vice president of the United States caps a journey almost as improbable as Barack Obama's. During seven terms as a U.S. senator from Delaware, Biden has never lived in Washington, instead commuting daily by train from Wilmington. In 1972, at age 29, he became the sixth-youngest senator ever elected, leading many to believe the White House was in his future.

But after two failed presidential campaigns — in 1988 and in the last election — Biden seemed fated to remain a Senate lifer.

Along the way he rose to become chairman of the Judiciary Committee and gained national prominence while leading the confirmation hearings of Supreme Court nominees Robert Bork and Clarence Thomas. He had also served twice as chairman of the Foreign Relations Committee.

Obama's limited time in the Senate and lack of international experience led to increased speculation that he would select Biden as his running mate to bridge the gap. "[Joe Biden is] a leader who sees clearly the challenges facing America in a changing world, with our security and standing set back by eight years of failed foreign policy," Obama said in introducing Biden as his selection on Aug. 23, 2008.

But the new president has yet to clarify the specific role Biden will play in the new administration. The appointment of Hillary Rodham Clinton as secretary of State all but ensures that Biden, despite his impressive résumé, will not be the point man on foreign policy as initially expected.

Nor does anyone expect him to emulate former Vice President Dick Cheney's muscular role. Upon taking office in 2001, Cheney demanded — and President George W. Bush approved — a mandate to give him access to "every table and every meeting," expressing his voice in "whatever area the vice president feels he wants to be active in," recalls former White House Chief of Staff Joshua B. Bolten.[1]

Cheney's push to expand presidential war-making authority is arguably his most lasting legacy, but he also served as a gatekeeper for Supreme Court nominees, editor of tax proposals and arbiter of budget appeals.

While most vice presidents arrive eager to expand the influence of their position, Biden faces the unusual conundrum of figuring out how to scale it back. "The only value of power is the effect, the efficacy of its use," he told *The New York Times*. "And all the power Cheney had did not result in effective outcomes." But without any direct constitutional authority in the executive branch, Biden does not want to return to the days when vice presidents were neither

finish kept him in the race, but he dropped out on Jan. 30 after finishing third in primaries in Florida and his birth state of South Carolina.

The one-on-one between Obama and Clinton continued through May. Clinton bested Obama in a series of supposedly "critical" late-season primaries — notably, Ohio and Pennsylvania — even as Obama pulled ahead in delegates thanks to caucus state victories and also-ran proportional-representation winnings from the primaries. He turned the most serious threat to his campaign — his relationship with the sometimes fiery black minister, Jeremiah Wright — into a plus of sorts with a stirring speech on racial justice delivered in Philadelphia on

March 18. With Clinton's "electability" arguments unavailing, Obama mathematically clinched the nomination on June 3 as the two split final primaries in Montana and South Dakota. Clinton withdrew four days later, promising to work hard for Obama's election.

With nearly three months before the convention, Obama went to Iraq and Europe to burnish his national security and foreign policy credentials. His 16-month timetable for withdrawal now essentially matched the Iraqi government's own position — weakening a Republican line of attack. An address to a huge and adoring crowd in Berlin underscored Obama's promise to raise U.S. standing in the world. The McCain campaign countered

seen nor heard. "I don't think the measure is whether or not I accrete the vestiges of power; it matters whether or not the president listens to me."[2]

And although he says he doesn't seek to wield as much influence as Cheney, many don't expect the loquacious Biden to follow Al Gore either, who in 1992 was assigned a defined portfolio by President Bill Clinton to work on environmental and technology matters. "I think his fundamental role is as a trusted counselor," said Obama senior adviser David Axelrod. "I think that when Obama selected him, he selected him to be a counselor and an adviser on a broad range of issues."[3]

And that's exactly how Biden — who at first balked at accepting the position — wants it. "I don't want to have a portfolio," Biden says. "I don't want to be the guy who handles U.S.-Russian relations or the guy who reinvents government."

"I want to be the last guy in the room on every important decision."

"It's irrelevant what the outside world perceives. What is relevant is whether or not I'm value-added," Biden contends. And very few debate his credentials for the position.

"I'm the most experienced vice president since anybody. Anybody ever serve 36 years as a United States senator?" he asks.[4]

But in all likelihood Biden's first move to Washington will surely be his last.

At age 66, he says he has no plans to pursue the presidency, or return to the Senate for that matter, in 2016 — the last full year of a possible second term for Obama. That suggests he'll truly serve Obama's ambitions rather than his own.

"This is in all probability, and hopefully, a worthy capstone in my career," he said.

— Darrell Dela Rosa

Newly sworn in Vice President Joseph R. Biden, his wife, Jill, and son Beau greet crowds during the Inaugural Parade.

[1] Barton Gellman and Jo Becker, " 'A Different Understanding With the President,' " *The Washington Post*, June 24, 2007, blog.washingtonpost.com/cheney/chapters/chapter_1.

[2] Peter Baker, "Biden Outlines Plans to Do More With Less Power," *The New York Times*, Jan. 14, 2009, www.nytimes.com/2009/01/15/us/politics/15biden.html?_r=1.

[3] Helene Cooper, "For Biden, No Portfolio but the Role of a Counselor," *The New York Times*, Nov. 25, 2008, www.nytimes.com/2008/11/26/us/politics/26biden.html.

[4] Baker, *op. cit.*

with an ad mocking Obama's celebrity status. On the eve of the convention, Obama picked Biden as his running mate. The selection won praise as sound, if safe. The four-day convention in Denver (Aug. 25-28) went off without a hitch. Obama's acceptance speech drew generally high marks, but some criticism for its length and predictable domestic-policy prescriptions.

McCain countered the next day by picking Alaska Gov. Sarah Palin as his running mate. The surprise selection energized the GOP base but raised questions among observers and voters about his judgment. For the rest of the campaign, the McCain camp tried but failed to find an Obama weak spot. Obama had already survived personal

attacks about ties to Rev. Wright, indicted Chicago developer Tony Rezko and one-time radical William Ayers. He had also fended off attacks for breaking his pledge to limit campaign spending by taking public funds. Improved ground conditions in Iraq shifted the contest from national security — McCain's strength — to the economy: Democratic turf. Obama held his own in three debates and used his financial advantage — he raised a record $742 million in all — to engage McCain not only in battleground states but also in supposedly safe GOP states.

By Election Day, the outcome was hardly in doubt. Any remaining uncertainty vanished when Virginia, Republican since 1968, went to Obama early in the evening. By 9:30, one blog had declared Obama the winner. The networks waited until the polls closed on the West Coast — 11 p.m. in the East — to declare Obama to be the 44th president of the United States. In Chicago's Grant Park, tens of thousands of supporters chanted "Yes, we can," as Obama strode on stage.

"If there is anyone out there," Obama began, "who still doubts that America is a place where all things are possible; who still wonders if the dream of our founders is alive in our time; who still questions the power of our democracy, tonight is your answer."[18]

A Team of Centrists?

President-elect Obama began the 76 days between election and inauguration by hitting nearly pitch-perfect notes in his dealings with official Washington — including President Bush and members of Congress — and with the public at large. Beginning with his first post-election session with reporters, Obama sounded both somber but hopeful in confronting what he continually referred to as the worst economic crisis in generations. He completed his selection of Cabinet appointees in record time before taking an end-of-December vacation with his family in Hawaii. Some discordant notes were sounded as Inauguration Day neared in January. But on the eve of the inauguration, polls showed Obama entering the Oval Office with unprecedented levels of personal popularity and hopeful support. (*See graph, p. 362.*)

Acknowledging the severity of the economic crisis, Obama started the announcement of Cabinet-level appointments on Nov. 24 by introducing an economic team that included New York Federal Reserve Bank President Timothy Geithner to be secretary of the Treasury. Geithner had been deeply involved in the Fed's moves in the financial bailout. Obama also named Summers, who had served as deputy undersecretary of the Treasury in the Clinton administration, as special White House assistant for economic policy.

A week later, Obama introduced a national security team that included Hillary Clinton as secretary of State and Gates as holdover Pentagon chief. Clinton accepted the post only after weighing the offer against continuing in the Senate with possibly enhanced visibility and influence. In addition, the appointment required former President Clinton to disclose donors to his post-presidential foundation to try to reduce potential conflicts of interest with his wife's new role.

Along with Gates, Obama also introduced Gen. Jones, a retired Marine commandant and former North Atlantic Treaty Organization supreme commander, as his national security adviser. He also said that he would nominate Holder, a former deputy attorney general, for attorney general; Gov. Janet Napolitano of Arizona for secretary of Homeland Security; and Susan E. Rice, a former assistant secretary of State, for ambassador to the United Nations with Cabinet rank. Holder was in line to be the first African-American to head the Justice Department.

Other Cabinet nominations followed in rapid succession: New Mexico Gov. Bill Richardson, like Clinton one of the contenders for the Democratic nomination, for Commerce; Gen. Eric Shinseki, a critic of Iraq War policies, for Veterans Affairs; and former Senate Democratic Leader Tom Daschle of South Dakota, for Health and Human Services and a new White House office as health reform czar.

Obama picked Shaun Donovan, commissioner of New York City's housing department, for Housing and Urban Development; outgoing Illinois Rep. Ray LaHood, a Republican, for Transportation; and Chicago public schools Commissioner Arne Duncan, a reformer with good relations with Chicago teacher unions, for Education. Steven Chu, a Nobel Prize-winning scientist and an advocate of measures to reduce global warming, was picked for Energy. Sen. Kenneth Salazar, a Colorado Democrat with a moderate record on environmental and land use issues, was tapped for Interior. Former Iowa Gov. Tom Vilsack, who had supported Clinton for the nomination, was chosen for Agriculture. And Rep. Hilda Solis, a California Democrat and daughter of a union family, was designated for Labor.

As Obama prepared to leave for Hawaii, some supporters were griping about the moderate cast of his selections. "We just hoped the political diversity would have been stronger," Tim Carpenter, executive director of Progressive Democrats of America, told Politico.com. But official Washington appeared to be giving him top marks. *The Washington Post* described the future Cabinet as dominated by "practical-minded centrists who have straddled big policy debates rather than staking out the strongest pro-reform positions."[19]

Obama arrived in Washington on Jan. 4 to enroll daughters Malia, 10, and Natasha ("Sasha"), 7, in the private Sidwell Friends School and begin two hectic work weeks before a long weekend of pre-inaugural events. By then, problems had begun to arise, including a corruption scandal over the selection of Obama's successor in the Senate; the withdrawal of one of his Cabinet nominees; and questions about several of his nominees for top posts.

The Senate seat controversy stemmed from a federal investigation of Illinois Gov. Rod Blagojevich that included tape-recorded comments by the Democratic chief executive that were widely depicted as attempting to sell the appointment for political contributions or other favors. In charging Blagojevich with corruption, U.S. Attorney Patrick Fitzgerald specifically cleared Obama of any involvement. But Obama had been forced to answer questions on the issue from Hawaii and had lined up with Senate Democratic Leader Harry Reid in promising not to seat any Blagojevich appointee. When Blagojevich went ahead and appointed former state Comptroller Roland Burris, an African-American, Reid initially resisted but eventually bowed to the fait accompli and welcomed Burris to the Senate.

Richardson had withdrawn from the Commerce post on Jan. 3 after citing a federal probe into a possible "pay for play" scandal in New Mexico.

Two other Cabinet nominees faced critical questions as Senate confirmation hearings got under way. Treasury Secretary-designate Geithner was disclosed to have failed to pay Social Security and Medicare taxes for several years and to have paid back taxes and interest only after being audited. Attorney General-designate Holder faced questions about his role in recommending that President Clinton pardon fugitive financier Marc Rich and in submitting a pardon application for members of the radical Puerto Rican independence movement FALN. Both seemed headed toward confirmation, however.

CURRENT SITUATION

Moving Quickly

Beginning with his first hours in office, President Obama is moving quickly to put his stamp on government policies by fulfilling campaign promises on such issues as government ethics, secrecy and counterterrorism. Along with the flurry of domestic actions, Obama opened initiatives on the diplomatic front by promising an active U.S. role to promote peace in the Middle East and naming

high-level special envoys for the Israeli-Palestinian dispute and the strategically important region of South Asia, including Afghanistan and Pakistan.

In the biggest news of his first days in office, Obama on Jan. 22 signed executive orders to close the Guantánamo prison camp within one year and to prohibit the use of "enhanced" interrogation techniques such as waterboarding by CIA agents or any other U.S. personnel. Human rights groups hailed the actions. "Today is the beginning of the end of that sorry chapter in our nation's history," said Elisa Massimino, executive director and CEO of Human Rights First.

Some Republican lawmakers, however, questioned the moves. "How does it make sense," House GOP Whip Eric Cantor asked, "to close down the Guantánamo facility before there is a clear plan to deal with the terrorists inside its walls?"

An earlier directive, signed late in the day on Jan. 20, ordered Defense Secretary Gates to halt for 120 days any of the military commission proceedings against the remaining 245 detainees at Guantánamo. Separately, Obama directed a review of the case against Ali Saleh Kahlah al-Marri, a U.S. resident and the only person designated as an enemy combatant being held in the U.S.

The ethics and information directives signed on Jan. 21 followed Obama's campaign pledges to limit the "revolving door" between government jobs and lobbyist work and to make government more transparent and accountable.

The new ethics rules bar any executive branch appointees from seeking lobbying jobs during Obama's administration. They also ban gifts from lobbyists to anyone in the administration. Good-government groups praised the new policies as the strictest ethics rules ever adopted. Fred Wertheimer, president of the open-government group Democracy 21, called them "a major step in setting a new tone and attitude for Washington."

On information policy, Obama superseded a Bush administration directive promising legal support for agencies seeking to resist disclosure of government records under the Freedom of Information Act. Instead, Obama called on all agencies to release information whenever possible. "For a long time now, there's been too much secrecy in this city," Obama said at a swearing-in ceremony for senior White House staff.

Obama also signed an executive order aimed at greater openness for presidential records following the

Should Congress and the president create a commission to investigate the Bush administration's counterterrorism policies?

YES

Frederick A. O. Schwarz Jr.
*Chief Counsel, Brennan Center for Justice,
New York University School of Law; co-author,
Unchecked and Unbalanced: Presidential
Power in a Time of Terror (New Press, 2008)*

Written for *CQ Researcher*, January 2009

In his inaugural address, President Obama rejected "as false the choice between our safety and our ideals." Throughout our history, seeking safety in times of crisis has often made it tempting to ignore the wise restraints that make us free and to rush into actions that do not serve the nation's long-term interests. (The Alien and Sedition Acts at the dawn of the republic and the herding of Japanese citizens into concentration camps early in World War II are among many historic examples.) After 9/11 we again overreacted to crisis, this time by descending into practices including torture, extraordinary rendition, warrantless wiretapping and indefinite detention. Each breached American values and thus made America less safe.

Our new president is taking steps to reject these actions. And some say this is all that is needed because we need to look forward. Others clamor for criminal prosecutions because to hold our heads high wrongdoers should be held to account.

But, to me, neither of these positions is right. Prosecution is not likely to be productive, and could well be unfair. At the same time, failure to learn more about how we went wrong poses two dangers: First, if we blind our eyes to the truth, we increase the risk of repetition when the next crisis comes.

Second, clearly and fairly assessing and reporting what went wrong — and right — in our reactions to 9/11 will honor America's commitment to openness and the rule of law. Committing ourselves to a full exploration is consistent with the ethos the new president articulated on his first day in office: "The way to make government responsible is to hold it accountable. And the way to make government accountable is to make it transparent."

For these two reasons, I have recommended that the president and Congress appoint an independent, nonpartisan commission to investigate national counterterrorism policies. This is the best way to achieve accountability and an understanding of how to design an effective counterterrorism policy that comports with fundamental values.

Shortly after his reelection in 1864, President Abraham Lincoln nicely articulated the necessity of learning from the past without seeking punishment: "Let us study the incidents of [recent history], as philosophy to learn wisdom from, and none of them as wrongs to be revenged."

NO

David B. Rivkin Jr. and Lee A. Casey
*Washington attorneys who served in the Justice
Department under Presidents Reagan and
George H. W. Bush*

Written for *CQ Researcher*, January 2009

A special commission would be both unnecessary and harmful. First, multiple congressional inquiries have already aired and analyzed all of the Bush administration's key legal and policy decisions. Indeed, whether through disclosures, leaks, media and/or congressional investigations, both the process and substance of the administration's war-related decisions have been publicized to an unprecedented extent. If any further inquiry into these policies is necessary, the normal congressional and executive branch investigatory tools are always available, including additional hearings.

Second, a special commission would be fundamentally unfair, beginning — as it would — with the proposition that the Bush policies represent systematic wrongdoing. The Bush policies were based upon well-established case law and reasonable legal extrapolation from the available authorities. Simply because the Supreme Court ultimately decided to change the legal landscape does not mean the Bush administration ignored the law; it did not. Moreover, although there have been many problems and certainly some abuses over the past seven years — Abu Ghraib being a case in point — these have been remarkably rare when compared with past armed conflicts and/or counterterrorism campaigns like the one Britain conducted in Northern Ireland.

A commission would also inevitably involve attacks on career officials in the intelligence community and the departments of Justice and Defense, not merely Bush political appointees. When combined with past investigations, the commission's work would inevitably burden, distract and demoralize the nation's intelligence capabilities. The end result would be the extension of a bureaucratic culture that already favors excessive caution and inaction among our key intelligence and law enforcement officials — the very developments, acknowledged by the 9/11 Commission, as contributing mightily to the analytical, legal and policy failures of 9/11.

Finally, a commission would warp our constitutional fabric and harm civil liberties. While many commissions have operated throughout American history, they have not focused on potential prosecutions. Such a private or quasi-governmental commission would not be constrained by the legal and constitutional limits on Congress and the executive branch, thus raising a host of important constitutional questions.

That the commission's supporters — so determined to vindicate the rights of enemy combatant detainees — seem untroubled by these issues is both ironic and terribly sad.

AT ISSUE

Will Obama's economic stimulus revive the U.S. economy?

YES Dean Baker
Co-director, Center for
Economic and Policy Research

NO J. D. Foster
Norman B. Ture Senior Fellow in
the Economics of Fiscal Policy,
The Heritage Foundation

Written for *CQ Researcher*, January 2009

Written for *CQ Researcher*, January 2009

President Obama's stimulus proposal is a very good start toward rescuing the economy. In assessing the plan, it is vitally important to recognize the seriousness of the downturn. The economy lost an average of more than 500,000 jobs a month in the last three months of 2008. In fact, the actual job loss could have been over 600,000 a month due to the way in which the Labor Department counts jobs in new firms that are not in its survey.

The recent announcements of job loss suggest that the rate of job loss may have accelerated even further. It is possible that we are now losing jobs at the rate of 700,000 a month. This is important, because people must understand the urgency of acting as quickly as possible.

With this in mind, the package being debated does a good job of getting money into the economy quickly. According to the projections of the Congressional Budget Office (CBO), 62 percent of the spending in the package will reach the economy before the end of 2010, with most of the rest coming in 2011. This money will be giving the economy a boost when we need it most.

At this point, there is considerable research on the impact of tax cuts, and the evidence suggests that they do not have nearly as much impact on the economy, primarily because a large portion of any tax cut is saved. According to Martin Feldstein, President Reagan's chief economist, just 10 percent of the tax cuts sent out last spring were spent. The rest was saved. Increased savings can be beneficial to household balance sheets, but savings will not boost the economy right now.

There will also be long-term benefits from President Obama's package. For example, the CBO projected we would save more than $90 billion on medical expenses over the next decade by computerizing medical records, which will be financed through the stimulus. In addition, weatherizing homes and offices and modernizing the electrical grid will substantially reduce our future energy use.

The Obama administration projects that this package will generate close to 4 million jobs, and several independent analysts have arrived at similar numbers. This will not bring the economy back to full employment, but it is still a huge improvement over doing nothing.

The cost of this bill sounds large, but it is important to remember that the need is large. If we were to just do nothing, the economy would continue to spiral downward, with the unemployment rate reaching double-digit levels in the near future.

President Barack Obama promises to create 3.5 million new jobs by the end of 2010, and that vow provides a clear measure by which to judge whether his policies work.

U.S. employment stood at about 113 million people in December 2008, so the Obama jobs pledge will be met if 116.5 million people are working by the end of 2010. Reaching this goal will require effective stimulus policies — and the only fiscal policy that can come close to reaching the goal is to cut marginal tax rates.

Obama's target for jobs creation was chosen carefully. Employment peaked at about 115.8 million jobs in November 2007. Obama's jobs pledge at that time was to create 2.5 million jobs, for a total of 116.5 million private sector jobs.

The November 2008 jobs report showed a half-million jobs lost, so his job-creating target rose by a half-million, affirming the 116.5 million target. Then last month's jobs report showed another half-million jobs lost, and the president raised the target again to its current 3.5 million total.

To stimulate the economy, Obama and congressional Democrats have focused on massive new spending programs. However, the federal budget deficit is likely to exceed $2.5 trillion over the next two years even before any stimulus is added. If deficit spending were truly stimulative, the economy would be at risk of overheating by now, not sliding deeper into recession.

Additional deficit spending won't be any more effective than the first $2 trillion, because government spending doesn't create additional demand in the economy. Deficit spending must be financed by borrowing, so while government spending increases demand, government borrowing reduces demand. Worse, since the government's likely to borrow between $3 trillion and $4 trillion over the next two years, the enormous waves of government debt will likely drive interest rates up. That would only prolong the recession and weaken the recovery.

An effective fiscal stimulus would defer the massive 2011 tax hike (higher tax rates on dividends and capital gains are scheduled to kick in), and also cut individual and corporate tax rates further to reduce the impediments to starting new businesses, hiring, working and investing.

To meet his goal, President Obama should junk his ideology and the wasteful spending that goes with it and focus on cutting marginal tax rates. That's the only way to hit his jobs creation target.

congressionally established five-year waiting period after any president leaves office. The order supersedes a Bush administration directive in 2001 by giving the incumbent president, not a former president, decision-making authority on whether to invoke executive privilege to prevent release of the former president's records.

On foreign policy, Obama on his first full day in office turned to the fragile cease-fire in Gaza by placing calls to four Mideast leaders: Egyptian President Hosni Mubarak, Israeli Prime Minister Ehud Olmert, Jordanian King Abdullah and Palestinian Authority President Mahmoud Abbas. Obama offered U.S. assistance to try to solidify the ceasefire that had been adopted over the Jan. 17-18 weekend by Israel and Hamas, the ruling party in Gaza.

Israel had begun an offensive against Hamas on Dec. 27 in an effort to halt cross-border rocket attacks into Israel by Hamas supporters. During the transition, Obama had limited himself to a brief statement regretting the loss of life on both sides. White House press secretary Robert Gibbs said Obama used the calls from the Oval Office to pledge U.S. support for consolidating the cease-fire by preventing the smuggling of arms into Hamas from neighboring Egypt. He also promised U.S. support for "a major reconstruction effort for Palestinians in Gaza," Gibbs said.

The next day, Obama took a 10-block ride to the State Department for Hillary Clinton's welcome ceremony as secretary following her 94-2 Senate confirmation on Jan. 21. As part of the event, Clinton announced the appointment of special envoys George Mitchell for the Middle East and Richard Holbrooke for Afghanistan and Pakistan.

In his remarks, Obama renewed support for a two-state solution: Israel and a Palestinian state "living side by side in peace and security." He also promised to refocus U.S. attention on what he called the "perilous" situation in Afghanistan, where he said violence had increased dramatically and a "deadly insurgency" had taken root.

Returning to domestic issues, Obama on Jan. 23 signed — as expected — an order to lift the so-called Mexico City policy prohibiting U.S. aid to any nongovernmental organizations abroad that provide abortion counseling or services. The memorandum instructed Secretary of State Clinton to lift what Obama called the "unwarranted" restrictions. The policy was first put in place by President Ronald Reagan in 1984, rescinded by President Clinton in 1993 and then reinstituted by President Bush in 2001.

After the weekend, Obama reversed another of Bush's policies on Jan. 26 by directing Environmental Protection Agency Administrator Lisa Jackson to reconsider the request by the state of California to adopt automobile emission standards stricter than those set under federal law. In a reversal of past practice, the Bush administration EPA had denied California's waiver request in December 2007. On the same day, Obama instructed Transportation Secretary Ray LaHood to tighten fuel efficiency standards for cars and light trucks beginning with 2011 model cars.

Working With Congress

President Obama is pressing Congress for quick action on an economic stimulus plan even as bipartisan support for a proposal remains elusive. Meanwhile, the new administration is struggling to find ways to make the financial bailout approved before Obama took office more effective in aiding distressed homeowners and unfreezing credit markets.

House Democrats moved ahead with an $825-billion stimulus package after the tax and spending elements won approval in separate, party-line votes by the House Ways and Means Committee on Jan. 22 and the House Appropriations Committee the day before. The full House was scheduled to vote on the package on Jan. 28 after deadline for this issue, but approval was assured given the Democrats' 256-178 majority in the chamber.

Obama used his first weekly address as president on Jan. 24 — now not only broadcast on radio but also posted online as video on YouTube and the White House Web site — to depict his American Recovery and Reinvestment Plan as critical to get the country out of an "unprecedented" economic crisis. The plan, he said, would "jump-start job creation as well as long-term economic growth." Without it, he warned, unemployment could reach double digits, economic output could fall $1 trillion short of capacity and many young Americans could be forced to forgo college or job training.

Without mentioning the tax and spending plan's minimum total cost, Obama detailed a long list of infrastructure improvements to be accomplished in energy, health care, education and transportation. He mentioned a $2,500 college tax credit but did not note other items in the $225 billion in tax breaks included in the plan — either his long-advocated $1,000 tax break for working families or the various business tax cuts added as sweeteners for Republicans.

Republicans, however, remained unconvinced. Replying to Obama's address, House Minority Leader John Boehner called the plan "chock-full of government programs and projects, most of which won't provide immediate relief to our ailing economy." On "Meet the Press" the next day, the Ohio lawmaker again called for more by way of tax cuts, criticized the job-creating potential of Obama's plan and warned of opposition from most House Republicans.

Appearing on another of the Sunday talk shows, McCain told "Fox News Sunday" host Chris Wallace, "I am opposed to most of the provisions in the bill. As it stands now, I would not support it."

On a second front, the principal members of Obama's economic team are assuring Congress of major changes to come in the second stage of the $700-billion financial rescue plan approved last fall. During confirmation hearings, Treasury Secretary-designate Geithner promised the Senate Finance Committee on Jan. 21 to expect "much more substantial action" to address the problem of troubled banks that has chilled both consumer and corporate credit markets since fall 2008.

Geithner's comments on the financial bailout were overshadowed by sharp questions from Republican senators about the nominee's tax problems while working for the International Monetary Fund. For several years, Geithner failed to pay Social Security and Medicare taxes, which the IMF — as an international institution — does not withhold from employees' pay as domestic employers do. Geithner repeatedly apologized for the mistake and pointed to his payment of back taxes plus interest totaling more than $40,000. In the end, the committee voted 18-5 to recommend confirmation; the full Senate followed suit on Jan. 26 in a 60-34 vote.*

On the bailout, Geithner said he would increase the transparency and accountability of the program once he assumed the virtually unfettered responsibility for dispensing the remaining $350 billion. He acknowledged criticisms that so far the program has benefited large financial institutions but done little for small businesses. He also promised to restrict dividends by companies that receive government help.

With many banks still holding billions in troubled assets on their balance sheets, speculation is increasing in Washington and in financial circles about dramatic action by the government. Possible moves include the creation of a government-run "bad bank" to buy distressed assets from financial institutions or even outright nationalization of one or more banks.

"People continue to be surprised by the poor condition of the banks," says Dean Baker, co-director of the Center for Economic and Policy Research, a liberal think tank in Washington. "Whatever plans they may have made a month ago might be seen as inadequate given the severity of the problem of the banking system."

With the stimulus package on the front burner, however, Obama went to Capitol Hill on Jan. 27 for separate meetings to lobby House and Senate Republicans to support the measure. The closed-door session with the full House GOP conference lasted an hour — slightly longer than scheduled, causing the president to be late for the start of the meeting on the other side of the Capitol with Republican senators.

In between meetings, Obama challenged GOP lawmakers to try to minimize partisan differences. "I don't expect 100 percent agreement from my Republican colleagues, but I do hope we can put politics aside," he said.

For their part, House Republican leaders expressed appreciation for the president's visit and his expressed willingness to compromise. But some renewed their opposition to the proposal in its current form. Rep. Tom Price of Georgia, chairman of the conservative House Republican Study Committee, said the proposal "remains rooted in a liberal, big-government ideology."

Obama's meeting with GOP senators came on the same day that the Senate Finance and Appropriations committees were marking up their versions of the stimulus package. The Senate was expected to vote on the proposal over the weekend, giving the two chambers two weeks to iron out their differences if the bill was to reach Obama's desk before the Presidents' Day recess.

OUTLOOK

Peril and Promise

One week after taking office, President Obama is getting high marks from experts on the presidency for carefully stage-managing his first policy initiatives while discreetly moving to set realistic expectations for the months ahead.

---

* Attorney General-designate Holder, Obama's other controversial Cabinet nominee, was expected to be confirmed by the full Senate on Jan. 29 or 30, after deadline for this issue, following the Senate Judiciary Committee's 17-2 vote on Jan. 28 to recommend confirmation.

"He's started out quite impressively," says Fred Greenstein, professor of politics emeritus at Princeton University in New Jersey and the dean of American scholars on the U.S. presidency. "So far, it's been a striking rollout week."

Other experts agree. "The Obama administration has met expectations for the first week," says Meena Bose, chair of the Peter S. Kalikow Center for the Study of the American Presidency at Hofstra University in Hempstead, N.Y. "There's been virtually no drama, which is an indication of how he intends to run his administration."

"The indications are all positive," says Bruce Buchanan, a professor of political science at the University of Texas in Austin and author of several books on the presidency. Like the others, Buchanan says Obama is holding on to popular support while striving either to win over or to neutralize Republicans on Capitol Hill.

The wider world outside Washington, however, is giving Obama no honeymoon in office. The U.S. economy is continuing to lag, while violence and unrest continue to simmer in three global hot spots: Gaza, Iraq and Afghanistan.

On the economy, Obama has initiated a daily briefing from senior adviser Summers in addition to the daily briefing on foreign policy and national security issues. "Frankly," Obama told congressional leaders on Jan. 23, "the news has not been good." The day before, the Commerce Department had reported that new-home construction fell to its slowest pace since reporting on monthly rates began in 1959. On the same day, new claims for unemployment benefits matched the highest level seen in a quarter-century.[20]

Meanwhile, leading U.S. policy makers were giving downbeat assessments of events in Afghanistan and Iraq. In testimony to the Senate Armed Services Committee, Defense chief Gates warned on Jan. 27 to expect "a long and difficult fight" in Afghanistan. A few days earlier, the outgoing U.S. ambassador to Iraq, Ryan Crocker, warned that what he called "a precipitous withdrawal" could jeopardize the country's stability and revive al Qaeda in Iraq. And special envoy Mitchell left Washington for the Mideast on Jan. 26, just as the fragile cease-fire between Hamas and Israel was jeopardized by the death of an Israeli soldier from a roadside bomb and an Israeli air strike in retaliation.

Obama continues to work at the problems with the same kind of message control that served him well in the election. After reaping a full day's worth of mostly favorable news coverage on the Guantánamo issue, the administration began directing laser-like attention to the economy

from Jan. 22 on. For example, the repeal of the Bush administration's ban on funding international groups that perform abortions was announced late on Friday, Jan. 23 — a dead zone for news coverage.

On foreign policy, Obama emphasized the Mitchell and Holbrooke appointments by personally going to the State Department for the announcements. And he underscored the inaugural's outreach to Muslims by granting his first formal television interview as president to the Arabic satellite television network Al Arabiya. Obama called for a new partnership with the Muslim world "based on mutual respect and mutual interest." One of his main tasks, he told the Dubai-based network in an interview aired on Jan. 27, is to communicate that "the Americans are not your enemy."[21]

Obama and his senior aides are also signaling to supporters that some of their agenda items will have to wait. In a pre-inauguration interview with *The Washington Post*, for example, he reiterated his support for a labor-backed bill to make it easier to unionize workers but downgraded it to a post-stimulus agenda item. Similarly, press secretary-designate Gibbs repeated Obama's support for repealing the military's "don't ask, don't tell" policy on homosexuals on the transition's Web site on Jan. 13, but the next day expanded on the answer: "Not everything will get done in the beginning," Gibbs said.[22]

Greenstein and Bose view Obama's inaugural address — which many observers faulted for rhetorical flatness — as a conscious, initial step to lower expectations about the pace of the promised "change we can believe in." Greenstein calls it a "get-down-to-work" address. Obama himself again evoked the inaugural's theme of determination in the face of adversity when he spoke to congressional leaders immediately following the address.

"What's happening today is not about me," Obama said at the joint congressional luncheon on Inauguration Day. "It is about the American people. They understand that we have arrived at a moment of great challenge for our nation, a time of peril, but also extraordinary promise."

"President Obama has done everything he can to tamp down this sense that he somehow walks on water," says Bose. "He has done everything he can to show that he is a man of substance.

"We have to recognize that these challenges aren't going to be met overnight and that we have to have confidence that we're going to meet them," she continues. "Now the question is, 'Can he govern? Can he show results?'"

NOTES

1. The text and video of the inaugural address are available on the redesigned White House Web site: www.whitehouse.gov. Some crowd reaction from Christopher O'Brien of CQ Press' College Division.

2. Quoted in Clea Benson, "An Economy in Foreclosure," *CQ Weekly*, Jan. 12, 2009.

3. Quoted in Aamer Madhani, "Will Obama Stick to Timetable?" *Chicago Tribune*, Nov. 6, 2008, p. 11.

4. Quoted in Shailagh Murray and Paul Kane, "Democratic Congress Shows It Will Not Bow to Obama," *The Washington Post*, Jan. 11, 2009, p. A5.

5. Norman Ornstein, "First Steps Toward 'Post-Partisanship' Show Promise," *Roll Call*, Jan. 14, 2009.

6. For a compact, continuously updated biography, see Barack Obama, www.biography.com. Background also drawn from Barack Obama, *Dreams from My Father: A Story of Race and Inheritance* (2004 ed.; originally published 1995). See also David Mendell, *Obama: From Promise to Power* (2007).

7. Quoted in Serge Kovaleski, "Obama's Organizing Years: Guiding Others and Finding Himself," *The New York Times*, July 7, 2008, p. A1.

8. Background drawn from Jody Kantor, "In Law School, Obama Found Political Voice," *The New York Times*, Jan. 28, 2007, sec. 1, p. 1.

9. Background drawn from Janny Scott, "The Story of Obama, Written by Obama," *The New York Times*, May 18, 2008, p. A1.

10. For a story on his mother's influence on Obama, see Amanda Ripley, "A Mother's Story," *Time*, April 21, 2008, p. 36.

11. See David Jackson and Ray Long, "Showing his bare knuckles: In first campaign, Obama revealed hard-edged, uncompromising side in eliminating party rivals," *Chicago Tribune*, April 4, 2007, p. 1; Rick Pearson and Ray Long, "Careful steps, looking ahead: After arriving in Springfield, Barack Obama proved cautious, but it was clear to many he had ambitions beyond the state Senate," *ibid.*, May 3, 2007, p. 1.

12. See Barack Obama, *The Audacity of Hope: Thoughts on Reclaiming the American Dream* (2006), pp. 105-107.

13. See David Mendell, "Obama routs Democratic foes; Ryan tops crowded GOP field," *Chicago Tribune*, March 17, 2004, p. 1.

14. For the full text of the 2,165-word speech, see http://obamaspeeches.com/002-Keynote-Address-at-the-2004-Democratic-National-Convention-Obama-Speech.htm. For Mendell's account, see *Obama, op. cit.*, pp. 272-285. Obama's conversation with Martin Nesbitt may have been reported first in David Bernstein, "The Speech," *Chicago Magazine*, July 2007; the anecdote is briefly repeated in Evan Thomas, *"A Long Time Coming": The Inspiring, Combative 2008 Campaign and the Historic Election of Barack Obama* (2009), p. 6. For the Post's account, see David S. Broder, "Democrats Focus on Healing Divisions," July 28, 2004, p. A1.

15. *CQ's Politics in America 2008* (110th Congress), www.cnn.com/video/#/video/world/2007/01/22/vause.obama.school.cnn.

16. Some background from Thomas, *op. cit.*

17. *Ibid.*, p. 9.

18. Many versions of the speech are posted on YouTube, including a posting of CNN's coverage.

19. Carpenter was quoted in Carrie Budoff Brown and Nia-Milaka Henderson, "Cabinet: Middle-of-the-roaders' dream?" *Politico*, Dec. 19, 2008; Alec MacGillis, "For Obama Cabinet, a Team of Moderates," *The Washington Post*, Dec. 20, 2008, p. A1.

20. See Kelly Evans, "Home Construction at Record Slow Pace," *The Wall Street Journal*, Jan. 23, 2009, p. A3.

21. See Paul Schemm, "Obama tells Arabic network US 'is not your enemy,'" The Associated Press, Jan. 27, 2009.

22. Obama quoted in Dan Eggen and Michael D. Shear, "The Effort to Roll Back Bush Policies Continues," *The Washington Post*, Jan. 27, 2009, p. A4; Gibbs quoted in, "Obama aide: Ending 'don't ask, don't tell' must wait," CNN.com, Jan. 15, 2009.

BIBLIOGRAPHY

Books by Barack Obama

Dreams from My Father: A Story of Race and Inheritance (Three Rivers Press, 2004; originally published by Times Books, 1995) is a literate, insightful memoir written in the three years after Obama's graduation from Harvard Law School. The three parts chronicle his "origins" from his birth through college, his three years as a community organizer in Chicago and his two-month pre-law school visit to his father's homeland, Kenya.

The Audacity of Hope: Thoughts on Reclaiming the American Dream (Crown, 2006) is a political manifesto written as Obama considered but had not definitively decided on a presidential campaign. The book opens with a critique of the "bitter partisanship" of current politics and an examination of "common values" that could underline "a new political consensus." Later chapters specifically focus on issues of faith and of race. Includes index.

Change We Can Believe In: Barack Obama's Plan to Renew America's Promise (Three Rivers Press, 2008), which includes a foreword by Obama, outlines steps for "reviving our economy," "investing in our prosperity," "rebuilding America's leadership" and "perfecting our union." Also includes texts of seven speeches from his declaration of candidacy on Feb. 7, 2007, to his July 24, 2008, address in Berlin.

Books About Barack Obama

The only objective, full-length biography is *Obama: From Promise to Power* (Amistad/Harper Collins, 2007) by David Mendell, the *Chicago Tribune* political reporter who began covering Obama in his first race for the U.S. Senate. An updated version was published in 2008 under the title *Obama: The Promise of Change.*

Two critical biographies appeared during the 2008 campaign: David Freddoso, *The Case Against Barack Obama: The Unlikely Rise and Unexamined Agenda of the Media's Favorite Candidate* (Regnery, 2008); and Jerome Corsi, *The Obama Nation: Leftist Politics and the Cult of Personality* (Threshold, 2008). Freddoso, a writer with National Review Online, wrote what one reviewer called a "fact-based critique" depicting Obama as "a fake reformer and a real liberal." Corsi, a conservative author and columnist best known for his

book *Unfit for Command* attacking Democratic presidential nominee John Kerry in 2004, came under fierce criticism from the Obama campaign and independent observers for undocumented allegations about Obama's background.

Two post-election books chronicle the 2008 campaign. Evan Thomas, *"A Long Time Coming": The Inspiring, Combative 2008 Campaign and the Historic Election of Barack Obama* (Public Affairs, 2009) is the seventh in *Newsweek's* quadrennial titles documenting presidential campaigns on the basis of reporting by a team of correspondents, with some reporting specifically not for publication until after the election. Chuck Todd and Sheldon Gawiser, *How Barack Obama Won: A State-by-State Guide to the Historic 2008 Presidential Election* (Vintage, 2009) gives an analytical overview of the campaign and election with detailed voting analyses of every state. A third title, *Obama: The Historic Journey*, is due for publication Feb. 16 by *The New York Times* and Callaway; the author is Jill Abramson, the *Times'* managing editor, in collaboration with the newspaper's reporters and editors.

Other books include John K. Wilson, *Barack Obama: The Improbable Quest* (Paradigm, 2008), an admiring analysis of Obama's political views and philosophy by a lawyer who recalls having been a student in Obama's class on racism and the law at the University of Chicago Law School; Paul Street, *Barack Obama and the Future of American Politics* (Paradigm, 2009), a critical depiction of Obama as a "power-conciliating centrist"; and Jabiri Asim, *What Obama Means: For Our Culture, Our Politics, Our Future* (Morrow, 2009) a depiction of Obama as creating a new style of racial politics — less confrontational than in the past but equally committed to social justice and more productive of results.

Articles

Purdum, Todd, "Raising Obama," *Vanity Fair*, **March 2008.**

The magazine's national editor, formerly a *New York Times* reporter, provided an insightful portrait of Obama midway through the 2008 primary season.

Von Drehle, David, "Person of the Year: Barack Obama: Why History Can't Wait," *Time,* **Dec. 29, 2008.**
Time's selection of Obama as person of the year includes an in-depth interview of the president-elect by Managing Editor Richard Stengel, Editor-at-large von Drehle and Time Inc. Editor-in-chief John Huey. The full text is at time.com/obamainterview.

On the Web

The Obama administration unveiled a redesigned White House Web site (www.whitehouse.gov) at 12:01 p.m. on Jan. 20, 2009 — even before President-elect Obama took the oath of office. The "Briefing Room" includes presidential announcements as well as a "Blog" sometimes being updated several times a day. "The Agenda" incorporates Obama's campaign positions, subject by subject. The site includes video of the president's speeches, including the inaugural address as well as the weekly presidential address — previously broadcast only on radio.

For More Information

American Enterprise Institute for Public Policy Research, 1150 17th St., N.W., Washington, DC 20036; (202) 862-5800; www.aei.org. Conservative think tank researching issues on government, economics, politics and social welfare.

Campaign for America's Future, 1825 K St., N.W., Suite 400, Washington, DC 20006; (202) 955-5665; www.ourfuture.org. Advocates progressive policies.

Center for American Progress, 1333 H St., N.W., 10th Floor, Washington, DC 20005; (202) 682-1611; www.americanprogress.org. Left-leaning think tank promoting a government that ensures opportunity for all Americans.

Center for Economic and Policy Research, 1611 Connecticut Ave., N.W., Suite 400, Washington, DC 20009; (202) 293-5380; www.cepr.net. Promotes open debate on key economic and social issues.

Center on Budget and Policy Priorities, 820 First St., N.E., Suite 510, Washington, DC 20002; (202) 408-1080; www.cbpp.org. Policy organization working on issues that affect low- and moderate-income families and individuals.

Concord Coalition, 1011 Arlington Blvd., Suite 300, Arlington, VA 22209; (703) 894-6222; www.concordcoalition.org. Nonpartisan, grassroots organization promoting responsible fiscal policy and spending.

Heritage Foundation, 214 Massachusetts Ave., N.E., Washington, DC 20002; (202) 546-4400; www.heritage.org. Conservative think tank promoting policies based on free enterprise, limited government and individual freedom.